Progress in Colloid & Polymer Science · Vol. 82

PROGRESS IN COLLOID & POLYMER SCIENCE
Editors: H.-G. Kilian (Ulm) and G. Lagaly (Kiel)

Volume 82 (1990)

Surfactants and Macromolecules: Self-Assembly at Interfaces and in Bulk

Guest Editors: B. Lindman (Lund),
J. B. Rosenholm (Abo)
and P. Stenius (Stockholm)

Springer-Verlag
Berlin Heidelberg GmbH

ISBN 978-3-662-15059-7 ISBN 978-3-7985-1688-5 (eBook)
DOI 10.1007/978-3-7985-1688-5
ISSN 0340-255-X

Originally published by Dr. Dietrich Steinkopff Verlag GmbH & Co. KG, Darmstadt in 1990
Softcover reprint of the hardcover 1st edition 1990

Chemistry editor: Dr. Maria Magdalene Nabbe; English editor: James Willis; Production: Holger Frey.

Type-Setting: Graphische Texterfassung, Hans Vilhard, D-6126 Brombachtal

Preface

This volume of Progress in Colloid & Polymer Science consists of papers presented at the 10th Scandinavian Symposium on Surface Chemistry, which was held at the Åbo Akademi University in Åbo (Turku), Finland, June 14—16, 1989. The first Symposium in the row was also held at Åbo Akademi University in 1962 with prof. em. Per Ekwall as the principal organizer.

The aim of this series of Sympsia is to collect reports on fundamental and applied research in "wet" surface and colloid chemistry. At the symposium 53 oral communications were given at two parallel sessions, whereof 6 plenary lectures and 13 invited lectures. Additionally 57 posters were presented at the Symposium. The presentations were attended by 195 delegates from 20 countries. This volume contains a collection of papers mainly given as oral communications. The papers are grouped under the following topics:

- Adsorption from solution
- Emulsions, foams, thin liquid films
- Self assembling systems
- Surface modification and interactions.

On behalf of the organizing committee, we would like to thank all the participants for their very active response for this Symposium volume, and for their very stimulating discussions during the sessions. A special thank is also due to the referees of the manuscripts as well as to the Advisory board consisting of: A. Almgren (S), K. S. Birdi (DK), A. Borseth (N), I. Danielsson (SF), J.-C. Eriksson (S), B. Hakulin (SF), N. Krogh (DK), K. Larsson (S), G. Lindblom (S), S. Nordberg (S), J. Sjöblom (N), O. Smidsrød (N), N.-G. Wannerberg (S).

Björn Lindman
Jarl B. Rosenholm
Per Stenius

Contents

Surface Modification and Interactions

Adsorption from Solution

Progress in Colloid & Polymer Science Progr Colloid Polym Sci 82:3—9 (1990)

Controlled adsorption of polymers and surfactants in particulate separation processes

B. M. Moudgil

Mineral Resources Research Center, Department of Materials Sciences and Engineering, University of Florida, Gainesville, Florida, USA

Abstract: Ultrapurification of fine powders and processing of low-grade, finely disseminated ores can be achieved by selectively flocculating one of the constituents (impurity or value component) and separating it from the bulk suspension by sedimentation or floc flotation. Controlled adsorption of polymers and/or surfactants is known to govern the separation efficiency. In this paper, the role of relative adsorption, as dictated by active sites constituting a given particulate surface, and its implications in the selective flocculation technology for ultrapurification of fine powders are discussed.

Key words: Polymer adsorption; ultrapurification; fine powders; active sites; heteroflocculation; selective flocculation

Introduction

Solid-solid separation in the fine particle size range is a critical materials problem in the ceramic and mineral industries. Separation of particulate impurities from the fine powders used to make advanced ceramic components is recognized as an important parameter in manufacturing defect-free, sintered products. However, in the mineral industry, the depletion of high-grade ores has led to the mining and processing of highly disseminated, low-grade, fine-grained ores. The economic recovery of values from finely ground ores is necessary to maintain a reliable raw material base for the metal, ceramic, and chemical industries.

Selective flocculation of the desired component from the mixture using polymeric and/or surfactant additives, followed by separation of the aggregated mass from the suspension by sedimentation or flotation has shown considerable promise in processing of fine particles. Flotation of fines in column flotation cells is another technique which is being developed to separate particles in the fine size range.

Selective adsorption of surfactants and polymers from solution governs the efficiency of separation achieved by flotation or selective flocculation technology. Both the magnitude of adsorption and the orientation of the adsorbed species are important parameters. In flotation, the amount of the surfactant adsorbed on a given substrate determines its water wetting characteristics, whereas in selective flocculation, the orientation of the adsorbed polymeric species is believed to strongly influence the properties of the aggregates formed and thereby, the overall process efficiency.

In the past, emphasis has been placed on achieving the selective adsorption of polymers and surfactants on desired substrates to obtain particulate separation. Although relative adsorption on different components of a mixture was considered to be important, it was the change in surface property (e.g., hydrophobicity) or some other physical phenomena (such as flocculation) which was monitored to assess the potential separation. The present discussion is limited to polymer adsorption as it applies to the efficiency of (fine) particulate separation by selective flocculation technology.

Active sites concept

Selective flocculation process is successfully applied on a commercial scale [1, 2] and is considered

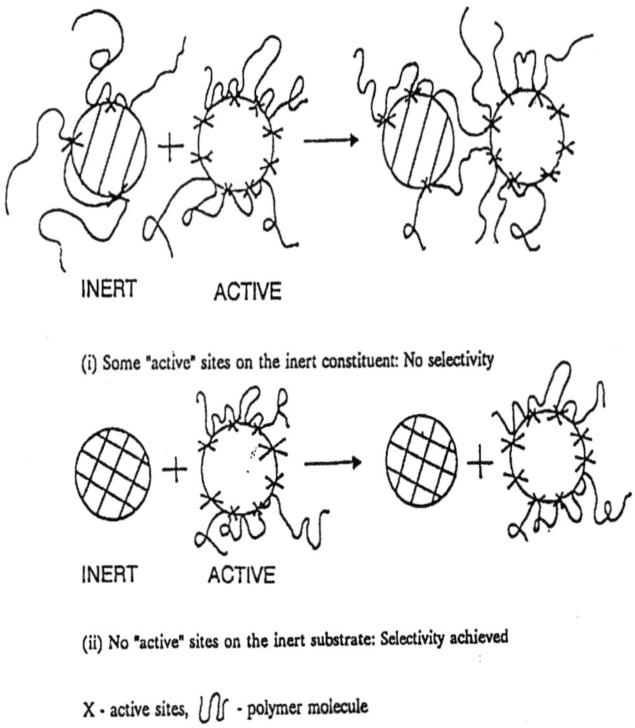

Fig. 1. Effect of active sites on heteroflocculation

to be a promising technique for processing of low-grade, fine-grained ores, and in ultrapurification of fine powders for advanced ceramic processing [3, 4]. Invariably, the selectivity predicted on the basis of *single component* flocculation tests is lost when mixed component tests are attempted. The loss in selectivity is attributed to i) heterocoagulation, ii) physical entrapment and entrainment, and iii) dissolved ion effect [5]. However, recently it has been shown that *heteroflocculation* between the inert and active constituent(s) of a mixture can also be a major factor in the loss of selectivity [6]. Heteroflocculation is schematically illustrated in Fig. 1; it arises as a result of adsorption of polymeric species also on the inert particles, which then are incorporated in the flocs of active species. Classification of an inert or active substrate is based on the degree of polymer (or surfactant) adsorption on a given component. Furthermore, the extent of adsorption on a given particle is attributed to the number of "active" sites to which the polymer or surfactant molecules attach. The ratio of "active" sites on the inert and active constituents along with feed composition governs the separation efficiency. In order to understand the role of active sites concept and to evaluate the effect of process parameters, a physical simulation of the flocculation process was attempted [7]. A major advantage of such a simulation was the ease with which active sites on different components could be varied, which otherwise is extremely difficult in real systems.

Physical simulation tests

In these tests, spherical wooden balls of 25-mm diameter, as shown in Fig. 2, were used as particles. Velcro pieces (4 mm^2) were glued on the wooden balls to simulate active sites. Thin 25-mm long strips of Velcro that attached only to the Velcro sites on the wooden balls were used to represent polymers. A given number of active components (balls with more velcro pieces glued onto the surface) was mixed with inert component and the polymer (Velcro strips) for 2 min in a cylindrical tank (0.45-m dia., 0.6-m length). At the end of the mixing, the number of aggregates, single balls, and free polymers were counted; structure of the aggregates formed was also analyzed.

Results of physical simulation studies with respect to effect of active sites and feed composition are presented in Figs. 3 and 4, respectively. It should be realized that, due to obvious differences in the simulated and real systems, only trends and not the absolute values need to be considered. An encouraging observation is that to achieve the desired separation total elimination of active sites from the inert constituent is not necessary. However, as the feed contains less and less of one of the components (e.g., active component), it becomes more and more difficult to separate that component from the bulk. In other words, ultrapurification is predicted to be difficult.

Simultaneously a mathematical model of the selective flocculation process was developed; it allowed cross-checking some of the predictions of physical simulation tests.

Mathematical model

In this model the entire floc growth process, including corrections for changes in floc density and floc mobility upon flocculation, are considered, which was not the case in earlier studies [8, 9]. To avoid the complications due to adsorption and floc-

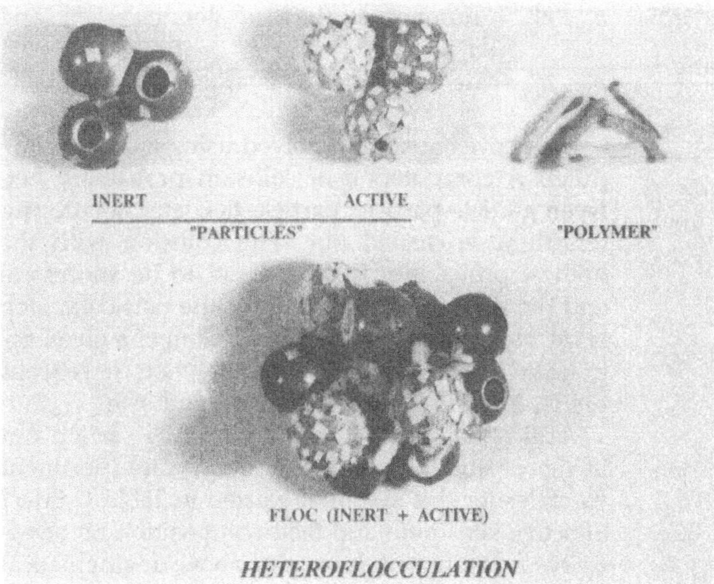

Fig. 2. Photo of wooden balls ("particles"), velcro pieces ("polymers") etc. used for physical simulation tests

culation occurrring at the same time, it is assumed that the rate of polymer adsorption is much faster than the kinetics of flocculation. Additionally, only binary collisions are considered since probability of binary collision events would be much higher than multibody collisions. The relevant rate equation for flocculation in a system consisting of particles i and j is given by

$$R_{ij} = F_{ij} E_{ij} P_{ij} \,, \tag{1}$$

where F_{ij} and E_{ij} represent collision frequency and efficiency factors, respectively, and P_{ij} is the sticking efficient that describes the probability of attachment of an adsorbed polymer to an active site upon collision. These terms are described by the following relationships:

$$F_{ij} = 1/6\,\bar{s}(d_i + d_j)^3(n_i + n_j)^2 \tag{2}$$

$$E_{ij} = \Phi_i\Theta_i(\Phi_j - \Phi_j\Theta_j) + \Phi_j\Theta_j(\Phi_i - \Phi_i\Theta_j) \tag{3}$$

$$P_{ij} = A_{ij}e^{-E_j/KT} + A_{ij}e^{-E_i/KT} \,, \tag{4}$$

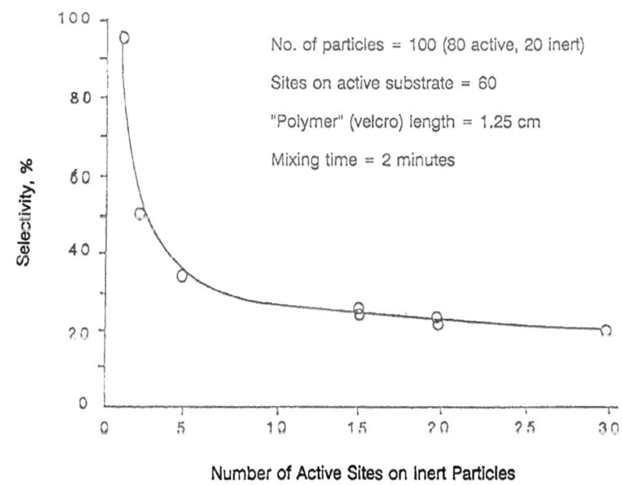

Fig. 3. Effect of nature of inert particle on selectivity in physical simulation tests

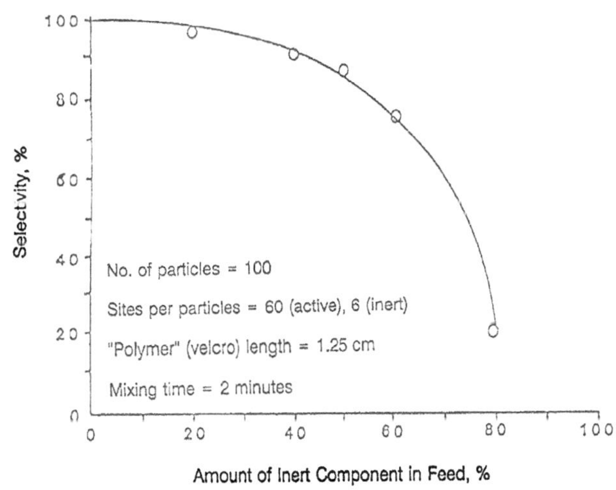

Fig. 4. Effect of feed composition on the selectivity in physical simulation tests

where $\bar{s}$-mean point velocity gradient and:

d_i, d_j = diameters of particles or flocs i and j, respectively;
n_i, n_j = number of particles or flocs of diameter d_i and d_j;
Φ_i, Φ_j = number of active sites on i and j, respectively;
Θ_i, Θ_j = fractional surface coverage of i and j, respectively;
A_{ij} and A_{ji} = frequency factors;
E_i, E_j = activation energy for polymer adsorption on i and j, respectively;
K = Boltzman constant;
T = absolute temperature.

Combining Eqs. (1)—(4) yields:

$$R_{ij} = 1/6\ \bar{s}(d_i + d_j)^3(n_i + n_j)^2[\Phi_i\Theta_i(\Phi_j - \Phi_j\Theta_j) + \Phi_j\Theta_j(\Phi_i - \Phi_i\Theta_i)] \times [A_{ij}e^{-E_j/KT} + A_{ji}e^{-E_i/KT}] \ . \quad (5)$$

However, it should be realized that as flocculation proceeds, mobility of the flocs will decrease, affecting E_{ij}. Additionally, accessible surface area for polymers to anchor on vacant active sites decreases with floc growth, which also affects E_{ij} [10]. As a result, the following corrections need to be taken into account, due to reduced mobility and surface area of the floc:

$$\text{Mobility}\ \alpha\ \frac{1}{\sqrt{\rho_f d_f}}\ , \quad \text{therefore,}\quad E_{ij}\ \alpha\ \frac{1}{\sqrt{\rho_f d_f}}\ ;$$

$$\text{Area of floc}\ \alpha\ \frac{1}{d_f}\ , \quad \text{therefore,}\quad E_{ij}\alpha\ \frac{1}{d_f}\ .$$

Incorporating the above corrections in (5) results in the following equation:

$$R_F = K'_{Fij}\frac{(n'_{Fij} + n''_{Fij})^2(d'_{Fij} + d''_{Fij})^3}{\rho'^{1/2}_{Fij}\rho''^{1/2}_{Fij}d'^{3/2}_{Fij}d''^{3/2}_{Fij}}E_{ij}P_{ij}\ , \quad (6)$$

in which

R_F = rate of aggregation of flocs consisting of two types of particles or flocs i and j;
K'_{Fij} = flocculation rate constant, depends on physical properties of medium and hydrodynamic conditions;
n'_{Fij}, n''_{Fij} = number of flocs of diameters d'_{Fij} and d''_{Fij};
ρ'_{Fij}, ρ''_{Fij} = densities of two interacting flocs.

The above equation is solved using a digital computer. A computation of collision probability between particle-particle, particle-floc, and floc-floc interactions is carried out. The collision with the highest probability is considered to be successful and the floc-size distribution for the particular step is calculated accordingly. Flocculation of a given aggregate is considered to occur when it reaches predetermined values of density and size.

Verification of the above model was carried out using a quartz-dolomite system. The pertinent variables for this system are listed in Table 1. Effect of active sites ratio and feed composition on selectivity as estimated from the above mathematical model are compared with the experimental data in Figs. 5 and 6, respectively. It is to be noted that similar trends in selectivity are predicted by physical simulation tests as well as the mathematical model.

Table 1. List of parameters employed in verification of the mathematical model

Parameter	Quartz	Dolomite
Active Sites Ratio, Φ	0.107	1.000
Fractional Surface Coverage, Θ	0.015	0.125
Diameter of Particle, d_p	0.53 µm	0.66 µm
Density of Particle, ρ_p	2.7 g/cc	2.8 g/cc

A decrease in selectivity as the amount of one of the components in the feed is reduced, as mentioned earlier, indicates the limitation of the selective flocculation technology particularly in ultrapurification of fine powders. However, it should be remembered that the criterion of separating the flocculated material from the suspension, both in physical simulation and mathematical model, was based on *gravity settling* of flocs. The other technique for separating flocs from suspension is floc flotation. Obviously, density and size of the flocs desired would be different, depending on whether the sedimentation or floc flotation method is used

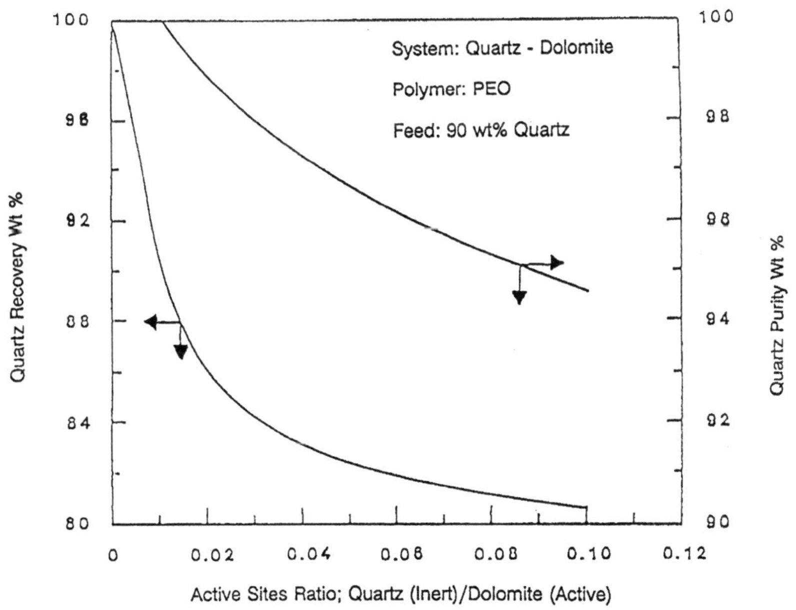

Fig. 5. Effect of active sites ratio on the selectivity (mathematical model)

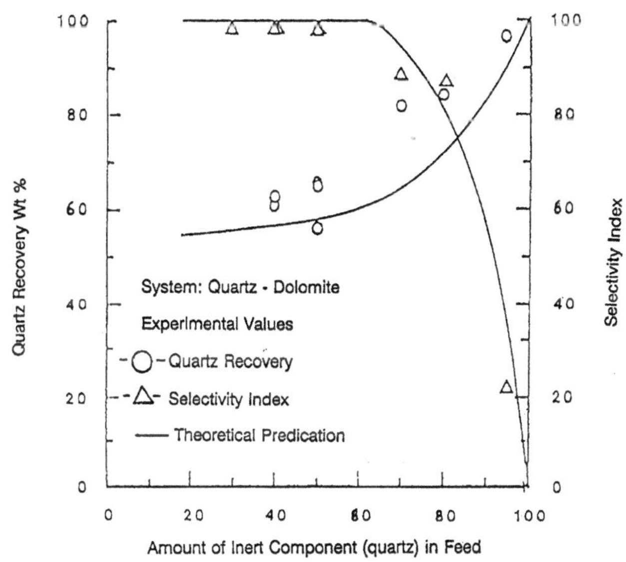

Fig. 6. Effect of feed composition on selectivity (mathematical model)

to separate the flocculated material from the suspension. Therefore, attempts to characterize flocs and eventually to control their properties were made.

Floc characteristics

Floc size and size distribution can be evaluated by observing the projected area of aggregates under an optical microscope or by in-situ measurements [11–13]. Density of a floc, however, is usually determined by measuring its buoyant density [11]. Considering that shape and porosity of aggregates vary widely, meaningful results from such measurements are not always possible. Therefore, a direct method of floc density determination was developed [14]. Additionally, in another technique employing a free settling equation, corrections for shape of the flocs along with turbulence were incorporated. These two techniques allowed a much more realistic determination of floc density measurements. It was observed that flocs can be characterized as fractals (see Fig. 7). The significance of this observation lies in the fact that it provides a unique basis for comparing flocs prepared under different experimental conditions.

It is evident that flocs of different densities can be separated from each other. This provides added flexibility in achieving the desired separation by selective flocculation. For example, if flocs of one component are of lower density they than can, perhaps, be separated from the denser flocs by the floc-flotation method. These concepts were applied to achieve ultrapurification of quartz contaminated with dolomite particles.

Ultrapurification of fine powders

Separation of dolomite (active constituent and an impurity) from quartz was attempted by floccu-

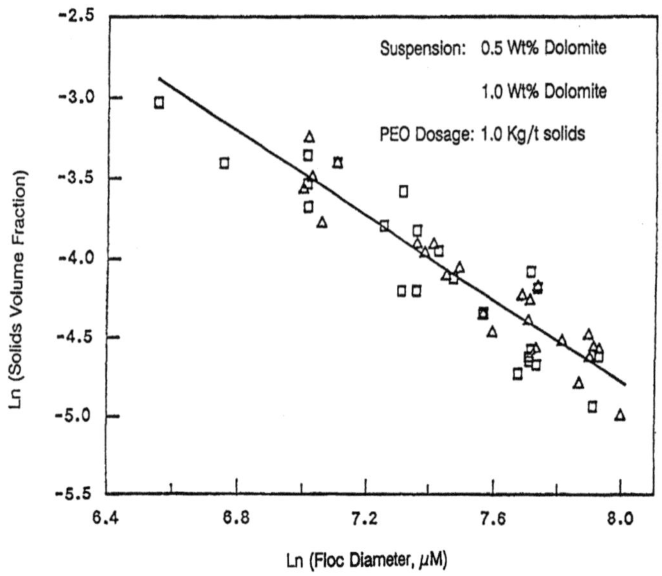

Fig. 7. Relationship between size and porosity of dolomite flocs (fractal dimension $D = 1.68$)

lating dolomite with polyethylene oxide (5000000 m.wt.), using sodium oleate as the coflocculant. Separation of the dolomite flocs from the quartz suspension was achieved by sedimentation as well as floc flotation. Results presented in Table 2 indicate the separation of dolomite from quartz. After several attempts it was clear that, although sedimentation technique is successful in acquiring higher purity, it was only possible at the expense of quartz recovery. It was also evident that since flocs of dolomite are less dense and small, floc flotation could therefore be a more efficient technique for separating the flocs from the suspension. Floc flotation was accomplished by bubbling nitrogen gas through a sparger fitted with a frit of medium porosity. It is clear from the results presented in Table 3 that ultrapurification of quartz with almost complete removal of dolomite at acceptable recovery levels is achievable by selective flocculation technique.

Table 2. Separation of dolomite flocs from quartz suspension by sedimentation. Particle Size: 15 μm; Flocculant: PEO 1 kg/t; Co-flocculant: Sodium oleate, 1 kg/t; Settling time: 250 s

Feed		Quartz product	
Quartz, %	Dolomite, %	Purity, %	Yield, %
95.0	5.0	98.94	72.1
99.0	1.0	99.22	86.0

Table 3. Removal of dolomite flocs from quartz suspension by floc flotation. Sodium oleate, 1 kg/t

Feed		Quartz product	
Quartz, %	Dolomite, %	Purity, %	Yield, %
95.0	5.0	98.49	82.2
99.0	1.0	99.25	87.5
99.75	0.25	99.90, 99.99	78,7, 80.9

Concluding remarks

Selective flocculation and froth flotation process efficiency is governed by the controlled adsorption of polymers and/or surfactant in the solid-liquid interfacial region. Until now, screening of the suitable polymers and surfactants, especially for selective flocculation, was based on conducting single-component flocculation tests followed by sedimentation or floc-flotation to separate the aggregates from the suspended material. Invariably, the selectivity predicted by the screening tests was lost when mixed component flocculation tests were attempted. The reasons for the loss of selectivity included heteroagulation, physical entrapment and entrainment, and dissolved ion effect. Recently, another factor termed heteroflocculation, has been determined to account for a significant selectivity loss. It is believed that controlling the adsorption on the specific component(s) below a certain level is a prerequisite for achieving the desired selectivity. The structure of the flocs formed, if manipulated properly, can also provide added flexibility in fine-particle separation technology.

Floc structure has been found to depend on the nature of the substrate, chemical nature of the polymer, and the amount of the polymer added. By judicious selection of the experimental parameters, it should be possible to generate flocs of desired properties.

It is evident from the above discussion that selective flocculation process holds a great promise for ultrapurification of fine powders and also for processing of finely disseminated, low-grade ores. Controlling the adsorption on various substrates, as mentioned earlier, is essential to minimize heteroflocculation. Poisoning the active sites, competitive adsorption of specific ions, and modification of the solvent power of the medium are some of the research areas which need to be investigated for reducing the polymer adsorption on the given component(s).

Acknowledgement

The author acknowledges financial support of this work by the Dow Chemical Company and the NSF PYI Award (MSM # 8352125).

References

1. Banks AF (1979) Selective Flocculation-Flotation of Slimes from Sylvinite ore. In: Somasundaran P, Arbiter N (eds) Beneficiation of Mineral Fines. AIME, New York, p 22
2. Colombo AF (1980) Selective Flocculation of Iron Bearing Materials. In: Somasundaran P (ed) Fine Particles Processing, AIME, New York, 53, pp 1034–1056
3. Moudgil BM, Shah BD (1987) Removal of Particulate Impurities from Ceramic Powders by Selective Flocculation, Advances in Ceramics, Vol 21: Ceramic Powder Science, The American Ceramic Society, Inc, pp 483–494
4. Moudgil BM, Vasudevan TV, McCombs A (1988) Selective Flocculation Method for Purification of Ceramic Powder, Ceramic Transactions: Ceramic Powder Science II, Vol 1, Messing GL, Fuller ER, Hansner H (eds) The American Ceramic Society, Westerville, Ohio, pp 372–379
5. Krishnan SV, Attia Y (1986) Floc Characteristics in Selective Flocculation of Fine Particles, Flocculation, Sedimentation and Consolidation, Moudgil BM, Somasundaran P (eds) Engineering Foundation, New York, pp 229–248
6. Moudgil BM, Shah BD, Soto HS (1987) Loss of Selectivity in Apatite-Dolomite Flocculation, Minerals and Metallurgical Processing, pp 27–31
7. Moudgil BM, McCombs A (1987) Physical Simulation of the Flocculation Process, Minerals and Metallurgical Processing, pp 151–155
8. Healy TW, LaMer VK (1962) J Phys Chem 66:1835–1838
9. Hogg R (1984) J Colloid Interface Sci 102:232–236
10. Moudgil BM, Vasudevan TV (1988) Factors Affecting Selectivity in Purification of Fine Powders by Selective Flocculation, Plumpton A (ed) Production and Processing of Fine Particles, Canadian Institute of Mining and Metallurgy
11. Matsumoto K, Mori Y (1975) Settling Velocity of Floc-New Measurement Method of Floc Density, J Chem Eng, Japan, Vol 8, pp 143–147
12. Dirican C (1981) The Structure and Growth of Aggregates in Flocculation, Thesis MS The Pennsylvania State University
13. Klimpel RC, Hogg R (1986) Effect of Flocculation on Agglomerate Structure, J Colloid Interface Sci, Vol 113, pp 121–131
14. Moudgil BM, Vasudevan TV (1989) Evaluation of Floc Properties for Dewatering Fine Particle Suspensions, Minerals and Metallurgical Processing, Vol 6, pp 142–145

Author's address:

Dr. B. M. Moudgil
Mineral Resources Research Center
Department of Materials Science & Engineering
University of Florida
Gainesville, Florida 32611, USA

Progress in Colloid & Polymer Science Progr Colloid Polym Sci 82:10—14 (1990)

Computer simulation of aggregation in adsorbed protein layers

M. Stenberg and H. Nygren

Department of Solid State Electronics, Chalmers University of Technology, Göteborg, Sweden and
Department of Histology, University of Göteborg, Göteborg, Sweden

Abstract: Some theoretical models describing the adsorption of proteins from solution are investigated. Computer simulations are carried out on a hexagonal lattice that allows lateral interactions between adsorbed protein molecules to be incorporated in the model. Diffusion limited adsorption determines the amount of particles at the surface and aggregation is allowed by lateral diffusion along the surface. The computer model can be characterized as a two-dimensional, stochastic, cellular automaton where the surface sites are arranged on a hexagonal lattice. The fate of every lattice site, or cell, is updated in discrete time steps. Different rules concerning nearest-neighbor interactions were tested, giving both irreversible and reversible diffusion-limited aggregation (DLA). — The fractal dimensions of individual clusters were extracted from simulated aggregates and found to be consistent with literature values on irreversible and reversible DLA in two dimensions. By comparing computer simulated aggregation to experimental electron micrographs of adsorbed ferritin layers some conclusions can be drawn concerning the adsorption of ferritin to hydrophobic quartz surfaces. The adsorbed protein molecules must, most probably, be mobile along the surface with a diffusion constant considerably lower ($5 \cdot 10^4$) than in the bulk solution. The initial aggregation could be well described by reversible DLA-clustering (at $\theta = 0.015$), but at higher surface coverages of proteins (at $\theta = 0.1$) the present model could not account for the low experimentally observed fractal dimension ($d_f = 1.05$).

Key words: Protein adsorption; two-dimensional aggregation; computer simulation; cellular automata

Introduction

It has long been known that proteins may very well adsorb and stick firmly to a surface introduced into a stable protein solution. It has recently been shown that during this adsorption process the surface actually induces aggregation of protein molecules at the surface and that these aggregates show fractal properties [1]. During the last few decades much theoretical and experimental work has been done on aggregation kinetics and fractal aggregates [2—4]. Computer simulations have been used to model both diffusion-limited aggregation (DLA) and reaction-limited aggregation (RLA) in one, two, or three dimensions. However, the process of aggregation at a surface in contact with a stable solution imposes boundary conditions for the aggregation process which, so far, have not been treated in the literature.

In this study we restrict our simulations to the simplified situation where the amount bound is limited by diffusion from the solution and also to different diffusion-limited aggregation models where the sticking coefficient is one. Both assumptions are motivated by experimental findings. The initial adsorption of ferritin at a hydrophobic surface has been found to be diffusion-limited [1]. Early attempts to measure the pair correlation function $g(r)$ for dilute surface concentrations of ferritin molecules failed to show any significant repulsive interactions between the adsorbed molecules [5]. Only attractive forces could be detected, thus indicating a sticking coefficient of one. Since experiments indicate reorganization of aggregates, we chose to study some models of reversible, diffusion-limited aggregation.

Since we are dealing with a system where nearest-neighbor interactions are important and where particles may suddenly appear at different sites at the surface, it is natural to use computer models known as cellular automata [6]. Our simulation methods will be close to that of two-dimensional, cellular, stochastic automata which have been used for studying two-dimensional phase transitions [7].

Simulation methods

The simulations were carried out for an experimental situation where a surface is introduced at time zero into a solution containing the adsorbing molecules at a homogeneous initial concentration. By assuming that adsorbed molecules do not leave the surface, we get the diffusion-limited adsorbed amount as

$$n_s = \frac{2}{\sqrt{\pi}} n_0 \sqrt{D_b t}\,, \tag{1}$$

where n_s is the surface concentration in number per unit area at time t, D_b is the diffusion constant in the bulk solution, and n_0 is the bulk concentration in number per unit volume. For a hexagonal lattice the maximum surface concentration in terms of number per unit area is

$$n_{sm} = \frac{1}{2\sqrt{3}\, r_0^2}\,, \tag{2}$$

where r_0 is the particle radius. The initial volume fraction φ_0 is

$$\varphi_0 = n_0 \frac{4}{3} \pi r_0^3\,, \tag{3}$$

and thus, for a hexagonal lattice the diffusion-limited surface coverage can be written

$$\theta = \frac{n_s}{n_{sm}} = \left(\frac{3}{\pi}\right)^{\frac{3}{2}} \varphi_0 \frac{\sqrt{D_0 t}}{r_0}\,. \tag{4}$$

Each position in the hexagonal lattice is updated in discrete time steps equal to the mean time needed for moving one lattice position by surface diffusion. In two dimensions the mean displacement Δr after a time Δt is

$$\Delta r = 2\sqrt{D_s \Delta t}\,, \tag{5}$$

where D_s is the surface diffusion constant.

With $\Delta r = 2\, r_0$ we get the time increment as

$$\Delta t = \frac{r_0^2}{D_s}\,. \tag{6}$$

After T iterations we get $t = T \cdot \Delta t$ and by combining Eqs. (4) and (6) we get the normalized relation

$$\theta(T) = C\sqrt{T}\,, \tag{7}$$

where the parameter C is related to initial concentration and the respective diffusion constants the relation by:

$$C = \left(\frac{3}{\pi}\right)^{\frac{3}{2}} \varphi_0 \sqrt{\frac{D_b}{D_s}}\,. \tag{8}$$

In the simulations we choose to specify c by giving the total number of iterations T_t and the corresponding final coverage θ_t so that

$$C = \frac{\theta_t}{\sqrt{T_t}}\,. \tag{9}$$

At iteration $T + 1$ we add new particles at random sites and we can use Eq. (7) to calculate the number of new particles. Thus, we start every itaration by increasing the coverage by an amount calculated from the relation:

$$\Delta\theta = \theta_t \sqrt{\frac{(T+1)}{T_t}} - \theta(T)\,. \tag{10}$$

After adjustment of the total number of particles every lattice-position is analyzed with respect to its neighbors. For every lattice-position a jumping direction is chosen at random and the jumping probability is determined from the neighbor distribution. In this work we have chosen to deal with two types of neighbors, defined by the jumping direction as shown in Fig. 1. Jumping with an A-type neighbor requires the creation of extra particle-liquid interface, whereas jumping with a B-type neighbor can be done by sliding along this neighbor. For example, by this distinction it is possible for dimers to have an increased mobility along the surface. In this study we have restricted the number of parameters to:

p_A = jumping probability for at least one A-type neighbor; and p_B = jumping probability for no A-type neighbor and at least one B-type neighbor.

The simulations were carried out on a 256 × 256 hexagonal lattice with periodic boundary conditions. The resulting aggregates were analyzed by measuring the radius of gyration R_g for an s cluster defined as

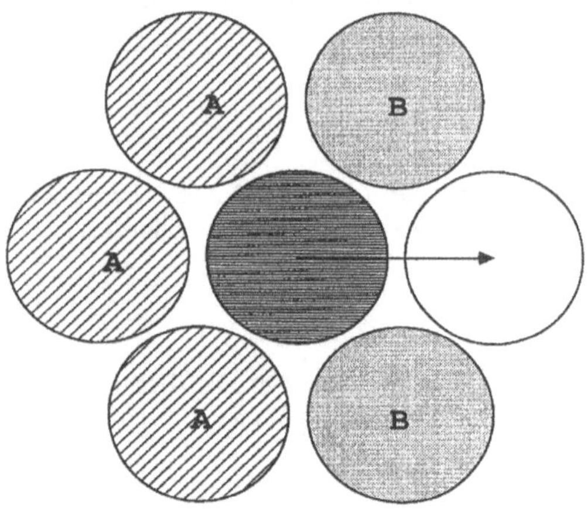

Fig. 1. Simulations are carried out on a hexagonal lattice. The fate of every lattice position is updated in discrete time steps. A jumping direction is chosed at random and the jumping probability is determined by the presence of neighbors. The present model can also distinguish between A- and B-type neighbors, as defined in the figure

$$R_g = \left(\sum_{i=1}^{s} r_i^2/s\right)^{\frac{1}{2}}, \tag{11}$$

where r_i is the distance of a site from the center of mass. Aveage values of R_g were then plotted against s and fitted to the form

$$R_g = \text{const.}\, s^{\rho}, \tag{12}$$

where, for scale-invariant clusters, the exponent ρ should be related to the fractal dimension d_f by $d_f = 1/\rho$.

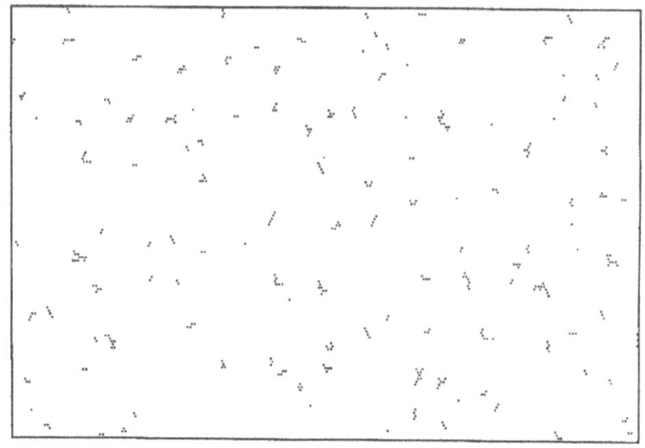

Fig. 2. Irreversible DLA-clusters generated after 2500 iterations at a coverage $\theta = 0.015$ (parameters: $C = 3 \cdot 10^{-4}$, $p_A = 0$ and $p_B = 0$)

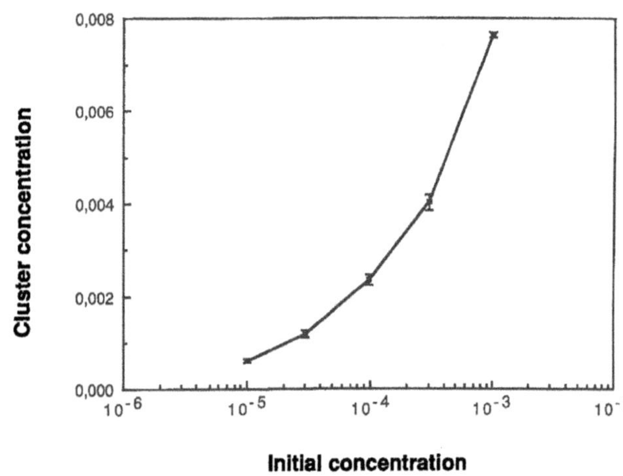

Fig. 3. Cluster concentration (n_c/n_{sm}) vs initial bulk concentration parameter c at a coverage of 0.015 for irreversible DLA-clusters. Mean and standard deviation from five different simulations are shown

Results and discussion

In Fig. 2 irreversible DLA-clusters are shown at a surface coverage of 0.015. In this case the particles are immobilized as soon as they find a neighbor ($p_A = 0$ and $p_B = 0$). After some initial clusters are formed the surface concentration of clusters n_c remains constant, because aggregation with existing clusters is more probable than the formation of new clusters. At a low surface coverage θ the cluster concentration is a function of the initial bulk concentration C. This relation is shown in Fig. 3 for a surface coverage of $\theta = 0.015$. Although this relation is different for reversible DLA, we can use it for a first estimation of the surface diffusion constant D_s. From measurements on ferritin [1] it was found that at an initial concentration of 10 µg/ml the cluster concentration n_c/n_{sm} was 0.0035 at a surface coverage of 0.015. Using a MW of 680000 for Fe-saturated ferritin and a radius r_0 of 5 nm, we get $\varphi_0 = 4.6 \cdot 10^{-6}$. From Fig. 3 we get a value of $C = 2 \cdot 10^{-4}$, and from Eq. (8) we can estimate that $D_s = 5 \cdot 10^{-4}\, D_b$.

In Fig. 4a irreversible DLA-clusters are shown at a surface coverage $\theta = 0.1$ generated at an initial concentration $C = 3.16 \cdot 10^{-4}$. From Fig. 5 we can see that the radius of gyration seems to follow Eq. (12) with $\rho = 0.582$ giving $d_f = 1/\rho = 1.71$. From

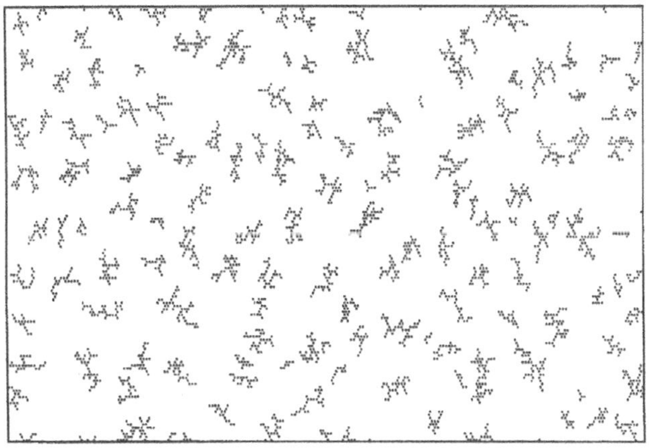

a

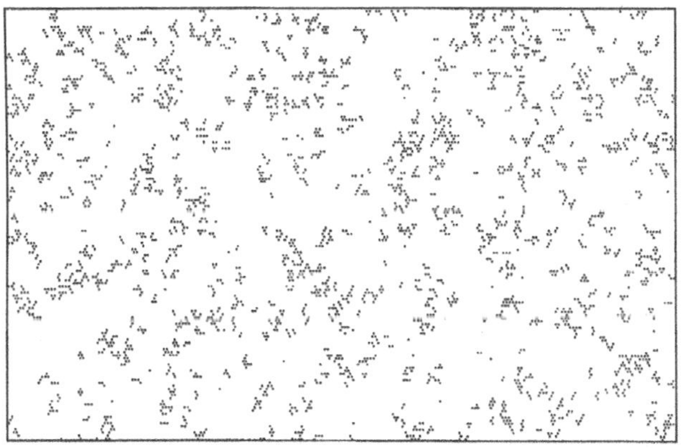

b

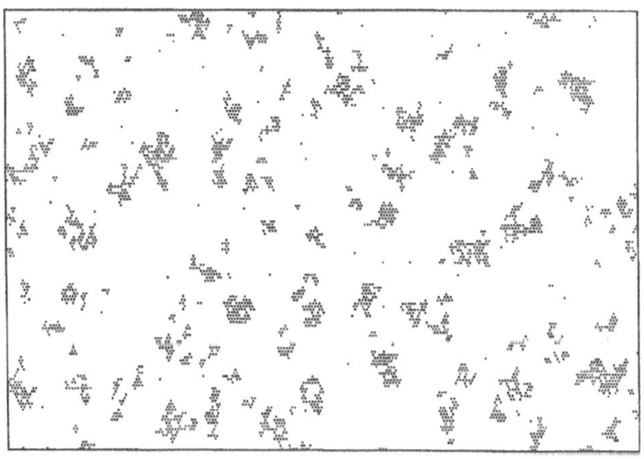

c

Fig. 4. DLA-clusters generated after 10^5 iterations at a coverage $\theta = 0.1$ ($C = 3.16 \cdot 10^{-4}$). a) irreversible DLA-cluster with $p_A = 0$ and $p_B = 0$; b) reversible DLA-cluster with $p_A = 2^{-7}$ and $p_B = 2^{-7}$; and c) reversible DLA-cluster with $p_A = 2^{-7}$ $p_B = 2^{-3}$

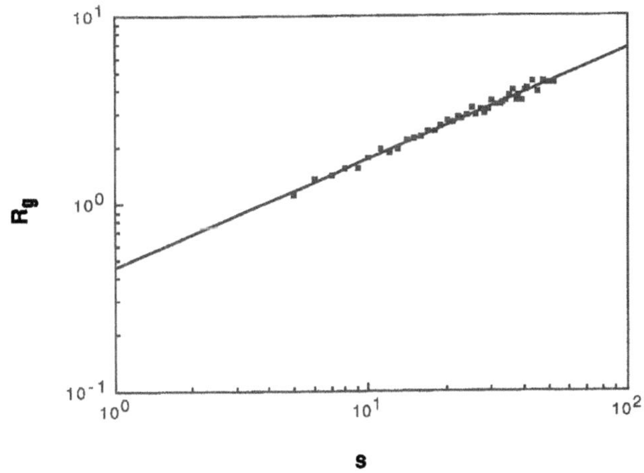

Fig. 5. A log-log plot of the radius of gyration R_g vs cluster mass s for irreversible DLA-clusters ($C = 3.16 \cdot 10^{-4}$, $p_A = 0$, $p_B = 0.1$, solid line $\rho = 0.582$, data from 272 clusters)

five different simulations, as in Fig. 5, we calculated the fractal dimension d_f to 1.70 ± 0.07. This value is consistent with the fractal dimension of irreversible DLA-clusters grown in two dimensions [8] (d_f = 1.70 ± 0.02 from radius of gyration). We conclude that our different boundary and grwoth conditions did not alter the fractal dimension, at least up to a coverage of 0.1. The cluster concentration is 0.0046 in Fig. 4a and has not changed significantly from $\theta = 0.015$ to $\theta = 0.1$, indicating that, so far, the different clusters have grown independently.

In Figs. 4b and c we can see what happens when aggregates are allowed to disintegrate. In Fig. 4b we do not differ between A- and B-type neighbors and the result of the disintegration is an increased number of clusters with a lower average mass. However, when the particles are allowed to slide along B-type neighbors with an increased probability, as in Fig. 4c, the result is a decreased number of clusters with a greater average mass. The increased mobility for smaller clusters dramatically changes the general appearance. In Fig. 6 we have analyzed the fractal dimensions. With no distinction between A- and B-type neighbors ($p_A = p_B$) we get a fractal dimension of d_f = 1.65 ± 0.08 (from five similar simulations as in Fig. 6). This number is consistent with earlier simulations of two-dimensional reversible DLA-clusters: d_f = 1.54 ± 0.08 in [9] and d_f = 1.57 ± 0.06 in [10]). When the difference between A- and B-type neighbors is considered, as in Fig. 4c, the fractal dimension changes, and for $p_B = 2^{-3}$ it increases to d_f = 1.70 ± 0.04, as shown in Fig. 6.

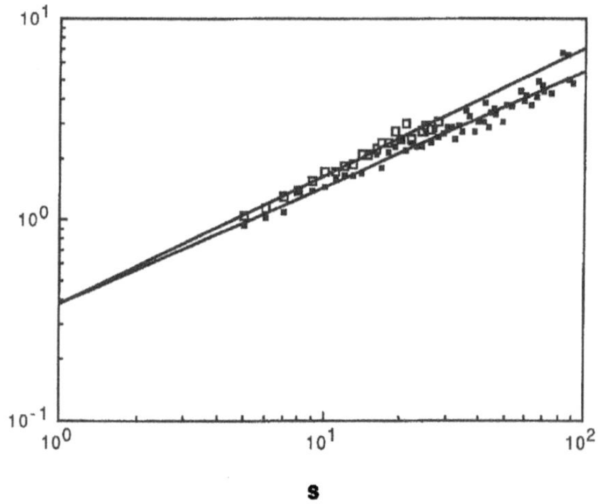

Fig. 6. A log-log plot of the radius of gyration R_g vs cluster mass s for reversible DLA-clusters ($C = 3.16 \cdot 10^{-4}$, $\theta = 0.1$): □) $p_A = 2^{-7}$ and $p_B = 2^{-7}$ (solid line $\rho = 0.637$, data from 504 clusters)

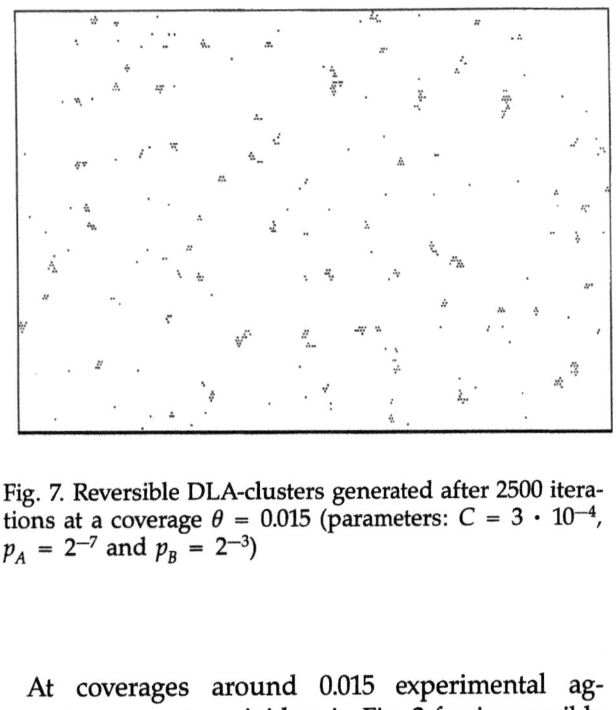

Fig. 7. Reversible DLA-clusters generated after 2500 iterations at a coverage $\theta = 0.015$ (parameters: $C = 3 \cdot 10^{-4}$, $p_A = 2^{-7}$ and $p_B = 2^{-3}$)

At coverages around 0.015 experimental aggregates are not as rigid as in Fig. 2 for irreversible DLA-clusters. Some type of reorganization must be considered. In an attempt to fit both the cluster concentration and the number of free single particles a good fit was obtained for a reversible DLA-cluster with $p_A = 2^{-7}$ and $p_B = 2^{-3}$, as seen in Fig. 7. These aggregates are also very similar to experimental aggregates, indicating that, at least in the initial stages, the ferritin adsorption induces reversible DLA-clustering according to the present model. For these parameters in the reversible DLA-clustering the cluster concentration vs initial concentration C is not very different from that of Fig. 3, indicating that the estimation of the surface diffusion constant can also be valid.

So far, we can see that these relatively simple models cannot account for the very low fractal dimension seen in experiments on ferritin aggregates at coverages around 0.1 ($d_f = 1.05$ in [1]). More sophisticated models are needed, perhaps involving more complex clustering of clusters. Lower fractal dimensions $d_f = 1.28 \pm 0.03$ have been reported in two dimensions by considering tip-to-tip clustering of clusters [10, 11]. The Fe-core of the ferritin molecule could be responsible for magnetic effects which may be important for the observed effects.

Acknowledgement

The present study was supported by a grant from the National Swedish Board for Technical Development (Grant No 89-468P).

References

1. Nygren H, Stenberg M (1990) Progr Colloid Polym Sci 82:15—18
2. Sander L (1986) Nature 322:789—793
3. Meakin P (1988) Adv Colloid Int Sci 28:249—331
4. Lin M, Lindsay H, Weitz D, Ball R, Klein R, Meakin P (1989) Nature 339:360—362
5. Stenberg M, Nygren H (1988) Bio Sci Abstr 1:257
6. Wolfram S (1984) Nature 311:419—424
7. Kinzel W (1985) Z Phys B 58:229—224
8. Witten T, Sander L (1981) Phys Rev Lett 47:1400—1403
9. Botet R, Jullien R (1985) Phys Rev Lett 55:1943—1946
10. Kolb M (1986) J Phys A Math Gen 19:L263—L268
11. Jullien R (1985) Phys Rev Lett 55:1697
12. Jullien R (1986) J Phys A Math Gen 19:2129—2136

Authors' address:

Dr. M. Stenberg
Department of Solid State Electronics
Chalmers University of Technology
412 96 Göteborg, Sweden

Surface-induced aggregation of ferritin at a liquid-solid interface

H. Nygren and M. Stenberg

Department of Histology University of Göteborg, and
Department of Solid State Eletronics Chalmers University of Technology, Göteborg, Sweden

Abstract: The kinetics of adsorption and the spatial distribution of adsorbed ferritin was studied by transmission electron microscopy (TEM) using a hydrophobic quartz grid as solid substrate. It was found that the initial adsorption process appears as a growth of molecular clusters at the liquid-solid interface. The rate of adsorption was limited by the mass transport of ferritin from the bulk to the liquid-solid interface. — Density-density correlation analysis $c(r)$ of the mass distribution of protein within the clusters by computer-aided image analysis showed a fractal dimension $d_f = 1.05$ at short distances (100 nm). The fractal dimension of the clusters is smaller than values obtained from theroetical models of two-dimensional diffusion-limited aggregation, indicating that the mass transport limitation of the adsorption process does not determine the structure of ferritin clusters. More complex models have to be used to account for the experimental observations.

Key words: Protein adsorption; solid phase; fractal; aggregation; electron microscopy

Introduction

The adsorption of proteins to hydrophobic surfaces is a rapid process, often diffusion-rate-limited, whereas the reverse reaction is orders of magnitude slower, effectively making the adsorption irreversible [1—5].

Examining the mechanism behind this effect we have studied the spatial distribution of adsorbed proteins by transmission electron microscopy (TEM). We report data on the kinetics of adsorption of ferritin onto a hydrophobic quartz surface, and on the structure of the adsorbed layer. Ferritin was chosen as model protein because of its iron nucleus which is easily seen in the electron microscope without further staining.

Experimental

Sample preparation: Hydrophobic, methyl silanized quartz grids were made as described [6]. Horse spleen ferritin (Fluka, Switzerland. Crystallized ×3) was dissolved in phosphate-buffered saline (PBS; 0.02 M phosphate buffer pH 7.2; 0.15 M Sodium chloride). Quartz grids were placed inverted on drops of PBS in a humified chamber. Ferritin solution was injected into the drops to a final concentration of 10 μg/ml. The adsorption was terminated by rinsing with PBS for 5 s. In some experiments, the grids were then incubated with 0.5% glutaraldehyde in PBS for 30 min. The grids were finally dried in a dry-air current.

Electron microscopy: The grids were examined in an electron microscope at 60 kV accelerating voltage. Electron micrographs were taken at ×20000 and copies were made at ×100000 magnification. The number of adsorbed molecules/m^2 was counted at 2—4 image fields (25—100 cm^2) for each picture.

Computer-aided image analysis: The micrographs were scanned, digitized, and analyzed by a computer with a technique similar to that used for analyzing computer-simulated aggregates [7, 8]. Whole pictures were scanned and digitized. A fast-Fourier-tranform was used to get the power spectrum, and an inverse transform was used to give the autocorrelation function. The inverse transform was averaged over all directions to give the radical correlation functions. For analyzing clusters an estimate of the density-density correlation function can be evaluated as described previously [9]. An estimate of the density-density correlation function $c(r)$ was evaluated as shown in Fig. 3. The correlation function $c(r)$ describes the probability of finding a second particle at a distance r in the same cluster and it is related to the scale invariance of the cluster by [10]:

$$c(r) = kr^{-(d-d_f)} , \quad (1)$$

where k is a constant, d is the dimension of space, and d_f is a fractal dimension characterizing the cluster.

Results and discussion

Electron micrographs of adsorbed ferritin are shown in Fig. 1. At a bulk concentration of ferritin of 10 µg/ml, the initial adsorption is seen as single molecules or clusters of 2—9 molecules at 2.5×10^9 sites/cm^2. The adsorption proceeds mainly as growth of the clusters formed initially (Figs. 1a—f). After 100 s the adsorbed ferritin is rarely found as single molecules, but is most commonly distributed as clusters of 10—100 molecules at 5×10^9 sites/cm^2 (Fig. 1f).

Heterogeneous distribution of adsorbed ferritin has been reported previously [11] on carbon surfaces and polymer surfaces and is, thus, not restricted to the quartz surfaces used in our experiments. The structure of the clusters is not altered by fixation with glutaraldehyde (Fig. 1d), indicating that the clusters are present in the wet state and not induced by the drying procedure.

A plot of the amount of bound protein vs time is shown in Fig. 2 together with a graph respresenting the maximal amount of ferritin that could reach the surface by diffusion [12, 13]. A careful analysis of the initial adsorption shows an acceleration of the rate of adsorption from 0.68×10^9 molecules/cm^2 s during 0—5 s to 2.3×10^9 molecules/cm^2 during 5—10 s. The acceleration is damped by the mass transport limitation after 10 s and the cluster growth is diffusion-rate-limited from 10 s to 100 s adsorption time. The diffusion rate limitation is probably due to depletion of protein from the bulk phase close to the surface [13]. The mass transport limited rate of adsorption entails that all molecules that could reach the surface by diffusion are adsorbed. The fact that diffusion-limited adsorption could be measured by TEM shows that there is no significant loss of adsorbed material during the rinsing procedure.

The process of surface-induced aggregation of protein molecules from a stable solution can intuitively be interpreted as a nucleation and growth phenomenon, related to heterogeneous precipitation of salts from metastable solutions. The fact that single molecules are seen adsorbed initially invalidates this interpretation, but it is still possible that the presence of adsorbed single molecules and small aggregates increases the number of sites available for stable adsorption on the surface, thus causing the initial acceleration of the adsorption process. In Fig. 3 measured density-density correlation $c(r)$ is shown for clusters appearing at the surface, as shown in Fig. 1f. From the slope in the logarithmic diagram we get a fractal dimension $d_f = 1.05$ at distances <100 nm. This dimension is close to that ($d_f = 1.2 \pm 0.15$) reported for experimental aggregation of silicon particles at an air-liquid interface [14]. The fractal dimension of the clusters can tell us something about the mechanism behind the formation of clusters of ferritin at the liquid-solid interface. Aggregation can proceed according to four principally different boundary conditions; diffusion-limited (sticking probability = 1) or reaction-limited (sticking probability <1), irreversible or reversible. The fractal dimension of ferritin clusters found in the present study is too low to reflect diffusion-limited aggregation (DLA) in two dimensions, as introduced by Witten and Sander [7], that is ($d_f = 1.67$), which describes irreversible growth of clusters with a sticking probability of 1 for the random walker. This leads to the conclusion that the diffusion-rate limitation of the adsorption process (Fig. 2) does not determine the structure of ferritin clusters. The aggregation of molecules at the surface must proceed according to other boundary conditions.

After searching for models of fractal growth using such boundary conditions for the aggregation process we found that the measured fractal dimension is lower than the values predicted by theoretical models describing cluster-cluster aggregation [15, 16], kinetic growth [17] and models describing growth of branched polymers [18]. The spatial distribution of molecules within the clusters of adsorbed ferritin is, thus, not in accordance with previously described computer simulations of particle aggregation. However, recently developed models for reversible DLA show promising results [19].

In conclusion, the initial adsorption of ferritin at a hydrophobic surface can be described as a

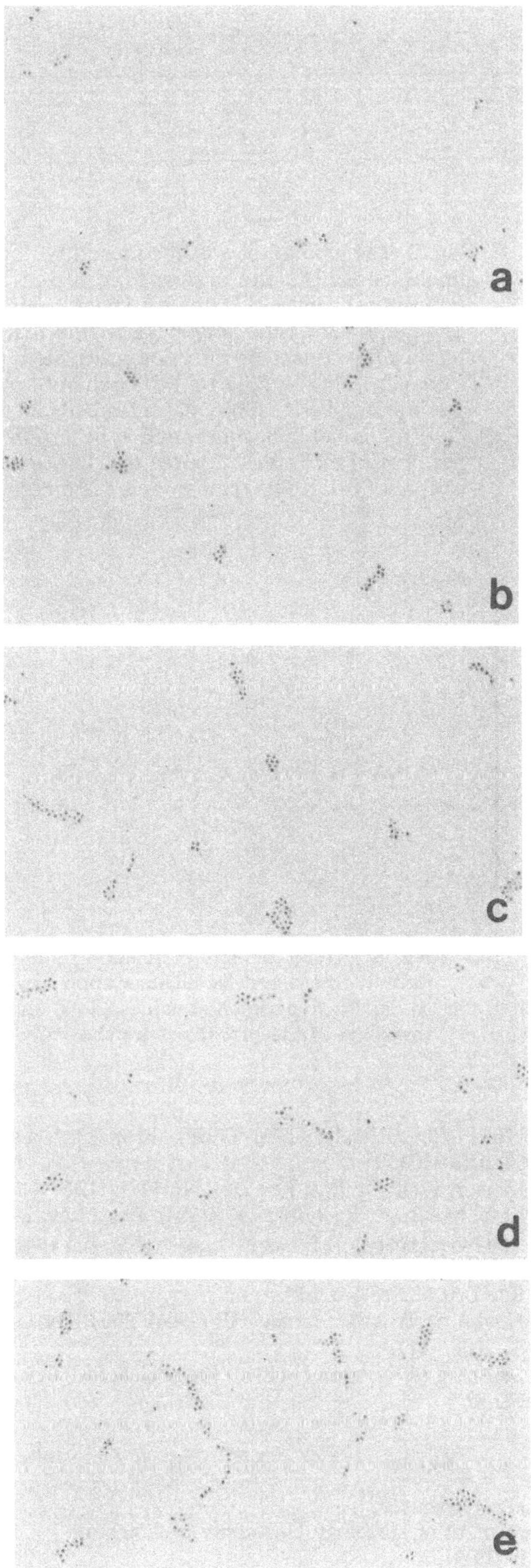

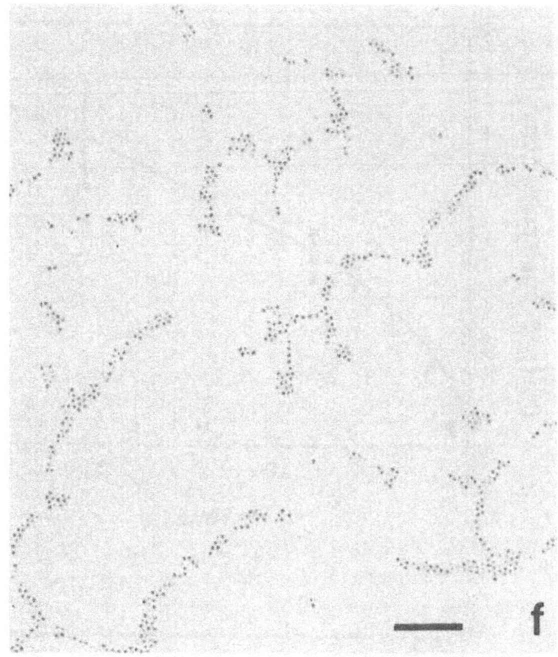

Fig. 1. Electron micrographs of ferritin adsorbed on methylized quartz from a solution containing 10 μg/ml of ferritin. Incubation time: a) 5 s; b) 10 s, c) 20 s, d) 25 s followed by fixation with glutaraldehyde; e) 40 s, and f) 100 s. Bar = 0.1 μm

surface-induced fractal aggregation which becomes mass-transport-limited, due to the formation of a concentration gradient of protein in the bulk phase. The adsorbing molecules are laterally mobile along the surface and the clusters are probably restructured during their formation. These experimental data may serve as a guideline for the development of theoretical models of protein adsorption.

Acknowledgement

The present study was supported by grants from the Swedish Medical Research Council (12X-06235) and The Board of Technical Development (STUF).

References

1. Gorman RR, Stoner GE, Catlin A (1971) J Phys Chem 75:2103–2107
2. Brynda E, Houska M, Lednicky F (1986) J Coll Int Sci 113:164–169

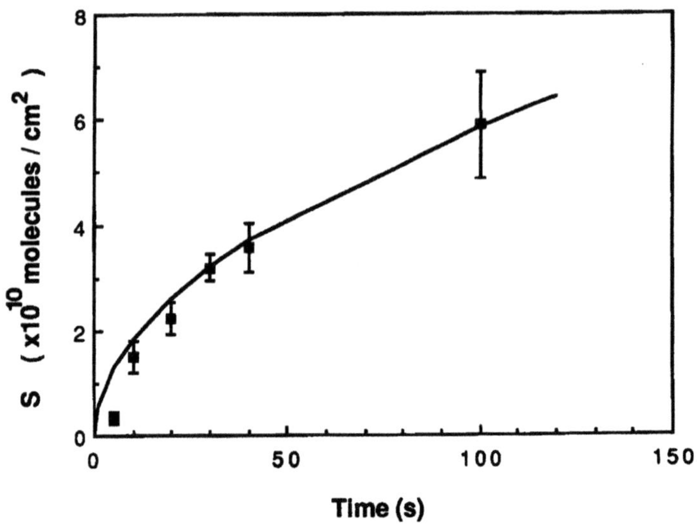

Fig. 2. The amount of adsorbed ferritin (molecules/cm²) plotte vs time of adsorption.Vertical bars = S.D.; n = 8 image fields. The graph shows the amount of ferritin that reaches the surface by diffusion, calculated from [12, 13] ($S = (2/\sqrt{\pi})C_0\sqrt{Dt}$), where S is the number of molecules adsorbed/cm²; C_0 is the initial bulk concentration of ferritin (number of molecules/ml); d is the diffusion constant (3.61×10^{-7} cm²/s), and t is time (s)

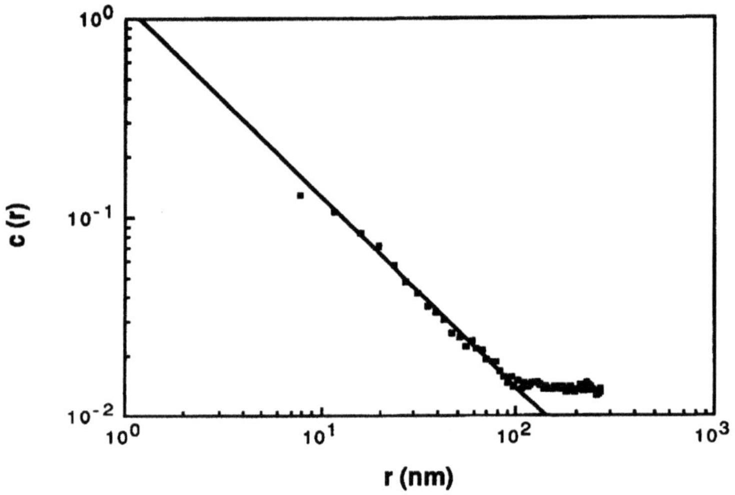

Fig. 3. Density-density correlation function $c(r)$ measured for clusters appearing in Fig. 1f. Experimental values (dots) and the slope of the plot (line) are shown

3. Norde W, MacRitchie F, Nowicka G, Lyklema J (1986) J Coll Int Sci 112:447—450
4. Chan BMC, Brash JL (1981) J Coll Int Sci 82:217—222
5. Wojcieshowskij P, Ten Hove P, Brash JL (1986) J Coll Int Sci 111:455—465
6. Stenberg M, Stemme G, Nygren H (1987) Stain Technol 62:231—236
7. Witten TA, Sander LM (1981) Phys Rev Lett 47:1400—1402
8. Witten TA, Sander LM (1981) Phys Rev B 27:5686—5694
9. Stenberg M, Nygren H (1987) Phys Rev Lett 59:1164—1166
10. Sander LM (1986) Nature 322:789—793
11. Feder J, Giaever I (1980) J Coll Int Sci 78:144—154
12. Trurnit HJ (1954) Arch Biochem 51:176—181
13. Stenberg M, Stiblert L, Nygren H (1986) J Theoret Biol 120:129—140
14. Hurd AJ, Schaefer DW (1985) Phys Rev Lett 54:1043—1046
15. Meakin P (1983) Phys Rev Lett 51:1119—1122
16. Kolb M, Botet R, Jullien R (1983) Phys Rev Lett 51:1123—1126
17. Majid I, Jan M, Coniglio A, Stanley HE (1984) Phys Rev Lett 53:1257—1260
18. Havlin S, Trus B, Stanley HE (1984) Phys Rev Lett 53:1288—1291
19. Stenberg M, Nygren H (1990) Progr Colloid Polymer Sci 82

Authors' address:

Dr. H. Nygren
Department of Histology University of Göteborg
P.O.B. 33031
40033 Göteborg, Sweden

Progress in Colloid & Polymer Science Progr Colloid Polym Sci 82:19—27 (1990)

Determination of the adsorption energy distribution function with the LOGA method

L. K. Koopal[1]), M. M. Nederlof[2]) and W. H. van Riemsdijk[2])

[1]) Department of Physical and Colloid Chemistry, Wageningen, The Netherlands
[2]) Department of Soil Science and Plant Nutrition, Wageningen, The Netherlands

Abstract: A method to determine the adsorption energy distribution function from adsorption data is presented. The overall isotherm on a heterogeneous surface is the summation of local adsorption contributions. The distribution function of the adsorption energy of affinity can be calculated, provided the overall isotherm and the local isotherm function are known. The LOGA method uses a special function to approximate the local isotherm, so that an analytical expression results for the energy distribution function. Results of the LOGA method are presented for "synthetic" overall isotherms based on various energy distribution functions and the Langmuir equation as local isotherm. Special attention is paid to the effect of experimental errors in the overall isotherm data. In this case pretreatment of the data is required. A smoothing spline (SP) procedure is recommended in which the extent of smoothing of the data corresponds with the experimental error in the data. A comparison of results obtained with the SP-LOGA method and the more sophisticated CAESAR method shows that the SP-LOGA result is satisfactory but is somewhat less precise than that obtained with CAESAR. In the presence of lateral interaction the expression for the LOGA distribution function depends on the type of heterogeneity. Expressions are given for both random and patchwise surfaces.

Key words: Determination of Adsorption energy distribution; determination of Affinity distribution function; Local isotherm approximation Local isotherm approximation by LOGA; comparison of "CAESAR" and "LOGA"

Introduction

Solid adsorbents will, in general, have non-uniform or heterogeneous surfaces. This heterogeneity can be the result of different crystal planes, crystal plane imperfections, different surface groups, amorphous solid structure, microporosity, etc. Understanding the adsorptive properties of solids, therefore, relies on the characterization of the surface heterogeneity with respect to adsorption. Such a characterization is especially useful for a comparison of similar but slightly different adsorbents. For instance, to study the effect of a temperature treatment on the adsorbent properties or to compare natural adsorbents which have a similar origin.

For the description of the adsorption on heterogeneous surfaces in general two extreme situations are considered: "random" and "patchwise" heterogeneity. For "random" heterogeneous surface the "equal-energy" sites are regularly or randomly distributed over the entire surface. On a "patchwise" surface the "equal-energy" sites are grouped together in patches that are assumed to behave independently, i.e., the boundary effects are neglected.

The description of the adsorption usually starts with the description of the adsorption on an ensemble of "equal-energy" sites as given by the local isotherm function. For the local isotherm, in principle, all adsorption equations derived for homogeneous surfaces can be applied. Through the local isotherm a distinction is made between interactions of the adsorbate molecules with the surface and those between the adsorbate molecules themselves. In this way the standard free energy of adsorption or the intrinsic affinity between adsorbate and surface is separated from the lateral interactions. For a patchwise heterogeneous surface the lateral interactions are thought to be restricted within the limits of a patch, whereas for a random surface these interactions are counted over the entire surface. In the absence of lateral interactions an ideal situation exists and the difference between patchwise and random heterogeneity vanishes.

Once an appropriate local isotherm is established the overall isotherm simply results as a weighted summation of the local adsorption contributions. For a discrete distribution of equal energy sites, we obtain for the overall adsorption:

$$\Gamma/\Gamma_m = \theta_t = \Sigma_i f_i \theta_i(K_i, x, \omega) \;, \tag{1a}$$

and for a continuous distribution

$$\Gamma/\Gamma_m = \theta_t = \int_{\Delta} \theta(K, x, \omega) f(\log K) d\log K \;, \tag{1b}$$

where Γ is the adsorption, Γ_m the monolayer adsorption, θ_t the overall coverage, θ_i or $\theta(K, x, \omega)$ the local isotherm equation with K_i or K the intrinsic affinity, x the mole fraction or relative pressure, ω the lateral interaction parameter, f_i or $f(\log K)$ the distribution of intrinsic affinities, and Δ is the range of $\log K$ values. Equation (1a) or (1b) is the starting point for the analysis.

When $\theta_t(x)$ or $\Gamma(x)$ is measured and for $\theta(K, x, \omega)$ one of the common local isotherm equations is assumed, f_i or $f(\log K)$ can be calculated by inverting the integral equation. With respect to this procedure it should be noted that there is always some arbitrariness in the choice of the local isotherm or in the way in which $\theta(K, x, \omega)$ and $f(\log K)$ are separated. Preferably, some independent information should be obtained about the magnitude of the lateral interactions.

Another problem with the inversion of Eq. (1) lies in the fact that Eq. (1b) is a so-called Fredholm integral equation of the first kind [1]. Inversion of Eq. (1b) to obtain $f(\log K)$ leads to instable solutions unless special precautions are taken. The most rigorous way is to use a special mathematical technique such as regularization [1–4] or Singular Value Decomposition [5, 6]. However, these methods are complicated and difficult to handle when only part of the overall isotherm is available. An alternative, but somewhat less rigorous way is to use a special function to approximate the local isotherm. Nederlof et al. [7] have shown that for well chosen functions the inversion of Eq. (1) can be done analytically. The most general approximation is a logarithmically symmetrical local isotherm or LOGA isotherm. It is the purpose of this paper to study this approximation in more detail by investigating the effect of experimental errors in θ_t and smoothing of θ_t on the calculated distribution function. Also, a comparison will be made with a result obtained with Singular Value Decomposition.

The LOGA method to obtain $f(\log K)$

In order to illustrate the LOGA method the Langmuir-equation is used as the local isotherm

$$\theta(K, x) = \frac{Kx}{1 + Kx} \;, \tag{2}$$

hence, lateral interactions are neglected for the moment. A simple, but poor approximation of the Langmuir isotherm is the Condensation Approximation (CA) introduced by Roginsky [8] and worked out by Hobson [9] and Harris [10]. The CA replaces Eq. (2) by a stepfunction:

$$\theta(K, x) = 0 \quad \text{for} \quad Kx \leqslant (Kx)^* \tag{3a}$$

$$\theta(K, x) = 1 \quad \text{for} \quad Kx > (Kx)^* \;, \tag{3b}$$

where $(Kx)^*$ is the intersection point between the true isotherm and the CA isotherm at $\theta = 0.5$. According to Eq. (2) at this point $(Kx)^* = 1$ or $x^* = 1/K^*$. Inserting Eq. (3) in Eq. (1b) gives

$$\theta_t(x^*) = \int_{\log K^*}^{\infty} f(\log K) d\log K \;. \tag{4}$$

Differentiation of Eq. (4) with respect to $\log x$ leads to the CA distribution function:

$$f_{CA}(\log K) = \frac{\partial \theta_t(x)}{\partial \log x} \tag{5a}$$

$$\log K = -\log x \;. \tag{5b}$$

The conversion of $\log x$ to $\log K$ follows from Eq. (2) for $\theta = 0.5$. Thus the CA distribution is simply obtained by taking the first derivative of the adsorption isotherm with respect to $\log x$, followed by substitution of $\log x = -\log K$ when Eq. (2) is the local isotherm. Evidently the CA method is rather crude in the case of Eq. (2) as local isotherm. A better result is obtained if an approximation is used which follows Eq. (2) more precisely. Some approximate functions and their corresponding distribution functions have been suggested in [7]. A good result can be obtained with the LOGA function

$$\theta(K,x) = \alpha(Kx)^{\beta} \qquad \text{for } Kx \leqslant (Kx)^* \tag{6a}$$

$$\theta(K,x) = 1 - \alpha(Kx)^{-\beta} \quad \text{for } Kx > (Kx)^* , \tag{6b}$$

where α, β, and the breakpoint value $(Kx)^*$ depend on the local isotherm equation. Note that Eq. (6), when plotted as $\theta(\log Kx)$, is symmetrical in $\theta = 0.5$. This symmetry point is used to determine the breakpoint value $(Kx)^*$. For the Langmuir equation as local isotherm we find $(Kx)^* = 1$. The distribution function corresponding with Eq. (6) and $(Kx)^* = 1$ is given by Eq. (7):

$$f_{LOGA}(\log K) = \frac{\partial\theta_t(x)}{\partial\log x} - \frac{0.189}{\beta^2}\frac{\partial^3\theta_t(x)}{(\partial\log x)^3} - \frac{0.189(2\alpha-1)}{\beta^2}\frac{d^2 f_{LOGA}(\log K)}{(d\log K)^2} \tag{7a}$$

$$\log K = -\log x . \tag{7b}$$

The conversion of $\log x$ to $\log K$ (Eq. (7b)) follows from the breakpoint condition, that is, by solving Eq. (2) for $\theta = 0.5$. Equation 7 is an implicit function from which $f(\log K)$ can be determined iteratively. Several suggestions have been made for α and β in case of the Langmuir equation as local isotherm. Some of the solutions are in fair agreement with previous work [11–17]. Note that for $\alpha = 0$, Eq. (6) reduces to Eq. (3), and Eq. (7) reduces to Eq. (5), the Condensation Approximation. Below we will consider mainly the case that $\alpha = 0.5$ and $\beta = 0.7$ for which Eq. (6) is a very good approximation of Eq. (2) (see Fig. 1). Moreover, for $\alpha = 0.5$ the last term on the RHS of Eq. (7a) vanishes so that no iterations are required for the calculation of $f(\log K)$. We denote this method as LOGA-1.

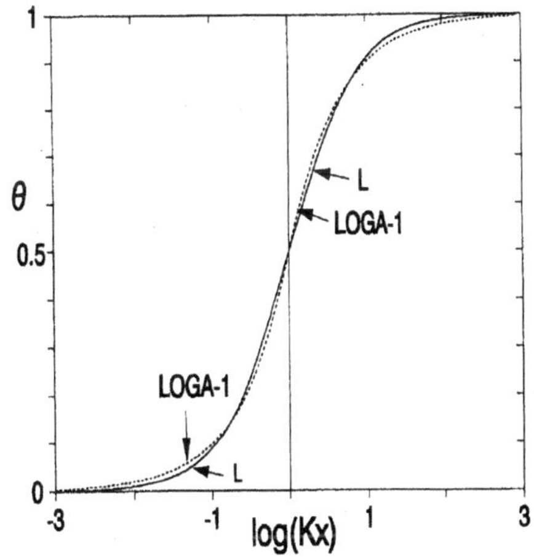

Fig. 1. The Langmuir isotherm and its LOGA approximation (see Eq. (6)). The LOGA parameters are $\alpha = 0.5$ and $\beta = 0.7$

Results and discussion

Exact data

In order to illustrate the potentials of the LOGA method, a set of synthetic overall adsorption isotherms is used so that the exact distribution functions are known. The synthetic isotherms and the corresponding distribution functions are shown in Figs. 2a and b, respectively. These figures also give an indication of the way in which surface heterogeneity affects the isotherm. Note that the logarithm of the mole fraction (or relative pressure) is plotted and that a dimensionless scale is obtained by using $\log \tilde{K}x$ instead of $\log x$, where $\tilde{K}$ is the average K value. The LOGA-1 distribution ($\alpha = 0.5$ and $\beta = 0.7$), corresponding to the Langmuir isotherm, can be written as

$$f_{LOGA\text{-}1}(\log K) = \frac{\partial\theta_t(x)}{\partial\log x} - 0.386\frac{\partial^3\theta_t(x)}{(\partial\log x)^3} \tag{8a}$$

$$\log K = -\log x . \tag{8b}$$

Results obtained for the three test isotherms are shown in Fig. 3. Sharp peaks are difficult to recover.

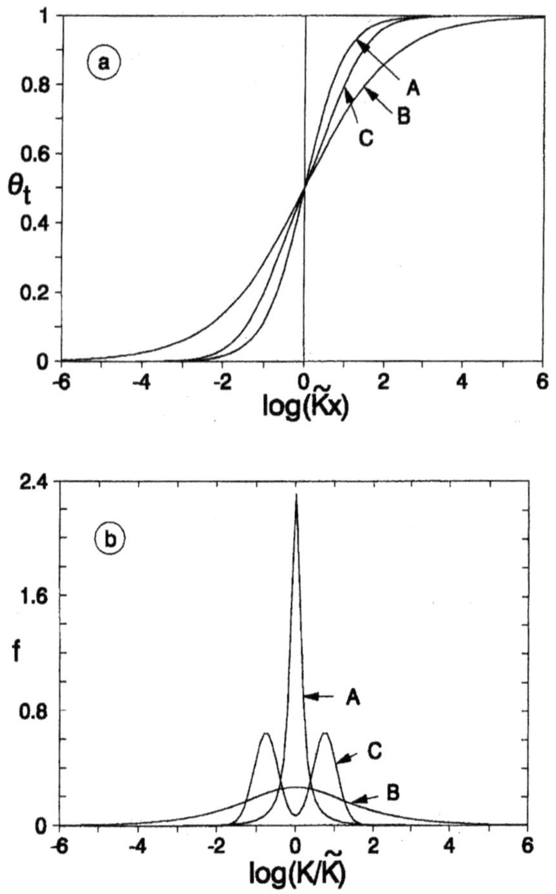

Fig. 2. a) Synthesized overall isotherms based on the Langmuir equation and three different distribution functions. b) The distribution functions on which the three isotherms are based

Figure 3a shows that a sharp peak with a "half width" of about 0.25 $\log K$ units results in a less sharp LOGA-1 peak with a "half width" of about one $\log K$ unit. The best result is obtained for the wide distribution function (see Fig. 3b). The bi-Gaussian distribution function (Fig. 3c) is also flattened, but the two peaks are reasonably well recovered.

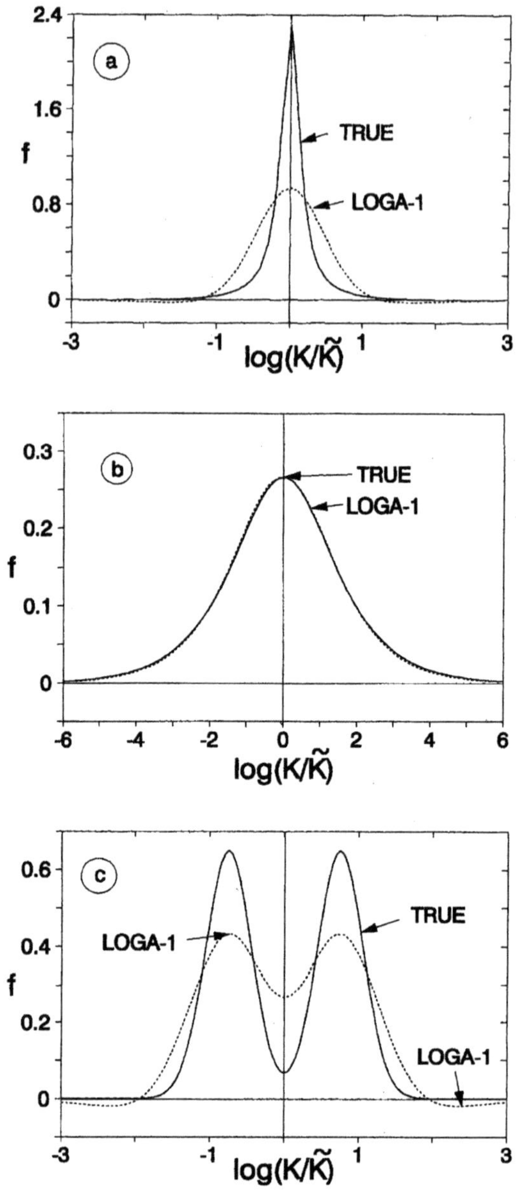

Fig. 3. Comparison of the LOGA-1 ($\alpha = 0.5$, $\beta = 0.7$) distributions with the true distribution functions. The error in the overall isotherm is negligible

Experimental errors and smoothing of θ_t

In reality the overall isotherm data will contain errors. The effect of an error of 5% in θ_t on the calculated distribution function is shown in Fig. 4 for the wide distribution; clearly, the result is very poor. In order to obtain more meaningfull results smoothing of the data is required. To this order we have used a smoothing spline procedure [18, 19] which provides continuous derivatives up to the fourth derivative (third order spline). The optimal solution is obtained when

$$\sum_{j=1}^{m} \left(\frac{\theta_{t,j} - \theta_{t,j}(sp)}{\theta_{t,j}(sp)} \right)^2 \Big/ \sigma_j^2 = m \, . \tag{9}$$

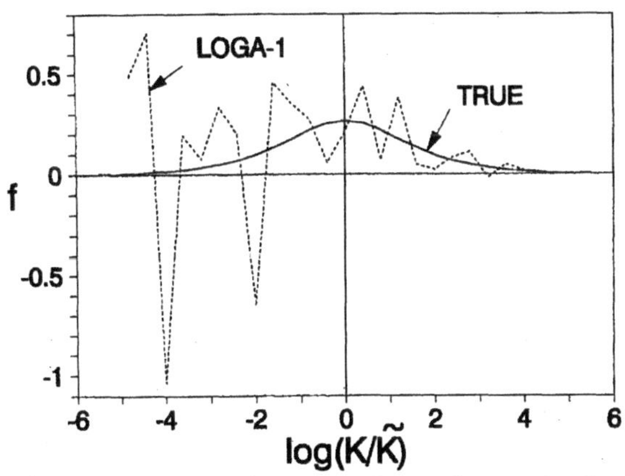

Fig. 4. The effect of an experimental error in θ_t on the LOGA-1 distribution. The error in θ_t is 5%. The result is compared with the true distribution function (isotherm B)

Where $\theta_{t,j}(sp)$ is the calculated value of $\theta_{t,j}$ after smoothing, and σ_j is the relative error in θ_t. The value of the smoothing parameter is adjusted until Eq. (9) is satisfied. A requirement for this procedure is that σ_j is known. In the absence of knowledge about σ_j a smoothing spline procedure in combination with cross validation [1, 20] can be used. In a forthcoming paper we will consider this aspect in more detail. Results obtained with the LOGA method after smoothing of the data will be indicated as SP-LOGA.

In Fig. 5 the SP-LOGA-1 results are shown for the three test functions. As expected, the wide distribution function again gives the best result (Fig. 5b). The applied smooting is adequate, the wild oscillations as shown in Fig. 4 have disappeared. Due to the error, the true distribution is somewhat less well approximated than in Fig. 3b, but the result is quite good. The sharp distribution function (Fig. 5a) and the bi-Gaussian distribution function (Fig. 5c) cannot be recovered very precisely, but fairly smooth distribution curves result. The isotherm data corresponding with the bi-Gaussian distribution function are, in this case, not good enough to distinguish the two peaks.

In general, Fig. 5 clearly indicates that with an error level of 5%, one can only poorly discriminate with respect to heterogeneity. For a description of such inaccurate results a simple Freundlich-type of equation [21] will often be adequate.

In practice, even excellent adsorption data still contain an error. As a final example an error level of 0.1% in θ_t is therefore assumed. The results obtained with the SP-LOGA-1 method using the three test isotherms are shown in Fig. 6. From these figures it follows that, in this case, the LOGA procedure allows a detailed heterogeneity analysis; the results are very close to those obtained in the absence of errors.

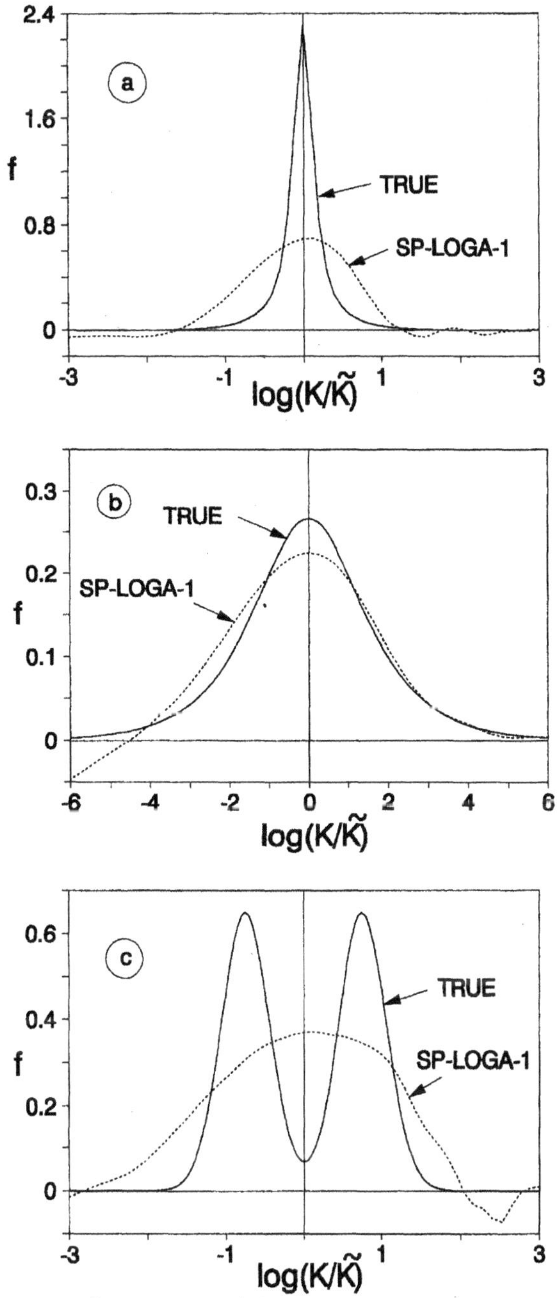

Fig. 5. Comparison of the SP-LOGA-1 distributions and the true distributions. The error in θ_t is 5%; pre-smoothing of the isotherm is done with the smoothing spline (SP) method (see text)

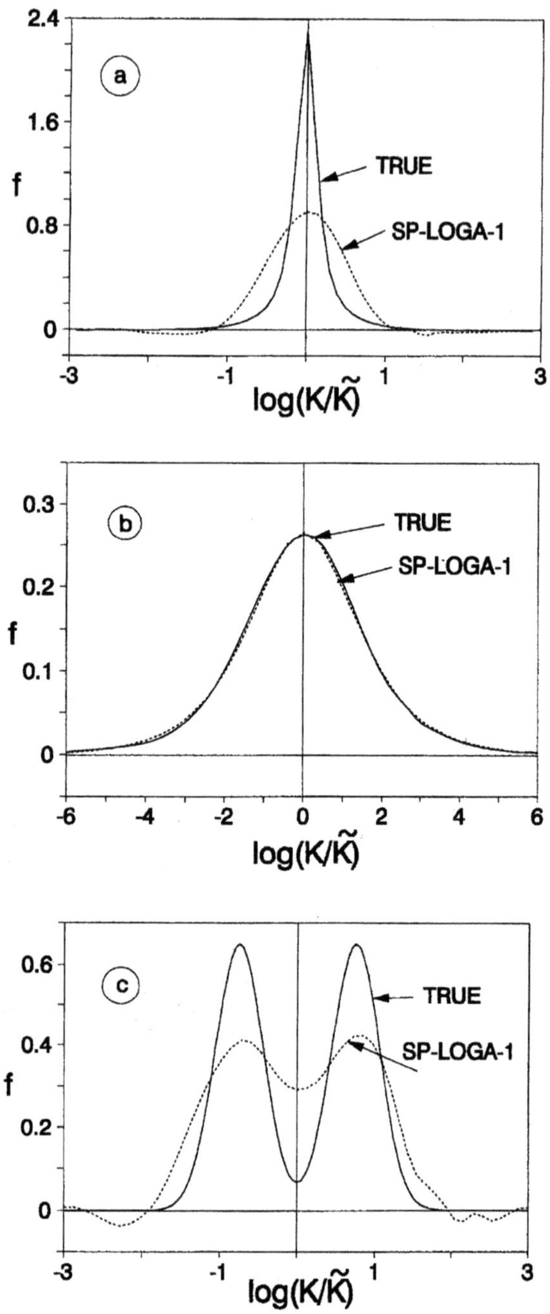

Fig. 6. Comparison of the SP-LOGA-1 distributions with the true distribution functions in the case of an error of 0.1% in θ_t

Comparison with CAESAR

In literature several numerical methods [1–6, 22, 23] have been presented for the calculation of the adsorption energy distribution function. One of these methods, CAESAR [5, 6] controls the ill posedness of Eq. (1b) by Singular Value Decomposition and acknowledges explicitly the presence of experimental errors. The two aspects make it one of the best methods presently available to calculate $f(\log K)$. The main advantage of the CAESAR method over the SP-LOGA method is that the "true" local isotherm equation is used. Comparison of results obtained with CAESAR and SP-LOGA-1 will show how close the latter method approximates the best possible result in practice. As data set for the comparison an overall isotherm based on the Langmuir equation as local isotherm and a bi-Gaussian distribution function is chosen. The distribution function is the same as that used to test CAESAR [5, 6]. The relative error in θ_t is taken to be 0.1%, representing a high quality data set.

The distribution functions calculated with this data set using CAESAR and SP-LOGA-1 are shown in Fig. 7, together with the true distribution function. The distribution obtained with CAESAR closely approximates the true distribution function, but the peak in the low affinity range (where the absolute errors in θ_t are the largest) is not fully recovered. This is related to the use of the Langmuir equation as local isotherm. A similar data set based on the Hill-de Boer equation including lateral attraction resulted in an accurate prediction of both peaks [6].

The SP-LOGA-1 distribution is less accurate. The approximation of the local isotherm leads to smoothing of the distribution function and a less

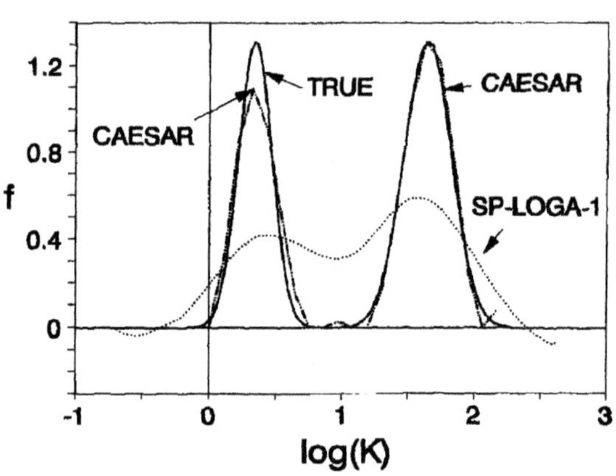

Fig. 7. Comparison of the result obtained with CAESAR [5, 6] and SP-LOGA-1. The experimental error in the isotherm (θ_t) is 0.1%. The Langmuir equation is used as local isotherm and the (true) distribution function is the same as that used in [6]

sharp distinction of the two peaks. The deviations from the true distribution are similar to those shown in Fig. 6c. A comparable result would have been obtained with the CAESAR method if (incorrectly) some lateral attraction had been incorporated [6]. For a final judgement of the SP-LOGA method it should be noted that the data set as used covers almost the entire range of θ_t values. The SP-LOGA method can also be used with an incomplete data set, whereas with CASESAR instabilities will occur in the distribution function when the window of θ_t is too small. Especially in the study of strongly heterogeneous natural materials, often only a rather small window of data is available [24].

Lateral interactions

So far the LOGA method has been used with the Langmuir equation as local isotherm, i.e., lateral interactions between the adsorbate molecules have been neglected. However, in the presence of lateral interactions the LOGA method can also be used. Preferably the Frumkin-Fowler-Guggenheim or FFG [25, 26] equation is chosen as local isotherm, because also in this case $\theta(\log Kx)$ is symmetrical in $\theta = 0.5$. The FFG equation can be written as:

$$\frac{\theta}{1-\theta} = Kx\exp(-\omega\theta^*) \,, \tag{10}$$

where ω is the lateral interaction parameter. For random heterogeneity $\theta^* = \theta_t$ and for patchwise surfaces $\theta^* = \theta$.

Random heterogeneity

With random heterogeneity the lateral interactions are counted independent of the type of sites surrounding a specific site. When the overall isotherm is known, θ^* ($= \theta_t$) in Eq. (10) is known. Hence, it is possible to calculate

$$x' = x\exp(-\omega\theta_t) \tag{11}$$

for each x and to replace $x\exp(-\omega\theta_t)$ in Eq. (10) by x'. In other words, for random surfaces Eq. (2) applies as local isotherm if x is replaced by x'. Consequently, the LOGA isotherm with $\alpha = 0.5$, $\beta = 0.7$ and breakpoint value $(Kx')^* = 1$ is an appropriate approximation of $\theta(\log Kx')$. The distribution function, denoted as f(LOGA-R-1), can be derived from Eq. (7) for $\alpha = 0.5$ and $\beta = 0.7$:

$$f_{LOGA\text{-}R\text{-}1}(\log K) = \frac{\partial\theta_t}{\partial\log x'} - 0.386\,\frac{\partial^3\theta_t}{(\partial\log x')^3} \tag{12a}$$

$$\log K = -\log x' = -\log x + 0.434\,\omega\theta_t \,. \tag{12b}$$

The conversion of $\log x$ to $\log K$ follows from the breakpoint condition, i.e. by substitution of the value $\theta = 0.5$ in the FFG equation. Hence, the only difference with Eq. (8) is the transformation of $\log x$ to $\log K$.

Patchwise heterogeneity

For patchwise surfaces and the FFG equation as local isotherm, θ^* in Eq. (10) equals θ. Equation (6) is now used to approximate the FFG equation. The breakpoint value $(Kx)^*$ is derived from the symmetry point at $\theta = 0.5$, this results in $(Kx)^* = \exp(\omega/2)$. The values of α and β depend on the value of ω. In Fig. 8 values of α and β are plotted as a function ω. The values are optimized by using a least sum of squares technique to minimize the difference between the LOGA and the true local isotherm.

Due to the fact that the breakpoint value is a function of ω the solution for the distribution function is no longer given by Eq. (7), but as:

$$\begin{aligned} f_{LOGA\text{-}P}(\log K) &= \frac{\partial\theta_t(x)}{\partial\log x} - \frac{0.189}{\beta^2}\,\frac{\partial^3\theta_t(x)}{(\partial\log x)^3} \\ &+ \frac{0.434}{\beta}\left\{2\alpha\sinh\frac{\beta\omega}{2}\right\}\frac{df_{LOGA\text{-}P}(\log K)}{d\log K} \\ &- \frac{0.189}{\beta^2}\left\{2\alpha\cosh\frac{\beta\omega}{2} - 1\right\}\frac{d^2 f_{LOGA\text{-}P}(\log K)}{(d\log K)^2} \end{aligned} \tag{13a}$$

$$\log K = -\log x + 0.434\,\omega/2 \tag{13b}$$

The derivation of Eq. (13) is similar to that of Eq. (7). Equation (13) is an implicit equation which cannot as easily be simplified as can Eq. (7), i.e.,

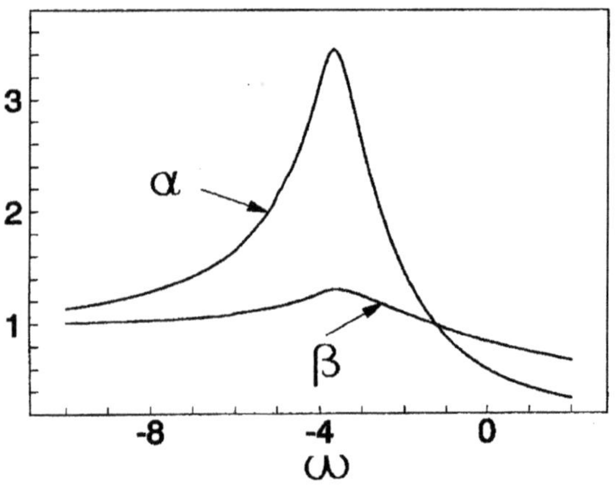

Fig. 8. Optimized values of α and β for negative values of ω

by taking $\alpha = 0.5$. This makes Eq. (13) less attractive for use in practice. For $\omega < -4$ the FFG isotherm shows a two-dimensional (2D) phase transition or 2D-condensation step around $\theta = 0.5$. For $\omega = -4.5$ the condensation step already covers a range of θ values from 0.225 to 0.775. Hence, for strong lateral attraction the Condensation Approximation is a reasonable approximation and Eq. (13) reduces to:

$$f_{CA\text{-}P}(\log K) = \frac{\partial \theta_t(x)}{\partial \log x} \tag{14a}$$

$$\log K = -\log x + 0.434\omega/2 \; . \tag{14b}$$

For $-4.5 < \omega < 0$ Eq. (14) can be used to simplify Eq. (13). Substitution of $f_{CA\text{-}P}(\log K)$ for $f_{LOGA\text{-}P}(\log K)$ on the RHS of Eq. (13) results in:

$$f^{(1)}_{LOGA\text{-}P}(\log K) = \frac{\partial \theta_t(x)}{\partial \log x} - \frac{0.434}{\beta} \times \left\{2\alpha \sinh \frac{\beta\omega}{2}\right\} \frac{\partial^2 \theta_t(x)}{d\log x^2} - \frac{0.189}{\beta^2} \left\{2\alpha \cosh \frac{\beta\omega}{2}\right\} \frac{\partial^3 \theta_t(x)}{(\partial \log x)^3} \tag{15a}$$

$$\log K = -\log x + 0.434\omega/2 \; . \tag{15b}$$

Equation (15) can be seen as a first iteration step in solving Eq. (13). Equation (15) is considerably more attractive than Eq. (13) and can be used for not too strong lateral attraction.

Conclusion

— The LOGA method offers a simple and reliable method to calculate the affinity distribution function, provided that the affinity distribution function is not a too-rapid varying function, and that the precision of the experimental data is not too low.

— In general, pre-smoothing of the isotherm data is required. To this aim a smoothing spline procedure is recommended. The optimal solution is selected on the basis of the experimental error in the data.

— In the presence of lateral interactions the LOGA method also applies, the LOGA expression for the distribution function depends on the type of heterogeneity.

Acknowledgements

This work was partially funded by both the Netherlands Integrated Soil Research Programme under contract number PCBB 8948 and by the EC Environmental Research Programme on Soil Quality under contract number EV4V-0100-NL (GDF).

References

1. Merz PH (1980) J Comput Phys 38(1):64—85
2. Twomey S (1962) J Ass Comp Math 10:97—101
3. House WA (1978) J Colloid Interface Sci 67(1):166—180
4. Brown LF, Travis BJ (1983) Chem Eng Sci 38(6):843—847
5. Koopal LK, Vos CH (1985) Colloid Surf 14:87—95
6. Vos CH, Koopal LK (1985) J Colloid Interface Sci 105(1):183—196
7. Nederlof MM, Van Riemsdijk WH, Koopal LK (1990) J Colloid Interface Sci 135:410—426
8. Roginsky SS (1965) Adsorption and Catalysis on heterogeneous surfaces. Academy of Sciences USSR, Moscow (Russian)
9. Hobson JP (1965) Canadian J Phys 43:1934; idem 43:1941
10. Harris LB (1968) Surface Sci 10:129—145
11. Cerofolini GF (1971) Surface Sci 24:391—403
12. Cerofolini GF (1974) Thin Solid Films 23:129—152
13. Hsu CC, Wojciechowski BW, Rudzinski W, Narkiewicz J (1978) J Colloid Interface Sci 67:292—303
14. Rudzinski W, Jagiello J, Grillet Y (1982) J Colloid Interface Sci 87:478—491

15. Ninomiya K, Ferry JD (1959) J Colloid Interface Sci 14:36—48
16. Hunston DL (1975) Anal Biochem 63:99—109
17. Thakur AK, Munson PJ, Hunston DL, Rodbard D (1980) Anal Biochem 103:240—254
18. Schoenberg IJ (1964) Proc NAS 52:947—950
19. Reinsch CM (1967) Numer Math 10:177—183; idem (1971) Numer Math 16:451—454
20. Craven P, Wahba G (1979) Numer Math 31:377—403
21. Sips R (1948) J Chem Phys 16(5):490—495; idem (1950) J Chem Phys 18(8):1024—1026
22. Sacher RS, Morrison ID (1979) J Colloid Interface Sci 70:153
23. Bräuer P, Fassler M, Jaroniec M (1985) Thin Solid Films 123:245—272
24. Buffle J (1988) Complexation reactions in aquatic chemistry: an analytical approach. Ellis Horwood Ltd, Chicester England
25. Frumkin A (1925) Z Phys Chem 116:466; idem (1964) J Electro Anal Chem 7:152
26. Fowler RH, Guggenheim EA (1949) Statistical thermodynamics, Cambridge University Press, Cambridge, England

Authors' address:

Dr. L. K. Koopal
Wageningen Agricultural University
Department of Physical and Colloid Chemistry
Dreijenplein 6
6703 HB Wageningen, The Netherlands

Adsorption of some common surfactants and polymers on TiO_2-pigments

M. Huldén and E. Sjöblom

Institute for Surface Chemistry, Stockholm, Sweden

Abstract: The adsorption of nonylphenol deca (oxyethyleneglycol) monoether (NPE_{10}), sodium dodecyl sulphate (SDS), polyacrylic acid sodium salt (PAA) and ethyl hydroxyethyl cellulose (EHEC) on TiO_2-pigment coated with different mineral oxides is reported. The adsorption isotherms are compared with data obtained by electrophoretic mobility measurements. Despite the limitations inherent in the mobility measurements, we show that this technique can be used to gain adequate information on the adsorption. — Coating of the TiO_2-pigment with different mineral oxides has a strong influence on the adsorption. This can be understood from the different acid/base characteristics of the coatings, e.g., TiO_2 coated with Al_2O_3 or (Al_2O_3 + SiO_2) results in a basic and an acidic surface, respectively. SDS, PAA, and EHEC adsorb on the basic pigment, but not on the acidic one. NPE_{10} does not adsorb on either pigment, nor does it adsorb on pigment stabilized with pre-adsorbed PAA. The competitive adsorption of PAA and EHEC was studied by the consecutive addition of the polymers to the pigment dispersion. The total amount of adsorbed polymer was always larger with two polymers present compared to one single polymer. EHEC does not significantly influence the adsorption of PAA, irrespectively of whether it is adsorbed before or after the PAA. Conversely, 45% of the preadsorbed EHEC desorbs upon the addition of PAA. When EHEC is added to a pigment with pre-adsorbed PAA, the final amount of EHEC on the surface is significantly lower than the amount adsorbed on the bare pigment.

Key words: Adsorption; polymer; surfactant; pigment; titanium dioxide; electrophoretic mobility

Introduction

In most practical cases, dispersions are complex systems containing more than one type of dispersed particles and a combination of different surfactants and polymers for the stabilization against flocculation. This may result in a competitive adsorption, i.e., the more surface-active components displace the less surface-active ones from the particle surface [1, 2]. In many cases this is detrimental to the stability and related properties [3, 4]. Waterborne paints are typical examples of such complex dispersions. To gain a better understanding of the behavior of these dispersions, we have studied the adsorption in model systems chosen with latex paints in view. This paper presents adsorption data on surface-treated titanium dioxide, one of the most widely used pigments in latex paints. Both surfactant and polymer adsorption is dealt with, including adsorption from solutions containing mixtures of adsorbates.

As adsorption studies are time-consuming and laborious, it can be adavantageous to combine them with electrokinetic measurements. The electrophoretic mobility of the particles, as a function of the adsorbate concentration in the solution, does in most cases give qualitative information on the adsorption [5]. According to Henry [6], the electrophoretic mobility u for non-conducting particles is given by:

$$\frac{\zeta \varepsilon}{1.5\,\eta} f(\kappa a)\,, \tag{1}$$

where ζ is the zeta potential, η is the solution viscosity, and ε is the dielectric constant. $f(\kappa a)$ is the Henry term which is dependent on a, the particle radius, and κ is the Debye-Hückel parameter; κ is influenced by the ionic strength.

When a nonionic surfactant adsorbs on a charged particle, the plane of shear is moved outwards from the particle surface and hence, the zeta potential and the electrophoretic mobility decrease. This is, of course, a very simplified picture of the relation between the adsorption and the electrokinetic data. Nevertheless, in the majority of cases we do get a significant change in the electrophoretic mobility upon adsorption. If the limitations inherent in the electrokinetic measurement are taken into consideration, this technique can be used to gain adequate information on the adsorption.

Materials and methods

All chemicals were used as supplied without further purification. The water was doubly distilled.

Surfactants

The nonylphenol deca (oxyethylene glycol) monoether, NPE_{10}, was of technical grade and supplied by Berol Nobel, Sweden. The sodium dodecylsulphate SDS was obtained from BDH Chemicals Ltd., England; The purity was stated to be >99%. The critical micelle concentrations (CMC) were determined by measuring the surface tension at room temperature as a function of surfactant concentration using a du Noüy tensiometer. CMC of NPE_{10} and SDS were found to be 0.058 mM and 8.7 mM, respectively; the corresponding literature values are 0.075 mM and 8.15—8.3 mM, respectively [25].

Polymers

Ethyl hydroxyethyl cellulose (EHEC) is a nonionic cellulose ether frequently used in latex paints as a thickener. According to the supplier, Berol Nobel AB, Sweden, the degree of substitution of ethyl groups and the molar substitution of ethylene oxide groups in this sample are both equal to 0.8. The molecular weight is about 80000 g/mol.

The sodium salt of polyacrylic acid (PAA) was obtained from Allied Colloids Ltd., England and from BASF, FRG. Both products are commercial pigment dispersants. The reported molecular weight of the PAA from Allied Colloids is about 2000 g/mol. The PAA from BASF has a molecular weight of 6000 g/mol as determined by viscosimetry in 1.5 M aqueous NaBr at 15 °C. The Mark-Houwink constants in this solvent are $K = 0.165$ ml/g and $a = 0.50$ [7]. The PAA from Allied Colloids was used in most of the electrokinetic measurements, and the one from BASF was used in the quantitative adsorption studies.

Pigments

The TiO_2-pigments were supplied by Kemira Vuorikemia, Finland. Rutile-type TiO_2 was coated with (Al_2O_3 + SiO_2) or Al_2O_3. The composition and mean particle size of the two pigments, here named A and B, are given in Table 1.

Table 1. The composition and size of the TiO_2-pigments

Pigment	Surface coating		Specific surface area[1]), $mg^2\ g^{-1}$	Mean diameter[2]), nm
	Al_2O_3	SiO_2		
A	2.84%	8.55%	52.5	538
B	3.17%	—	16.5	283

[1]) BET area, nitrogen;
[2]) Photon Correlation Spectroscopy, assuming spherical particles.

The charge properties of the pigment surfaces were characterized by measuring the electrophoretic mobility as a function of pH. As can be seen in Fig. 1, the two pigments differ much in this respect. A comparison of the curves with similar measurements on pure mineral oxides [5, 8] shows that the surface of pigment A is very similar to pure silica. The charge characteristics of pigment B falls between those of TiO_2 and Al_2O_3. The isoelectric point is 7.5 compared to a value of 6 reported for TiO_2 and 9.0—9.5 for Al_2O_3. The main difference between the two TiO_2-pigments is that pigment A has an acidic surface, whereas pigment B has a basic surface.

Electrophoretic mobility

The electrophoretic mobility was determined by measuring Doppler effects in the scattered light using a Malvern Zetasizer II instrument. The pigment concentration was 0.01%. In the measurements of pigment mobility as a function of pH, the electrolyte was 10^{-2} M KNO_3 and the pH was adjusted with solutions of KOH and HNO_3. In the qualitative determinations of the adsorption the electrolyte was 10^{-2} M $NaHCO_3$, which, at the same time, buffered the dispersions to pH 8.6.

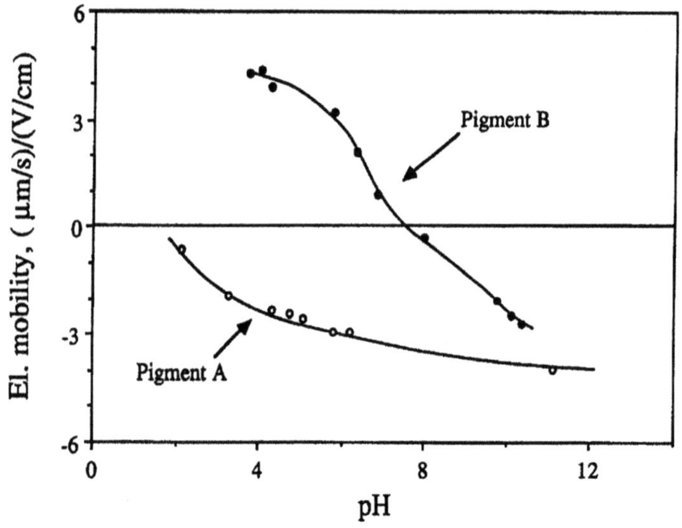

Fig. 1. The electrophoretic mobility of TiO_2-pigments as a function of pH in 10^{-2} M KNO_3. Pigment A is coated with (Al_2O_3 + SiO_2) (○) and pigment B with Al_2O_3 (●)

Adsorption isotherms

The adsorption isotherms were determined at room temperature. A batch procedure was used. Pigment dispersions of 10 (w/w)% were mixed with surfactant/polymer solutions of known concentration. The surfactant adsorption was determined without adjustment of pH or electrolyte concentration. The pH of 10% dispersions of pigments A and B are 9.4 and 7.0, respectively. In order to reduce electrostatic effects, polymer adsorption was measured in the presence of 0.1 M NaCl and with pH adjusted to 8 or 8.6 with NaOH/HCl. At this pH, pigment B is very weakly negative while pigment A carries a large negative charge (Fig. 1). The equilibration time for the surfactant adsorption was about 3 h, and for polymer adsorption, at least 17 h. After centrifugation, the residual adsorbate in the supernatant was determined. The concentrations of NPE_{10} and PAA were determined spectrophotometrically at 275 and 210 nm, respectively. The SDS concentration was determined by the methylene blue procedure of Longwell and Maniece [9]. EHEC was analyzed by a modified version of the Antron method developed by Black [10].

Competitive adsorption

The competitive adsorption of two adsorbates on pigment B was studied by adding the adsorbates to the pigment dispersion consecutively. First, the pigment was dispersed in a solution containing enough of adsorbate (1) to ensure maximum adsorption. After 1 day of equilibration, solutions were added containing adsorbate (2) in varying concentrations and adsorbate (1) in a concentration equal to the equilibrium concentration obtained in the first stage. In this way desorption of adsorbate (1) due to dilution was avoided. After equilibration for 1 additional day, the adsorbed amount and the electrophoretic mobility were determined. In the presence of EHEC the concentration of PAA cannot be determined spectrophotometrically. Therefore, in these experiments a potentiometric titration method was used. After adjustment of pH to 3.5 with HCl, the solutions were titrated with 0.01 M NaOH. A linear relation was found between the consumption of NaOH and the PAA concentration. The Antron-method used to analyze EHEC is not disturbed by the concentration of PAA present in these studies.

Results and discussion

Surfactant adsorption

The electrophoretic mobility of pigment A was independent of the amount of added NPE_{10} up to concentrations of at least 0.35 mM. This indicates that no adsorption takes place. This is also confirmed by the quantitative adsorption measurements. Neither does this surfactant adsorb on pigment B. The adsorption behavior of a surfactant to a solid is affected, not only by its affinity to the solid, but also by its affinity to the medium. When both the medium and the surface are polar, there has to be some specific interaction between the pigment surface groups and the surfactant in order to attain adsorption. For these pigments, hydrogen bonding between silanol- or aluminium hydroxyl groups on the pigment surface and the oxygen atoms in the ethyleneoxide groups could be envisaged. However, the gain in free energy is probably small as it only involves the replacement of a similar

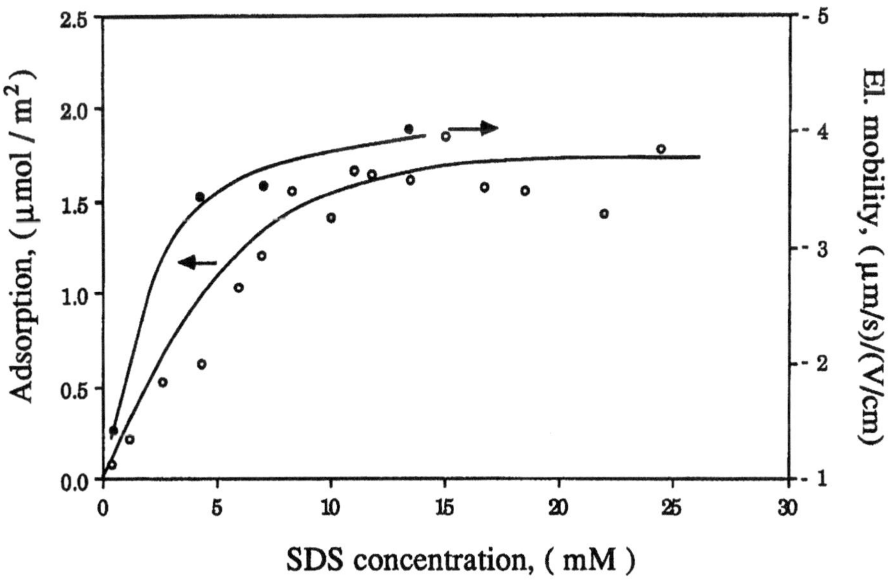

Fig. 2. Adsorption of SDS on pigment B. Comparison of the adsorption isotherm (○) with electrophoretic mobility data (●); pH = 7

bond between the surface and water. In other studies on the adsorption of nonionic ethoxylated surfactants on similar oxide surfaces, examples are given where adsorption takes place [11—16], as well as where no adsorption can be detected [15—16]. Lawrence et al. [17] report a very low adsorption of nonionic ethoxylated surfactants on alumina and alumina-coated silica. They attribute this fact to the formation of surfactant: Al^{3+} complexes, which are stable relative to the adsorbed state. However, contrary to their findings, we could not detect any higher levels of Al^{3+} in the water phase of the pigment dispersions containing NPE_{10}.

SDS does not adsorb on the acidic pigment A. This can be explained by the high negative charge of pigment A, resulting in an electrostatic repulsion between the pigment surface and the surfactant. However, SDS adsorbs on pigment B and there is a good correlation between the electrophoretic mobility data and the adsorption isotherm (Fig. 2). Since pH at this measurement is lower than the isoelectric point of pigment B, the adsorption is favored by the electrostatic attraction between the positively charged surface and the anionic surfactant. At a low surface coverage the charge of the particles is reversed and further adsorption is due to lateral interactions between the hydrophobic parts of the surfactant in aggregates formed on the surface. The isotherm levels close to the CMC of the surfactant and the maximum surface coverage corresponds to an average area per molecule of 0.98 nm^2. This is about twice the molecular area of a close-packed SDS monolayer. Observations of the sedimentation rate of the pigment revealed a very fast sedimentation in the concentration range 1.7 to 5.8 mM of SDS. This corresponds to an adsorbed amount of 30—60% of the maximum adsorption. At very low and high surface coverages the sedimentation rate is low. These observations reflect a change of the surface from hydrophilic to weakly hydrophobic as the first SDS molecules adsorb with the anionic head groups oriented towards the surface and a subsequent change back to a hydrophilic surface as additional surfactant molecules adsorb with the anionic group towards the water phase. The low surface coverage implies adsorption in isolated clusters or hemimicelles, rather than as a complete bilayer.

Polymer adsorption

The electrophoretic mobility of both pigments decrease with increasing concentration of EHEC (Fig. 3). At these EHEC concentrations the viscosity of the medium is increased, which in addition to a reduction in charge may be the cause of a reduced

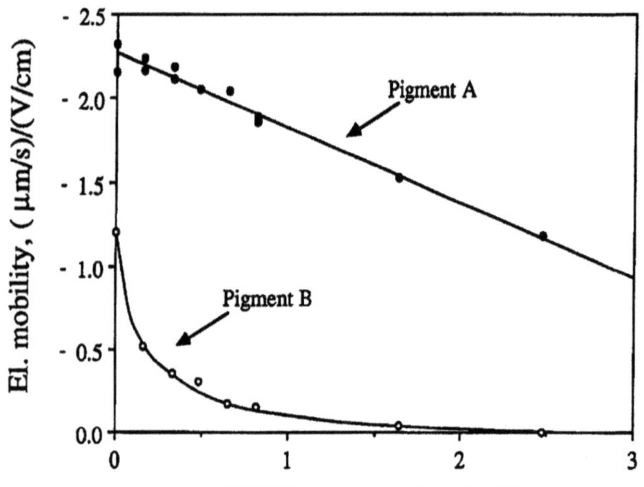

Fig. 3. The electrophoretic mobility of pigment A (●) and pigment B (○) as a function of EHEC concentration in solution. pH = 8.6 and $[NaHCO_3] = 10^{-2}$ M

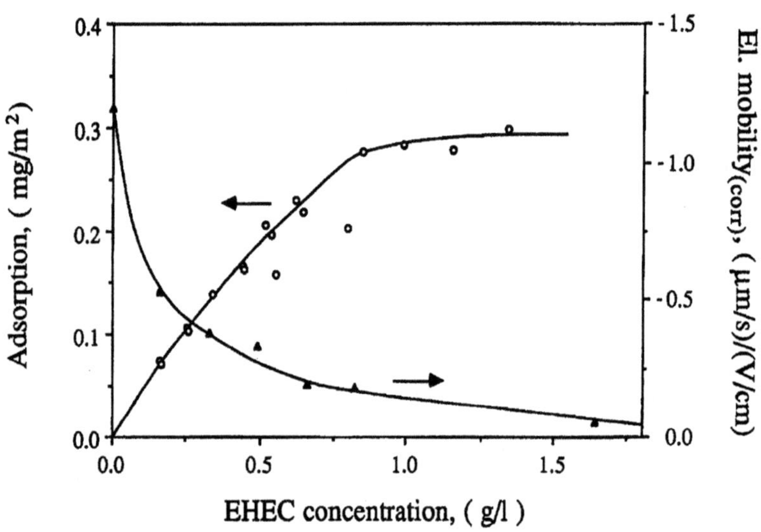

Fig. 4. Adsorption of EHEC on pigment B. Comparison of the adsorption isotherm (○) with electrophoretic mobility data, corrected for the influence of the viscosity (△)

mobility according to Eq. (1). The way in which the mobility decreases differs for the two pigments. For pigment B the mobility is strongly decreased in a narrow concentration range, while the decrease is linear for pigment A. This indicates that EHEC adsorbs on pigment B, but not on pigment A. Adsorption measurements confirmed this conclusion. Thus, the decrease in electrophoretic mobility of pigment A is entirely due to the increase in viscosity with increasing EHEC concentration. The relation between u and $\eta(c_{EHEC})$ obtained for pigment A can be used to correct the electrophoretic mobility of pigment B, so that it only reflects the change in the zeta potential upon adsorption. As can be seen from Fig. 4 there is a good agreement between the adsorption isotherm and the corrected electrophoretic mobility data. At an EHEC concentration of about 1 g/l the adsorption levels off and reaches the value 0.30 mg/m^2.

Figure 5 shows the change in electrophoretic mobility of the pigments as a function of concentration of the sodium salt of polyacryl acid (PAA). The mobility of pigment A is unaffected by the PAA concentration, indicating the absence of adsorption. The surface of pigment B is also negatively charged at the pH of this measurement, but not to the same extent as the acidic surface of pigment A. A sharp increase in the mobility is obtained at very low

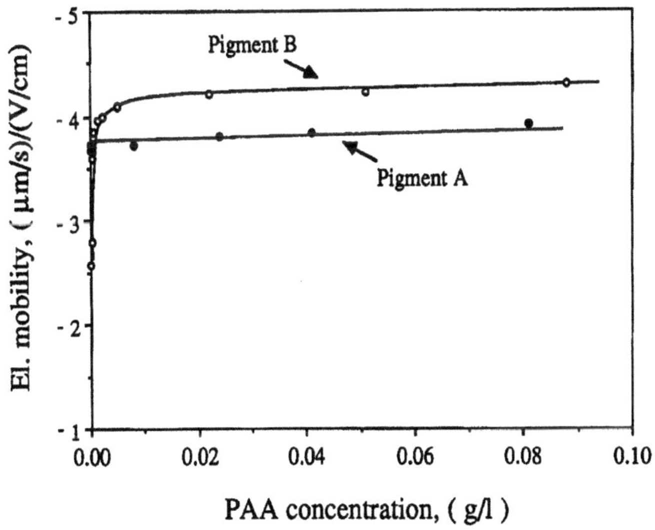

Fig. 5. The electrophoretic mobility of pigment A (●) and pigment B (○) as a function of PAA concentration in solution; pH = 8.6 and $[NaHCO_3]$ = 10^{-2} M

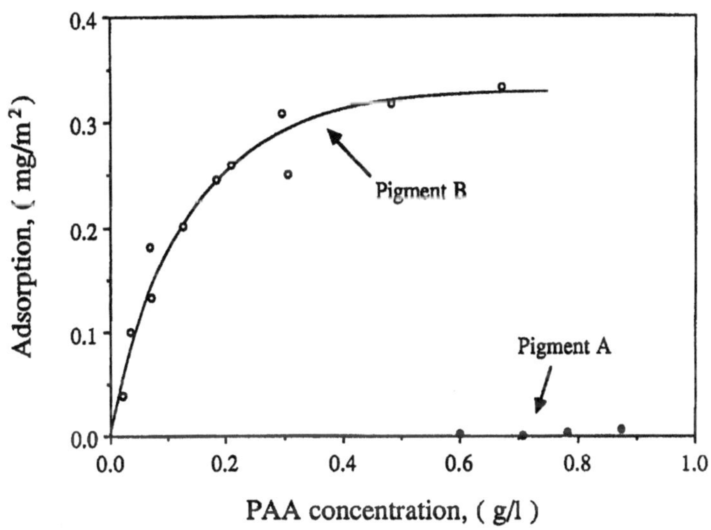

Fig. 6. Adsorption of PAA on pigment A (●) and pigment B (○); pH = 8.6 and [NaCl] = 0.1 M

PAA concentrations and a constant value is reached at about 0.01 g/l PAA in solution. It is obvious that the polyelectrolyte adsorbs on pigment B, despite the fact that both the adsorbate and the adsorbent are negatively charged. Adsorption measurements confirmed that the PAA adsorbs on the basic pigment, but not on the acidic one (Fig. 6). The absence of adsorption on the acidic surface is explained by the electrostatic repulsion between the surface and the polyelectrolyte. PAA does, however, adsorb on the negative surface of pigment B. The adsorption of a polyelectrolyte to a like-charged surface is possible when the electrostatic repulsion is weaker than the other contributions to the adsorption energy. This is clearly shown by Hesslink's theoretical model of polyelectrolyte adsorption [18] and by a number of experimental investigations, e.g., [19–21]. One conceivable contribution to the adsorption energy is hydrogen bonding between hydroxyl groups on the pigment surface and carboxyl groups in the polyelectrolyte, i.e., a strong acid/base interaction between these functional groups.

A comparison of the adsorption isotherms in Figs. 4 and 6 shows that the adsorption of PAA reaches a plateau level at a much lower polymer concentration than EHEC, signifying that the adsorption free energy is greater for PAA. However, the maximum adsorbed amount is almost equal for the two polymers at these conditions of salinity and pH.

The agreement between electrokinetic data and the adsorption isotherm for PAA is not as good as for EHEC. The electrophoretic mobility reaches a constant value at very low PAA concentrations, far from the concentrations at which maximum adsorption is attained. The reason for this is probably that polyelectrolyte adsorption effects the electrophoretic mobility, both by shift in the shear plane and by addition of charge to the surface. The shift in shear plane counteracts the effect of added charge on the mobility. It should also be noted that the surface area to solution volume ratio is much smaller in the electrokinetic measurements compared to the adsorption measurements. For polydisperse polymers the shape of the adsorption isotherm is affected by the adsorbent content. This behavior is explained in terms of a preferential adsorption of larger molecules over smaller ones [22, 23]. The degree of polydispersity of the PAA is not known. However, for polymers of low molecular weight, like the PAA used in this study, the polydispersity is usually low and, hence, the affect of adsorbent content should be of minor importance.

Competitive adsorption

Nonionic surfactants, e.g., ethoxylated nonylphenols are often used as wetting agents to aid the pigment dispersion in water-borne paints [24]. However, NPE_{10} does not adsorb on either bare pigment, as stated earlier, or on pigment saturated with pre-adsorbed PAA (Fig. 7).

The competitive adsorption of EHEC and PAA was studied by consecutive addition of the polymers to the pigment dispersion. A slight decrease in the electrophoretic mobility (corrected for viscosity effects) is observed upon the addition of EHEC to a pigment dispersion stabilized with pre-adsorbed PAA (Fig. 8). However, the change in mobility is much smaller than when EHEC is added to bare pigment (c.f. Fig. 4). From the electrokinetic data, we may conclude that EHEC adsorbs on the PAA covered pigment, but to a much lower degree than on the bare pigment. Adsorption measurements confirm this conclusion and give additional information (Fig. 9). PAA does not desorb from the surface when EHEC is adsorbed. The maximum amount of EHEC adsorbed on a PAA-covered pigment is 40% of the plateau adsorption on bare pigment.

If the order of polymer addition is reversed a similar result is obtained. The sharp increase in negative electrophoretic mobility on addition of PAA to the pigment dispersion, with pre-adsorbed EHEC, indicates an adsorption of the polyelectrolyte (Fig. 10). As shown in Fig. 11, the adsorbed

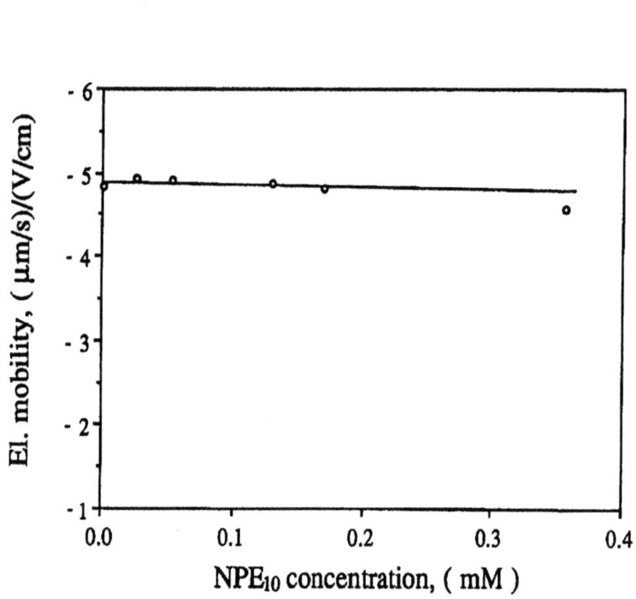

Fig. 7. The electrophoretic mobility of pigment B, prestabilized by PAA, as a function of NPE_{10} concentration in solution. [PAA] = 0.2 g/l, pH = 8.6 and $[NaHCO_3] = 10^{-2}$ M

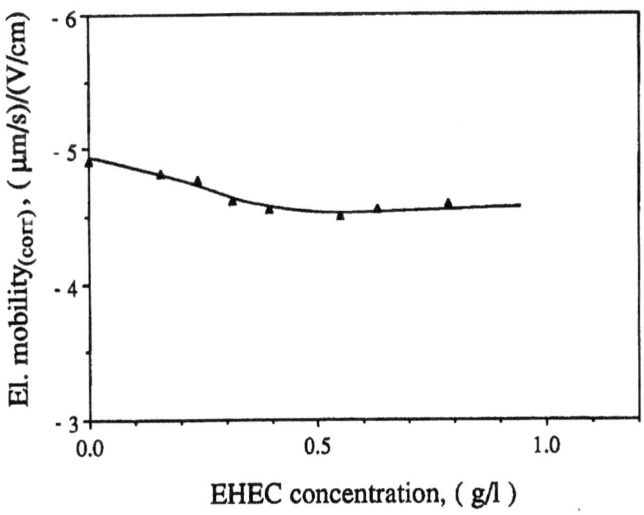

Fig. 8. The electrophoretic mobility (corrected for viscosity effects) of pigment B, prestabilized by PAA as a function of EHEC concentration in solution; [PAA] = 0.15 g/l, pH = 8.6 and $[NaHCO_3] = 10^{-2}$ M

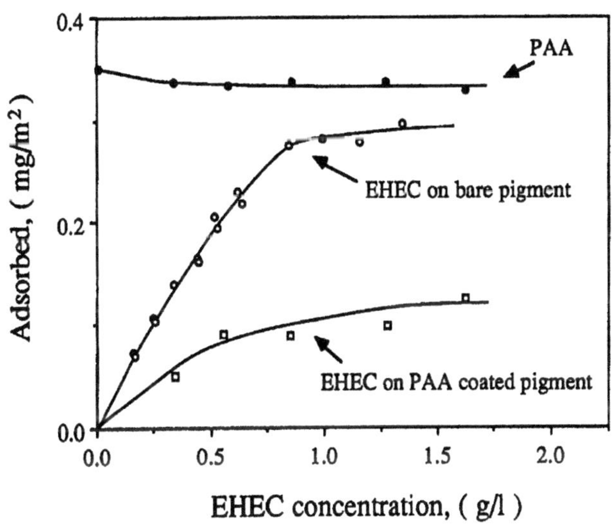

Fig. 9. Variation in the amount of adsorbed polymer on pigment B as a function of EHEC concentration in solution. The pigment is prestabilized by PAA (●); □ = EHEC; ○ = adsorption of EHEC on bare pigment; pH = 8 and [NaCl] = 0.1 M

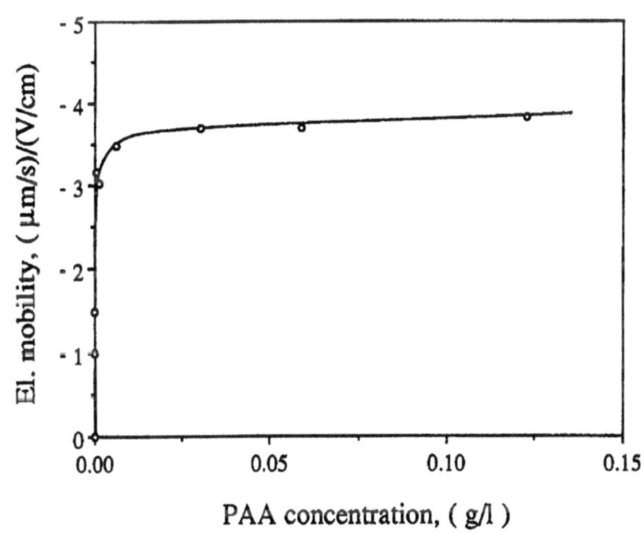

Fig. 10. The electrophoretic mobility of pigment B, prestabilized by EHEC, as a function of PAA concentration in solution; [EHEC] = 1.56 g/l, pH = 8.6, and $[NaHCO_3] = 10^{-2}$ M

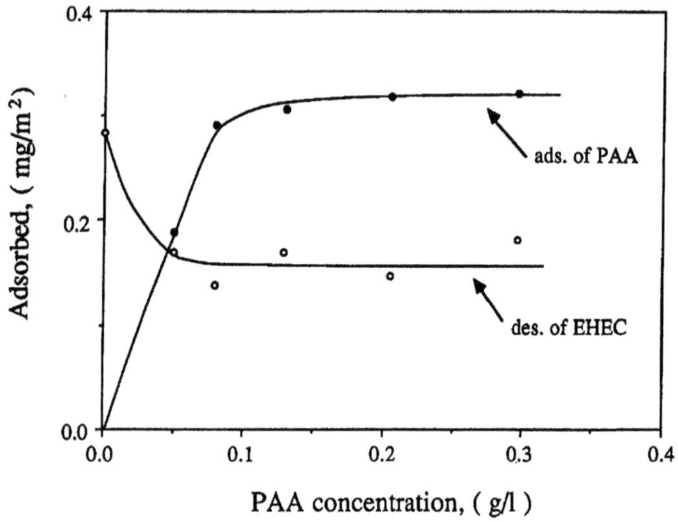

Fig. 11. Change in the amount of adsorbed polymer on pigment B, as a function of PAA concentration in solution. The pigment is prestabilized by EHEC. ● = PAA; ○ = EHEC; pH = 8 and [NaCl] = 0.1

amount of PAA on the EHEC-covered pigment is equal to the amount of PAA adsorbed on bare pigment. On adsorption of PAA about 45% of the pre-adsorbed EHEC desorbs.

The ability of PAA to displace part of the EHEC (while the reverse is not the case) is in agreement with the conclusion drawn from the individual adsorption isotherms, namely, that the adsorption energy to pigment B is lower for EHEC than for PAA. Other studies of competitive adsorption show that a polymer with a lower adsorption energy is displaced by a polymer with a larger adsorption energy [26, 27].

The final amount of EHEC at the surface is to some extent dependent on the order of addition. If EHEC is the polymer added first, the final amount of EHEC on the surface is 30—40% higher than in the case when EHEC is added after PAA. As the observed difference is relatively small and as we have made no experiments so far to investigate the time-dependence of the adsorption, not too great importance should be attached to the observed difference. However, it deserves mention that Cszempesz and Rohrsetzer, who studied competitive adsorption between different nonionic polymers, reported even larger effects of the order of polymer addition [28].

An interesting point to note is that, regardless of the order of addition, the total amount of adsorbed polymer is larger with two polymers present than that compared to with one single polymer. This is not the case if the only difference between the polymers is the molecular weight [26, 27]. A plausible explanation to this enhancement in the adsorbed amount is as follows: The chemical potential of a polymer in solution increases in the presence of another polymer, and this augments the driving force for adsorption. At the same time the effect is not counteracted by a large increase in the chemical potentials of the polymers on the surface, which is the case if the polymers are mixed at the surface. It would be most interesting to gain more information of the polymer conformation at the surface.

Conclusions

The comparison of electrokinetic measurements with adsorption isotherms demonstrated, in many cases, good agreement. If the limitations inherent in the electrophoretic mobility measurements are taken into consideration, this technique can be used to gain adequate information on the adsorption.

Coating of TiO_2-pigment with different mineral oxides, as expected, has a strong influence on the adsorption. This can be understood in view of the different acid/base characteristics of the coatings, e.g., TiO_2 coated with Al_2O_3 (pigment B) and Al_2O_3 + SiO_2 (pigment A) result in a basic and an acidic surface, respectively. SDS, PAA, and EHEC adsorb on this basic pigment, but not on the acidic one. NPE_{10} does not adsorb on either pigment, nor does it adsorb on pigment stabilized with pre-adsorbed PAA. The competitive adsorption of PAA and EHEC was studied by consecutive addition of the polymers to the pigment dispersion. The adsorption of PAA is not affected by the presence of EHEC, irrespective of whether it is added before or after the PAA. However, on adsorption of PAA, about 45% of the pre-adsorbed EHEC desorbs. When EHEC adsorbs on preadsorbed PAA, the final amount of EHEC on the surface is lower than the amount adsorbed on the bare pigment. The results of the competitive adsorption are in accordance with the conclusion that could be drawn from the individual adsorption isotherms of the polymers, namely, that PAA has a higher affinity than EHEC to the surface of pigment B. An interesting result is that the total amount of adsorbed polymer is larger with two polymers present compared to one single polymer.

Acknowledgement

This work was supported by Nordisk Industriefond and five Scandinavian companies; Jotun A/S and Norsk Hydro A/S (Norway), Kemira Oy (Finland), Becker Industrifärg AB and Berol Nobel AB (Sweden). The model pigments were made by Mr. Tuomo Losoi (Kemira).

References

1. Kronberg B, Lindström M, Stenius P (1986) ACS Symp Ser 311:225—240
2. Järnström L, Ström G, Stenius P (1987) Tappi J 70:101—107
3. Kronberg B, Kuortti J, Stenius P (1986) Colloids and Surfaces 18:411—425
4. Stenius P, Kuortti J, Kronberg B (1984) Tappi J 67:56—59
5. Hunter RJ (1981) Zeta potential in colloid science. Academic Press, London
6. Henry DC (1931) Proc roy soc A 133:106
7. Kurata M, Tsunashima Y, Iwama M, Kamada K (1975) In: Brandrup J, Immergut EH (eds) Polymer Handbook. Wiley, New York, p IV—9
8. Kumano I (1984) In: Kitahara A, Watanabe A (eds) Electrical phenomena at interfaces. Marcel Dekker, New York, pp 356—363
9. Waters J, Taylor CG (1977) In: Cross J (ed) Anionic Surfactants — Chemical Analysis. Marcel Dekker, New York, pp 211—212
10. Black HC Jr (1951) Anal Chem 23:1972
11. Levitz P, Van Damme H (1986) J Phys Chem 90:1302—1310
12. Gellan A, Rochester CH (1985) J Chem Soc, Faraday Trans 1 81:2235—2245
13. Furlong DN, Aston JR (1982) Colloids and Surfaces 4:121—129
14. Rouquerol J, Partyka S, Rouquerol F (1982) In: Rouquerol J, Sing KSW (eds) Adsorption at the gas-solid and liquid-solid interface. Elsevier Scientific Publishing Company, Amsterdam, pp 69—74
15. Scamehorn JF, Schechter RS, Wade WH (1981) J Colloid Interface Sci 85:494—501
16. Fukushima S, Kumagai S (1973) J Colloid Interface Sci 42:539—544
17. Lawrence SA, Pilc JA, Readman JR, Sermon PA (1987) J Chem Soc, Chem Commun 1035—1036
18. Hesslink FT (1983) In: Parfitt GD, Rochester CH (eds) Adsorption from Solution at the Solid/Liquid Interface. Academic Press, London, pp 377—412
19. Mäkinen PO, Losoi T, Kohonen A (1988) In: 12th Congress of the Federation of Scandinavian Paint and Varnish Technologists, Helsinki, pp 8.1—8.18
20. Morrison Jr WH (1985) J Coatings Technol 57:55—65
21. Girod G, Lamarche JM, Foissy A (1988) J Colloid Interface Sci 121:265—272
22. Koopal LK (1981) J Colloid Interface Sci 83:116—129
23. Hlady V, Lyklema J, Fleer GJ (1982) J Colloid Interface Sci 87:395—406
24. Lein MM, Brakke B, Keltz G, Kiezulas MP, Leavy CM, Marderosian R, Withington D (1983) J Coatings Technol 55:81—90
25. Mukerjee P, Mysels KJ (1971) Critical Micelle Concentrations of Aqueous Surfactant Systems, National Standard Reference Data Series, Washington
26. Ramachandran R, Somasundaran P (1987) J Colloid Interface Sci 120:184—188
27. Wright JA, Harrop R, Williams PA, Pass G, Robb ID (1987) Colloids and Surfaces 24:249—258
28. Csempesz F, Rohrsetzer S (1984) Colloids and Surfaces 11:173—186

Authors' address:

Margareta Huldén
Institute for Surface Chemistry
Box 5607
11486 Stockholm, Sweden

Progress in Colloid & Polymer Science Progr Colloid Polym Sci 82:38—42 (1990)

A calorimetric investigation of the adsorption of octylamine on titanium dioxide from aqueous solutions*)

A. B. A. Pettersson and J. B. Rosenholm

Department of Physical Chemistry, Åbo Akademi, Åbo, Finland

Abstract: Enthalpies of displacement, as well as adsorption isotherms, have been measured for the adsorption of aliphatic octylamine on TiO_2 from aqueous solutions at 298 K, using sorption microcalorimetry. The adsorbent studied was the rutile form of TiO_2. The pH of the aqueous octylamine solutions were natural. The results show that the adsorption isotherm is sigmoid. At the critical micelle concentration (CMC) the surface is not close-packed, but the adsorbate molecules are strongly adsorbed. Above CMC the preferential adsorption continues. However, the differential enthalpy of displacement is very weak. The surface acid-base properties of the rutile used were measured by titration of the surface with butylamine in benzene and trichloroacetic acid also in benzene, using a series of Hammet indicators. The density of the acid sites on the rutile, as determined by titration, is greater than the density of monolayer coverage of octylamine on rutile obtained by calorimetric measurements. The adsorption up to CMC may thus be explained by an acid-base interaction.

Key words: Rutile; octylamine; displacement enthalpy; adsorption isotherm; adsorption density; surface acidity; surface basicity

1. Introduction

Considerable interest has been focused on TiO_2, due to its suitability in many commercial applications (principally as white pigment in paint, paper, textiles, ceramics, plastics, and rubber). Scientifically, the availability and use of two abundant crystal forms (anatase and rutile) with distinct differences in their physico-chemical properties opens an option to relate crystal structure and chemical reactivity. Pigments treated with aqueous solutions of cationic surfactants become lipophilic so that they suspend more readily in an oily dispersion [1—4]. It was thus considered worthwhile to undertake a systematic study of the properties of TiO_2 in different media. In this work only rutile was studied.

Rutile is known to have a strong surface heterogeneity, the amphoteric surface consisting of both basic and acidic sites at the three crystal faces (110, 100 and 101). According to the literature there are two forms of adsorbed molecular water present on rutile surfaces under ambient conditions. One form is hydrogen bonded to surface hydroxyl ions and the other is held as a coordinating ligand to surface cations. The latter may undergo subsequent dissociation to hydroxo ligands. Basic amine compounds can adsorb strongly on rutile surfaces through coordination with surface cations such as Lewis acidic Ti^{4+}, or through hydrogen bond formation with certain surface hydroxylic species after displacing the aqueous or hydroxo ligands [1, 5].

Adsorption from solution is a relatively complex phenomenon; it depends on the nature of solute-solvent interactions in the solution phase and in the interfacial region, as well as on their interactions with the adsorbent [6].

The mechanism of the adsorption of cationic surfactants on mineral or inorganic substrates has been infrequently investigated. On negatively charged hydrophilic surfaces suspended in water, cationic

*) Poster presented during the 10th Scandinavian Symposium on Surface Chemistry, Åbo, Finland, June 14—16, 1989.

surfactants are known to adsorb, leaving the hydrophobic groups of the surfactant in contact with the aqueous solution. The adsorption thus renders the surface hydrophobic. At higher coverage, attractive chain-chain lateral interactions are believed to give rise to two-dimensional aggregates on the surface. These are commonly referred to as "hemimicelles". The hydrophobic moiety of the surfactant plays a significant role in hemimicellization. For octylamine the hydrocarbon chain is rather short. Many questions remain open concerning the hemimicellization [7—11].

Octylamine, soluble to a limited degree in water, forms an isotropic solution in the concentration range studied at 298 K. The molecularly dispersed octylamine has been found to consist of hydrates such as the hexahydrates $(C_8H_{17}NH_2)_2 \cdot 3\ H_2O$ and $C_8H_{17}NH_2 \cdot 6\ H_2O$ [12, 13].

According to literature, a typical isoelectric point for rutile is at about pH = 5 [14—16]. Because of the unshared pair of electrons on the nitrogen atom, octylamine is a strong proton acceptor and electron pair donor and is, therefore, basic in character. Starting from the approximate IEP, the octylamine tend hence to be adsorbed to a greater extent at increased pH's, when the potential-determining OH^- ions make the titania surface more negatively charged [7].

2. Experimental

Materials and methods

The displacement enthalpies were measured using a LKB 2107 sorption microcalorimetry system. The adsorption isotherms (reduced surface excess) were measured by the dynamic method described by Noll and Burchfield [17] using monitored refractive index of the supernatant stream. Tests made prior to the calorimetric measurements showed that the response of the refractive index detector was proportional to the amount of octylamine in solution. The detector used was a Knauer differential refractometer.

The rutile was from Kemira Oy, Pori, Finland; it was produced by the sulfate process using Al^{3+} ions as doping agent. The samples came directly from the calcining furnace without further treatment. The rutile was crushed with a mortar, sieved, and washed free of soluble salts. The water used was double distilled, subsequently purified with a Milli-Q water system and then degassed. The minimum resistivity of the water was 10^7 ohm · cm, and it was used in all experiments. After the washing, the TiO_2 was carefully dried. Because the particles were larger than the primary particles, the titania was suspected to be porous. Interpretation of x-ray diffraction pattern of the substrate, using Ni-filtered copper *Kα* radiation, indicated that the samples consisted of pure rutile.

The 1-octylamine used was Genamin 8 R 100 D by Hoechst, and it was used without further purification.

Surface acidity and basicity

The acidity (the number of acid sites per area) at different acid strength (the ability of acid sites to convert an adsorbed basic Hammet inciator into its conjugate acid) of the acidic sites on the rutile was estimated by the widely used titration method described by Johnson [18] and Benesi [19]. The titrator was a 0.1 N solution of n-butylamine (Fluka, puriss.) in benzene (Merck, p.a. grade). When the indicator used undergoes a color change, the value of the Hammet acidity function H_0 (measure of the acid strength of the sites) is equal to or lower than the pK_a of the indicator. The amount of acid sites of the powder added is the number of mole n-butylamine consumed in the titrator. The change of color was visually recorded. The basicity was measured using the method described by Yamanaka and Tanabe [20, 21], and gives the basicity on a common scale using the same basic indicators as for measuring acidic properties. The titrator was a 0.1 N solution of trichloroacetic acid (Merck, p.a. grade) in benzene. The value of the H_0 (measure of the basic strength of the sites) now being equal to or grater than the pK_a of the indicator. The basic indicators used with respective pK_a values were: neutral red (+6.8, Merck), methyl red (+4.8, Merck), phenylazonaphthylamine (+4.0, Aldrich), benzeneazodiphenylamine (+1.5, Aldrich) and crystal violet (+0.8, Merck) [22, 23].

There has been a lot of debate about whether the method gives the sum of Brønsted and Lewis acid sites or only Brønsted surface acid groups, because of steric hindrance of the indicator molecules to adsorb onto the sites. There are also reports that surface sites other than the acidic or basic are able to react with the indicators. These are serious drawbacks to the method. It may, however, be useful as an easy and quick way of screening surface acidities and basicities, when the surface is as amphoteric as it is for rutile [24, 25].

3. Results and discussion

The mean value of BET(N_2) surface area and the specific area measured with a calorimetric method [26] for rutile were 8.73 m^2/g. Because the latter method gave a slightly larger area, we consider the pores of the titania particles to be macroscopic. The determined mean specific area reflects the surface available for adsorption from solution, and hence it was used in the calculations. The enthalpy of immersion when wetting rutile with water at 298 K was —525 mJ/m^2. The outgassing temperature was

453 K. Assuming the cross-sectional area of a water molecule to be 0.1054 nm^2 [27] (9.5 H_2O molecules/nm^2), our wetting enthalpy gives a molar wetting enthalpy of —33.3 kJ/mol (monolayer).

The pH's of the aqueous solutions varied between 8.9 and 10.25 depending on octylamine concentration. The ionization constant K_b for octylamine in water is $4.5 \cdot 10^{-4}$, and pK_a for the conjugate acid BH^+ is about 10.7. The data refers to the concentration range of 0.0008 M—0.016 M [28]. The pH's of the solutions indicate that most of the octylamine is protonated.

The CMC for octylamine could not be found in literature. As an ammonium salt the CMC was found to be $1.75 \cdot 10^{-1}$ M [29]. According to the supplier, the CMC for the octylamine used is at about 10^{-5} on the molefraction scale. The interpretation of the results of our investigations are based upon the latter value. For many systems it is known that the adsorption attains saturation at equilibrium concentrations which approximate the CMC for the surfactant.

Table 1 gives the acidities and basicities (reported as the number of acidic or basic sites/area) of the rutile surface at their different acid and base strength. Because the pK_a's of the surface acidic sites are more than 1—2 units less than the pK_a of the conjugated acid of octylamine, strong adsorption may occur [1]. The rutile had an overall slightly basic surface property.

Table 1. Amount of surface acidic and basic sites at different acid and base strength of rutile used

	$H_0 \leqslant 1.5$	$H_0 \leqslant 4.0$	$H_0 \leqslant 4.8$	$H_0 \leqslant 6.8$
Acidity/sites/nm^2	0	0.04	0.63	1.79
	$H_0 \geqslant 1.5$	$H_0 \geqslant 4.0$	$H_0 \geqslant 4.8$	$H_0 \geqslant 6.8$
Basicity/sites/nm^2	3.46	0	0	0

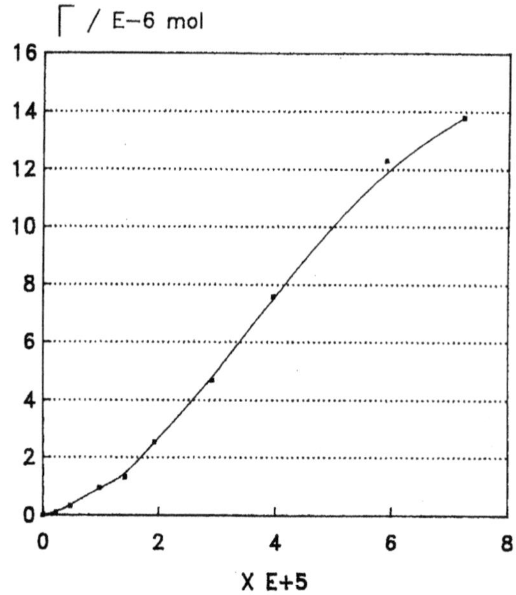

Fig. 1. Surface excess for octylamine at rutile/water interface versus mole fraction octylamine at 298 K

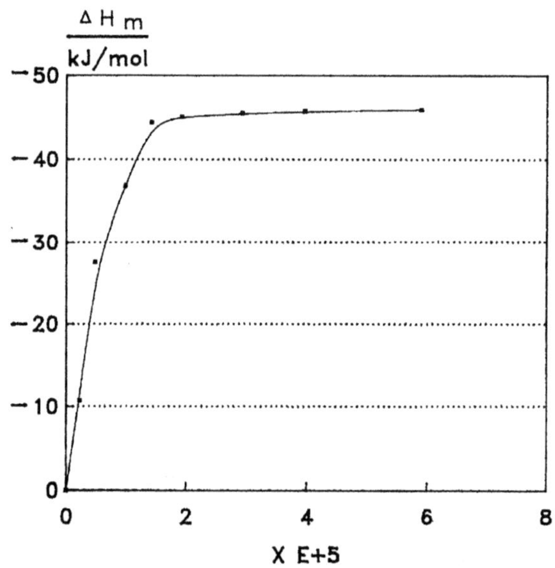

Fig. 2. Enthalpy of displacement of octylamine at rutile/water interface versus mole fraction octylamine at 298 K

Figure 1 shows the total surface excess for octylamine at rutile/water interface of 0.2427 g powder, as a function of the mole fraction of octylamine in bulk solution at 298 K. According to the figure, the preferential adsorption of octylamine at rutile/water interface is of Brunauer's type V adsorption isotherm. The sigmoid type isotherms has frequently been applied to solution adsorption, as multilayer adsorption or indication of porous adsorbent. The latter causing a capillary effect.

Figure 2 shows the exothermic enthalpy of displacement of water by octylamine at rutile/water interface as the function of the mole fraction of octylamine in bulk solution at 298 K. The low CMC for octylamine is close to the breakpoint towards the plateau. According to Fig. 1, the adsorption continues above CMC, but the differential enthalpy of displacement is very weak.

Figure 3 shows the exothermic enthalpy of displacement of water by octylamine at rutile/water interface as a function of surface concentration of octylamine at 298 K. According to Fig. 3, a monomolecular coverage of very strongly adsorbed molecules is obtained at only 0.5 octylamine molecules/nm^2, the exothermic enthalpy of displacement being at about —45 kJ/mol. Since vertically close-packed, aliphatic amines require 4.3 molecules/nm^2 space, it is realized that the basic octylamine adsorbs strongly only to certain surface sites. The high energies in Fig. 3 indicate hydrogen bonding up to at least CMC [30].

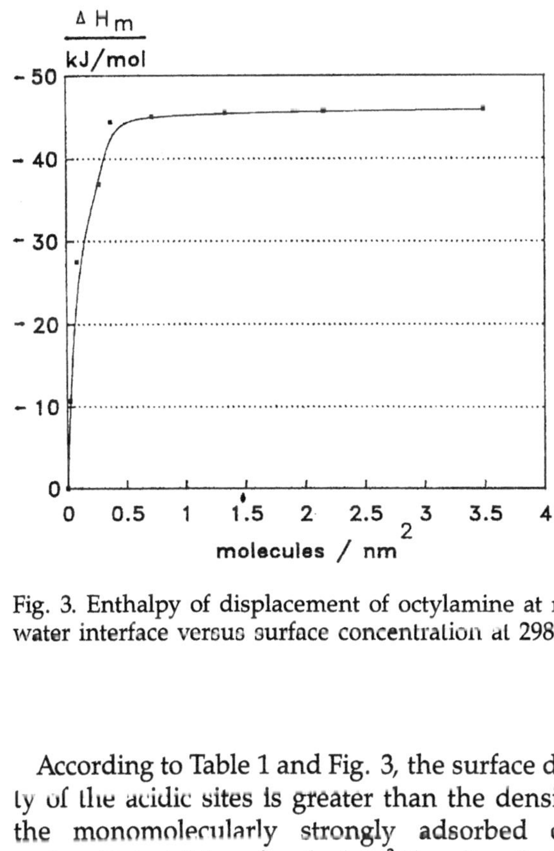

Fig. 3. Enthalpy of displacement of octylamine at rutile/water interface versus surface concentration at 298 K

According to Table 1 and Fig. 3, the surface density of the acidic sites is greater than the density of the monomolecularly strongly adsorbed octylamine. But at 0.5 molecules/nm^2 the sites that produce strong bonds are occupied. The adsorption of octylamine in the strongly bonded region may thus be explained by an acid-base interaction.

Much of the adsorption above CMC is somewhat confusing. If close-packed monolayer is achieved (far above CMC), 4.3 molecules/nm^2 gives a calculated surface excess of about $1.5 \cdot 10^{-5}$ mol. This may be the plateau (outside the measurements) of saturated adsorption at about the equilibrium mole fraction octylamine of 10^{-4}.

For long chain ions such as amines the adsorption mechanism is evidently obscured by the complicated solution behavior of these species.

One far-fetched explanation may be some kind of energetically weak phase separation induced by the interactions with the solid surface and the high pH. Such phase separation could be expected to lead to dependence of adsorption on solution.

Acknowledgements

The authors wish to thank The Academy of Finland for support of this work, and Kemira Oy, Finland, for supplying The TiO_2 samples and for financial support to A.P.

References

1. Solomon DH, Hawthorne DG (1983) Chemistry of Pigments and Fillers, John Wiley & Sons, New York
2. Jefferies HD (1986) In: Parfitt GD (ed) Dispersion of Powders in Liquids — Whit Special Reference to Pigments. Elsevier Applied Science Publishers, London, pp 395—469
3. Gutcho MH (1980) Inorganic Pigments, Noyes Data Corporation, Park Ridge
4. Black W (1986) In: Parfitt GD (ed) Dispersion of Powders in Liquids — With Special Reference to Pigments, Elsevier Applied Science Publishers, London, pp 149—201
5. Jones P, Hockey JA (1972) Trans Faraday Soc 68:907
6. Adamson AW (1976) Physical Chemistry of Surfaces, John Wiley & Sons, New York
7. Ginn ME (1970) In: Jungermann E (ed) Cationic Surfactants, Surfactant Science Series, Vol. IV, Marcel Dekker Inc., New York, pp 341—367
8. Aveyard R (1987) In: Tadros TF (ed) Solid/Liquid Dispersions, Academic Press, London pp 111—129
9. Smith RW, Akhtar S (1976) In: Fuerstenau MC (ed) Flotation — A. M. Gaudin Memorial Volume, Volume 1, American Institute of Mining, Metallurgical and Petroleum Engineers Inc., New York, pp 87—116
10. Hough DB, Rendall HM (1983) In: Parfitt GD, Rochester CH (eds) Adsorption from Solution at the Solid/Liquid Interface, Academic Press, London, pp 247—319

11. Somasundaran P, Ramachandran R (1988) In: Wasan DT, Ginn ME, Shah DO (eds) Surfactants in Chemical/Process Engineering. Marcel Dekker Inc., New York, pp 195—235
12. Ralston AW (1948) Fatty Acids and Their Derivatives, John Wiley & Sons, New York, pp 650—667
13. Ralston AW, Hoerr CW, Hoffmann EJ (1942) J Am Chem Soc 64:1516
14. Furlong DN, Parfitt GD (1978) J Colloid Interface Sci 65:548
15. Cornell RM, Posner AM, Quirk JP (1975) J Colloid Interface Sci 53:6
16. Parfitt GD (1976) Pure & Appl Chem, Vol 48, Pergamon Press, pp 415—418
17. Noll LA, Burchfield TE (1982) Department of Energy/Bartlesville Energy Technology Center/RI-82/7, Oklahoma
18. Johnson O (1955) J Phys Chem 59:827
19. Benesi HA (1957) J Phys Chem 61:970
20. Yamanaka T, Tanabe K (1975) J Phys Chem 79:2409
21. Yamanaka T, Tanabe K (1976) J Phys Chem 80:1723
22. Tanabe K (1970) Solid Acids and Bases, Academic Press, New York
23. Tanabe K (1981) In: Anderson JR, Boudart M (eds) Catalysis — Science and Technology, Volume 2, Springer-Verlag, Heidelberg, pp 231—273
24. Benesi HA, Winquist BHC (1978) Adv Catal, Volume 27, pp 97—182
25. Hodgkin JH, Hawthorne DG, Swift JD, Solomon DH (September 10, 1974) Commonwealth Scientific and Industrial Research Organization, U.S. Patent 3,834,923
26. Groszek AJ (1966) Chemistry and Industry 15:1754
27. Day RE, Parfitt GD, Peacock J (1974) J Colloid Interface Sci 46:17
28. Hoerr CW, McCorkle MR, Ralston AW (1943) J Amer Chem Soc 65:328
29. Mukerjee P, Mysels KJ (1970) Critical Micelle Concentrations of Aqueous Surfactant Systems, Nat Stand Ref Data Ser — Nat Bur Stand 36, Washington
30. Arsentiev VA, Leja S (1976) In: Kerker M (ed) Colloid and Interf Sci Vol V, Academic Press Inc., London, pp 251—270

Authors' address:

M. Sc. (Tech.) A. B. A. Pettersson and
Prof. J. B. Rosenholm
Department of Physical Chemistry
Åbo Akademi, Porthansgatan 3—5
20500 Åbo, Finland

 Progr Colloid Polym Sci 82:43—51 (1990)

Influence of calcium and phosphate ions on the adsorption of partially hydrolyzed polyacrylamides on TiO_2 and $CaCO_3$

Y. El Attar Sofi, C. Carquille, J. M. Lamarche and A. Foissy

Laboratoire Electrochimie des Solides, U.A. CNRS, Université de Franche-Comté, Besançon, France

Abstract: The adsorption of neutral and 16% hydrolyzed polymers having similar molecular weights was compared on calcium carbonate and titanium dioxide. A detailed analysis of the measurements shows a lack of coherence between adsorption trends and electrical interactions between surfaces and molecules. The same coverage was measured on two oppositely charged surfaces, but in similar dispersing media (containing similar pH, ionic strength, and calcium concentration). Calcium ions considerably increase the uptake of hydrolyzed polymers due to the complexation and incidence on molecular overlaps in the interface. Phosphate ions decrease the adsorption of neutral and hydrolyzed polymers to zero on both minerals. This phenomena can be explained by the preferential adsorption of phosphates on the sites otherwise available for polymer binding. Results are highly dependent on the experimental procedure.

Key words: Adsorption; polyacrylamides; TiO_2; $CaCO_3$; calcium influence; phosphate influence; surface charge density

Introduction

Polyelectrolyte adsorption is highly dependent on such parameters as surface and polymer charge densities, pH, and ionic strength, all of which have an effect on the electrical interactions between the substrate and the molecule. This adsorption, to a lesser extent, also depends on the solvent quality, non-electrical binding energy, kinetic effects, experimental procedure, and possibly other factors, which all contribute to the complexity of this subject and provide ambiguous and diverging experimental results.

Some studies have been carried out on the adsorption of the polyacrylamide group on minerals in aqueous media. Much of this research has been industrially-aimed: the adsorption on cellulose for papermaking [1], on asbestos for floculation and filtration applications [2], and on silica and iron oxide for floculation in the presence of divalent ions [3]. Some authors, in a more general scope, have correlated the uptake on titanium dioxide to the variation of charge density and molecular conformation [4], showing an increase in adsorption with ionic strength [5], and the influence of both a complexing ion (Ca^{2+}) and a surfactant [6]. More recent studies have emphasized the electrical interactions between polyacrylic acid and several mineral oxides [7], compared the adsorption of neutral polyacrylamide on silica and silicon carbide [8], and analyzed the electrical interactions between hydrolyzed polyacrylamides and a modified silica surface [9]. Surprisingly enough, some of these studies show a difference of as much as tenfold in the adsorption measurements in apparently quite similar experimental conditions [8, 9].

The present study will elaborate on previous investigations concerning the adsorption of polyacrylic acid and polyacrylamides on titanium dioxyde [10, 11] and calcium carbonate [12], and will mainly deal with the adsorption of partially hydrolyzed molecules on both of these minerals, placing particular emphasis on the influence of ions having a strong interaction with the surface sites and macromolecules.

Materials

All materials used in this study have been previously described, namely titanium dioxide [13, 14], calcium carbonate [12, 15], and polyacrylamides [10, 15]. Consequently, only the main characteristics relevant to the present paper will be given.

Titanium dioxide

Titanium dioxide is a commercial product (P 25, Degussa, FRG). The crystalline form is anatase and the specific area is 53 ± 2 mg^2 g^{-1} (B.E.T. and immersion calorimetry). In spite of this value, the grain size measured by light scattering after 5 min sonication in a solution of polyacrylic acid is about 0.4 µm. the particles are then probably aggregated. The zero point of charge and the isoelectric point are both close to pH 6.5 in sodium chloride and potassium nitrate solutions. Calcium ions are highly adsorbant on the surface, consequently, the particles remain positive up to pH 11 in a solution containing 10^{-2} M Ca^{++} and 5 wt % titanium dioxide.

Calcium carbonate

Calcium carbonate is a precipitated calcite (SOCAL U1, Solvay, Belgium) with a specific surface area of 22 m^2 g^{-1}. The isoelectric point is at pH 9.5 in normally aerated solutions. It has been shown that the electrophoretic mobility is actually related to the calcium content of the solution by Nernst's Law [16]. The particles are positive at any pH < 11 in solutions containing at least 10^{-3} M Ca^{++} ions.

Polyacrylamides

Polyacrylamides are commercial linear compounds (Hoechst, FRG) with a mean molecular weight of 2 × 10^6. The polydispersity of polymers is high enough. The degree of hydrolysis can be potentiometrically controlled: The so-called neutral polymer (hereafter referred to as PAM) is less than 0.3% hydrolyzed. The hydrolysis ratio of the other polymer (hereafter called HPAM) is 16% hydrolyzed. Some properties of the aqueous solutions of these polymers are given in [10], especially their intrinsic viscosity as a function of pH and salt concentration.

Experimental procedure

It is a well-known fact that adsorption of high molecular weight polymers depends on such experimental steps as the order and rate of adding reactants, mixing procedure, solid concentration, and possibly other technical details. This adsorption is both the result of a very slow desorption rate leading to an apparent state of nonequilibrium between the adsorbed and free polymer molecules [17], and the molecular weight distribution [18]. A specific experiment will be described to further illustrate this point after explaining the general experimental procedure used in the present study.

The solid was dispersed with a magnetic stirrer in 25 ml of a solution prepared at a given pH and salt concentration (altering the pH concentration accordingly before adding the polymer). The solid concentration was 0.2 wt %. A series of 50-ml solutions was prepared containing varying amounts of polymer at the same pH and salt concentration as the solid dispersions. 25 ml of each solution was discarded for analytical purposes. The solid dispersions and the polymer solutions were stirred slowly with a magnetic rod for at least 24 h. Polymer solutions were then very slowly added to the respective dispersions using a burette, and the mixtures were kept for a 24-h period on an oscillating table before measurements were taken. The remaining polymer concentration after solid centrifugation was measured in the solution using a carbon analyzer (Dohrmann D.C. 80, Xertex, Santa Clara, California). The amount adsorbed was calculated using the initial and residual concentrations.

Experimental results

Adsorption isotherms in KNO_3 solutions are of the high-affinity type in both titanium dioxide and calcium carbonate, and closely resemble those obtained previously in NaCl solutions [10]. Only the variation of the adsorbed amount at the plateau level shall be considered here.

pH and ionic strength influence

Figures 1 and 2 show the HPAM adsorption as a function of pH and three ionic strengths, respectively, on titanium dioxide and calcium carbonate. Adsorption increases with an increase of ionic strength at pH 8 on both substrates. At pH below 6 for titanium dioxide the adsorption is independent of the ionic strength.

Neither titanium dioxide nor calcium carbonate show pH dependence above pH 8, but adsorption increases considerably on titanium dioxide between pH 8 (about 20 mg · mg^{-1}) and pH 4 (about 60 mg · g^{-1}).

Figure 3 shows that the adsorption of the unhydrolyzed polymer is also slightly dependent on ionic strength above pH 8 and that adsorption increases with acidity. The same level of adsorption is obtained for both polymers on titanium dioxide below pH 4.

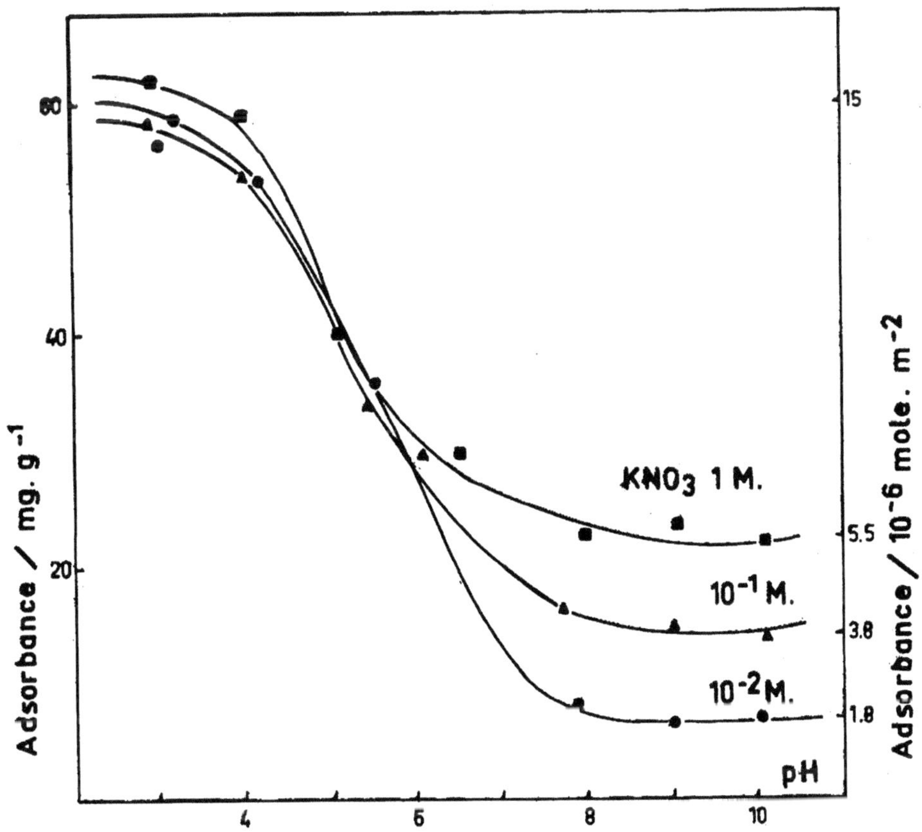

Fig. 1. Adsorption of HPAM on TiO_2 in KNO_3

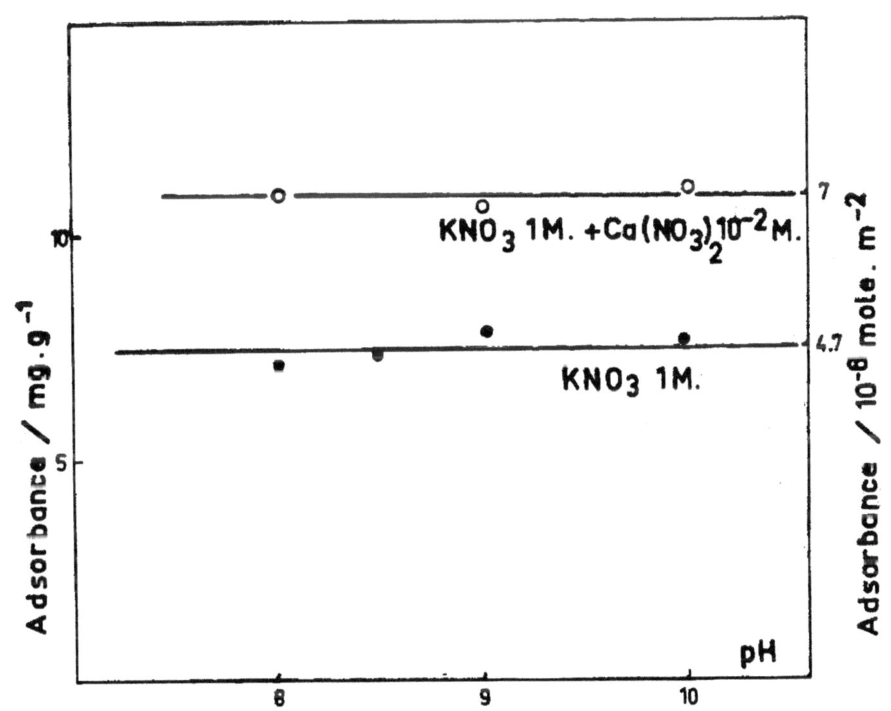

Fig. 2. Adsorption of HPAM on $CaCO_3$

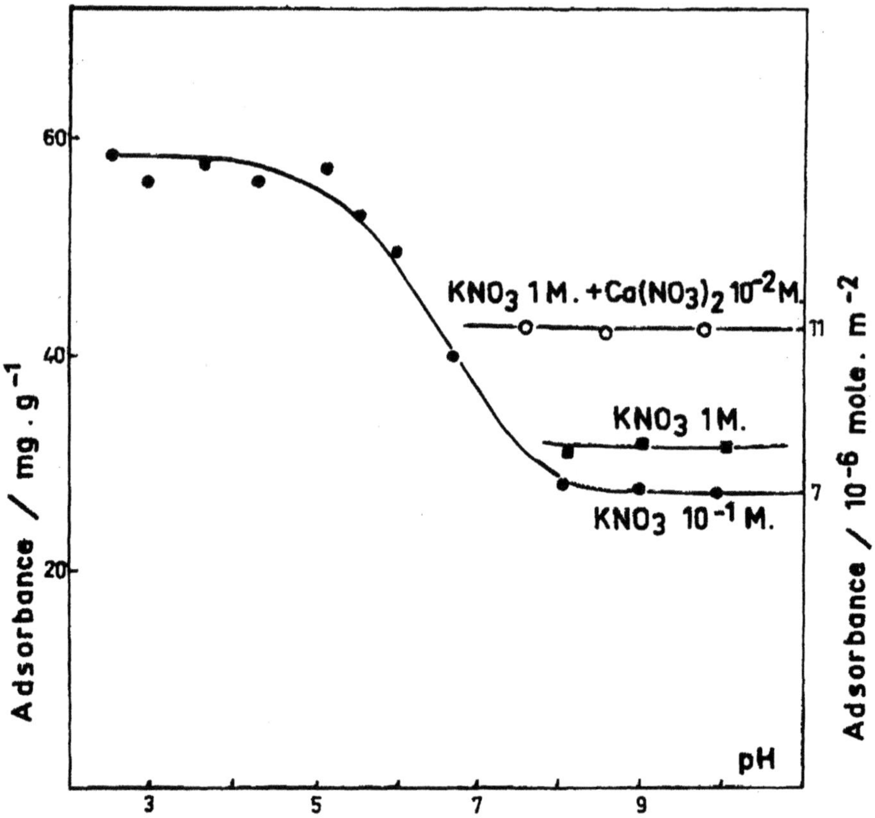

Fig. 3. Adsorption of PAM (unhydrolyzed polyacrylamide) on TiO_2

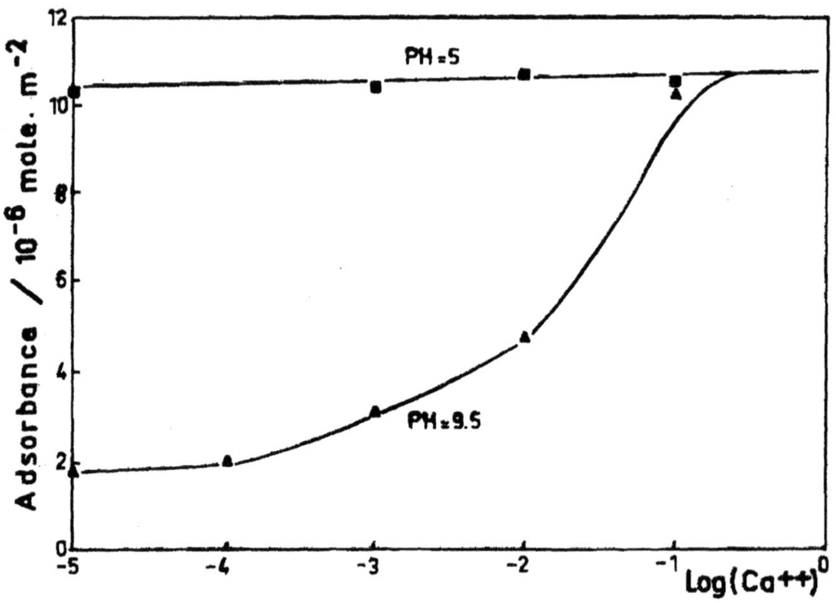

Fig. 4. Adsorption of HPAM on TiO_2 vs Ca^{++} content

Calcium ion influence

Figure 4 shows that an addition of calcium ions greatly increases HPAM adsorption at pH 9.5. Another interesting feature is the adsorption level which is the same at both pH 9.5 and pH 5 when the calcium concentration is above 10^{-1} M.

An appreciable increase in HPAM adsorption on calcium carbonate at pH 9.5 can also be observed, becoming even greater when ionic strength is lower (Fig. 5).

An equal level of adsorption was measured in surface density units on titanium dioxide and calcium carbonate at the highest surface density: 10^{-5} M · m^{-2} at pH 9.5. This measurement is also very close to that of the PAM adsorption in the same conditions.

Phosphate ion influence

Phosphate ions have the opposite influence on HPAM adsorption on both minerals. Examples of this phenomenon for titanium dioxide are given in Fig. 6 (pH 5, KNO_3 1 M) and for calcium carbonate in Fig. 7 (pH 9.5, KNO_3 1 M). The adsorption level is zero with a phosphate concentration of about 10^{-1} M × l^{-1} when using the general experimental procedure (referred to as "method a" in the figures). The same results are obtained at pH 9.5 for titanium dioxide. There is no adsorption of the unhydrolyzed polymer on either mineral due to the presence of phosphate ions when performed in similar experimental conditions.

Experimental procedure influence

As mentioned previously, discrepancies may occur in results from apparently similar experimental procedures, and even more so between different authors. In this experiment, for example, the adsorption of HPAM was carried out in the presence of calcium ions using a procedure (method b) which differs from the one generally used in the following manner: instead of adding the calcium salt at the chosen concentration in both the suspension and the polymer solution before mixing, a double concentration of salt was added with the solid in the suspension and no calcium salt was added to the polymer. Thus, the total chemical composition is similar, but, as shown in Fig. 8, the adsorption is about half of that measured previously. Indeed, Fig. 8 shows that adsorption increases slowly until it reaches an adsorption level close to that obtained using method a.

A similar phenomenon is shown in solutions containing phosphate ions. The curves entitled "method b" in Figs. 6 and 7 provide measurements when all of these ions are added together with the polymer solutions and similarly, when none of them are added with the solid before mixing. No experiments have been performed to measure the effects of ageing.

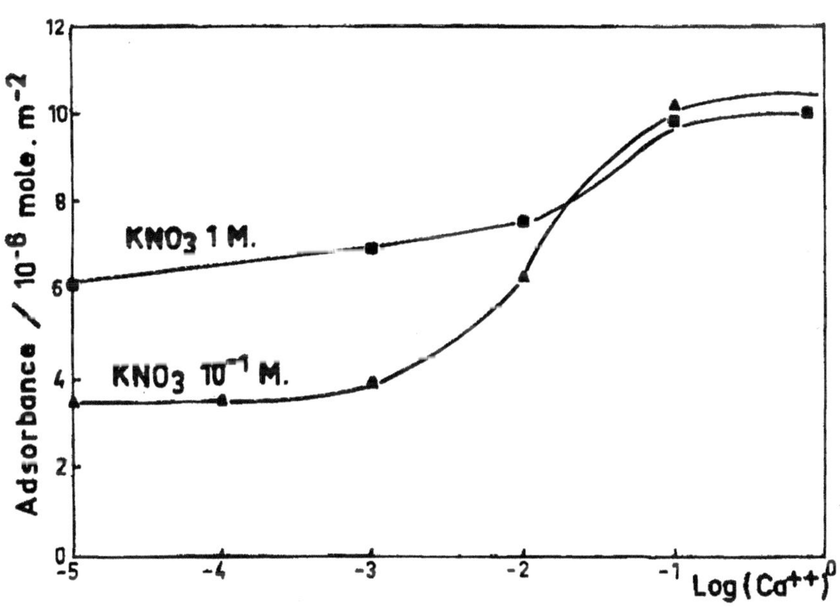

Fig. 5. Adsorption of HPAM on $CaCO_3$ vs Ca^{++} content at pH 9.5

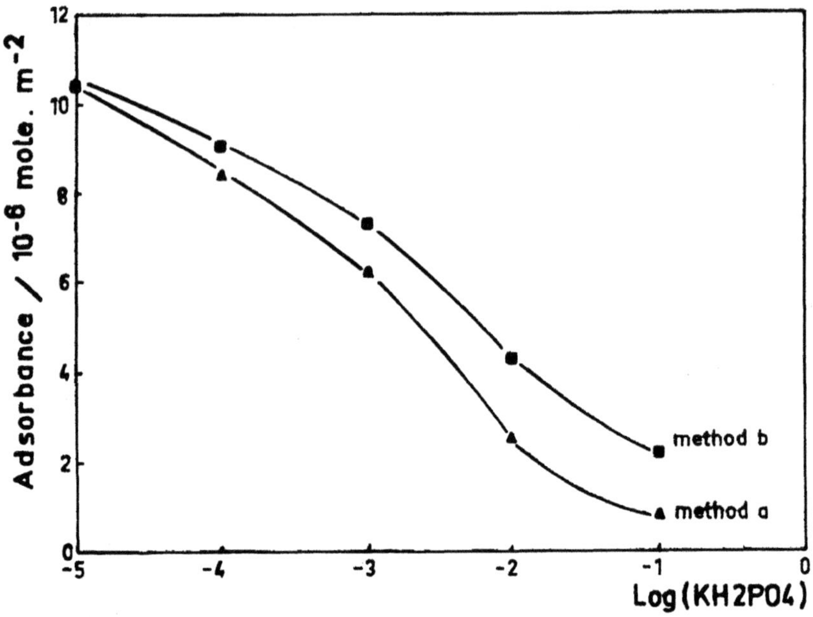

Fig. 6. Adsorption of HPAM on TiO_2 vs phosphate content at pH 5, KNO_3 1 M; two experimental procedures

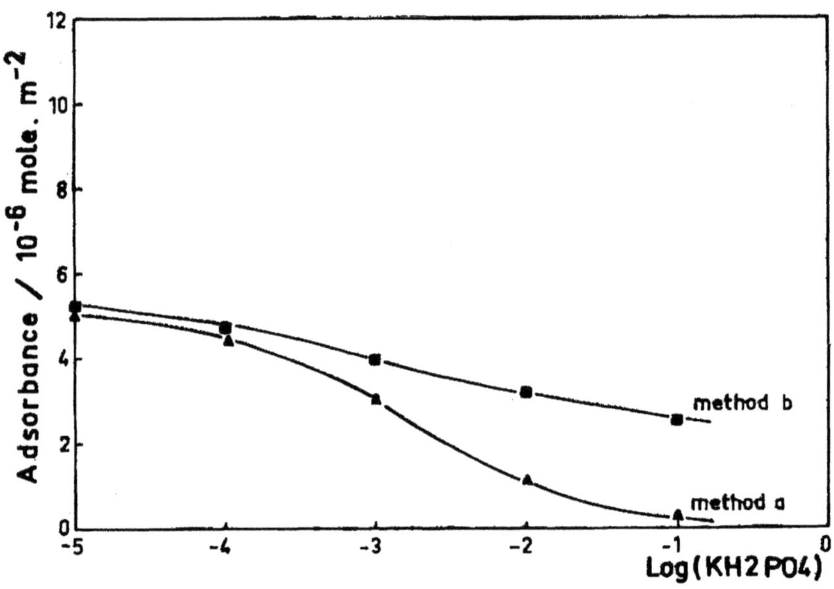

Fig. 7. Adsorption of HPAM $CaCO_3$ vs phosphate content at pH 9.5, KNO_3 1 M; two experimental procedures

Discussion

The variations in polyelectrolyte adsorption as a function of the solution's pH and saline composition, qualitatively-speaking, are a result of the influence of these two parameters on the electrical interactions between the molecules and the mineral surface and, sometimes, on the molecular conformation in the adsorption layer, as is the case with polyacrylic acid.

Gebhardt et al. [7] attribute the increased adsorption in acidic media to the formation of increasingly attractive forces between the positively charged oxide surfaces (silica, hematite, rutile) and the negatively charged polymer. Lecourtier and Chauveteau [8] use this same reasoning to explain

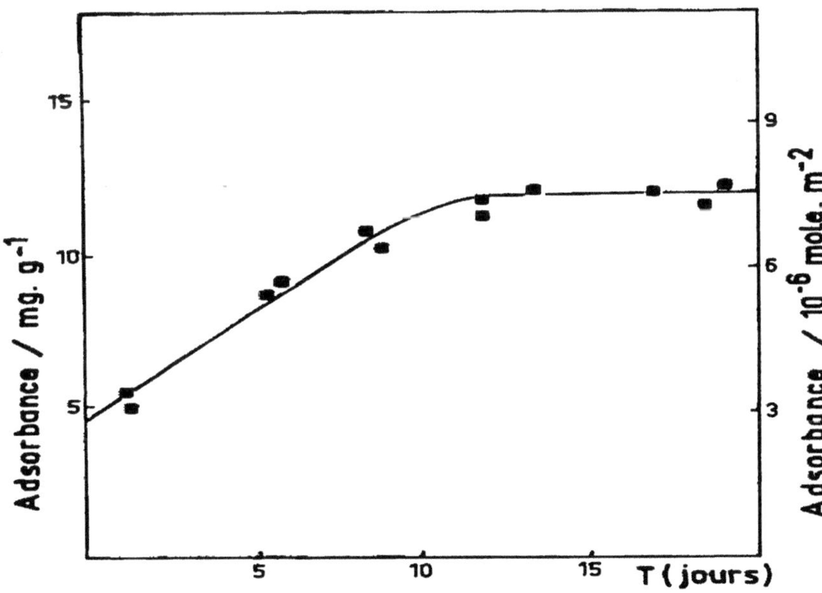

Fig. 8. Adsorption of HPAM on $CaCO_3$ vs aging of system, method b (see text)

the increased adsorption of hydrolyzed polyacrylamides on silicium carbide as a result of electrostatic repulsion screening when ionic strength is increased. Such arguments are consistent with the general decrease in HPAM adsorption on TiO_2 as the pH increases to approximately pH 6.5 (Fig. 1). In fact, the average surface charge reverses from a positive to a negative value in this pH range [15]. When taking a closer look at the adsorption variations in Figs. 1 and 2, however, some unjustified trends based on electrostatic interactions between the mineral surface and polymer can be observed:

a) adsorption on titanium dioxide does not change above pH 8, whereas the surface density of the negative sites increases considerably;

b) HPAM adsorption on $CaCO_3$ does not depend on pH (Fig. 2) but the particle charge is reversed around pH 8;

c) HPAM adsorption on TiO_2 in acidic medium does not depend on pH, although one would expect much variation as the surface becomes more positively charged and the polymer less negatively charged when the pH decreases below pH 4 or 5. On this same point, Jean-Chronberg [9] found a very high adsorption level for HPAM around the isoelectric pH on a silica surface modified by grafted Al-OH functions (—Si—O-Al-OH).

d) the absence of adsorption in the presence of phosphate ions cannot solely result from electrostatic repulsions since the same phenomenon occurs with unhydrolyzed polymers as well.

To explain the differences between the adsorption variations and those of surface-polymer electrostatic interactions, other factors involved in the adsorption reaction must be considered. We now intend to show how the amount adsorbed in the experiments we performed essentially depends on the interactions between ionized polymer groups, i.e., size and entanglement feasibility, as well as the presence of specific adsorption sites.

pH and ionic strength dependence

An interesting analogy can be made between the values obtained with both minerals when considering the adsorption in molar density.

The surface adsorption is the same, approximately 6×10^{-6} M $\cdot$ m^{-2} on TiO_2 and $CaCO_3$ in alkali media ($\cong$ pH 8) with high ionic strength. This value is also very close to the adsorption measured for a similar polymer on sand and silicon carbide by Chauveteau et al.: 5.8×10^{-6} $\cdot$ m^{-2} [8]. In these experiments, however, the surface charge is quite different, being positive on $CaCO_3$, slightly negative on TiO_2, and highly negative on SiC. This phenomenon can be explained by the fact that the surface coverage is limited mostly by the molecular size of the polyelectrolyte and electrical forces of repulsion between adsorbed molecules. These factors, in turn, are dependent on the dispersion medium parameters: pH and salt concentration.

Considering this, the adsorbed amount variation as a function of pH and ionic strength shows some amount of consistency. Adsorption remains constant above pH 8, although the surface charge increases. There is no change in the molecular size and intermolecular distance as the pH increases, due to a fully ionized polymer. On the contrary, the HPAM remains uncharged below pH 4, its adsorption thus being independent of pH and ionic strength and quantitatively resembling that of PAM. Finally, the significant decrease in adsorption from pH 4 to pH 8 might be a consequence of the ionization of carboxylic groups leading to molecular stretching and intermolecular repulsions.

The adsorption dependence of PAm on pH still remains unexplained. Since adsorption slightly depends on ionic strength and calcium ion concentration at a higher pH, this phenomenon is suspected to originate in the few carboxylic functions observed by potentiometric titration.

Calcium ion concentration dependence

Calcium ions provoke a greater increase of HPAm adsorption in alkali media (pH 9.5) than in monovalent sodium and potassium solutions. In the first case, coverage reaches a maximum of 10^{-5} M · m^{-2} in 10^{-2} M Ca^{++} solutions, but in the second case, only reaches 6 × 10^{-6} M · m^{-2} in 10^{-1} M K^{+} solutions.

In calcium solutions the adsorption maximum is again the same for both minerals. This shows that the value depends more on the molecular properties in the solid-liquid interface than on the surface electro-chemical characteristics. Surely, calcium ions do modify the surface sites. They strongly adsorb on both minerals as a reversal of charge occurs in solutions containig 10^{-4} M Ca^{++} for a $CaCO_3$ surface and 10^{-3} M · Ca^{++} for a TiO_2 surface. The calcium concentration for such a transformation is, however, one order of magnitude lower than for a significant increase in adsorption (10^{-2} M · Ca^{++}).

The experiment using two different mixing procedures, but having the same final composition (Fig. 8) also differentiates between the respective roles of adsorbed and "in the bulk" calcium ions. In method a, the polymer is added free of calcium; it then adsorbs in a more stretched conformation on the positive surface than in method b where it is formerly in contact with the complexing calcium ions. In the first case (method a), the adsorbed molecules slowly change their conformation as they interact with calcium. The reduction of electrical repulsions and molecular size allows more adsorption to take place and the uptake eventually reaches the same value as in method b.

Phosphate ion concentration dependence

The decrease and, possibly, the suppression of PAM and HPAM adsorptions in the presence of phosphate ions give an insight into the binding of the polymer on the surface. As shown in Fig. 9, phosphate ions adsorb strongly on titanium dioxide. An adsorption plateau close to 1.8 μ · mole × m^{-2} is measured with an equilibrium concentration of about 2 × 10^{-3} mole · dm^{-3}. In a previous study using chloride counterions we have shown that the polymer molecules displace the ions, so as to occupy the surface sites. In the present case the binding with the phosphate ions is stronger than with the polymer functional group, then no displacement occurs. This is shown in Fig. 9 by the similarity of both adsorption isotherms, with and without the addition of polymer. The difference between both experimental methods in Figs. 6 and 7 is due to a kinetic effect. In method b, where the phosphate ions are added together with the polymer molecules, some of these molecules are adsorbed as they reach the surface sites prior to the phosphate ions. Eventually, however, they are displaced by the ions, giving an adsorption similar to that obtained with method a (Fig. 8).

Conclusion

A systematic measurement of the adsorption of neutral and partially hydrolyzed polyacrylamides has been taken on TiO_2 and $CaCO_3$ as a function of experimental procedure, pH, ionic strength, and calcium and phosphate concentrations. Comparisons between the adsorption on both substrates, as well as trends based on the parameters, show that adsorption is limited mostly due to the molecular size and intermolecular forces which, in turn, depend on the polymer charge density.

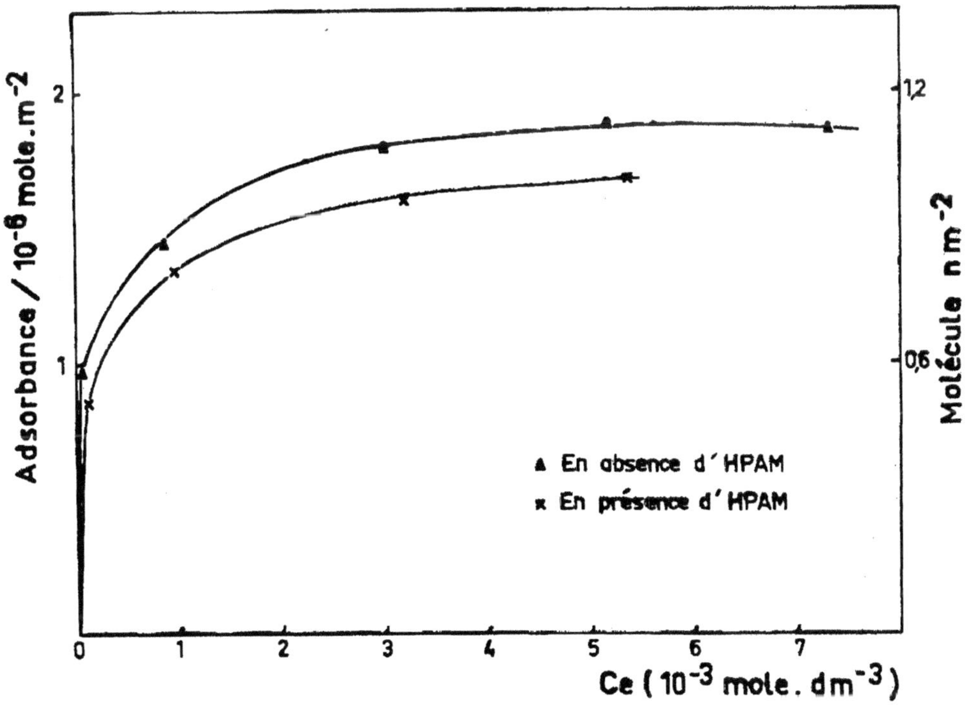

Fig. 9. Adsorption of phosphate ions on TiO_2 vs KH_2PO_4 at 9.5: Δ without HPAM, and X with HPAM

Variations in the mineral surface charge in the present experiment do not seem to be a dominant factor. The suppression of adsorption when adding phosphate ions gives some insight into the binding reaction of the polymer to the surface sites. Phosphate ions, unlike chloride or nitrate, bind strongly enough to the site so as to not be displaced by the polymer.

Acknowledgements

This work was supported by the CNRS-ARTEP through its PIRSEM project on the conversation of natural resources, and by the Insitut Français du Pétrole.

References

1. Howard G, Lyth Hudson F, West J (1977) Appl Polym Symp 61:389
2. Klimenko O, Lapin V (1980) Kolloid Z 42, 5:968
3. Slater R, Clark J, Kitchener J (1969) Ceram Soc Proc 13:1
4. Chong P, Curthoys G (1979) Int Miner Process 5:335
5. Dodson P, Somasundaran P (1984) J Colloid Interface Sci 97:481
6. Bocquenet Y, Siffert B (1984) Colloids Surf 9:147
7. Gebhardt J, Furstenau D (1983) Colloids Surf 7:221
8. Chauveteau G, Lecourtier J (1985) Colloque RCP-CNRS "Rhéologie des solutions de polymères", Sainte Croix aux Mines
9. Jean-Chronberg A (1988) Thesis, Univ Strasbourg, France
10. Girod G, Lamarche J, Foissy A (1988) J Colloid Interface Sci 121, 1:265
11. Foissy A, El Attar A, Lamarche J (1983) J Colloid Interface Sci 96, 1:275
12. Lamarche J, Persello J, Foissy A (1983) Ind Eng Chem Prod Res Dev 22, 1:123
13. Foissy A, M'Pandou A, Lamarche J, Jaffrezic-Renault N (1982) Colloids Surf 5:123
14. Foissy A (1985) Thesis, Univ Franche-Comté, Besançon, France
15. Lamarche J (1989) Thesis, Univ Franche-Comtè, Besançon, France
16. Pierre A, Lamarche J, Mercier R, Foissy A, Persello J (1989) submitted J Dispersion Sci and Tech
17. Van der Schee H (1984) Ph. D. Thesis, Agricultural Univ, Wageningen, Netherlands
18. Koopal L (1981) J Colloid Interface Sci 83, 1:116

Authors' address:

A. Foissy
Laboratoire Electrochimie des Solides
U.F.R. des Sciences et Techniques
25030 Besançon, Cedex, France

Progress in Colloid & Polymer Science Progr Colloid Polym Sci 82:52—61 (1990)

Retention & thermodynamics of homologous series in reversed-phase liquid chromatography

S. I. Andersen and K. S. Birdi

Fysisk-Kemisk Institut, Technical University of Denmark, Lyngby, Denmark

Abstract: The retention mechanism in reversed-phase liquid chromatography of homologous series of n-alkanes, n-alkylbenzenes, and polyaromatic hydrocarbons (PAH) on a octadecyl-bonded phase has been studied at temperatures ranging from 10° to 45°C. Pure single component eluents, e.g., ethanol and methanol were used to avoid preferential adsorption of the components in the commonly used binary water-alcohol eluents. The retention data was investigated according to the solvophobic theory of Horvath et al. (1976) J. Chrom. 125:129. Enthalpy H^0 and entropy S^0 of adsorption was calculated from Van't Hoff plots. The relationship between the capacity factor and the carbon number n in the homologous series was investigated, showing a specific constant contribution to the free energy of adsorption per incremental carbon group in each series at constant T. This indicates that adsorbates are present in a flat orientation on the surface of the adsorbent. The effect of solubility of the solutes in the mobile phase is investigated. Enthalpy-entropy compensation with the same compensation temperature for each series implies that the same mechanism governs the retention of n-alkanes, alkylbenzenes, and PAHs.

Key words: Non-aqueous RPLC; hydrocarbons; adsorption; retention; thermodynamics

Introduction

n-Octadecyl-bonded silica (ODS) gels are widely used as packing materials in liquid chromatography. Even though much attention has been given to the studies of these alkyl-bonded materials the actual mechanism of retention at molecular scale has not yet been revealed [1, 2]. Most of the work reported in literature is concerned with the retention using binary mobile phases with one of the constituents being water. The retention of the solutes in the liquid chromatography is determined by the properties of the chromatographic system, namely the sorbate, the type of the sorbent and the eluent composition. In reversed-phase liquid chromatography, the retention of hydrocarbons is determined by nonspecific interactions of solute and eluent with the sorbent, as well as by the specific and nonspecific interactions between solute and eluent [3]. The studies carried out with binary eluent mixtures complicates the understanding of the retention due to changes caused by preferential or excess adsorption of one of the eluent components on the stationary phase [4].

In this work, we have investigated the retention of different hydrocarbons; n-alkanes, n-alkylbenzenes, and polyaromatics at various temperatures in order to study the thermodynamics of retention. Pure ethanol and methanol were used as eluents in order to avoid the possibility of preferential adsorption from the commonly used binary alcohol — water eluents [5]. Such a preferential adsorption will be difficult to estimate, and with the present lack of knowledge about the nature and role of the bonded alkyl stationary phase, we do not know what the effect will be upon the retention [4, 6]. Using two different eluents we can investigate the effect of eluent on retention. The enthalpy of retention or adsorption is calculated from Van't Hoff plots, and the existence of enthalpy-entropy compensations for retention of the various types of solutes is analyzed.

Experimental

Apparatus

The liquid chromatography used consisted of a LKB 2150 HPLC pump, a Rhodyne 7125 injector, and a LKB 2142 refractive index detector equipped with a Radiometer Servograph REC 51 recorder, (Denmark). A 250 × 4 mm i.d. LiChrospher 100-RP-18 (5 μm) column was used. The column was thermostated and the temperature controlled by a HETO (Denmark) thermostat water bath ±0.2 K. 20 μl 0.1% (w/w) solute in ethanol or methanol was injected. All solutes were analyzed three or more times, and standard deviations were less than 1.0% for all measurements. All measurements were performed isocratically in the temperature range of 10°—50°C at a flow of 0.50 ml/min.

Chemicals

All solutes investigated are listed in Table 1. These were all used as received and had purities >99%. Most chemicals were obtained from Fluka or Merck.

Ethanol was obtained from DDSF (Denmark) and methanol (LiChrosolv) from Merck (FRG). The eluents were degassed by sonification prior to use.

Column dead-time determination: In literature the importance of correct dead-time determination has been stressed, since calculations of capacity factors and selectivity are based upon the dead time. Different methods that involve mathematical linearization of retention times of homologues [7] or retention times of unretained reagents has been proposed.

The dead-time t_0 of the column was measured from the retention time of 2:7 hydroxynaphthalene, as well as by the method of homologous series by Grobler and Balisz [8]. Because the retention time of the naphthol had no temperature dependence and was shorter than the dead-time obtained from linearization of n-alkane retention data, the value of the naphthol was used as t_0 in the calculations. The lower value of the naphthol could be caused by what has been called the negative adsorption [9], but using this value gave good correlations between alkane carbon numbers and retention times in agreement with literature [10]. The magnitude of t_0 as obtained in ethanol (258 s) and methanol (268 s) systems were different, which is in accord with literature [11].

Table 1. List of eluted compounds

No.	Solute	No.	Solute
1	2:7 Naphthol	16	Octylbenzene
2	Naphthalene	17	Nonylbenzene
3	1-Methylnaphthalene	18	Decylbenzene
4	2-Methylnaphthalene	19	Dodecylbenzene
5	Anthracene	20	n-Pentane
6	9-methylanthracene	21	n-Hexane
7	Phenanthrene	22	n-Heptane
8	Pyrene	23	n-Octane
9	Chrysene	24	n-Nonane
10	Perylene	25	n-Decane
11	Tetraline	26	n-Dodecane
12	Decaline	27	n-Tetradecane
13	Pentylbenzene	28	n-Hexadecane
14	Hexylbenzene	29	n-Octadecane
15	Heptylbenzene	30	n-Eicosane

Theoretical

The Van't Hoff equation gives:

$$-\ln K = \Delta G^0/RT = \Delta H^0/RT - \Delta S^0/R \,, \tag{1}$$

where K is the retention equilibrium constant, ΔG^0 is the Gibbs free energy of adsorption, ΔH^0 the enthalpy, and ΔS^0 the entropy of adsorption. The magnitudes of ΔH^0 and ΔS^0 can be calculated from Van't Hoff plots of $\ln K$ vs $1/T$.

K can be replaced by the capacity factor k' according to the relation $k' = \Phi K$ [13, 14], where Φ is the column constant or phase ratio, and $k' = (t_R - t_0)/t_0$. The magnitude of the capacity factor can be calculated from t_R. The column constant Φ in reversed phase chromatography using octadecyl-bonded stationary phases can be difficult to determine, because of the lack of knowledge of the retention behavior of these bonded column materials [12]. It has been claimed that the column constant may change with changing eluents due to the incorporation of eluent molecules in the stationary phase [11].

Inserting the capacity factor in Eq. (1) gives

$$\begin{aligned}\ln k' &= \ln \Phi - \Delta G^0/RT \\ &= -\Delta H^0/RT + \Delta S^0/R + \ln \Phi \,.\end{aligned} \tag{2}$$

The Van't Hoff plot according to Eq. (2) gives $\Delta H^0/R$ as the slope, but makes the determination of ΔS^0 difficult because the intercept will have the value of $\Delta S^0/R + \ln \Phi$. Assuming Φ to be constant when the eluent is not changed the intercept value can be used to indicate the trend of the entropy.

The Gibbs free energy of the transfer of the molecule, in liquid chromatography, from solution to the surface of the stationary phase can approximately be expressed as the sum of the adsorption energies of structural fragments of a molecule. In homologous series this will be given as

$$\Delta G^0 = \Delta G_0^0 + n\Delta G_i^0 , \quad (3)$$

where ΔG_0^0 is the adsorption energy of the first member in the series, ΔG_i^0 is the adsorption energy for one incrmental group and n the number of incremental groups [9]. In the case of n-alkane chains the incremental group is the methylene group CH_2. From Eqs. (2) and (3), the capacity factor can be expressed as

$$\ln k' = a + bn , \quad (4)$$

where $a = \ln\varphi - \Delta G_0^0/RT$ and $b = -\Delta G_i^0/RT$. From the slope b the adsorption energy of one incremental group in the homologue can be calculated as follows:

$$-b = \Delta G_i^0/RT = \Delta H_i^0/RT - \Delta S_i^0/R ; \quad (5)$$

the enthalpy and entropy of adsorption of the incremental group can be obtained from plots of b vs $1/T$. The magnitude of ΔH_i^0 can also be estimated from linearization of ΔH^0 vs n.

The above relations allow us to analyze the thermodynamics of retention of different homologues.

stant in the temperature range investigated. Data points below 17 °C were not included in the regression analysis of the alkanes, because of the possible phase transition of the stationary phase at a temperature about 17 °C [15]. When this phase transition occurs the adsorption and retention mechanism may change. In literature the effect of the phase transition is reported to be more pronounced as the chain length of the n-alkanes increases [15], this is also seen from our data (Fig. 1).

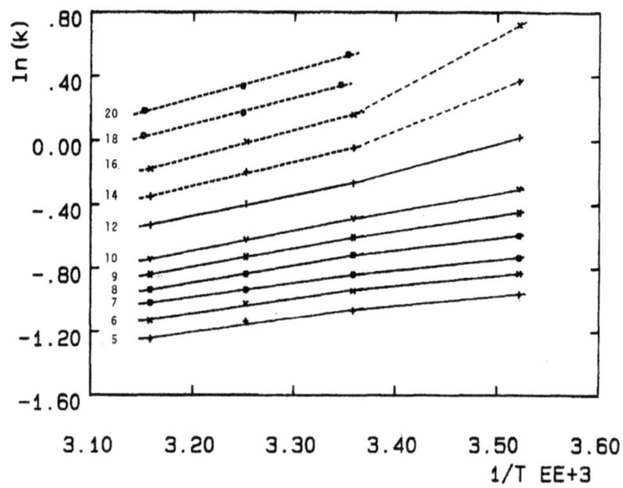

Fig. 1. Van't Hoff plot of $\ln k'$ vs $1/T$ for n-alkanes eluted with ethanol on ODS. Carbon numbers indicated. Note the apparent change in retention mechanism below 17 °C (3.44E—3 1/K)

Results and discussion

Most of the solutes investigated are clearly classified members of homologous series, but for the polyaromatic hydrocarbons (PAH) the classification procedure can be difficult due to the differences in condensation and number of aromatic rings. A few solutes such as tetraline and decaline are also difficult to place in any of the three groups of n-alkanes, alkylbenzenes, and PAHs. A few methyl-substituted PAHs have also been investigated.

Van't Hoff plots according to Eq. (2) of $\ln k'$ vs $1/T$ yield straight lines with intercept $\Delta S^0/R + \ln\Phi$ and slope $-\Delta H^0/R$ (Figs. 1—3). Most solutes gave regression coefficients better than 0.99 in agreement with the regular retention theory [13, 14] (Table 2). This indicated that the enthalpy of retention is con-

System: n-Alkanes and n-alkylbenzenes

Equations (3) and (4) are well established in chromatography and linearity in $\ln k'$ vs n plots have been reported for n-alkanes (n = carbon number), n-methylesters and n-alkylbenzenes (n = carbons in chain) in different eluents [10]. The linearity has been shown to be valid only as long as the carbon number of the homologues are less than that of the bonded alkyl chain in the stationary phase [11]. In literature [9, 10] some doubt has been expressed about the thermodynamic reason for the existence of linear $\ln k'$ vs n relations. Kazakevitch et al. [9] showed mathematically that the dependence of $\ln K$, and not of $\ln k'$, upon n in the n-alkanes should be linear, but that for $n > 4$ the $\ln k'$ vs n relationship was approximately linear. The deviation for short-chain molecules can be seen as an effect of dif-

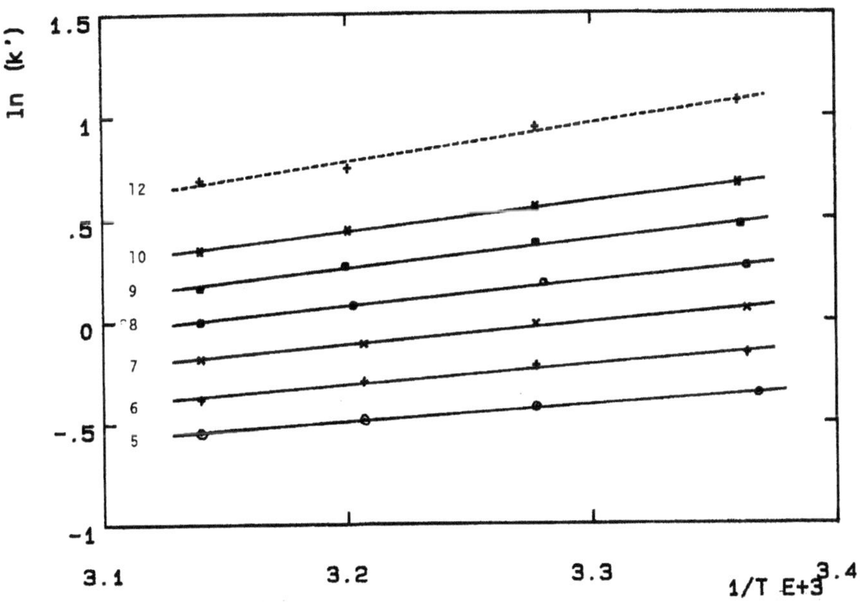

Fig. 2. Van't Hoff plot of $\ln k'$ vs $1/T$ for n-alkylbenzenes eluted with methanol. Carbon number of alkyl chain indicated

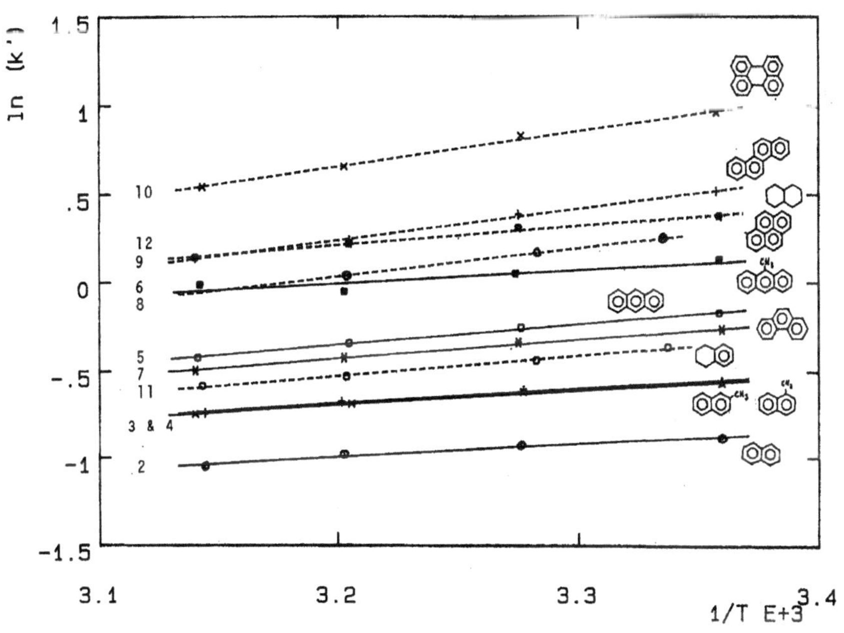

Fig. 3. Van't Hoff plot of $\ln k'$ vs $1/T$ for polyaromatic hydrocarbons and hydroaromatics: tetraline and decaline, eluted with methanol. Structures and compound number indicated

ferences in the adsorption energy of the methyl and the methylene groups. As the chain increases above $n = 4$ the contributions from the incremental group become constant and Eq. (4) becomes valid. The ΔG_0^0 term in Eq. (3) contains the difference in adsorption energy between a methylene chain containing n carbons and a n-alkane containing n carbons.

Calculated values of the coefficients a and b in Eq. (4) are given in Table 3 for n-alkanes and n-alkylbenzenes compared with values taken from literature [11, 15]. Methylene has been used as incremental group for both solute types. The magnitude of b (i.e., the Gibbs free energy of adsorption per methylene) are in good agreement with literature, while magnitudes of a which is

Table 2. Results from Van't Hoff calculations for retention in ethanol and methanol. Numbers refer to Table 1

Solute No	Ethanol			Methanol		
	ΔH^0 kJ/mol	$\ln \Phi - \Delta S^0/R$	r	ΔH^0 kJ/mol	$\ln \Phi - \Delta S^0/R$	r
1						
2	3.709	−3.069	0.66	6.571	−3.527	0.9842
3	7.927	−4.124	0.9846	7.005	−3.378	0.9937
4	6.091	−3.379	0.9560	6.768	−3.295	0.9955
5	9.727	−4.563	0.9551	9.676	−4.076	0.9982
6	10.675	−4.744	0.9305	10.368	−4.049	0.9952
7	8.339	−4.425	0.9259	8.882	−3.847	0.9977
8	9.378	−4.461	0.9755	13.272	−5.074	0.9996
9	12.488	−5.148	0.9997	14.948	−5.515	0.9987
10	14.466	−5.904	0.9938	16.558	−5.716	0.9972
11	8.524	−4.643	0.88	9.480	−4.175	0.9963
12	8.294	−4.121	0.91	9.120	−3.299	0.9958
13	6.393	−3.506	0.9997	7.135	−3.240	0.9962
14	6.982	−3.635	0.9983	8.124	−3.434	0.9904
15	8.347	−4.073	0.9999	9.041	−3.594	0.9957
16	9.120	−4.272	0.9965	10.038	−3.790	0.9976
17	8.888	−4.074	0.9992	11.486	−4.163	0.9948
18	9.320	−4.133	0.9998	12.203	−4.259	0.9995
19	11.457	−4.774	0.9995	15.435	−5.162	0.9892
20	6.308	−3.626	0.9889	6.334	−3.170	0.9768
21	6.666	−3.647	0.9918	7.132	−3.317	0.9937
22	7.379	−3.822	0.9999	8.749	−3.731	0.9929
23	8.997	−4.355	0.9999	9.699	−3.906	0.9877
24	9.494	−4.445	0.9999	10.415	−3.998	0.9865
25	10.624	−4.780	0.9999	11.320	−4.154	0.9916
26	11.031	−4.719	0.9991	14.528	−5.021	0.9932
27	12.575	−5.124	0.9992	17.215	−5.701	0.9960
28	13.903	−5.455	0.9994	19.579	−6.266	0.9959
29	14.038	−5.313	0.9965	22.077	−6.885	0.9967
30	15.441	−5.699	0.9965	24.433	−7.460	0.9980

Table 3. Coefficients calculated from the relation $\ln k' = a + bn$, 25 °C. Reference temperatures if different from 25 °C are given in ( ). For n-alkanes and alkylbenzenes n = carbon number in chain

Solute type	Eluent	a	b	r
n-Alkane	MeOH	−1.6831	0.2077	0.9998
	EtOH	−1.6183	0.1114	0.9998
Ref. [21]	MeOH	−1.4688	0.1971	0.999
Ref. [7]	EtOH	−1.1849	0.0932 (22)	0.9999
Akylbenzene	MeOH	−1.3778	0.2042	0.9999
	EtOH	−1.4813	0.1107	0.9999
Ref. [21]	MeOH	−1.1444	0.1917	0.999
Ref. [7]	EtOH	−1.4855	0.1109	0.9998

related to the phase ratio Φ differ as expected, due to differences in columns and, thereby, in phase ratios. The fact that magnitudes of b as found here are the same as reported in literature from experiments on other ODS columns with the same eluents, may imply that the incremental free-energy of retention per methylene group is almost constant on ODS, and that the well-known differences between different columns are mostly related to the term a. The possibility also exists that the columns used in literature and ours have nearly the same characteristics.

The magnitude of b was found to be similar for n-alkane and n-alkylbenzene in the same eluent, implying that the adsorption takes place in the same energetic manner. Differences in the magnitude of term a are caused by the difference in ΔG_0^0 for the two solute types. This can be related to solute-eluent interactions or may as well be caused by the procedure used to calculate the coefficient. The magnitude of a-coefficient is comprised of three contributions: one from the extrapolation, another from the functional group, and finally, $\ln \Phi$ [15].

The magnitude of b is lower for ethanol than for methanol, indicating that the solutes are less strongly retained when the solubility in the mobile phase increases, making the adsorbed state less preferential.

In order to confirm the validity of the equations the isothermal data of n-alkanes (n = 5 to 16) were linearized according to Eq. (4) to minimize deviations due to errors in dead-time determination and experimental deviations. Estimated $\ln k'$ values were then used in Van't Hoff plots to calculate new $\Delta H^0/R$ values, giving a constant contribution to ΔH^0 for each incremental methylene group. The linear ΔH^0 vs n relationship has been reported in literature for n-alkanols [30]. In Table 4 coefficients in the relation $\Delta H^0 = A + Bn$ are given for n-alkanes and n-alkylbenzenes, where B is the incremental contribution to the enthalpy of retention. In Fig. 4 ΔH^0 vs n dependencies for n-alkanes and n-alkylbenzenes for both raw and linearized data are given. This obvious connection between the Van't Hoff equation and the $\ln k'$ vs n relation proves the existence of a thermodynamical basis for Eq. (4) and the retention mechanism. The intersection of the correlation lines around n = 5—6 in Fig. 4 may indicate a change in the retention mechanism or dependence on n occurring below n = 5 in accord with the calculations of Kazakevitch et al. [9] and the results of Hanai et al. [20] (in acetonitrile/water

Table 4. Linear dependency of ΔH^0 on carbon number of n-alkanes and n-alkylbenzenes. $\Delta H^0 = A + Bn$ (kJ/mol)

Solute	Eluent	A	B	r
n-alkane	MeOH	—0.225	1.232	0.9985
	EtOH	3.629	0.612	0.9852
n-alkyl-	MeOH	1.047	1.232	0.9935
Benzenes	EtOH	3.221	0.666	0.9661

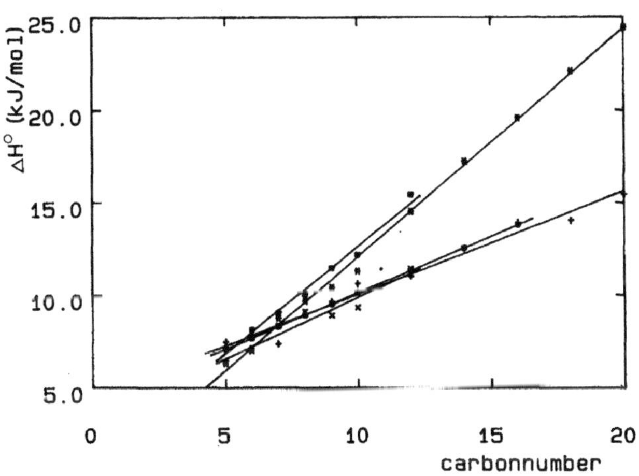

Fig. 4. Enthalpy of retention vs carbon number n of alkyl chains of n-alkane (+/EtOH, */MeOH) and n-alkylbenzenes (×EtOH, @/MeOH) eluted with ethanol and methanol. Linear regressions according to $\Delta H^0 = A + Bn$. For alkane/EtOH data (o) linearized according to $\ln k' = a + bn$ is shown

eluents). At constant temperature the incremental group has a specific contribution to the change in Gibbs free energy of adsorption. The incremental contribution is merely the same for methylene groups in both solute types, decreasing when going from methanol to ethanol. This constant increment per methylene group in n-alkanes and n-alkylbenzenes indicates that the molecules are adsorbed in a flat position on the stationary phase [16], and that no differences exist in the adsorption of a CH_2-group, whether it is a member of n-alkane or n-alkylbenzene. Recently, dependencies on surface area and planarity has been reported for a vast number of hydrocarbons and substituted hydrocarbons [1, 15]. In the investigated range of carbon numbers (n = 5—20) there is no evidence of deviations from the linear ΔH^0 dependence upon n, indicating that the solute are adsorbed in the same flat manner, independent of the carbon number. For alkyl chains longer than the ODS, deviations could be expected in accordance with the findings of Möckel and Freyholdt [11], but even for eicosane ($n = 20$) the linearity persists in our data. The adsorption mechanism may be different for shorter ($n < 5$) alkyl chains, as mentioned above. Apparently, the chain must have a certain length before the incremental contribution for each CH_2-group is constant. This could be an effect caused by the size of the molecule.

System: Polyaromatic hydrocarbons

The influence of the size, shape, and structure of polyaromatic hydrocarbons on their reaction in RPLC has been widely investigated. Different descriptions have been used in the correlation of retention time, i.e., carbon number (n), connectivity indices [18], shape factors [19], Van der Waals volumes [18, 20, 21], length-to-breadth ratios [18, 22], and combinations of these. Inspite of all this work, no universal description has been reported. Some descriptions are satisfactory with isomers and others are not. The dependency on these descriptions are not as easy to interpret as the $\ln k'$ vs n relationship of the alkyl chains. Often the main explanation will be entropic due to differences in the ordering of the adsorbed molecules. In Table 5, correlations for the relationship of capacity factors upon two of these descriptions are given; the Van der Waal volume and the correlation factor F = number of double bonds + primary and secondary carbons — 0.5 number of non-aromatic rings [18], both having regression coefficients better than 0.97. The Van der Waals volume relationship indicates again a flat adsorption on the stationary phase. The dependency on F (for PAHs the number of double bonds) is not easy to interpret, but can be seen as an indicator of the size and condensation of the PAH system. Nevertheless, the elution order cannot simply be related to either of the above used descriptions. An example is the retention order of isomers like anthracene and phenanthrene. The retention of anthracene is always stronger than that of phenanthrene [21]. This difference in $\ln k'$ for isomers has not been totally explained in literature. T. Hanai et al. [20] explained the lack of correlation between Van der Waals volume and $\ln k'$ for the two solutes to be due to changes in the volume of one of the solutes during the chromatographic process. This is not in accordance with Jinno et al. [18] who

Table 5. Polyaromatic Hydrocarbons. Dependencies of ΔH^0 on Van der Waals volume and correlation factor F. $\Delta H^0 = A + Bn$

Eluent	Descriptor	A	B	r
MeOH	Corr. F	−4.8965	3.0299	0.9763
EtOH		−6.1135	2.0606	0.9761
MeOH	VdW	−7.0437	0.1747	0.9680
EtOH		−8.4347	0.1694	0.9826

stated that the difference in the shape of the two solutes made the difference in retention. In literature the length-to-breadth ratio (L/B) has been used to correlate retention of isomeric PAHs, eluting in the order of increasing L/B [22]. This could indicate an entropic phenomena. The ΔS^0 values of anthracene calculated from the Van't Hoff plots are numerically larger than the values for phenanthrene. The anthracene molecules will be more ordered in the adsorbed state than the phenanthrene molecules, due to the more extended and stretched shape of the molecule. The effect of the solubility of the solute in the eluent on the retention behavior cannot be neglected. Locke [23] stated that the basis for selectivity was the relative solubility of each PAH in the mobile phase. According to Shinoda [24], the solubility of phenanthrene in ethanol at 25 °C is about 15 times larger than the solubility of anthracene. This indicates that the solute-eluent interactions give rise to the differences in retention times for chemically alike molecules. As mentioned above, the change of eluent had great effect on the incremental Gibbs free energy of retention.

The difference in capacity factors in the two eluents shows the effect of solubility on the retention. Going from ethanol to methanol the mobile phase environment becomes more hostile as the solubility of every solute investigated decreases, and the retention increases. This is seen as well in the increase of the enthalpy of retention and the incremental ΔH_i^0 (= B) contribution. For all the above-used descriptors, n, F, and VdW, the incremental contribution increases from ethanol to methanol in accord with the increased retention in methanol. The fixed contribution ΔH_0^0 (= A) increases as well, reflecting the increased hostility of the eluent and making the adsorbed state more preferable. The entropic term shows no significant trend in going from one eluent to the other. For n-alkanes and alkylbenzenes the entropy becomes more negative as the chain length increases and the solubility decreases, indicating that the solutes are preferentially retained on the stationary phase as the alkyl chain becomes longer [6]. All descriptors used in this work are size-dependent parameters, and linear dependencies of ΔH^0 upon these parameters support the idea of a flat conformation during adsorption and retention.

Enthalpy-entropy compensation

An extra-thermodynamic approach to the analysis of physicochemical data is based on the enthalpy-entropy compensation phenomena. When the changes in the enthalpy are compensated by changes in the entropy term they give rise to constant Gibbs free energies [25]; this gives rise to linear dependencies of ΔH^0 on ΔS^0. As the phase ratio Φ is not known, the entropy cannot be calculated directly. In the following calculations on the compensation phenomena we have assumed Φ to be constant, allowing us to investigate the phenomena from plots of ΔH^0 vs $\Delta S^0/R + \ln\Phi$.

According to the Gibbs relation for the free energy change the slope of the compensation line has the dimension of temperature and is termed the compensation temperature [25, 26]. This phenomena is seen in many different association processes like hydrogen bonding [27, 28] and micelle formation [26], and has also been reported for reversed-phase chromatography [25, 29]. The mechanism of the phenomenon is not well understood in chromatography [25, 14], but could be explained as was implied by Pimentell and McClellan [27] for hydrogen-bonding systems: "A higher value of ΔH^0 implies stronger bonding hence, greater order leading to a larger value of ΔS^0."

The compensation temperature T_c is expected to be related to the mechanism of adsorption, i.e., different compensation temperatures indicate differences in adsorption mechanism [14]. In Fig. 5 plots of ΔH^0 vs $\Delta S^0/R + \ln\Phi$ yield straight lines for both eluents, although a certain scattering of the data point can be seen. In these plots the slope has the value of R times T_c. The compensation line calculated for all the solutes investigated is shown and compensation temperatures are given in Table 6, indicating no significant difference in the compensation temperature and, hence, no difference in adsorption mechanism of the different solutes. Deviations are more likely produced by the procedure of the Van't Hoff calculation. This is probably the reason for the position of the tetraline-ethanol data point calculated from a very poor $\ln k'$ vs $1/T$ plot correlation ($r = 0.92$). Another explanation could be the non-planar conformation of the

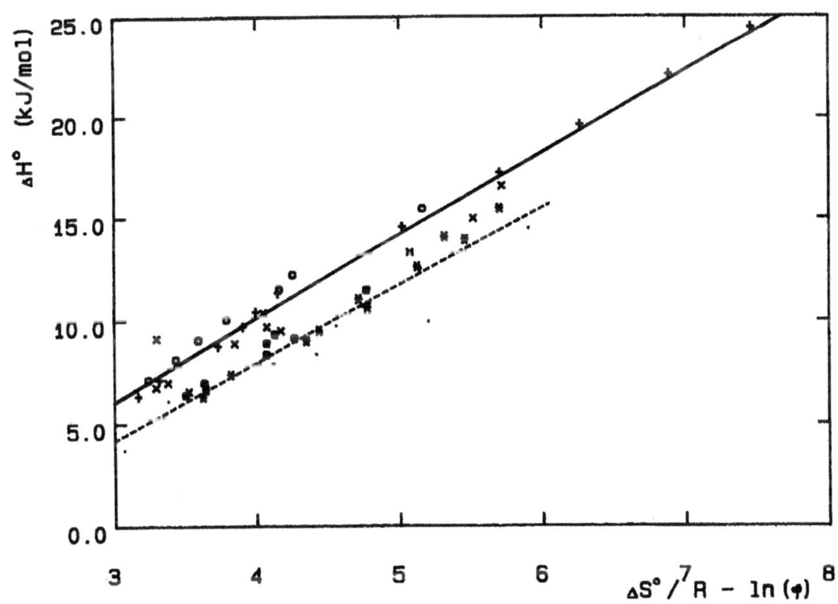

Fig. 5. ΔH^0 vs $\Delta S^0/R - \ln\Phi$. Enthalpy-entropy compensation line calculated for all compounds eluted in ethanol, ----, and methanol, —. n-alkanes */EtOH, +/MeOH; n-alkylbenzenes @/EtOH, o/MeOH; PAHs, •/EtOH, ×/MeOH

Table 6. Compensation temperatures T_c from plots of ΔH^0 vs $\Delta S^0/R + \ln\Phi$

	MeOH		EtOH	
	T_c	r	T_c	r
All solutes	489	0.9865	476	0.9823
n-alkanes	502	0.9992	515	0.9939
n-alkylbenzenes	519	0.9958	472	0.9867
PAH	499	0.9937	471	0.9921
PAH + Subst. PAH	475	0.9907	439*)	0.9720

*) In ethanol the Van't Hoff plots for the substituted PAHs and the PAHs have poor regression coefficients which affect the compensation temperature.

tetraline molecule, hindering the adsorption from being flat as it is for the other solutes investigated.

The equality of retention mechanism for the different solute groups is interesting considering the difference in structure and the possibility of increased interaction between solute, eluent, and stationary phase when the solute contains aromatic rings. The compensation phenomena does not give any evidence of whether the interactions are directly with the reversed-phase (solvophobic theory) or with solvent molecules adsorbed on the surface of the reversed-phase (sorption theory) [29].

Compensation temperatures in methanol and ethanol are merely the same, indicating the same retention mechanism in both eluents (Table 6). Compensation temperatures as found here are numerically of the same order of magnitude as the values reported elsewhere [14, 30, 31]. Deviations can be ascribed to the calculation procedure and variations in the data used. The difference in positions of the compensation lines in Fig. 5 is too large to be caused only by the difference in phase ratios when using ethanol and methanol. This must be explained as an increase in entropy caused by the increased solubility of the solutions in ethanol relative to the solubility in methanol. That T_c is nearly the same for both eluents can be a result of the eluents being members of the same homologous series of alcohols and, hence, no radical differences exist in the interactions between eluent and solvent, and eluent and stationary phase.

Conclusions

The thermodynamics of the retention of different hydrocarbons on n-octadecyl bonded silica using ethanol and methanol have been investigated using reversed-phased liquid chromatography. Based on the enthalpy-entropy compensation phenomena the overall mechanism of retention is seen to be the same in ethanol and methanol for all solutes investigated. The increased retention in methanol relative to ethanol is governed by the decrease of solubility in the mobile phase, making the adsorption onto the stationary phase more preferable. Differences in retention of isomers are caused only by differences in solubility, and not as reported [20] by changes in the Van der Waal volume during the chromatographic process.

The well-known equation $\ln k' = a + bn$ is shown to be based upon the adsorption of molecules in a flat position on the stationary phase. The slope representing the incremental Gibbs free energy of adsorption per incremental methylene group is dependent on the eluent. As the solubility of the solute in the eluent increases when going from methanol to ethanol the b-value decreases. The magnitudes of the b-coefficient, as found in this work, are in agreement with values reported in literature.

The $\ln k'$ vs n relationship is shown to be an effect of the linear dependency of the enthalpy of retention on the number of carbons in the n-alkyl chains in n-alkanes and n-alkylbenzenes.

The main conclusion is that the retention mechanisms in pure alcohol eluents are governed by the differences in solubility of the different solutes.

References

1. Jinno K, Shimurs S, Tanaka N, Kimata K, Fetzer JC, Biggs WR (1989) Chromatographia 27:285
2. Yamaguchi J, Hanai T (1989) Chromatographia 27:371
3. Lanin SN, Nikitin YS (1988) Chromatographia 25:272
4. Zhu PL (1986) Chromatographia 21:229
5. Chattoraj DK, Birdi KS (1984) Adsorption & the Gibbs Surface Excess, Plenum Press, New York
6. Stalcup AM, Matire DE, Wise SA (1988) J Chromatog 442:1
7. Wainwright MS, Haken JK (1980) J Chromatogr 184:1
8. Grobler A, Balizs G (1974) J Chromatogr Sci 12:37
9. Kazakevitch YV, Eltekov YA (1988) Chromatographia 25:965

10. Colin H, Krstukoi'c AM, Gonnord Mf, Guiochon G, Yun Z, Janders P (1983) Chromatographia 17:7
11. Möckel HJ, Freyholdt T (1983) Chromatographia 17:215
12. Colin H, Guiochon G (1978) J Chromatogr 158:183
13. Horváth C, Melander WR (1983) In: Heftmann E (ed) Journal of Chromatography Library 22A pp 27—103
14. Melander WR, Horváth C (1980) In: Horváth C (ed) HPLC Advances and Perspectives vol 2, Academic press pp 114—279
15. Möckel HJ, Welter G, Melzer H (1987) J Chromatogr 388:255
16. Katz S, Gray DG (1981) J Colloid Interface Sci 82:318
17. Sleight RB (1974) J Chromatogr 83:31
18. Jinno K, Kawasaki K (1983) Chromatographia 17:445
19. Wise SA, Bonnett WA, Guenther FR, May WE (1981) J Chrom Sci 19:457
20. Hanai T, Jukurogi A, Hubert J (1984) Chromatographia 19:266
21. Hanai T, Hubert J (1984) J Chromatogr 290:197
22. Wise SA, May WE (1983) Anal Chem 55:1479
23. Locke DC (1974) J Chrom Sci 12:433
24. Shinoda K (1978) Principles of Solution and Solubility. Marcel Dekker, New York, pp 204
25. Melander W, Cambell DE, Horváth C (1978) J Chromatogr 158:21
26. Singh HN, Daleem SM, Singh RP, Birdi KS (1980) J Phys Chem 84:2191
27. Pimentell GC, McClellan AL (1960) The Hydrogen Bond, Freeman & Co, pp 219—221
28. Birdi KS, Køhler K, Andersen SI (1988) XIX international Conf. on Solution Chemistry, August 15—18 Lund, Sweden
29. Jinno K, Ozaki N (1984) J Liquid Chrom 7:877
30. Vigh G, Varga-Puchony (1980) J Chromatogr 196:1
31. Opperhuizen A, Sinnige TL, Van der Steen JMD, Hutzinger O (1987) J Chromatogr 388:51

Authors' address:

K. S. Birdi
Fysisk-Kemisk Institut
Technical University of Denmark
Buildn 206
2800 Lyngby, Denmark

Emulsions, Foams, Thin Liquid Films

 Progr Colloid Polym Sci 82:65—75 (1990)

Competitive adsorption of food macromolecules and surfactants at the oil-water interface*)

E. Dickinson, S. R. Euston and C. M. Woskett

Procter Department of Food Science, University of Leeds, England

Abstract: Competitive adsorption between macromolecules and low-molecular-weight surfactants is an important factor affecting the formation and stability of food emulsions. We report experimental measurements and computational results relating to the displacement of disordered proteins from the oil-water interface by small-molecule amphiphiles. A Monte Carlo simulation of polymer chains + small displacer molecules shows how the average configurations of adsorbed macromolecules (expressed in terms of trains, loops and tails) are dependent on displacer concentration and the strength of polymer-displacer interactions. Measurements of interfacial tension and interfacial viscosity indicate distinct differences in behaviour between nonionic (noninteracting) surfactants and ionic (interacting) surfactants in mixtures with casein(ate) or gelatin.

Key words: Competitive adsorption; protein-surfactant interaction; computer simulation; interfacial tension; interfacial viscosity; food emulsions

Introduction

Food emulsions contain two classes of molecules which have a strong tendency to adsorb at the oil-water interface: amphiphilic macromolecules (mainly proteins) and small-molecule surfactants (lipids and their derivatives). The distribution of these two classes of molecules between the droplet surface and the two bulk phases is an important factor controlling the formation, stability and texture of edible oil-in-water emulsions such as salad dressing or ice cream [1]. This distribution is affected by competitive adsorption between macromolecules and surfactants at the oil-water interface, and by the nature of macromolecule-surfactant interactions, both at the interface and in the bulk aqueous phase. In real food systems, the situation is complicated by the fact that there is invariably a mixture of several different individual proteins which are adsorbing competitively [2], as well as various low-molecular-weight surfactants, both oil-soluble and water-soluble, which are competing with one another for binding sites on the proteins and at the interface. As a step towards understanding the real complex systems, it is convenient to consider model systems containing just a single protein and a single small-molecule surfactant. The information currently available in the literature on such model protein-surfactant system has recently been reviewed [3].

Proteins perform two important roles in the manufacture and stabilization of emulsions. They produce a lowering of the tension at the newly formed oil-water interface at the instant of emulsification, and they confer long-term stability by forming a thick macromolecular protective layer around the surface of the dispersed droplets. The first role is favoured by protein molecules which are small, hydrophobic and disordered. The proteinaceous emulsifying agent sodium caseinate is predominantly composed of such molecules (mainly α_{s1}-casein and β-casein). Another disordered protein is gelatin, but, compared with caseinate, the molecules are larger, more hydrophilic, and much more polydisperse. Gelatin is less surface-active than caseinate, but once adsorbed it gives thicker, stronger films. It is with these two disordered but

*) Plenary lecture presentation at the 10th Scandinavian Symposium on Surface Chemistry, 14—16 June 1989, Turku, Finland.

contrasting food proteins that we shall be predominantly concerned in this paper. Previous research at Leeds has considered the surface and emulsifying properties of the individual caseins [4—7], the competitive adsorption from mixtures of α_{s1}-casein and β-casein [8], and the competitive and cooperative adsorption from mixtures of caseinate and gelatin [9—12].

The presence of low-molecular-weight surfactants can effect the formation and stabilization of food emulsions and foams in a number of different ways. 1) During homogenization, smaller emulsion droplets are produced due to the more rapid reduction in interfacial tension than with protein alone [13]. 2) Displacement from the interface of protein involved in bridging flocculation results in dispersion of droplets and a lower emulsion viscosity [14]. 3) Interaction with adsorbed protein produces a stronger and thicker adsorbed layer [15, 16]. 4) Additional stabilization is provided by liquid crystals at oil-water or air-water interfaces [17] or in the continuous phase between emulsion droplets [18, 19]. 5) Emulsions are destabilized by enhancement of globule clumping by partial disruption of protein interfacial layers during whipping or air incorporation, e.g. during ice-cream manufacture [20]. Food emulsifiers generally consist of blends of small-molecule surfactants; commercial "monoglycerides", for instance, are mixtures of mono-, di- and triglycerides, accompanied by small amounts of free fatty acids. In studies on model systems, therefore, it is convenient to use single pure (non-food) surfactants in preference to real food emulsifiers.

The problem of competitive adsorption has been considered theoretically by two different statistical approaches [21]: firstly, using continuum models in which adsorbate molecules are represented as spheres of various sizes, and secondly, using lattice-based models in which each polymer segment or small molecule occupies a single lattice site. In an analytical approach of the second type, Cohen Stuart and coworkers have described [22] competitive adsorption at a plane wall from a system of polymer + surfactant + solvent with adsorbate-adsorbent interactions represented by parameters of the Flory-Huggins type. Based on a simple one-layer approximation, an analytic expression was obtained [22] for the so-called critical surfactant concentration at which polymer displacement is complete. However, the complexity of protein structure and interactions is such that analytic theories are unlikely to provide a complete description of protein + surfactant competitive adsorption. One way of bridging the gap between current idealized theories of competitive adsorption and the molecular complexity of real systems is by computer simulation [23, 24].

In what follows, we describe numerical results for a recently developed computer simulation model of competitive adsorption between polymer chains and small displacer molecules on a three-dimensional tetrahedral lattice. Then, experimental results are presented for interfacial behaviour of mixtures of disordered protein + nonionic surfactant and disordered protein + ionic surfactant. It is convenient to discuss the nonionic and ionic surfactants separately since the former (noninteracting surfactants) have little affinity for proteins, whereas the latter (interacting surfactants) bind strongly to proteins [3].

Computer simulation of competitive adsorption

The Monte Carlo simulation model consists of an assembly of linear homopolymers (50 segments each) + small displacer molecules (two segments each) confined to a three-dimensional lattice in a basic simulation cell which is periodic in the x- and y-directions, and bounded by two parallel plane interfaces in the z-direction. Polymer chains are defined as self-avoiding walks on the tetrahedral lattice. Changes in chain conformation are generated as described by Lal and coworkers [25, 26] by a combination of random configurational changes along four-bond sequences of the lattice chain and random chain regeneration at the ends. The total number of lattice sites is set at 108800, of which 6400 (i.e., 6% of the total) are surface lattice sites. Technical details of the Monte Carlo algorithm may be found elsewhere [24].

Over and above the excluded volume effects, interactions between pairs of segments are limited to those between nearest neighbours (segments occupying adjacent lattice sites). The polymer-polymer segment interaction energy is set at $u_{pp} = 0$, corresponding to a solution of polymer in an athermal solvent. The displacer-displacer segment interaction energy is set at $u_{dd} = 2.0\ kT$, where k is Boltzmann's constant and T is temperature. The polymer-displacer energy is varied within the range $-1.0 \leqslant u_{pd}/kT \leqslant 3.0$. Transfer of a polymer segment from bulk to surface is associated with an adsorption energy of $u_p^{ads} = -0.75\ kT$, and for a

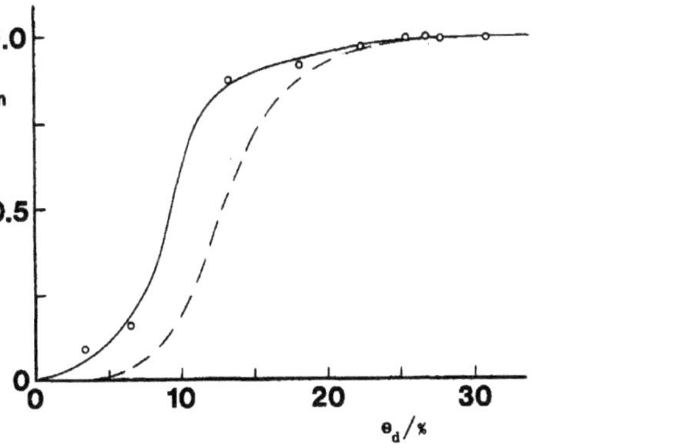

Fig. 1. Displacement of polymer segments from a planar interface by small molecules from a Monte Carlo simulation. The fractional change in the proportion of train segments (f_{train}) at constant u_{pd} = 1.5 kT is plotted as a function of displacer surface coverage θ_d. The limiting value f_{train} = −1.0 corresponds to complete loss of train segments from the interface. The dashed line refers to a single polymer molecule at the interface

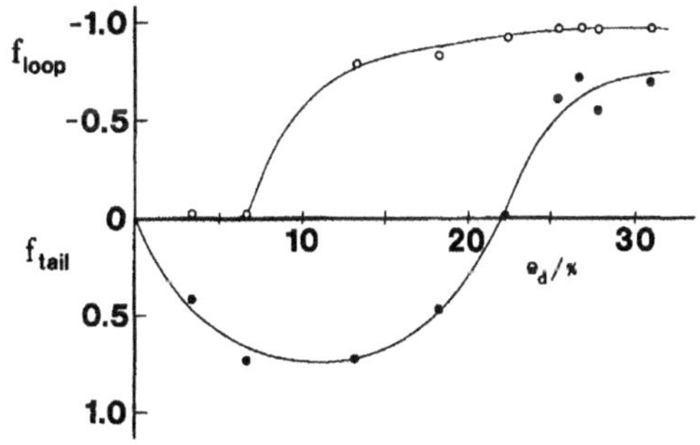

Fig. 2. Fractional changes in the proportions of loop segments (f_{loop}) and tail segments (f_{tail}) of simulated polymer at constant u_{pd} = 1.5 kT as a function of displacer surface coverage θ_d: ○ = loop segments; ● = tail segments. The limiting values f_{loop} = −1.0 and f_{tail} = −1.0 correspond, respectively, to complete loss of loop and tail segements from the interface. A positive value of f_{tail} implies an increased number of tail segments as compared with the situation in the absence of displacer

displacer segment the energy is u_d^{ads} = −3.0 kT. With 30 polymer molecules each of 50 segments, the overall volume fraction (site fraction) of polymer is φ_p = (50 × 30)/108800 = 1.38%; this would correspond to a polymer surface coverage of θ_p = (50 × 30)/6400 = 23.4% (maximum) if every single polymer segment were to occupy a surface lattice site. Displacer volume fraction is varied between φ_d = 0 (pure polymer) and φ_d = 10.6%. A segment on adsorbed polymer molecule is classified according to whether it occupies a surface lattice site (a *train* segment), or whether it forms part of a dangling chain region attached to the surface at both ends (a *loop* segment) or at just one end (a *tail* segment).

In simulations carried out in the absence of displacer, the polymer has on average 43% of its total segments in trains (θ_p = 10.1%), 30% in loops, 15% in tails, and 12% unadsorbed. The effect of adding displacer on the proportions of the three types of polymer segments is indicated in Figs. 1 and 2 for the case of a constant polymer-displacer segment interaction energy u_{pd} = 1.5 kT. Figure 1 shows the fractional change in the proportion of train segments as a function of the displacer surface coverage θ_d. We see that there is little displacement of train segment at low displacer surface coverages ($\theta_d \lesssim$ 5%), but that displacement is more-or-less complete for $\theta_d \gtrsim$ 25%. Also given in Fig. 1 are the corresponding results for an isolated polymer molecule ($\varphi_p \to 0$). We see that the isolated polymer train segments are rather more difficult to displace (i. e., a higher displacer surface coverage is required for the same fractional change in the proportional of train segments).

Figure 2 shows fractional changes in the proportions of loop and tail segments as a function of displacer surface coverage θ_d. At low coverages the tail fraction increases by an amount roughly equal to the loss of train segments (Fig. 1), but the loop fraction remains relatively unchanged. The number of tail segments reaches a maximum at $\theta_d \sim$ 10% and it remains greater than for the polymer alone for all displacer coverages $\theta_d \lesssim$ 22%. We note that the tail fraction never reaches zero in Fig. 2, even at $\theta_d \sim$ 30% when the train fraction is very low.

A pictorial illustration of the effect of displacer concentration on polymer configuration is provided by the two-dimensional projections of individual 50-segment chains in Fig. 3. Picture (a) shows an adsorbed polymer in the absence of displacer. Picture (b) shows an adsorbed polymer in the presence of a low concentration of displacer (φ_d = 0.24%) which is not significantly different from that in the absence of displacer. Picture (c) shows an adsorbed polymer at higher displacer concentration (φ_d = 0.49%) with a long extended tail. Picture (d) shows a polymer molecule at φ_d = 3.9% that is completely displaced from the interface.

Figure 4 shows the fractional change in the proportion of train segments as a function of the polymer-displacer segment interaction energy u_{pd} for a constant displacer concentration φ_d = 2.73%.

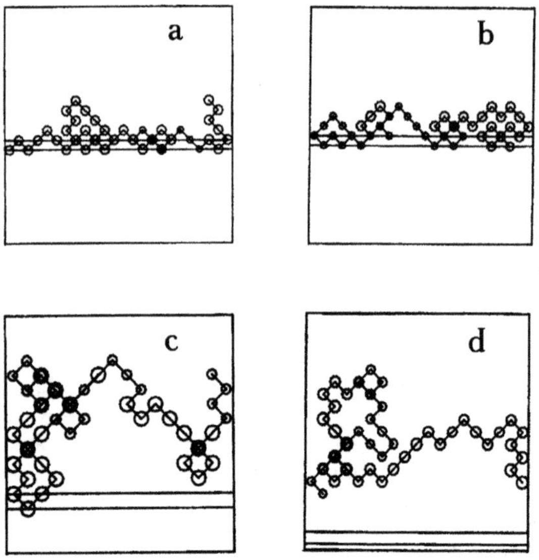

Fig. 3. Individual simulated polymer chains depicted as two-dimensional projections onto the xz-plane. At constant u_{pd} = 1.5 kT, the four pictures refer to different displacer concentrations: (a) $\theta_d = 0$, (b) $\theta_d = 0.24\%$, (c) $\theta_d = 0.49\%$, (d) $\theta_d = 3.9\%$. The interface is represented by a pair of parallel lines pointing in the x-direction. For clarity, no displacer molecules are represented in pictures (b), (c) or (d)

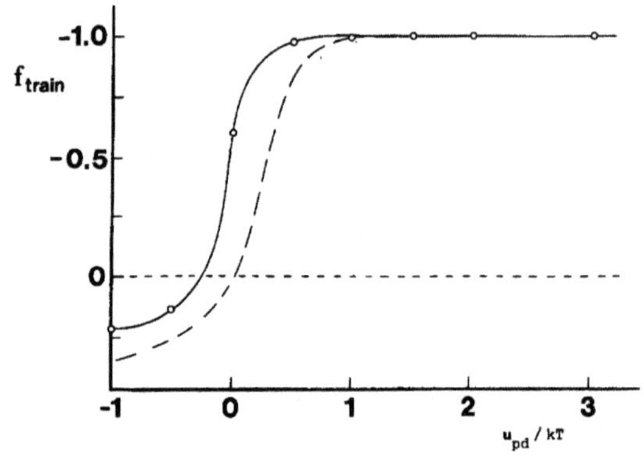

Fig. 4. Effect of polymer-displacer interaction energy u_{pd} on the fractional change in the proportion of train segements (f_{train}) at constant θ_d = 2.73%. The value f_{train} = 0.0 represents the train fraction in the absence of displacer. The dashed line refers to a single polymer molecule at the interface

For polymer-displacer attraction ($u_{pd} < 0$), there is an increased amount of polymer at the interface due to weak complex formation between adsorbed displacer molecules and polymer train segments. At $u_{pd} \sim 0$, the train fraction is rather sensitive to the

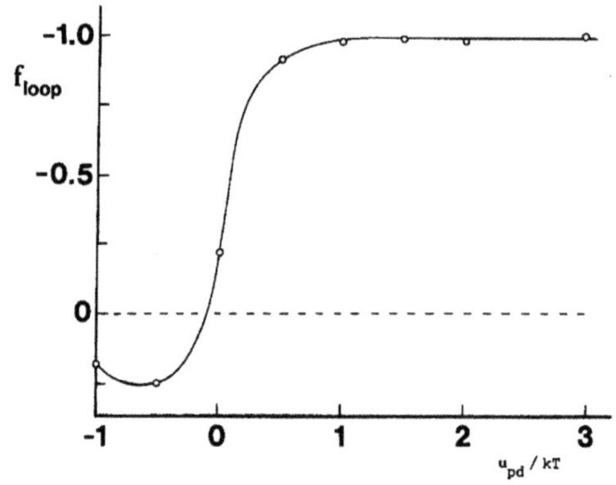

Fig. 5. Effect of polymer-displacer interaction energy u_{pd} on the fractional change in the proportion of loop segments (f_{loop}) at constant φ_d = 2.73%

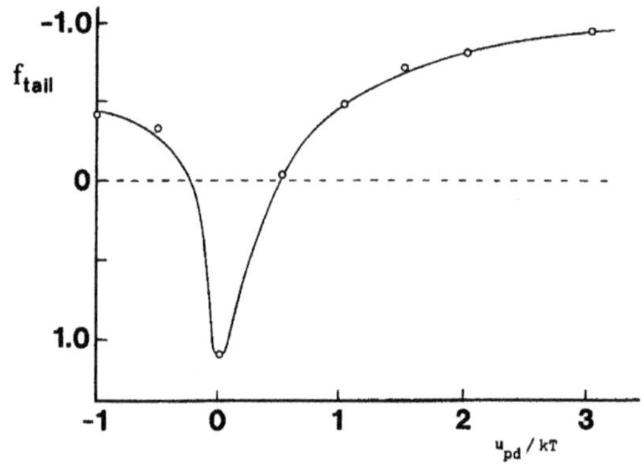

Fig. 6. Effect of polymer-displacer interaction energy u_{pd} on the fractional change in the proportion of tail segments (f_{tail}) at constant θ_d = 2.73%

strength of the polymer-displacer interaction. The effect of u_{pd} on the loop fraction follows a similar trend as illustrated in Fig. 5. With attractive interactions ($u_{pd} < 0$) there is a slight increase in loop formation; with no interaction ($u_{pd} = 0$) there is a slight decrease; and with repulsive interactions ($u_{pd} > 0$) there is a rapid loss of loop segments. At u_{pd} = —1 kT, interfacial polymer-displacer complex formation is sufficiently strong to draw more segments into trains, thereby reducing the loop fraction. Figure 6 shows the effect of u_{pd} on the tail fraction. We see that the presence of the displacer (φ_d = 2.73%) is to reduce the proportion of tail segments, except at $u_{pd} = 0$, where the proportion of tail segments is twice that in the absence of displacer. The reason for this is that the condition $u_{pd} = 0$ corresponds to a situation of partial displacement with long, dangling polymer tails. Moving to an attractive interaction (u_{pd} = —0.5 kT) causes tails to be converted to trains under the influence of interfacial complex formation. Conversely, the introduction of a repulsive interaction (u_{pd} = 0.5 kT) further increases displacement with the associated loss of tail segments as whole polymer molecules are removed from the interface.

To summarize the results from the Monte Carlo simulations, we can make two general statements. Firstly, it is clear that displacer surface coverage is a parameter of crucial importance in determining the extent of polymer displacement. Displacer surface coverage may be changed by adjusting the total displacer concentration as described above, or by adjusting the displacer adsorption energy u_d^{ads} relative to the polymer adsorption energy u_p^{ads} as described elsewhere [24]. Secondly, it is clear that the adsorbed polymer configuration (as expressed by the proportions of trains, loops and tails) is very sensitive to small changes in polymer-displacer interactions, either attractive or repulsive.

To make this Monte Carlo model a more realistic representation of protein + surfactant systems, we ought to ascribe different adsorption and interaction energies to the two ends of the displacer molecule. An alternative Monte Carlo model incorporating orientationally-dependent surfactant-interface and surfactant-macromolecule interactions has recently been described [23]. Figure 7 shows a "snapshot" of the instantaneous positions of the particles in the basic simulation cell of this two-dimensional system of macromolecules (large circles) + surfactant molecules (little circles). Strong surfactant-macromolecule binding leads to a tendency for individual surfactant-macromolecule complexes to be repelled from one another and also from the interface once it is covered with a surfactant monolayer. The bound surfactant molecules lead to an entropic barrier preventing individual macromolecules from getting close enough to the interface to adsorb.

Experiments on protein + nonionic surfactant systems

As an illustration of the competitive adsorption between disordered protein and noninteracting surfactant at the oil-water interface, we report interfacial measurements for caseinate + dioxyethylene dodecyl ether ($C_{12}E_2$) and caseinate + octaoxyethylene dodecyl ether ($C_{12}E_8$). The two oxyethylene surfactants form a contrasting pair insofar as $C_{12}E_2$ is predominantly oil-soluble, whereas $C_{12}E_8$ is water-soluble. High purity surfactant

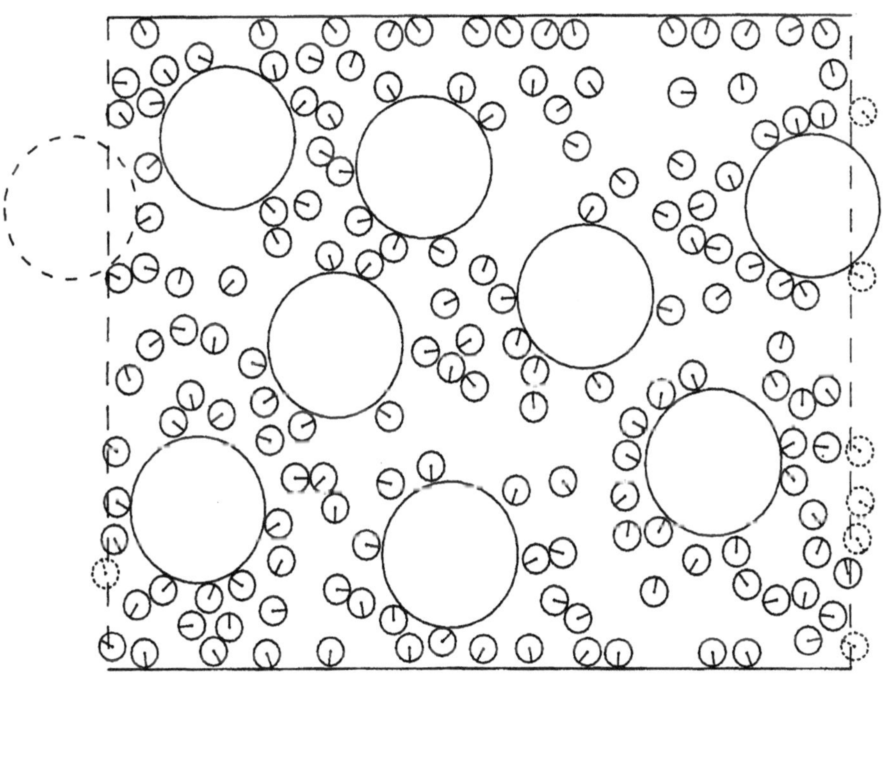

Fig. 7. Representation of instantaneous particle positions in a two-dimensional Monte Carlo simulation of competitive adsorption of macromolecules and surfactants. The basic rectangular simulation cell has hard, impermeable, adsorbing walls at the top and bottom, and permeable walls with periodic boundary conditions to the left and right. The lines on the small particles (surfactants) indicate the directions of orientationally-dependent attractive interactions with the adsorbing walls or the large particles (macromolecules). For the case shown, there are approximately seven surfactants bound to each macromolecule (on average). (See [23] for further details.)

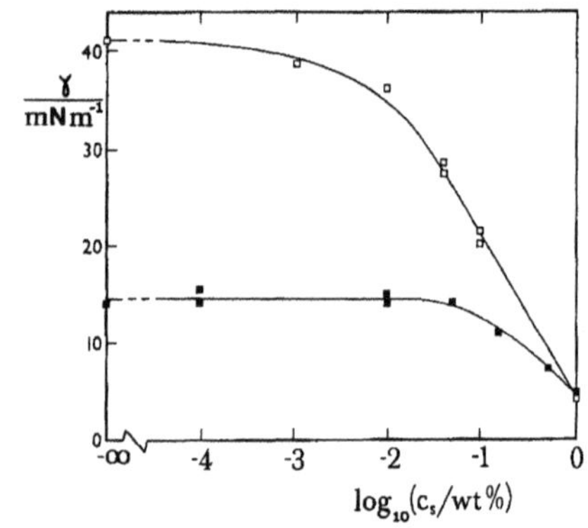

Fig. 8. Interfacial tension γ of caseinate + $C_{12}E_2$ at the *n*-tetradecane-water interface (20 mM imidazole buffer, pH 7, 25 °C) as a function of the logarithm of the surfactant concentration c_s: □ = no protein; ■ = 0.1 wt% protein

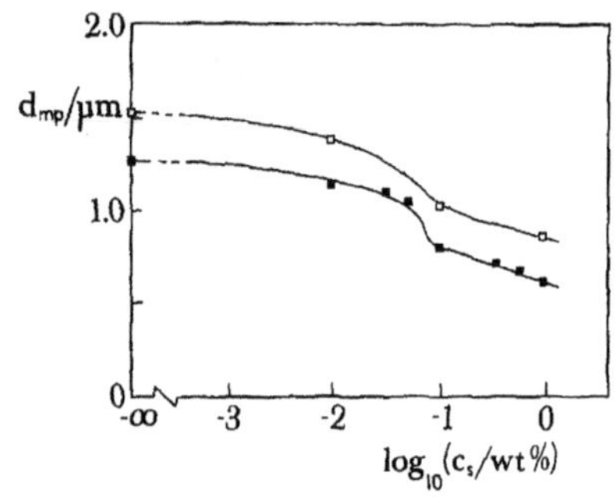

Fig. 9. Effect of $C_{12}E_2$ on droplet size in caseinate-stabilized n-tetradecane-in-water emulsions (20 wt%, pH 7) prepared by single-stage, high-pressure homogenization (300 bar). The most-probable droplet diameter d_{mp} from the Coulter counter particle-size distribution is plotted against the logarithm of the surfactant concentration c_s: □ = 0.3 wt% protein; ■ = 0.5 wt% protein

samples were obtained from the Nikko Company, Japan. Commercial sodium caseinate was obtained from the Scottish Milk Marketing Board.

Figure 8 shows steady-state interfacial data for caseinate + $C_{12}E_2$ at the n-tetradecane-water interface (pH 7, 25 °C) measured by the pendant drop method. The tension is always lower in the presence of 0.1 wt% caseinate than in the absence of protein, but the difference becomes less as the $C_{12}E_2$ concentration increases. At c_s = 1 wt%, the steady-state tension is the same within experimental error (±0.5 mN m^{-1}) as that measured in the absence of protein. Hence, we may infer that, at c_s ~ 1 wt%, the primary oil-water interface is composed of a surfactant monolayer, with the caseinate completely displaced from the interface, or possibly residing in a secondary adsorbed layer. In relation to emulsion formation, the lower tension at the oil-water interface in the presence of added surfactant leads to smaller average droplet sizes in protein-stabilized systems. For n-tetradecane-in-water emulsions (20 wt% oil, pH 7) prepared by high-pressure homogenization (20 °C, 300 bar), Fig. 9 shows a plot of the most probable (median) droplet diameter as a function of the $C_{12}E_2$ concentration c_s in the oil phase. The effect of added surfactant is to shift the droplet-size distribution measured by the Coulter counter to smaller diameters. This effect is most marked for $c_s \gtrsim 0.05$ wt%, corresponding to the $C_{12}E_2$ concentration where the equilibrium tension in the mixed system is suggestive of the onset of protein displacement from the oil-water interface (see Fig. 8). Addition of 1 wt% $C_{12}E_2$ leads to a reduction in the average emulsion droplet size by about a factor of 2. As expected, consistently smaller droplets are obtained with 0.5 wt% caseinate than with 0.3 wt% caseinate over the whole range of surfactant concentration. At constant protein content, the emulsions containing most added surfactant are the most stable to creaming on storage at 25 °C [13].

Figure 10 shows steady-state interfacial data for caseinate + $C_{12}E_8$ at the n-tetradecane-water interface (pH 7, 25 °C) measured by the static Wilhelmy plate method. As with $C_{12}E_2$ (see Fig. 8), the tension is always less (within experimental error) in the presence of protein, but the difference becomes reduced as c_s increases. The two curves coincide at a molar $C_{12}E_8$ concentration of $c_s \approx 1.0 \times 10^{-4}$ M, and beyond this point the protein must presumably be completely displaced from the primary interfacial layer because the surface excess concentration of surfactant (1.27 mg m^{-2}) and the corresponding limiting area per molecule (0.70 nm^2) are the same in each case. Similar behaviour to that shown in Fig. 10 has also been found [27] with the disordered protein gelatin and the globular protein lysozyme. With gelatin + $C_{12}E_8$, the tension curves in the absence and presence of gelation (0.4 wt%) become coincident at a molar surfactant concentration of $c_s \approx 1.0 \times 10^{-5}$ M. This is consistent with the predictions of simple analytic theory [28] and computer simulation [23, 24] that a greater bulk concentration of surfactant is required to achieve the same

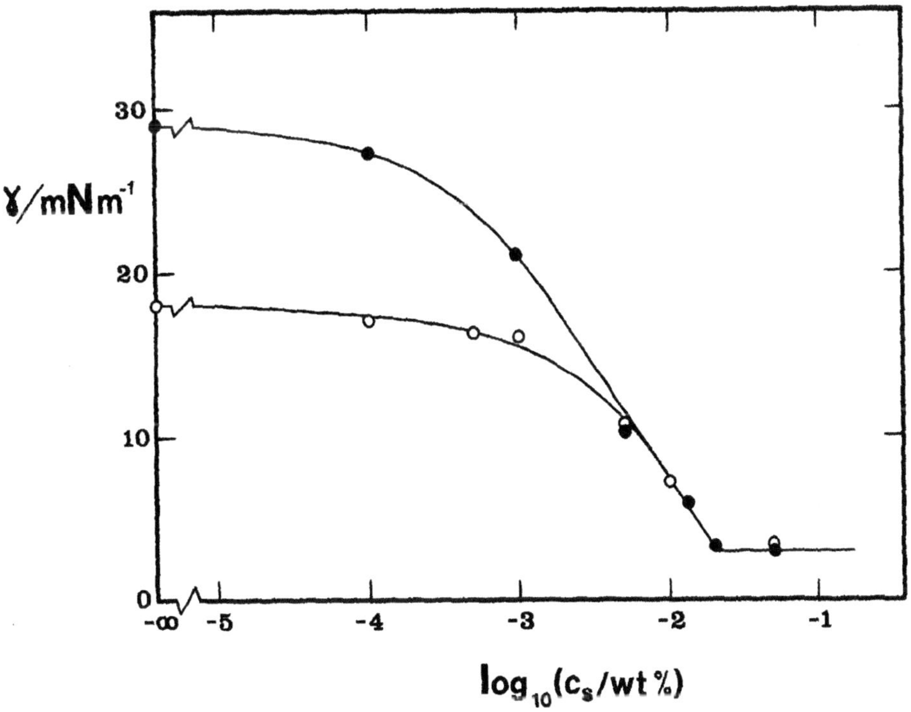

Fig. 10. Interfacial tension γ of caseinate + $C_{12}E_8$ at the n-tetradecane-water interface (20 mM imidazole buffer, pH 7, 25°C) as a function of the logarithm of the surfactant concentration c_s: ● = no protein; ○ = 0.1 wt% protein

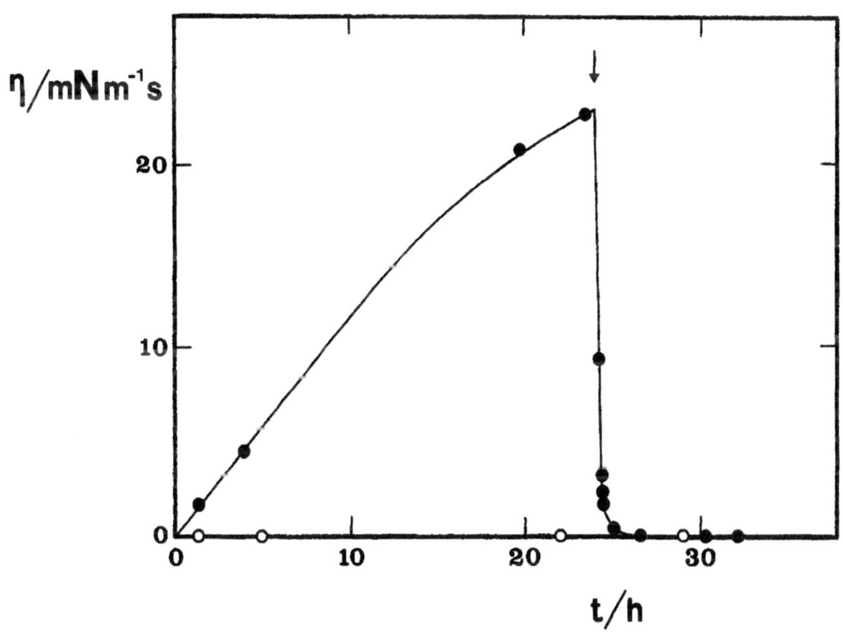

Fig. 11. Surface shear viscosity η of caseinate + $C_{12}E_8$ at the n-tetradecane-water interface (20 mM imidazole buffer, pH 7, 25°C) as a function of the adsorption time t: ○ = 0.1 wt% caseinate + 10^{-4} wt% $C_{12}E_8$ together at t = 0; ● = 0.1 wt% caseinate present at t = 0, 10^{-4} wt% $C_{12}E_8$ added (↓) at t = 24 h

equivalent surface coverage for an increase in protein surface activity (caseinate instead of gelatin).

Additional information on the nature of the interfacial adsorbed film in systems containing protein + surfactant may be obtained from surface rheology measurements. Figure 11 shows surface shear viscosity measurements at the n-tetradecane-water interface (pH 7, 25 °C) determined using a Couette-type torsion-wire surface viscometer [10]. In the absence of surfactant, the film adsorbed from the caseinate solution (0.1 wt%) develops a surface viscosity of ca. 20 mN m^{-1} s after 24 h. The presence of $C_{12}E_8$ in the aqueous phase at a concentration of $c_s = 10^{-4}$ wt% is sufficient, however, to prevent the formation of an adsorbed film with any measurable surface viscosity (<0.2 mN m^{-1} s). Alternatively, addition of the same concentration of surfactant below a caseinate film aged for 24 h leads to a rapid fall in interfacial viscosity down to values which are too low to be detected. Since the tension data in Fig. 10 would seem to indicate that caseinate is predominant at the interface at $c_s = 10^{-4}$ wt%, we must conclude from the surface viscosity data that the surfactant disrupts interactions between adsorbed protein molecules without significantly displacing protein segements from the interface. The situation may be similar to that in the simulation at low displacer surface coverage (see Fig. 1), where the proportion of polymer train segments is relatively unaffected by displacer adsorption. Assuming that train segments contribute mostly to the surface free energy, this would imply little effect on the interfacial tension at these low surfactant surface coverages, but a substantial effect on the intermolecular hydrogen-bonding interactions between adsorbed protein molecules which may be mainly responsible for the resistance to shear deformation in the adsorbed film.

The results reported above are generally consistent with the view of earlier workers that nonionic surfactants have only a weak affinity for water-soluble proteins. Weak complexes, if they do form, will tend to be rather surface-*in*active, since hydrophobic areas of protein and nonionic surfactant come together in the binding [3]. Nonionic food surfactants may give more complex behaviour than oxyethylene surfactants, although general trends with monoglycerides seem similar to those reported here. With ovalbumin + 1-monocaproin, for instance, it has been reported [29] that there is no discernible protein-surfactant interaction based on an experimental surface tension isotherm very similar in form to that shown in Fig. 10. And, in another study with casein + glycerol monostearate, it was found [30] that the surface pressure isotherm in the high-pressure region is essentially the same as for the monoglyceride alone, suggesting protein displacement from the interface. There appears to be little previous work on the surface rheology of systems of protein + nonionic surfactants. The closest work known to the authors is that of MacRitchie [31] for spread films of polyalanine + lecithin, where it was found that, as the mixed film is compressed, the surface viscosity first increases normally, but then goes through a maximum and decreases with further increase in surface pressure (cf. Fig. 11).

Experiments on protein + ionic surfactant systems

The strong binding of many anionic and cationic surfactants to protein is well known. Two types of protein-surfactant interaction are commonly distinguished [3]: high-affinity specific binding at low surfactant concentrations, and cooperative nonspecific binding at high concentrations. The high-affinity binding is due to specific electrostatic and hydrophobic interactions between surfactant molecules and sites on the protein. In the cooperative binding region, the driving force for complex formation is similar to that for the cooperative association of surfactant molecules into micelles in the absence of protein. The main effect of protein-surfactant binding on the competitive adsorption is to reduce the amount of surfactant available to saturate the interface. This means that it takes more ionic surfactant to displace the same protein from the interface.

To illustrate the competitive adsorption between disordered protein and interacting surfactant at the oil-water interface, we report interfacial measurements for gelatin + sodium dodecyl sulphate (SDS) at the n-tetradecane-water interface (pH 7, 25 °C). Figure 12 shows tension data as a function of SDS concentration c_s for three gelatin concentrations (0.01 wt%, 0.1 wt% and 0.4 wt%), as well as for the pure surfactant. At the highest SDS concentration (well above the critical micelle concentration (CMC)), the tension for gelatin + SDS is essentially the same as that for SDS alone, suggesting that the protein is completely displaced from the interface. At all lower surfactant concentrations, however, the situation is complex, with the form of the curves

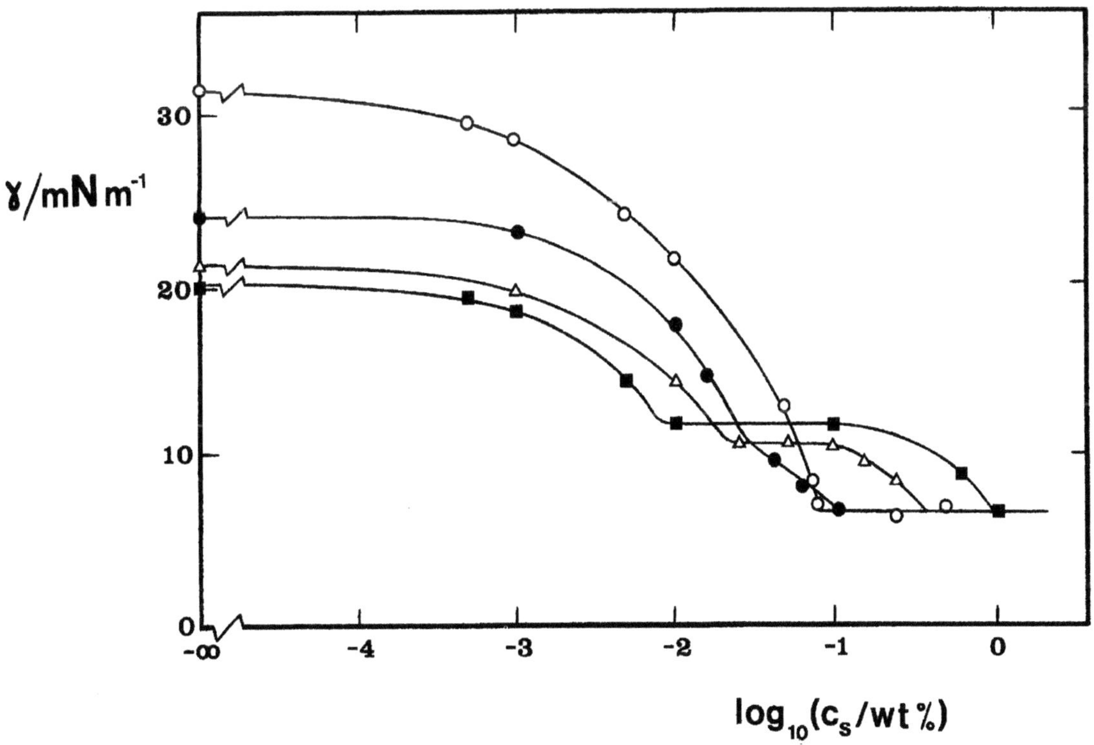

Fig. 12. Interfacial tension γ of gelatin + SDS at the n-tetradecane-water interface (5 mM phosphate buffer, pH 7, 25 °C) as a function of the logarithm of the surfactant concentration c_s: ○ = no protein; ● = 0.01 wt% protein; △ = 0.1 wt% protein; ■ = 0.4 wt% protein

dependent on the bulk gelatin concentration. At gelatin concentrations of 0.1 wt% and 0.4 wt%, there is a distinct plateau region in the plot of the tension against logarithm of surfactant concentration. This plateau region extends over a considerable range of SDS concentration in the vicinity of the pure surfactant CMC. Presumably, in this plateau region, it is more favourable in free energy terms for surfactant molecules to bind cooperatively onto the protein than it is for them to displace hydrophobic protein train segments from the interface. The end of the plateau region corresponds roughly to the surfactant concentration at which free SDS molecules begin to displace protein-surfactant complexes from the interface. At the lowest gelatin concentration (0.01 wt%), there is no discernible plateau region, but there is still a cross-over to tensions slightly greater than those for pure surfactant near the CMC.

We can use the difference in CMC for pure SDS and gelatin + SDS to estimate the extent of protein-surfactant binding at high surfactant concentrations. The data at the three gelatin concentrations in Fig. 12 are consistent with a value of $(6 \pm 1) \times 10^3$ molecules SDS per molecule of gelatin (based on a gelatin molecular weight of 7.7×10^5 daltons determined by light scattering). Within the experimental uncertainty, this value is consistent with the binding number obtained by equilibrium dialysis [27].

The general form of the behaviour of the interfacial tensions in Fig. 12 is in agreement with the recent results of Wüstneck and Wärnheim [32] for gelatin + SDS at the toluene-water interface. The results are interpreted in terms of a protein-surfactant complex that is more surface-active than either of the individual components by themselves. Maximum values of the surface rheological parameters in gelatin + SDS at the air-water interface are found [15] at SDS concentrations correspond to the beginning of the plateau region in the plot of tension against logarithm of surfactant concentration. Above the CMC, as surfactant displaces gelatin from the interface, it is found [15] that the surface viscosity drops by several orders of magnitude (cf. Fig. 11), eventually reaching very low steady values at the highest surfactant concentrations. Foam films stabilized by gelatin-SDS complexes have thicknesses intermediate between those for gelatin or SDS alone [16].

Qualitatively similar behaviour to that outlined above for gelatin + SDS has also been observed with other ionic surfactants, cationic as well as anionic, though the detailed shape of the plot of tension against logarithm of surfactant concentration appears to be dependent on the nature of both the surfactant and the protein [3]. In summary, we note two main differences between the ionic (interacting) and the nonionic (noninteracting) surfactants. Firstly, more of the ionic surfactant is required to displace protein from the interface, since some of it is used to bind to the protein and is therefore not available to act as a displacer. Secondly, at concentrations below that required for protein displacement, the formation of protein-surfactant complexes at the interface leads to a substantial increase in surface rheological properties of the adsorbed film as compared with those of the pure protein film.

Concluding remarks

We have seen that many of the features of competitive adsorption of food macromolecules and small-molecules surfactants can be described in terms of a computer simulation of systems of linear homopolymers + small displacer molecules. Though the displacer molecules simulated so far are chemically rather simplistic in comparison to real surfactant molecules, the general effects of polymer-displacer interactions on the competitive adsorption appear to follow those found experimentally for disordered protein + surfactant.

It remains to be seen whether steady-state measurements made at the planar oil-water interface are of direct applicability in predicting competitive adsorption behaviour at the droplet surface in food oil-in-water emulsions. In a recent study of the effect of low-molecular-weight surfactants on the properties of cream liqueurs (dairy based emulsions containing alcohol), it was found [33] that the extent of displacement of caseinate from the oil droplet surface was similar for the two types of surfactant: commercial glycerol monostearate (nonionic, oil-soluble) and sodium stearoyl lactylate (anionic, water-soluble). Even at the highest emulsifier concentration studied (2 wt%), the displacement was not complete; about one-third of the total amount of protein was retained in the interfacial layer around the droplets. Other studies have shown that mixed mono- and diglyerides, as occur in the commercial glycerol monostearate, exhibit interfacial complex formation with caseinate at the oil-water interface [34]; and so it appears that this (mixed) emulsifier is not behaving simply as a nonionic, noninteracting surfactant (e.g., caseinate + $C_{12}E_2$). Possibly, the primary interfacial layer contains only low-molecular-weight surfactant (and ethanol), with the interacting protein adsorbed in adjacent secondary layers and/or liquid crystalline structures. In this way, the emulsion droplets would still be stabilized partially by a protective layer of sterically stabilizing macromolecules even in the presence of added emulsifier.

Finally, we note that in most food emulsions competitive adsorption occurs between different macromolecules, as well as between macromolecules and small molecules. One significant difference between these two types of competition is the timescale: protein displacement by surfactant occurs much more rapidly than the displacement of one protein by another. This is mainly because of the larger surfactant diffusion coefficient, both in bulk solution and at the fluid interface.

Acknowledgements

S. R. E. acknowledges receipt of a Studentship from the Ministry of Agriculture, Fisheries and Food. C. M. W. acknowledges receipt of an AFRC Studentship in conjunction with Unilever Research (Colworth House).

References

1. Dickinson E, Stainsby G (1982) Colloids in Food. Applied Science, London
2. Dickinson E (1986) Food Hydrocolloids 1:3–23
3. Dickinson E, Woskett CM (1989) In: Bee RD, Richmond P, Mingins J (ed) Food Colloids. Royal Society of Chemistry, London, pp 74–96
4. Dickinson E, Robson EW, Stainsby G (1983) J Chem Soc Faraday Trans I 79:2937–2952
5. Dalgleish DG, Dickinson E, Whyman RH (1985) J Colloid Interface Sci 108:174–179
6. Dickinson E, Whyman RH, Dalgleish DG (1987) In: Dickinson E (ed) Food Emulsions and Foams. Royal Society of Chemistry, London, pp 40–51
7. Dickinson E (1989) J Dairy Res 56:471–477
8. Dickinson E, Rolfe SE, Dalgleish DG (1988) Food Hydrocolloids 2:397–405

9. Chesworth SM, Dickinson E, Searle A, Stainsby G (1985) Lebensm Wiss Technol 18:230—232
10. Dickinson E, Murray BS, Stainsby G (1985) J Colloid Interface Sci 106:259—262
11. Castle J, Dickinson E, Murray BS, Stainsby G (1987) ACS Symp Ser 343:118—134
12. Dickinson E, Murray A, Murray BS, Stainsby G (1987) In: Dickinson E (ed) Food Emulsions and Foams. Royal Society of Chemistry, London, pp 86—99
13. Dickinson E, Mauffret A, Rolfe SE, Woskett CM (1989) J Soc Dairy Technol 42:18—22
14. Darling DF, Birkett RJ (1987) In: Dickinson E (ed) Food Emulsions and Foams. Royal Society of Chemistry, London, pp 1—29
15. Wüstneck R, Hermel H, Kretzschmar G (1984) Colloid Polym Sci 262:827—832
16. Wüstneck R, Müller H-J (1986) Colloid Polym Sci 264:97-102
17. Friberg SE, Solans C (1986) Langmuir 2:121—126
18. Barry BW (1971) Rheol Acta 10:96—105
19. Barry BW (1975) Adv Colloid Interface Sci 5:37—75
20. Berger KG (1976) Dairy Industries International 41:236—243
21. Dickinson E, Lal M (1980) Adv Molec Rel Int Proc 17:1—87
22. Cohen Stuart MA, Fleer GJ, Scheutjens JMHM (1984) J Colloid Interface Sci 97:515—525
23. Dickinson E (1988) Molec Phys 65:895—908
24. Dickinson E, Euston SR (1989) Molec Phys 68: 407—421
25. Clark AT, Lal M, Turpin MA, Richardson KA (1975) Faraday Discuss Chem Soc 59:189—195
26. Clark AT, Lal M (1978) J Chem Soc Faraday Trans II 74:1857—1869
27. Woskett CM (1989) Ph. D. thesis, University of Leeds
28. de Feijter JA, Benjamins J, Tamboer M (1987) Colloids Surf 27:243—266
29. Ericsson B, Hegg PO (1985) Prog Colloid Polym Sci 70:92—95
30. Paquin P, Britten M, Laliberté M-F, Boulet M (1987) ACS Symp Ser 343:677—686
31. MacRitchie F (1970) J Macromol Sci Chem A4:1169—1176
32. Wüstneck R, Wärnheim T (1988) Colloid Polym Sci 266:926—929
33. Dickinson E, Narhan SK, Stainsby G (1989) J Food Sci 54:77—81
34. Doxastakis G, Sherman P (1986) Colloid Polym Sci 264:254—259

Authors' address:

E. Dickinson
Procter Department of Food Science
University of Leeds
Leeds LS2 9JT, England

Progress in Colloid & Polymer Science Progr Colloid Polym Sci 82:76—81 (1990)

Spreading of oil drops under a solvent vapor: influence of Marangoni effect

P. Carles and A. M. Cazabat

Physique de la Matière Condensée, Collège de France, Paris, France

Abstract: The spreading dynamics of a silicone oil (PDMS) drop in contact with the vapor of a solvent (hexane, cyclohexane, decaline) were studied. The observed behaviors are completely different from the ones that are expected and indeed observed in ideal systems. First, the contamination induces a surface tension gradient which may counteract or accelerate the spreading, depending on its direction. In a second stage, the surface of the drop becomes more homogeneous and the drop may either resume spreading or retract, again depending on the surface tension of the solvent.

Key words: Spreading; wetting; Marangoni effect; contamination; surface tension

Introduction

Wetting phenomena belong to everyday life and have been known for a long time. Recently, much effort has been developed to understanding them, which is now sufficiently achieved in ideal situations (pure, non-volatile, homogeneous liquids spreading on smooth, homogeneous surfaces) [1—3]. Yet, in practical situations, systems are non-ideal [4], and their behavior may differ notably from the "normal" ones.

Of particular interest is the influence of surface tension gradients on the dynamics of spreading [4—8]. A theoretical analysis [5] has predicted that they could not overcome adverse wettability, whereas a study of spreading of mixtures [6] has shown they actually could. Such gradients are often encountered, e.g., during the spreading of volatile mixtures [6] or with some kinds of contamination [4, 9]. We have undertaken to study the spreading of oil droplets under an atmosphere that contains a volatile solvent of the oil, and we observe phenomena that can be explained in terms of surface tension gradients (Marangoni effect [7]).

Experimental

The spreading experiments are performed on microscope glass slides. The slides, used once only, are cleaned in the following way: they are left for 1 h in hot sulfochromic acid, and then rinsed, first with tap water, then three times with tridistilled water. They remain in tridistilled water until just before use, when they are soaked for 1 h in 0.1 N sodium hydroxyde solution, and then for another 10 min in tridistilled water, before being rinsed with methanol and allowed to dry. This cleaning procedure gives homogeneous hydrophilic surfaces.

The spreading fluid is a silicone oil, polydimethylsiloxane (PDMS), trimethylsiloxy terminated. It is assumed that this oil does not react with the glass surface. It is available with various viscosities, typically between 10^{-3} and a few thousand Pa.s, corresponding to different molecular weights. We use rather light, but not volatile species; viscosity is between 0.02 and 10 Pa.s, and surface tension is between 20.6 and 21.5 mN · m^{-1}, depending on the molecular weight. Because of its low surface tension, PDMS is usually rather insensitive to contamination, except in some particular cases, like the one we have investigated. The spreading dynamics of pure PDMS on glass were the subject of a previous study [3].

We use different volatile solvents of the PDMS, viz. hexane, cyclohexane, and decaline. Results with decaline were reported in a previous paper [9], and we will deal here mainly with hexane and cyclohexane. These liquids are good solvents of the PDMS, well above the theta point at room temperature, and they mainly differ in their surface tension: hexane's is 18.4 mN · m^{-1}, cyclohexane's is 25.5 mN · m^{-1}, and decaline's is 26.3 mN · m^{-1}. Cyclohexane and decaline do not wet glass, whereas hexane does.

Experimental procedure

On a cleaned glass slide, we place a drop of PDMS of about one microliter, the spreading dynamics of which we follow for a few minutes: this allows the calculation of the drop volume, from the known laws of spreading of PDMS on glass (see next section). Then we bring a large reservoir of solvent, enclosed in a box, so that from that moment on the experiment proceeds in a closed cell, isolated from the outer atmosphere (Fig. 1). The reservoir has a circular shape and is laid around the drop so that the revolution symmetry is not broken.

The experiments are performed under a simple video set-up: the drop is lit with a parallel beam. Using strioscopy, the contrast is enhanced so as to make the contact line clearly visible. Its evolution is recorded on a video tape in real time, and later the radius of the wetted spot is measured and plotted against time.

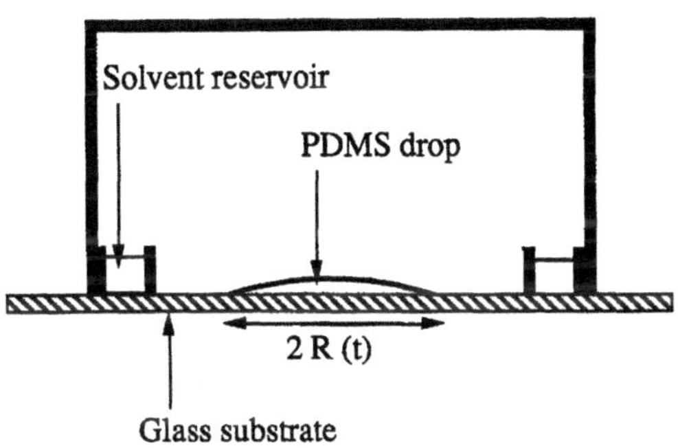

Fig. 1. Schematic representation of the experimental system

"Normal" laws

Let us first recall briefly the spreading behaviors that are expected for a pure, non-volatile liquid spreading on a smooth homogeneous surface. This "ideal" case has been extensively studied and discussed [1–3], both from theoretical and experimental points of view, and we shall only point out a few results.

Both gravity and capillary forces drive the spreading and, roughly speaking, their importance depends on the current radius of the wetted spot R, compared with the capillary length $\kappa^{-1} = \sqrt{\frac{\gamma}{\rho g}}$, where γ is the surface tension of the liquid, ρ is its specific gravity, and g is gravity's acceleration.

If $R > \kappa^{-1}$, the main driving force for the spreading is gravity and the radius $R(t)$ is expected to scale as

$$R(t) \propto \Omega^{3/8}\left(\frac{\rho g t}{\eta}\right)^{1/8}. \quad (1)$$

If $R < \kappa^{-1}$, the main driving force is capillarity and one expects

$$R(t) \propto \Omega^{3/10}\left(\frac{\gamma t}{\eta}\right)^{1/10}. \quad (2)$$

In these formulae, Ω is the drop volume, η the viscosity of the oil, and $\propto$ means "proportional to".

It has been shown [3] that the spreading of PDMS on glass follows these laws and that the proportionality coefficient is close to 1, but the transition between capillarity- and gravity-driven spreading does not simply take place near $R = \kappa^{-1}$. However, as mentioned before, the curve $R(t)$ provides a way to measure the drop volume.

One may wonder why the spreading parameter S does not appear in these dynamics. The macroscopic drop is actually preceded by a precursor film [1, 2, 10] and, thus, the liquid spreads on a film of itself. However, in the case we study, the precursor is little developed and we expect it to play only a minor role (see the following).

Spreading under solvent vapor

The "normal" laws we have just discussed are not valid when we deal with spreading in non-ideal conditions. The spreading under a solvent vapor is a good example of the effect of contamination. The observed phenomena depend on the solvent.

With cyclohexane (Fig. 2) the drop first spreads very quickly, under laws that have nothing to do with Eqs. (1) or (2), which correspond to slow spreading. Then, after say 2 min, the spreading slows down, the contact line becomes irregular and the drop retracts. During this retraction phase the drop has an irregular, not circular shape, and the

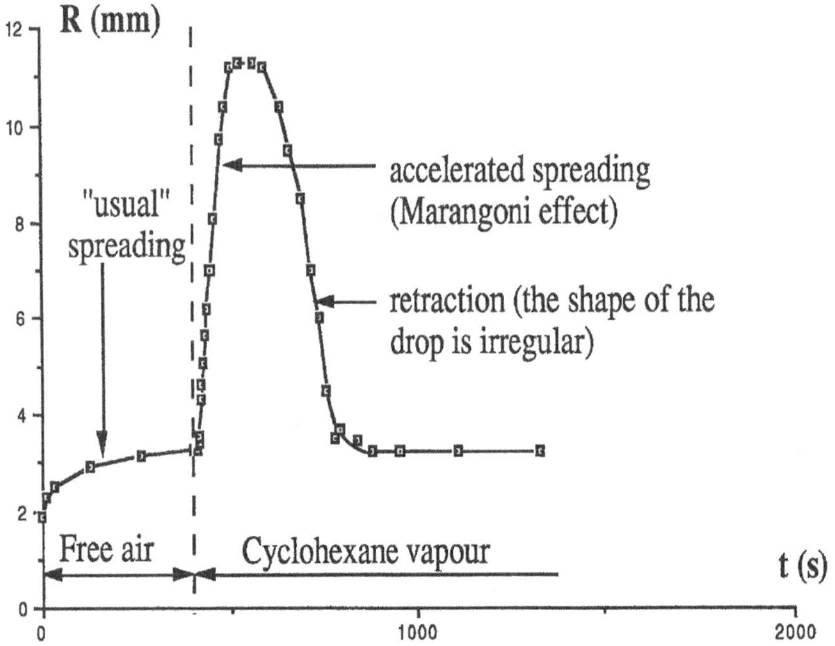

Fig. 2. Spreading of a PDMS drop (η = 0.02 Pa · s, Ω = 0.7 10^{-9} m^3) under cyclohexane vapor

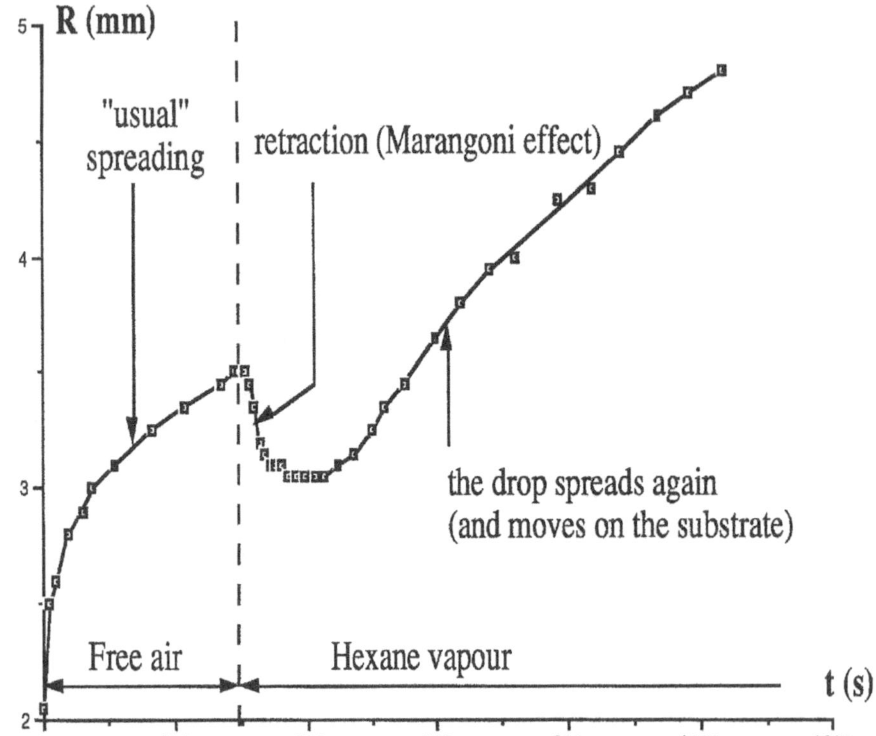

Fig. 3. Spreading of a PDMS drop (η = 0.02 Pa · s, Ω = 0.9 10^{-9} m^3) under hexane vapor

radius $R(t)$ we plot is only of a typical size and not very precise. Eventually, the drop reaches an equilibrium state, again with a circular shape.

With hexane (Fig. 3) the phenomena are somewhat reversed: first the drop retracts, then it resumes spreading, but spreads faster than in "normal" spreading, and its center of mass drifts on the glass plate. During all these stages, the drop remains circular.

With decaline [9] the phenomena resemble those with cyclohexane. We do observe accelerated spreading in the beginning. With small drops, we then observe the same behavior as that reported with cyclohexane, i.e., retraction of the drop — but with larger drops, we have only evidenced, instead of the retraction, a further spreading with instabilities of the contact line.

Discussion

The liquid is, of course, a pure liquid only in the beginning — as soon as it comes in contact with the solvent vapor it becomes a mixture. As a matter of fact, many parameters vary during our experiments: the temperature, viscosity, specific gravity, and surface tension become functions of the space-time coordinates. Moreover, the volume of the drop increases and there is, possibly, adsorption of solvent on the glass substrate. What allows the analysis is the fact that those changes are qualitatively the same with both hexane and cyclohexane, except for the change in surface tension: in that case the sign of the effect is reversed. That is why our observations give evidence of the presence of surface tension effects.

Because of the inhomogeneous composition of the liquid the surface tension of the drop is not homogeneous, which induces a shear stress on the surface, and, consequently, a motion of the liquid towards high surface tension regions (Marangoni effect [7]). The surface tension might also vary because of temperature inhomogeneities, due to heats of condensation, but such an effect cannot explain the difference between hexane and cyclohexane. Besides, the diffusion coefficients for temperature are much higher than those for concentration, so that the temperature is expected to be, as a first order approximation, constant in the drop. So we shall discuss the results in terms of composition inhomogeneities.

We shall not discuss our observations in terms of precursor film, because at short times, when the drop comes in contact with the solvent vapor, this film is barely developed (it is less than 1 millimeter long). So, at least in the case of accelerated spreading, the drop catches up to its precursor very quickly. Generally speaking, we cannot exclude a dramatic influence of the contamination on the precursor film [4]. But, as we have not yet looked at our drops on a microscopic scale, it seems unnecessary, for the time being, to discuss our results with microscopic arguments.

First stage

Let us deal first with what we shall call the first stage, namely the accelerated spreading under cyclohexane or decaline and the retraction under hexane. Both of these effects can be explained in terms of Marangoni effect if we assume the concentration of solvent to be higher near the edge of the drop. Indeed, in such an hypothesis, the surface tension under cyclohexane or decaline vapor would be higher near the edge, thus inducing a movement of liquid in the direction of spreading, whereas, on the other hand, under hexane vapor it would be higher near the center of the drop, thereby creating a movement of liquid opposing the spreading and leading to the retraction. So we must explain why the concentration of solvent should be higher near the edge of the drop.

But first a point deserves a particular discussion: what is the thickness of the layer, the composition of which determines the surface tension? In the case of Van der Waals interactions, it is clear that the surface tension depends on an average of concentrations, taken over, say, 1000 Å [9]. With this lengthscale, it is necessary to take into account the whole concentration profile in the drop to know the local surface tension. But, in fact, on the average the predominant terms come from the first layers, say the first 20 Å. Besides, if there are shorter range forces, we may expect an even stronger contribution of the first layers to the surface tension. Therefore, we can consider [11] that the surface tension only depends on the surface composition of the liquid (from a macroscopic point of view: this surface composition may include a few layers).

At the interface between liquid and vapor there is equilibrium between the compositions of the vapor and the liquid phases: because of diffusion, which is a very fast process at short distances, the first layers of the liquid are in equilibrium with the first layers of the vapor phase. We can draw, therefore, a first conclusion: although diffusion is much faster in the vapor phase than in the liquid, we cannot consider, as a first approximation, that the vapor phase composition is homogeneous: this would lead to an homogeneous surface tension.The PDMS drop acts like a pump, and locally depletes the vapor phase from its solvent.

There are several mechanisms that can explain that the vapor phase should be less depleted near the edge of the drop, leading to a higher interface concentration there. First, it may simply be due to the geometry of our experiments: the solvent is provided by a reservoir which is closer to the edge of the drop than to its center, which ensures that the vapor phase is less easily depleted near the edge. Second [11], it may be due to the presence of a permanent depletion zone: the drop can be surrounded by a rather homogeneously depleted atmosphere, except near its edge, which arrives — when it advances due to spreading — in regions that have not yet been depleted. There is a last possible explanation [9]: near the edge, the drop is thinner and acts as a less efficient pump than near the center.

Those three effects probably take place at the same time. We have undertaken a systematic study of the influence of the oil viscosity η, of the drop volume Ω, and of its dynamic contact angle at the moment when it comes in contact with the solvent vapor, but no clear feature has emerged yet.

Late stage

Let us now deal with the late stages of the phenomena, i.e., retraction under cyclohexane and spreading under hexane. We have here to strongly distinguish between these two types of behaviors.

We have already mentioned that cyclohexane does not wet glass. Thus, when the drop has become rich in solvent, it must be non-wetting. Besides, with time, its composition becomes more homogeneous and the surface tension gradients weaken. However, when there is a competition between Marangoni effect and wettability, there is, at present, no quantitative criterion available to predict which one will overcome [6]. Still, we can interpret the retraction as a dewetting phenomenon, and the final state obtained as a non-wetting drop in equilibrium, with a finite contact angle.

With hexane, the phenomena are different. Because of its low surface tension, hexane adsorbs on glass, and it forms a thin film on the substrate. It also adsorbs at the interface between liquid and gas, and the surface of the drop is almost pure hexane. We thus just have a drop, the surface of which is nearly pure hexane, spreading on a film of hexane. The spreading is faster than expected because the adsorbed film is liquid and because of the other effects (increase of volume, decrease of viscosity) we have neglected. The overall movement is due to a slight tilt of the glass plate.

This interpretation deserves a short remark: because of its rather high surface tension, the cyclohexane concentration should be lower at the interface than in the bulk, and it may seem difficult to get a non-wetting drop. But cyclohexane is present in the vapor phase, and this imposes the high surface concentration.

The vapor pressure of decaline is much lower than that of cyclohexane, which can explain why it is difficult to make "large" drops retract: it takes a long time before they have adsorbed enough decaline to become non-wetting and to have a rather homogeneous composition. We have not yet checked whether, with large enough PDMS drops under cyclohexane vapor, instabilities of the contact line do appear.

Conclusions

We have shown evidence of the role of Marangoni effect in the spreading of oil drops in presence of solvent vapor. Our results are still preliminary, but they are intended to lead to a better understanding of contamination effects.

Acknowledgements

Fruitful discussions with P. G. de Gennes, F. Heslot, and F. Van Voorst Vader are gratefully acknowledged.

References

1. De Gennes PG (1985) Rev Mod Phys 57:827—863
2. Derjaguin BV, Churaev NV, Muller VM (1987) Surface Forces, Consultants Bureau, New York and London, 440 pp, and references therein; Lopez J, Miller C, Ruckenstein E (1976) J Colloid Interface Sci 56:460—468; Joanny JF (1985) thesis, Paris

3. Cazabat AM, Cohen Stuart MA (1986) J Phys Chem 90:5845—5849
4. Bascom WD, Cottington RL, Singleterry CR (1964) In: Contact angle, wettability and adhesion, Advances in Chemistry Series 43, ACS, pp 355—379
5. Neogi P (1985) J Colloid Interface Sci 105:94—101
6. Pesach D, Marmur A (1987) Langmuir 3:519—524
7. Thomson J (1855) Phil Mag S 4, 10:330; Marangoni C (1871) Nuovo Cimento 5:239; Scriven LE, Sternling CV (1960) Nature 187:186—188
8. Troian SM, Wu XL, Safran SA (1989) Phys Rev Lett 66:1496—1499; Joanny JF (1989) J Colloid Interface Sci 128:407—415; Brochard F (1989) Langmuir 5:432—438
9. Carles P, Cazabat AM (1989) Colloids and Surfaces 41:97—105
10. Hardy WB (1919) Phil Mag S 6, 38:49—55
11. De Gennes PG, personal communication

Authors' address:

P. Carles
Physique de la Matière Condensée
Collège de France
11, place Marcelin Berthelot
75231 Paris Cedex 05
France

The dynamics of ultrathin films

A. M. Cazabat, F. Heslot, N. Fraysse, P. Carles and P. Levinson

Collège de France, Physique de la matière condensée, Paris, France

Abstract: The dynamics of thin and ultrathin wetting films have been studied experimentally by ellipsometry. The results are analysed in reference to the hydrodynamic theory of the Gennes in the case of thin films, i.e., films whose thickness is large compared to molecular sizes. — For ultrathin films, molecular effects become dominant, for which no theory is available yet. We propose a qualitative analysis in terms of the two-dimensional film pressure which accounts satisfactorily for the experimental findings.

Key words: Wetting; thin films; film pressure; spreading; monomolecular films

Introduction

When a liquid spreads spontaneously on a smooth solid surface, a microscopic precursor film grows at the edge of the macroscopic liquid front [1—4]. A convenient situation for the theoretical analysis is the case of non-volatile liquids, i.e., when the transport of molecules through the gaseous phase can be neglected [4].

Three length scales must be distinguished for such an analysis: the *macroscopic* part (if any) corresponds to thicknesses above the micrometer. It is a continuous medium with the properties of a bulk liquid which will be described by hydrodynamics.

Smaller thicknesses correspond to the microscopic range, where the film no longer has the bulk properties. This range is conveniently described by introducing the disjoining pressure π [1]. For a film of thickness z

$$\pi(z) = -\frac{1}{v_0}\,[\mu_F(z) - \mu_B]\ .$$

Here, v_0 is the molecular volume, $\mu_F(z)$ is the chemical potential of liquid molecules in the film, μ_B is its value in the bulk.

In the microscopic range, one has to distinguish between a *mesoscopic* [5] range, for thicknesses much larger than the molecular size (say, $z \geqslant 20$ Å) and a *molecular* range. In the mesoscopic part the medium is a continuum which can be treated by inserting in hydrodynamic equations the long range part of $\pi(z)$, described by the DLVO theory. In the molecular part, the coarse-grained structure of the film becomes apparent. Few attempts have been made to describe this situation [6, 7] where the short range part of $\pi(z)$ controls the dynamical behavior of the film.

Microscopic precursor films of various liquids have been studied by ellipsometry, both in the mesoscopic and in the molecular ranges [8—10]. In the following, we shall recall briefly the main results of these experiments. In the mesoscopic range, the hydrodynamic treatment can be used. In the molecular range, we shall propose an analysis in terms of two-dimensional film pressure P rather than disjoining pressure π.

Experimental findings

We have studied wetting films of various liquids spreading on silicon wafers covered with natural oxide (silica layer thickness ≈ 20 Å); the (very low) residual roughness of these surfaces plays no role in the thickness measurements [12] and is expected to have no significant effect on the film dynamics [13]. The measured thickness is the total thickness of the silica layer (baseline) plus the film (and possibly, a contaminant layer). Measurements are done before and after the film deposition, and the film thickness is calculated as the difference between the two measurements (each of them being a map of the surface).

The liquids are:

- polydimethylsiloxane, methyl terminated (PDMS), a light silicone oil with molecular weight 2400 daltons, viscosity:

$$\eta = 20 \times 10^{-3} \text{ Pa s} .$$

The mass distribution is somewhat wide (polydispersity index 1.7) but no secondary peak appears in the GPC analysis (courtesy of S. Boileau). Lighter species (cyclic molecules) do not contribute to the observed phenomena.

- squalane (SQ), a saturated branched alkane molecule

$$C_{30}H_{62}; \quad \eta = 21 \times 10^{-3} \text{ Pa s};$$

- tetrakis (2-ethylhexoxy) silane (TK), a star-like molecule with a central silicon atom and four aliphatic arms, molecular weight 545 daltons, $\eta = 6.8 \times 10^{-3}$ Pa s.

The ellipsometric set-up has a thickness resolution of 0.2 Å (including the stability of the baseline) and spatial resolution of 200 μm × 1 mm. Contamination is reduced by constant flow of dry nitrogen. It can be monitored during the experiments by measuring the thickness on the bare surface (baseline = silica layer + contaminants). This thickness is found to vary by less than 0.2 Å per week.

Two experimental situations have been investigated:

- First, the capillary rise geometry: the macroscopic front is static at long times. The film grows from the edge of the meniscus along the vertical wall. The wafer is vertical and its bottom part is put into contact with the reservoir.
- Second, tiny drops with no macroscopic part. The maximum thickness of the drops is less than 200 Å.

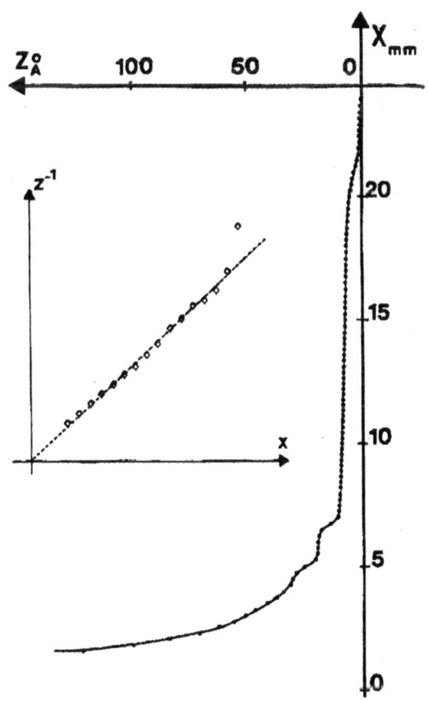

Fig. 1. Elllipsometric profile of a film of TK in the capillary rise geometry; measurement after 300 h. Inset: plot of the inverse z^{-1} vs x-coordinate in the mesoscopic range ($x \leqslant 5$ mm)

Capillary rise geometry

A typical ellipsometric profile for a film of TK is given in Fig. 1. Here, the characteristic features of the mesoscopic scale ($z \geqslant 25$ Å) and the molecular scale are obvious.

In the mesoscopic part, the film profile is mainly determined by the residual velocity of the meniscus. If the long-range part of $\pi(z)$ is dominated by Van der Waals interactions, the hydrodynamic theory of de Gennes predicts [4] that the film profile scales like x^{-1}, where x is the longitudinal coordinate. This prediction is well obeyed (see insert, Fig. 1), but has a restricted validity (200 Å — 25 Å typically). Similar results are obtained for PDMS.

In the molecular part the layering is clearly visible. The height of the successive steps is 1 molecular diameter. The same phenomenon, although less visible, occurs also for PDMS. Structuration is not observed in squalane.

The length of the film is found to obey the diffusion-like formula:

$$L(t) = \sqrt{D_F t} \, ,$$

with

$$D_F(\text{PDMS}) = 2.8 \times 10^{-10} \text{ m}^2 \text{ s}^{-1} \qquad [8]$$

$$D_F(\text{TK}) = 1.65 \times 10^{-10} \text{ m}^2 \text{ s}^{-1} .$$

The edge of the film smoothes out with time, due to two-dimensional diffusion of molecules on the surface. This smoothing also obeys a diffusion-like formula. If a is the width of the edge (a poorly defined quantity, indeed) one observes

$$a(t) \approx \sqrt{D_m t} \, .$$

The molecular diffusion coefficient D_m is in the 10^{-11} m² · s⁻¹ range.

Tiny drops

The molecular diffusion is more conveniently studied by investigating tiny drops, whose overall thickness falls below 1 molecular diameter at long times. Here the ellipsometric signal gives the density of probability of molecules on the surface. At long times, a Gaussian probability is observed, from which a value of D_m can be deduced:

$$D_m(\text{PDMS}) = 6 \pm 2 \times 10^{-11}\ \text{m}^2\ \text{s}^{-1} \qquad [8]$$

$$D_m(\text{SQ}) = 7 \times 10^{-11}\ \text{m}^2\ \text{s}^{-1} \qquad [9]$$

Discussion

The "mesocopic" part of our films is driven by a macroscopic moving edge (the meniscus needs a infinite time to reach its equilibrium position!). It is the adiabatic case of the hydrodynamic theory by de Gennes [4, 14] which accounts satisfactorily for our results.

No theory is available for the molecular part, i.e., the two first layers that represent the main contribution to the film length. Here a hydrodynamic description of friction does not hold. A better way is to use a two-dimensional approach and to introduce the film pressure P.

A two-dimensional analysis

Let c be the number of film molecules per unit area of the solid. The film thickness is just

$$z = cv_0 \, .$$

The film pressure $P(c)$ is related to the disjoining pressure $\pi(z)$ by

$$P(z) = -z\pi(z) + \int_0^z \pi(h)dh \, .$$

Let us consider a non-uniform film. The pressure gradient verifies

$$\frac{\partial c}{\partial t} + \frac{\partial}{\partial x}\left[\frac{1}{a}\frac{dP}{dc}\frac{\partial c}{\partial x}\right] = 0 \, .$$

Here a is a molecular coefficient, that is the inverse of a mobility.

This diffusion-like equation can be also expressed as

$$\begin{cases} \dfrac{\partial c}{\partial t} + \dfrac{\partial}{\partial x}\left[D(c)\dfrac{\partial c}{\partial x}\right] = 0 \\ D(c) = 1/a \ \ dP/dc \, . \end{cases}$$

The diffusion coefficient $D(c)$ is proportional to the derivative of the pressure with respect to concentration.

At very low c, P is given by the two-dimensional perfect gas law

$$P = kTc \, .$$

In this range, the diffusion coefficient has a simple expression:

$$D(c) = D_0 = \frac{kT}{a}: \text{ that is, the usual Stokes-Einstein}$$

relation. Here D_0 is the diffusion coefficient of an isolated molecule on the solid surface.

The case of PDMS

Film pressure data on solids are not available, but many studies on liquid substrates have been published.

Let us assume that some similarity exists between film pressures of PDMS on silica (our experimental case) and on water (where data are available).

The most striking feature of PDMS film pressures on water is their step-like character [15—16]. A schematic representation is given in Fig. 2. Such a behavior reflects the oscillatory character of the short range part of the disjoining pressure $\pi(z)$ [17, 18].

Very roughly, the film behavior will be described by three diffusion coefficients (see Fig. 2)

$$\begin{aligned} c &< c_1 & D &= D_1 \\ c_1 \leqslant c &\leqslant c_2 & D &= D_2 \\ c_2 &< c & D &= D_3 \, , \end{aligned}$$

with $D_2 > D_3 \gg D_1$.

The edge of such a film (connected to a reservoir) will evolve according to the schematic representa-

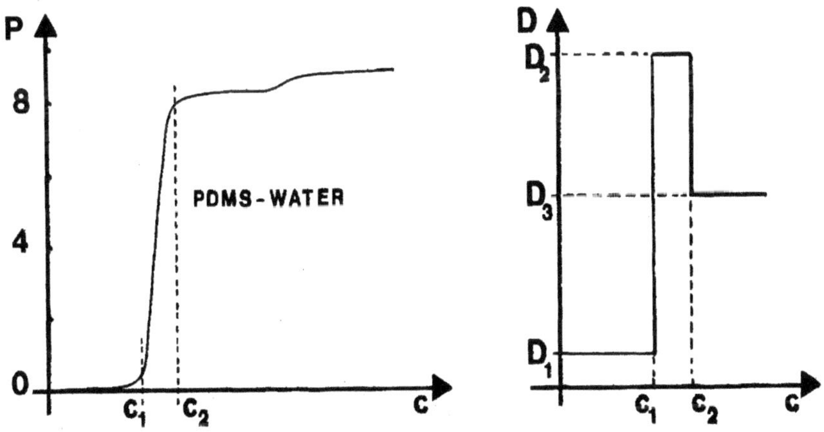

Fig. 2. Film pressure $P(c)$ of PDMS, water (schematic), and corresponding diffusion coefficients $D(c)$:

$$D(c) = \frac{1}{a}\frac{dP}{dc} ;$$

c is the surface concentration of molecules. P is given in 10^{-3} N m^{-1} units

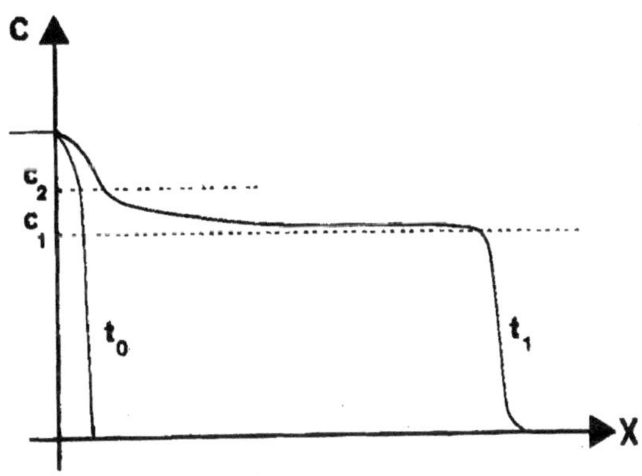

Fig. 3. Evolution of a film profile (schematic)

tion of Fig. 3. The similarity with ellipsometric profile of Fig. 1 is obvious, with:

$$D_m \equiv D_1$$
$$D_F \equiv D_2 .$$

The dynamics of the second layer could be treated similarly, in the reference frame of the first layer.

Conclusion

The dynamics of thin and ultrathin wetting films have been investigated by ellipsometry. Thin films are conveniently described by the hydrodynamic approach of de Gennes. For ultrathin films, the two-dimensional film pressure is a relevant parameter. A diffusion-like equation governing the film dynamics has been obtained. It allows to predict the dynamics of the film if its statics properties are known.

Acknowledgements

Fruitful discussions with PG. de Gennes, G. Findenegg, Y. Larher, J. M. di Meglio, J. Scheutjens, and F. Van Voorst Vader are gratefully acknowledged.

References

1. Deryaguin BV, Churaev NV, Muller VM (1987) Surfaces forces — Consultant Bureau — Plenum Press, New York and London, and references
2. Churaev NV, Starov VM, Deryaguin BV (1982) J Coll Int Sci 89:16—24
3. Teketzke GF (1983) PhD Thesis — University of Minnesota
4. de Gennes PG (1985) Rev Mod Phys 57:827—863
5. de Gennes PG (1988) Unpublished lecture at the Collège de France — Paris . "Quelques problèmes interfaciaux"
6. Abraham DB, Collet P, de Coninck J, Dunlop F (1990) Phys Rev Lett 65:195—198
7. Summ BD, Yushchenko VS, Shchukin ED (1987) Colloids and surfaces 27:43—55
8. Heslot F, Cazabat AM, Fraysse N (1989) J Phys Condens Matter 1:5793—5798
9. Heslot F, Cazabat AM, Levinson P (1989) Phys Rev Lett 62:1286—1289
10. Heslot F, Fraysse N, Cazabat AM (1989) Nature 338:640—642
11. Marty O, Drake B, Gould S, Hansma PK (1988) J Vac Sci Technol A6:2089—2092

12. Cazabat AM (1988) Lectures at the Summer School — Les Houches —. To be published: Liquids at the interfaces. Gordon and Breach
13. Andelman D, Joanny JF, Robbins MO (1988) Europhys Lett 7:731—736 and private discussions
14. Joanny JF, de Gennes PG (1986) J Phys Paris 298:121—127
15. Jarvis NL (1969) J Coll Int Sci 29:647—657
16. Granick S (1985) Macromolecules 18:1597—1602
17. Israelachvili JN, Mc Guiggan P, Homola AM (1988) Science 240:189—191
18. Horn RG, Israelachvili JN (1988) Macromolecules 21:2836—2841

Authors' address:

A. M. Cazabat
Collège de France
Physique de la matière condensée
11 place Marcelin Berthelot
75231 Paris Cedex O5, France

Drainage of foam films in the presence of nonionic micelles

A. D. Nikolov, D. T. Wasan[1]), N. D. Denkov, P. A. Kralchevsky and I. B. Ivanov

Laboratory of Thermodynamics and Physicochemical Hydrodynamics, University of Sofia, Faculty of Chemistry, Sofia, Bulgaria
[1]) Department of Chemical Engineering, Illinois Institute of Technology, Chicago, Illinois, USA

Abstract: The stratification (step-wise thinning) of horizontal foam films formed from micelar solutions of nonionic surfactants is studied experimentally. The effects of film area, temperature, oil solubilization, surfactant and electrolyte concentrations are investigated. The data support the colloid crystal model of stratification, according to which the step-wise thinning is due to a layer-by-layer destruction of an ordered structure of spherical nonionic micelles inside the film. The analysis of the data shows that the main factor governing the stratification is the volume fraction of the micelles. This fact leads to an unified interpretation of the observed effects, both with nonionic and with anionic surfactant solutions.

Key words: Foam films; nonionic surfactants; stratifying films; micellar structuring; steric stabilization

Introduction

At present the nonionic surfactants are the subject of intensive research and development with continuously growing fields of application. Their wetting, foaming, dispersing or emulsifying properties are widely utilized in practice. At equal molar concentrations the nonionic surfactants, as a rule, exhibit greater surface activity and lower critical micellar concentrations (CMC) than the ionic surfactants [1, 2].

It was recently observed [3] that foam films formed from micellar solutions of nonionic surfactants exhibit step-wise thinning. This phenomenon, called ,,stratification" was observed long ago with anionic surfactants [4, 5]. The step-wise thinning of films from solutions of ionic micelles was explained [3, 6, 7] as a layer-by-layer destruction of a structure of spherical micelles inside the film. The occurrence of the stratification depends strongly on the interaction between the micelles and on their concentration in the solution.

The aim of the present paper is a comprehensive experimental study of the occurrence of stratification with nonionic surfactant solutions. Special attention is paid to the influence of different factors (surfactant concentration, temperature, film area, solubilization, etc.) on the step-wise thinning. The results are compared with similar data for anionic surfactants. Correlations are found between the micellar aggregation in the surfactant solution and the stratification and stability of the films. The results indicate how one can control (increase or decrease) the stability of microheterogeneous systems containing nonionic surfactant micelles in the disperse medium.

Experimental materials and technique

a) Materials

In our experiments we used a commercial nonionic surfactants Enordet AE 1215-9.4 and AE 1215-30 (products of Shell Co.). Both surfactants are ethoxylated alcohols with straight paraffinic chain containing between 12 and 15 carbon atoms and with a hydrophylic chain containing, on the average, 9.4 and 30 ethylene-oxide groups, respectively. The data of Mitchell et al. [8] for ethoxylated alcohols obtained by small-angle neutron scattering show that these nonionic surfactants in aqueous solutions form liquid crystal phase only at concentrations higher than 30 wt % at room temperature. In our experiments we

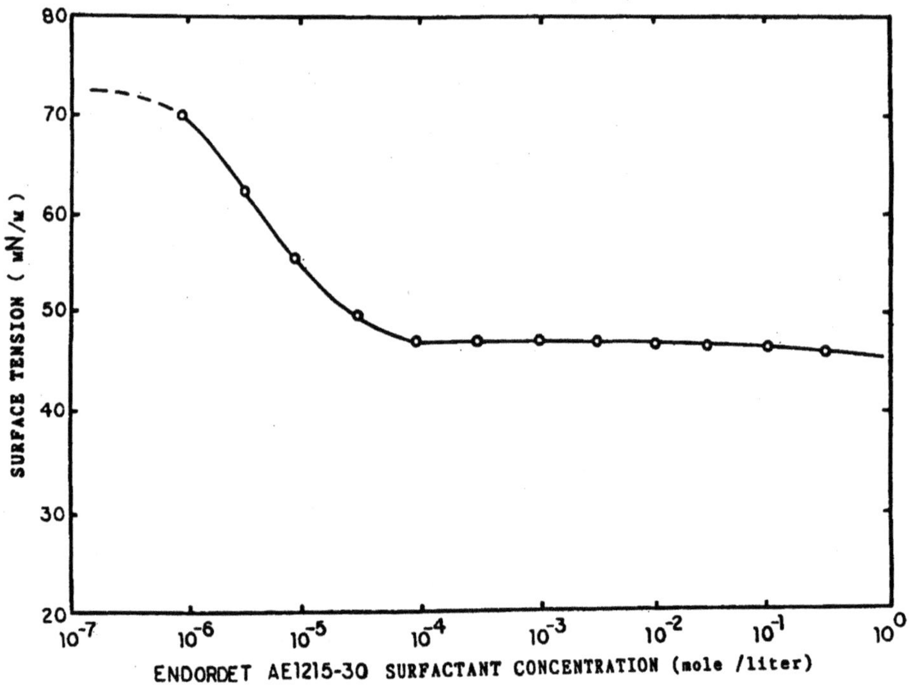

Fig. 1. Plot for the experimental data for the surface tension of an Enordet AE 1215-30 solution vs the surfactant concentration at 25 °C

used surfactant solutions of concentrations above CMC, but below 10 wt %. In this concentration-range spherical micelles are formed in the solution (see [1, 2, 8]).

Figure 1 shows surface tension isotherm at 25 °C for Enordet AE 1215-30, measured by the Wilhelmy plate technique. There is no minimum on the isotherm, which is an indication of the absence of surface contaminations.

b) Measurements of micellar size and distribution

The available light-scattering data [9] show that at low concentration and room temperature the micellar solutions of nonionic surfactants behave as a solution of hard spheres with a volume fraction that is larger than the surfactant weight fraction. This fact indicates that at low micellar concentrations the surfactant molecules are strongly hydrated. In the case of micellization in solutions of polyoxyethylene surfactants, like those studied by us, the ethylene-oxide groups interact with the water molecules and form a shell enveloping the micellar core, the latter consisting mostly of the hydrocarbon phase [10]. The outer diameter of this shell in fact determines the size of the effective hard sphere characterizing the interaction between micelles.

The micellar size and the distribution in the nonionic surfactant solutions studied by us have been measured by Malvern 4700 dynamic light-scattering technique. All measurements were done by Malvern Instruments Inc., Southborough, Mass., USA.

c) Measurements of foam film thickness

Horizontal microscopic foam films were formed from the studied solutions of nonionic surfactants in a vertical cylindrical capillary. The latter has a hydrophylic inner wall of radius 1.5 mm. The films were formed by sucking out the liquid from a biconcave drop inside the capillary through an orifice in the capillary wall. The same technique was applied previously [3, 6] for studying foam films formed from micellar solutions of anionic surfactants. The horizontal film was observed from above by means of a microscope with illumination through the objective. We used monochromatic light of wavelength $\lambda = 546$ nm. Stratification is observed, both with anionic and nonionic surfactants.

After the film is formed, it immediately starts to decrease its thickness. When the latter becomes smaller than about 104 nm (this thickness corresponds to the last interference maximum) step-wise changes in the film thickness can be observed. The film remains for a few seconds in a metastable state with almost constant thickness. Then one or several dark spots (thinner than the rest of the film) appear and rapidly increase their area. Soon the spots cover the whole film and the film stays for several seconds in the new metastable state. Then new darker spots appear and spread, and then a subsequent metastable state is reached. Eventually the film reaches a stable state and step-wise transitions no longer occur.

The thickness of the film during the step-wise thinning was recorded using reflected light interferometry technique. The light reflected from a portion of the film is conducted to a photomultiplier by means of a fine optic fiber. The resulting photocurrent is recorded as a function of time. The experimental setup is described in details in [6]. A typical interferogram is presented in Fig. 2. The minimum and the maximum intensities of the reflected light correspond to the last interference minimum and maximum. The instantaneous film thickness can be calculated from the instantaneous intensity of the reflected light as reported in [6]. The calculated heights of the steps are also shown in the interferogram.

The steps in the interferogram correspond to the metastable states of the film. Their width is the lifetime of the respective metastable state. The irregularity of some of the upper steps (Fig. 2) is due to the consecutive passage of several similar dark spots through the scope field of the optic fiber.

Interferograms obtained at different experimental conditions are shown and discussed below.

Experimental results

Preliminary experiments showed that the stratification of films formed from aqueous solutions of nonionic surfactants depend on several factors: film area, temperature, surfactant concentration, concentration of added electrolyte, oil solubilization, etc. In the following, we will separately consider the role of each factor, holding the other physical conditions constant.

a) Effect of film area

Since the capillary used is cylindrical, the horizontal films have circular shape. The radius r_c of the three-phase contact line encircling the film can be controlled by changing the capillary pressure as explained in [6]. The two interferograms shown in Figs. 2 and 3 represent the step-wise thinning of foam films formed from 0.052 mol/l (8 wt %) solution of Enordet AE 30 at temperature 32 °C. The difference between the figures is due to the different film radii: r_c = 0.4 mm for Fig. 2, and r_c = 0.2 mm for Fig. 3. Six transitions are observed with the larger film, whereas their number is only 4 with the smaller one. (As explained above, the upper transitions sometimes are not very distinct in the interferograms, but they are visually well observed.) In both cases the stratification starts when the film thickness becomes around 70 nm and the mean height of a step is the same and equal to 9.7 nm. However, the thinning of the smaller film stops at thickness 34 nm (vs 14 nm for the larger one) and the last two transitions (observed with the larger film) do not take place. Therefore, the decrease of the film area can stop the stratification. In the case illustrated in Fig. 3 the final stable film contains two layers of micelles, which are in diffusional equilibrium with the bulk micellar solution in the biconcave meniscus surrounding the film.

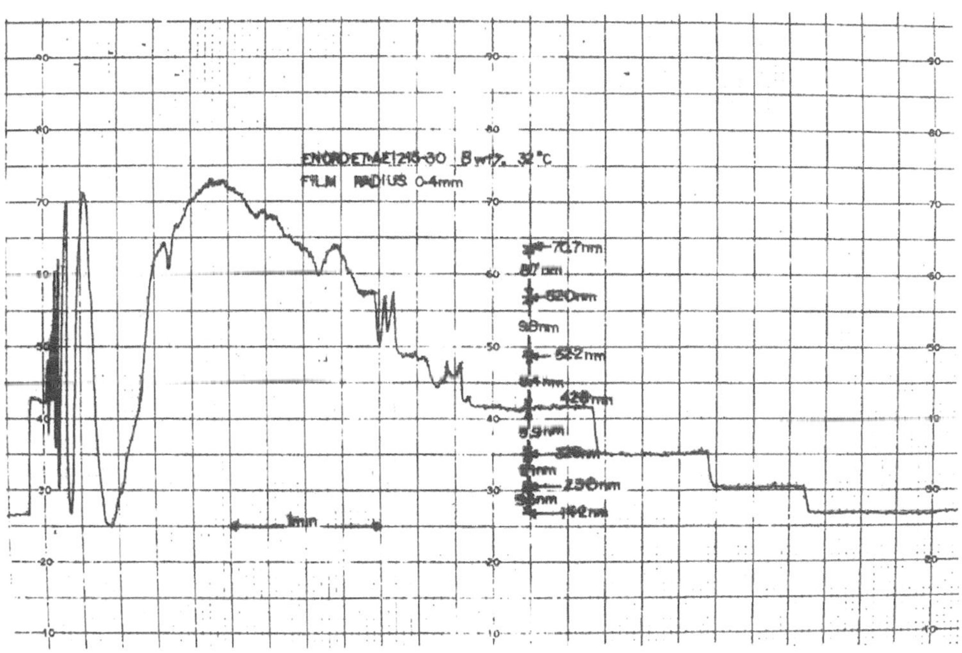

Fig. 2. Photocurrent vs time in the interferogram of thinning oof the Enordet AE 1215-30 film at surfactant concentration 5.2 × 10^{-2} mol/l at 32 °C and film radius 0.4 mm

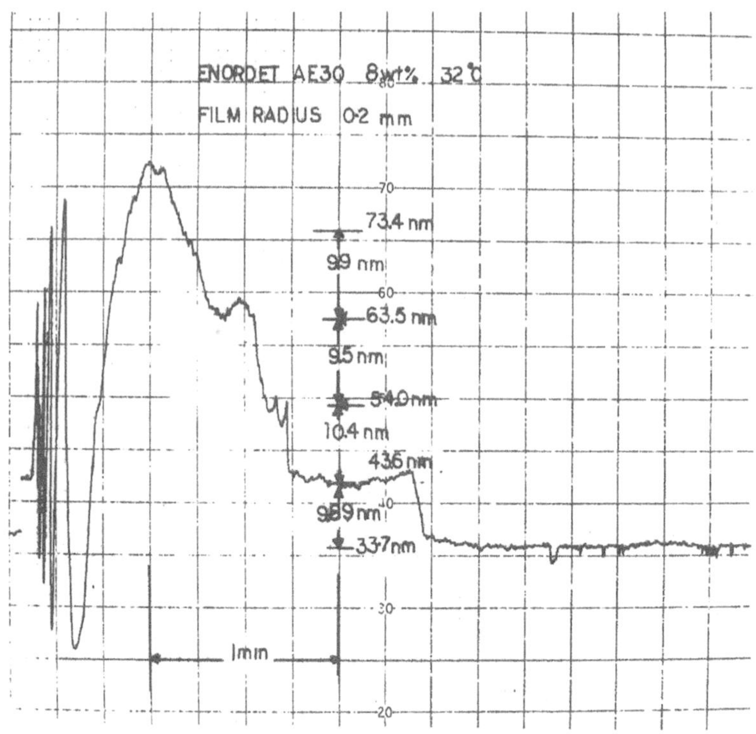

Fig. 3. Photocurrent vs time in the interferogram of thinning of the Enordet AE 1215-30 film at surfactant concentration 5.2 × 10^{-2} mol/l at 32 °C and film radius 0.2 mm

A similar effect was recently observed with stratifying films formed from anionic surfactant solutions. In [11] a diffusive-osmotic mechanism was proposed for interpretation of the data for formation and expansion of spots in the stratifying films. According to this mechanism the appearance of a spot is due to condensation of vacancies in the micellar structure inside the film. The theoretical model predicts that there is a minimum size of the (stable) spots in the film. When the film area is small enough, the total amount of vacancies in the film can be smaller than the one needed for creation of a stable spot. Then spots cannot appear and the stratification stops.

We expect that a similar explanation can be given also for the case of nonionic surfactants, because of the existence of analogous diffusive-osmotic effects. The development of such a model for nonionic surfactants needs, however, additional studies, both theoretical and experimental.

b) Effect of the temperature

A decrease of the temperature, similar to the decrease in the film area, can prevent the occurrence of the last few-step-wise transitions. The stratification stops at a larger film thickness. Figure 4 represents an interferogram obtained under the same conditions as those for Fig. 2 (r_c = 0.4 mm, 8 wt % solution of Enordet AE 1215-30), except for the temperature, which was 26 °C instead of 32 °C. One sees that the number of step-wise transitions is reduced to 2 at 26 °C, as opposed to 6 at 32 °C. The thickness of the final stable film is 14 nm at 32 °C, whereas it is considerably greater — 49 nm — at 26 °C. Hence, the decrease of the temperature leads to increase of the film stability and thick stable films can be obtained.

The effect of temperature is larger with solutions of nonionic surfactants having lower degree of ethoxylation. Figure 5 represents the interferogram of a thinning film formed from 2 wt % aqueous solution of Enordet AE 1215-9.4 (with an average number of 9.4 ethylene-oxyde groups per molecule) with r_c = 0.3 mm at 24 °C. One sees that with the surfactant with shorter polyoxyethylene chain step-wise thinning is not observed, the final thickness is larger and the drainage is slower.

The velocity of drainage of the stratifying films depends strongly on the temperature. Let us define

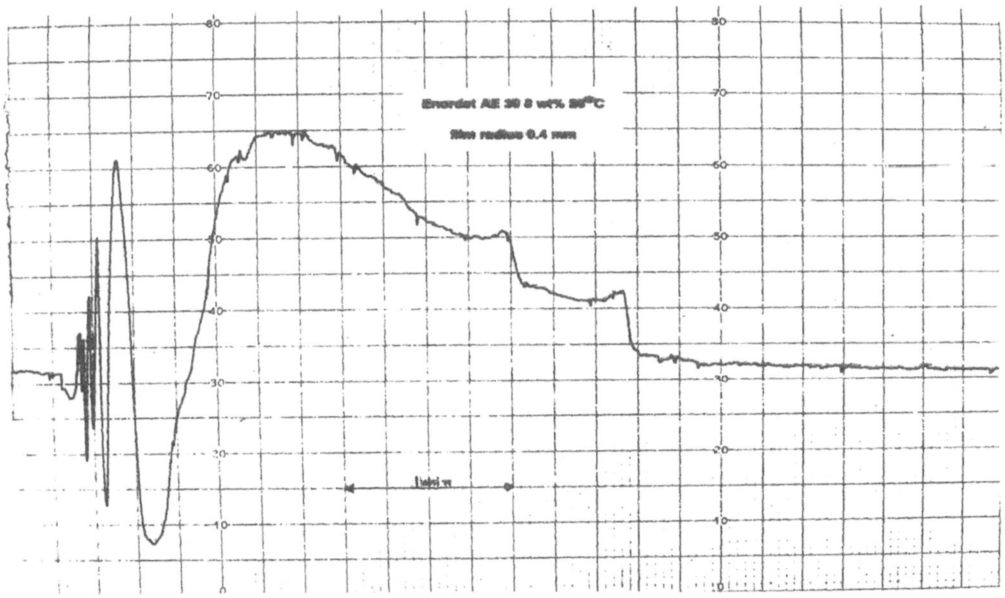

Fig. 4. Photocurrent vs time in the interferogram of thinning of the Enordet AE 1215-30 film at surfactant concentration 5.2×10^{-2} mol/l at 26 °C and film radius 0.4 mm

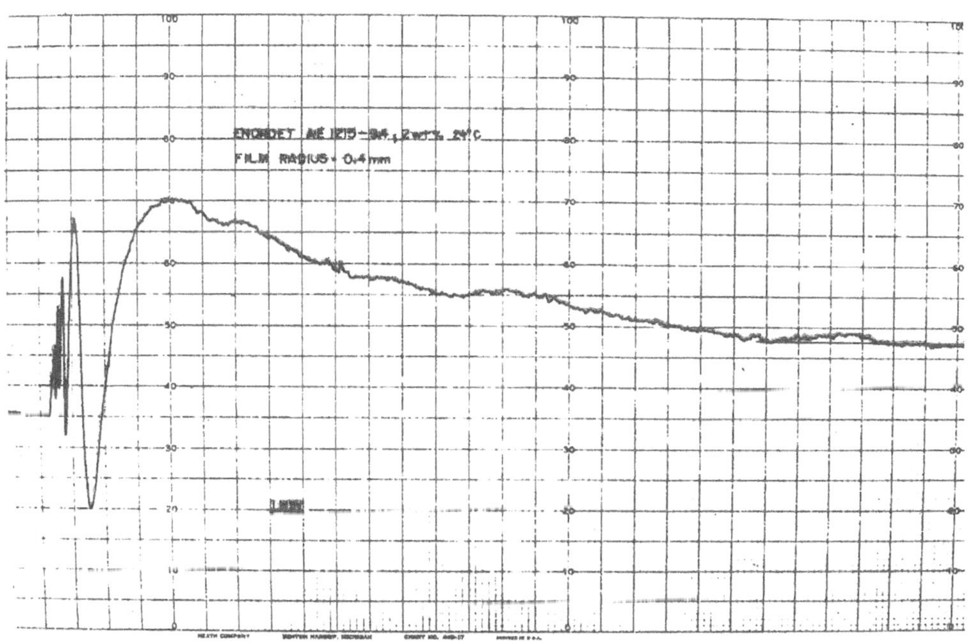

Fig. 5. Photocurrent vs time in the interferogram of thinning of the Enordet AE 1215-9.4 film at surfactant concentration 3.2×10^{-2} mol/l at 24 °C and film radius 0.4 mm

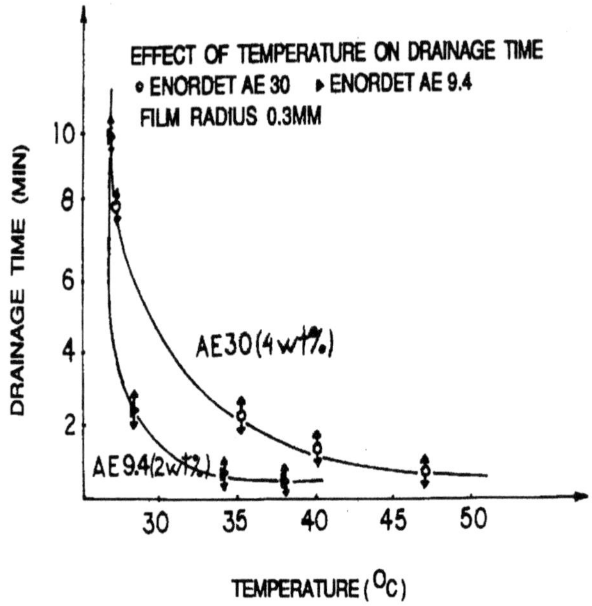

Fig. 6. Effect of temperature on drainage time (○ Enordet AE 1215-30, ◆ Enordet AE 1215-9.4). Film radius 0.3 mm

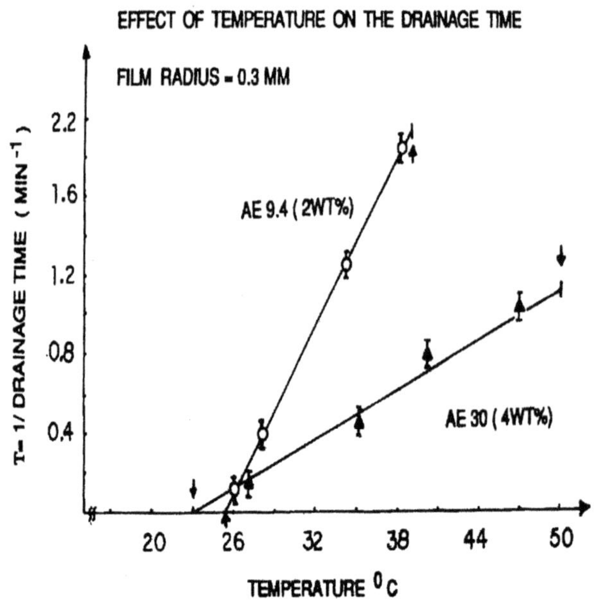

Fig. 7. Effect of temperature on drainage time (○ Enordet AE 1215-30, ◆ Enordet AE 1215-9.4). Film radius 0.3 mm

drainage time of the film by the time elapsed between the appearance of the last interference maximum (it is well seen in the interferograms and corresponds to film thickness about 104 nm) and the moment when the final stable thickness is reached. The data for the dependence of the drainage time so defined on the temperature are shown in Fig. 6 for Enordet AE 1215-30 at concentration 4 wt % (upper curve) and Enordet AE 1215-9.4 at concentration 2 wt % (lower curve). Both curves refer to film radii 0.3 mm. For temperatures below 30 °C the drainage time sharply increases and the step-wise transitions are either few or there are none. The final film thickness in this temperature range is about 60 nm or greater and increases with the decrease of temperature.

At temperatures higher than 30 °C pronounced stratification is observed. All possible consecutive step-wise transitions occur and the final film thickness is about 15 nm and probably corresponds to a film that does not contain any micelles.

At temperatures higher than 35 °C the step-wise transitions become irregular: two or more transitions can occur at once. At even higher temperatures the step-wise transitions are absent: the final stable state is directly reached. At temperatures close to the surfactant cloud point (but below it) the films rupture when they reach the critical thickness (stable states do not exist).

It is interesting to note that the dependences in Fig. 6, become linear when the reciprocal drainage time is plotted as a function of temperature — see Fig. 7.

c) Effect of surfactant concentration

Due to the strong effect of the film area and the temperature on the stratification, we studied the role of the surfactant concentration at fixed film radius, r_c = 0.3 mm, and at constant temperature 27 °C. Data for stratifying films at different surfactant concentrations are presented in Table 1. The film thicknesses correponding to the different steps were calculated from the experimental interferograms. Both for Enordet AE 1215-9.4 and for Enordet AE 1215-30 the number of the step-wise transitions increases with the increase of the surfactant concentration. For the lowest concentrations studied only one step-wise transition was observed. Five to six transitions occurred at higher concentrations. The surfactant with longer polyoxyethylene chain (Enordet 1215-30) exhibits more step-wise transitions for a given molar concentration.

Table 1. Experimental data for stratifying films from aqueous solutions of Enordet AE

		Enordet AE 1215-9.4			Enordet AE 1215-30		
Surfactant concentration (mol/l)		0.008	0.064	0.130	0.026	0.052	0.100
Final film thickness		17.3	16.8	16.0	16.1	15.0	14.5
Transition number and respective film thickness (nm)	1	32.1	31.0	29.9	27.1	25.5	24.4
	2	—	45.0	42.8	—	36.5	34.4
	3	—	59.4	56.2	—	47.2	43.4
	4	—	74.0	69.2	—	57.7	53.3
	5	—	—	82.9	—	68.7	63.3
	6	—	—	—	—	—	72.5
Mean height of a step (nm)		15.0 ±0.6	14.2 ±0.4	13.4 ±0.4	11.2 ±0.6	10.8 ±0.5	9.9 ±0.4
Micelle diameter (nm)		15.0 ±1.5	13.7 ±1.5	—	10.6 ±2.1	—	—

To check the hypothesis that the thickness of a stratifying film is determined by the number of the micellar layers inside it, we compared the mean height of a step with the diameter of the micelles (the last two rows of Table 1). The mean diameter of the micelles was measured by means of Malvern technique as explained above. One sees that the mean values of the micellar diameter and the height of the step coincide in the framework of the experimental accuracy. This fact implies that the stepwise thinning can really be interpreted as a layer-by-layer destruction of a micellar structure inside the stratifying film. Analogous experiments with thinning films containing anionic surfactant micelles or latex particles reveal [3, 6, 7] that the stratification is typical for films containing almost monodisperse spherical colloidal particles.

We will note also that the micellar diameters as well as the heights of the steps presented in Table 1 exhibit a weak tendency to decrease when increasing the nonionic surfactant concentration.

d) Other factors affecting the stratification

The stratification of foam films, formed from solutions of *anionic* micellar surfactant solutions, is strongly depressed by addition of neutral electrolyte [6, 7]. This effect is due to the shrinking of the double electric layer around the micelles. Although nonionic micelles have no electric layers, our experiments with *nonionic* surfactants still showed that neutral electrolyte can suppress the stratification.

In our experiments we added NaCl to aqueous solutions of Enordet AE 1215-9.4, —20, and —30 (nonionic surfactants containing 9.4, 20, and 30 ethylene oxyde groups, respectively). It turned out that the minimum concentration of NaCl needed for complete suppression of the stratification was dependent on the number of ethylene oxyde (EO) groups. That is why we presented the experimental data in Fig. 8 as a diagram showing the number of EO groups vs NaCl concentration. The line divides the regions of stratifying and nonstratifying films. The high values of the NaCl concentration indicate that the effect of neutral electrolyte on stratification is connected with dehydratation of the micellar EO groups (salting out).

We also studied the effect of pH. We found out that variation of pH between 2 and 10 does not affect the stratification if the ionic strength is held constant (about 10^{-2} mol/l).

To investigate the role of the solubilization of oil on the step-wise transitions, we used 5.2×10^{-2} mol/l solution of Enordet AE 1215-30 preequilibrated for 48 h with n-decane. It is known that under these conditions the decane is solubilized

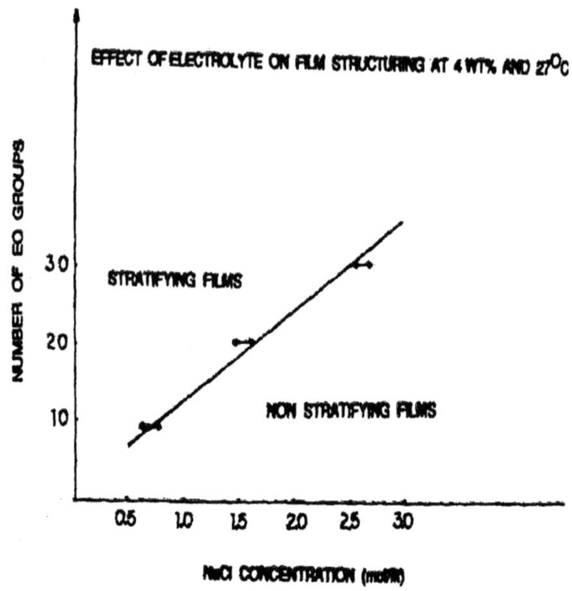

Fig. 8. Effect of electrolyte on film stratification at 27°C and 4 wt % surfactant concentration

and swollen micelles are formed [18]. Interferograms of the step-wise thinning before and after solubilization of decane are compared in Fig. 9a, b. Although the number of steps is the same, the lifetimes are shorter for the solutions containing swollen micelles, and the time for draining to the final film thickness was, respectively, 5.3 and 3.7 min. In addition, the steps in Fig. 9b are not as distinct as those in Fig. 9a. These facts suggest that the solubilization leads to less stable micellar structuring.

Discussion

As mentioned above, a colloid crystal model of stratifying films was proposed in [3, 6, 7, 11]. According to this model the step-wise thinning is due to the layer-by-layer destruction of the structure formed by spherical colloidal particles inside the film. An argument in favor of this model is the fact that the height of a step is approximately equal to the colloid particle diameter; (see Table 1 for nonionic micelles and [3, 6, 7] for anionic micelles and polystyrene latex particles). The effective micellar diameter of anionic micelles was defined in [6] as

$$2R_\kappa = 2(R_p + 1/\kappa) \,, \tag{1}$$

where R_p is the particle (micellar) radius and $1/\kappa$ is the effective thickness of the double electric layer around the particle.

According to the colloid crystal model any factor suppressing formation of particle ordering inside the film will also restrain the stratification. The parameter determining the disorder-order transition in a system of hard spheres is the volume fraction φ of the particles [20]. Then one can expect that *any factor decreasing the volume fraction of the colloidal particles will depress the particle ordering inside the film and hence — the stratification.* Our aim below is to demonstrate that such a conclusion is supported by the experimental data, both for anionic and nonionic surfactants.

For anionic surfactants the effective volume fraction of the micelle is

$$\varphi = C_{\mathrm{mic}} \left(\frac{4}{3} \pi R_\kappa^3 \right) , \tag{2}$$

where C_{mic} is the number density of the micelles in the solution. For nonionic surfactants the micellar volume fraction is simply

$$\varphi = C_{\mathrm{mic}} \left(\frac{4}{3} \pi R_p^3 \right) . \tag{2}$$

Let us first briefly consider the case of *anionic* surfactants. In [3, 6, 7] stratification with micellar solutions of sodium dodecylsulfate (NaDS) was studied. Both NaDS and NaCl are strong electrolytes. However, addition of NaDS or NaCl to the surfactant solution leads to quite different effects on stratification. The increase of NaDS concentration increases the number of the step-wise transitions and the time for drainage of the film [6]. This correlates with the fact that the addition of NaDS increases C_{mic} and, thereby, the micellar volume fraction φ, calculated by means of Eqs. (1) and (2). Hence, this favors the micellar structuring inside the thinning film.

The addition of NaCl to the surfactant solution leads to the exact opposite effect: the number and the regularity of the step-wise transitions decreases and the drainage of the stratifying film accelerates. The influence of added neutral electrolyte on the micellar volume fraction φ can be analyzed with the help of the experimental results of Sigter and Mysels [12]. These data and the calculated values of

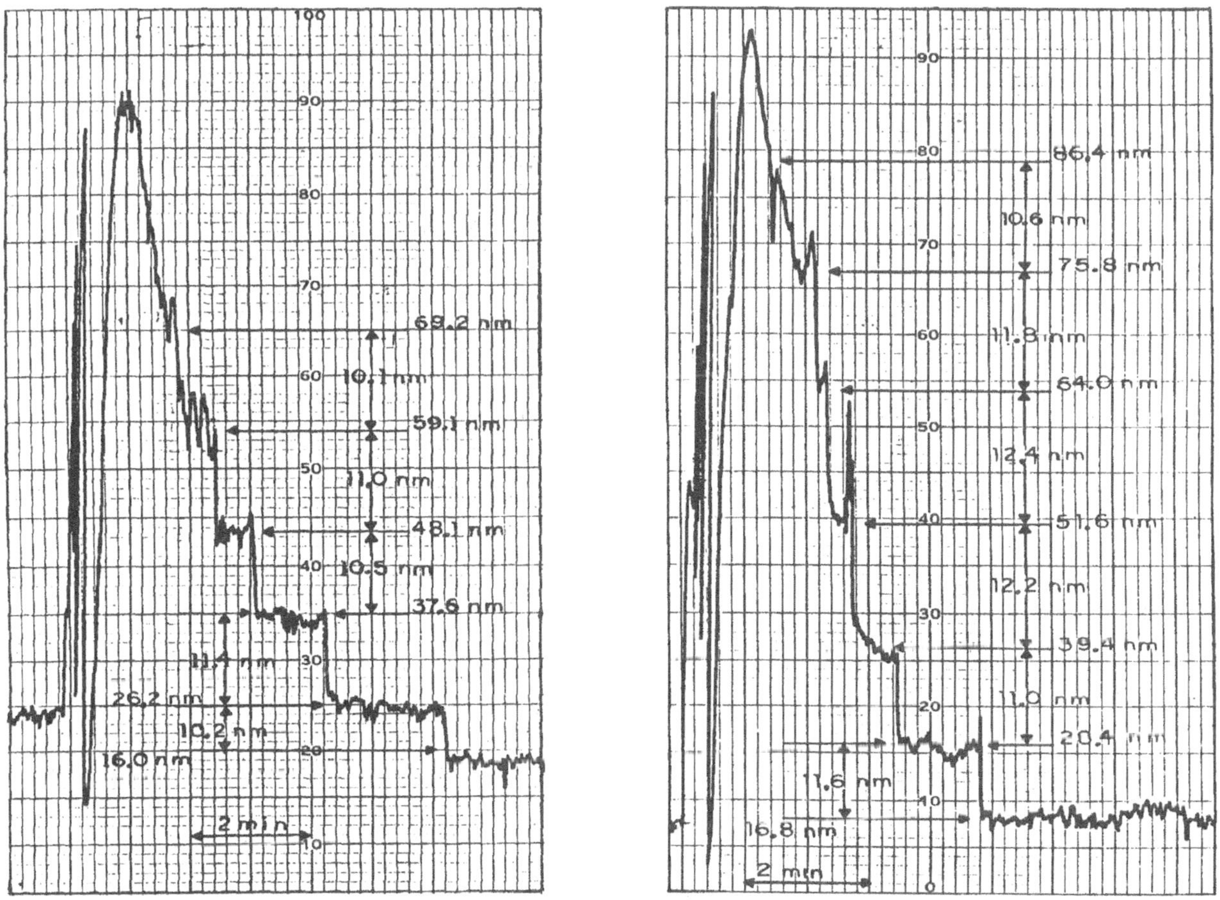

a **b**

Fig. 9. a) Photocurrent vs time in the interferogram of thinning of the Enordet AE 1215-30 film at surfactant concentration 5.2×10^{-2} mol/l. b) The same as in a) but the solution is preequilibrated with n-decane

Table 2. Effect of concentration of NaCl on the properties of micellar solutions of sodium dodecyl sulfate

C_{NaCl} (mol/l)	CMC (mol/l)	ν_a	R_p (nm)	u	$C_{mic} \times 10^4$ (mol/l)	δ_l (nm)	$1/\kappa$ (nm)	$2R_\kappa$ (nm)	φ %
0	8.1×10^{-3}	80	2.15	0.29	2.7	18.3	2.9	10.0	9
0.01	5.3×10^{-3}	89	2.1	0.28	2.8	18.1	1.6	8.8	5
0.03	3.1×10^{-3}	99	2.30	0.29	2.7	18.3	1.6	7.9	3
0.05	2.3×10^{-3}	104	2.34	0.30	2.7	18.3	1.3	7.3	3

some other parameters are presented in Table 2. The micellar concentration is defined as

$$C_{mic} = (C_{NaDS} - CMC)/\nu_a \ , \quad (4)$$

where C_{NaDS} is the total surfactant concentration and ν_a is the number of surfactant molecules aggregated in a micelle. δ_l in Table 2 is the mean distance between two micelles in the surfactant solution, proportional to the cubic root of C_{mic}. The effective thickness of the double electric layer around the micelle $1/\kappa$ is calculated from the equation

$$\kappa^2 = 4\pi e^2 C_{ion}/(\varepsilon kT) \ ,$$

where e is the electron charge, ε is the dielectric permittivity of the solution, T is temperature, k is Boltzmann constant, and

$$C_{ion} = 2C_{NaCl} + a\nu_a C_{mic} + 2\ (CMC) \ ,$$

with a being the degree of electrolytic dissociation of the surface ionizable groups of the micelles. The values of R_κ and φ in Table 2 are calculated by means of Eqs. (1) and (2). One can see that the increase of NaCl concentration decreases CMC, but increases the aggregation number ν_a of the micelles. As a result, the values of C_{mic} and δ_l are almost constant. However, the addition of NaCl decreases $1/\kappa$, R_κ, and thereby, φ. The reduction of the effective micellar volume fraction φ by the added electrolyte correlates with the observed depression of the stratification.

Now, let us do a similar analysis of the data for stepwise thinning of films formed from *nonionic* surfactant solutions.

The influence of temperature on stratification can also be explained by the change of the volume fraction φ of the nonionic micelles in the solution. (In this case φ must be determined from Eq. (3).) The data shown in Table 3 are calculated by using the experimental results of Nakagava et al. [13] for the nonionic surfactant $C_{10}H_{21}O(CH_2CH_2O)_{12}CH_3$, which has physico-chemical behavior close to that of Enordet AE 1215-9.4. The volume of a micelle V_p and the effective micellar radius R_p are calculated from the second virial coefficient A_2 in the equation for the osmotic pressure

$$\Pi = C_w RT(1/M_p + A_2 C_w) \ , \quad (5)$$

where C_w is the weight concentration of the micelles, and M_p is the molecular mass of the micelles. In accordance with [2], we assumed that

Table 3. Effect of temperature on the properties of micellar solutions of $C_{10}H_{21}O(CH_2CH_2O)_{12}CH_3$

T °C	CMC × 10^3 (mol/l)	M_p (kg/mol)	ν_a	C_{mic} × 10^4 (mol/l)	A_2 × 10^4	V_p × 10^{26} (m^3)	φ %
10	2.5	32.9	47	7.4	2.55	11.5	5.2
29	1.6	37.1	53	6.7	2.06	11.8	4.8
51	1.2	45.5	65	5.2	1.36	11.7	3.9

Table 4. Effect of NaCl on the properties of micellar solutions of $C_{13}H_{27}O(CH_2CH_2O)_{10}CH_3$

C_{NaCl} (mol/l)	ν_a	M_p (kg/mol)	C_{mic} × 10^4 (mol/l)
0	92	58.7	8.53
0.3	110	70.0	7.11
0.5	141	90.3	5.69

Table 5. Effect of NaCl on the properties of micellar solutions of $C_{13}H_{27}O(CH_2CH_2O)_{22}CH_3$

C_{NaCl} (mol/l)	ν_a	M_p (kg/mol)	C_{mic} × 10^4 (mol/l)
0	28	32.5	7.7
0.3	29	33.5	7.4
0.5	28	32.5	7.4

$$V_p = M_p^2 A_2 / 4N_A \,, \qquad (6)$$

where N_A is Avogadro's number. One sees that in spite of the increase of the aggregation number ν_a with the temperature, the volume of a micelle does not change significantly. However, the decrease of the micellar concentration C_{mic} leads to decrease of the volume fraction φ of the micelles. This explains why the stratification becomes less pronounced and even disappears entirely when the temperature is raised. On the contrary, when the temperature is sufficiently low, the micellar structure in the film becomes so stable that the film can exist in equilibrium with the bulk surfactant solution. As a result, stable thick films are formed (see Figs. 4 and 5). Such films can contain several micellar layers. It should be noted that the above findings correlate well with the very pronounced effect of temperature on the steric stabilization (see [14, 15]).

An analogous explanation can be given to the effect of neutral electrolyte (e.g., NaCl) on the stratification of thinning films from nonionic surfactant solutions (see Fig. 8). Table 4 contains data for $C_{13}H_{27}O(CH_2CH_2O)_{10}CH_3$, obtained by Becher [16]. One sees that the increase of C_{NaCl} leads to increase of the aggregation number ν_a and to decrease of the micellar concentration C_{mic}. Note that the relative change of C_{mic} without electrolyte and with 0.5 mol/l NaCl in Table 4 are the same as in Table 3 for 10°C and 51°C. Unfortunately, data for the second virial coefficient allowing the use of Eq. (6) for calculation of V_p are not avaialable, so that we could not calculate the micellar volume fraction φ. Nevertheless, it is known (see, e.g., [17]) that the salting-out action of the electrolyte, similar to the action of the temperature, leads to a partial dehydration of the ethylene-oxide chain and reduces the volume of the micelles. Therefore, one can infer that, in this case, the observed depression of stratification can also be explained by reduction of the micellar volume fraction φ. In favor of such an assumption are the data presented in Table 5. They show that the dehydration effect due to the added electrolyte is considerably smaller when the ethylene-oxyde chain is longer. This result is in agreement with our data in Fig. 8. Accordingly, for 30 ethylene-oxide units a high concentration of 2.5 mol/l NaCl is needed to entirely suppress the stratification.

To interpret our experimental results for the effect of solubilization on the step-wise transitions, we used the data of Nakagava et al. [18] for aqueous solutions of $C_{10}H_{21}O(CH_2CH_2O)_8CH_3$ preequilibrated with decane. By using Eq. (5), we calculated the volume fraction φ of the micelles (see Table 6). Similar to the increase of the temperature, the solubilization leads to an increase of the aggregation number ν_a and to reduction of the second virial coefficient A_2 and of the micellar number density C_{mic}. The volume of a micelle V_p and the respective volume fraction φ also decrease, which is consonant with the suppression of the step-wise thinning established in Fig. 9.

Concluding remarks

The present paper is the first experimental study of stratifying films formed from nonionic surfactant solutions. We investigated the effect of film area, temperature, solubilization, surfactant, and salt concentration on the occurrence of the step-wise transitions during the thinning of the film.

Table 6. Effect of solubilization on the properties of micellar solutions of $C_{10}H_{21}O(CH_2CH_2O)_8CH_3$ preequilibrated with n decane

oil/surfactant (wt %)	ν_a	M_p (kg/mol)	$C_{mic} \times 10^3$ (mol/l)	$A_2 \times 10^3$	$V_p \times 10^{26}$ (m^3)	φ %
0	83	43.1	5.73	1.69	13.0	7.4
1.32	87	45.6	5.47	1.13	9.8	5.4
2.33	90	47.8	5.29	0.98	9.3	4.9
3.16	94	50.5	5.06	0.91	9.6	4.8
3.78	96	52.0	4.96	0.68	7.6	3.8
4.93	105	57.5	4.53	0.66	9.1	4.1

The experimental results support the colloid crystal model of the stratification proposed in [3, 6]. According to this model, the step-wise thinning of films from surfactant solutions is due to a layer-by-layer destruction of a structure of spherical micelles formed in the film. We established experimentally that the mean height of a step and the diameter of a nonionic micelle are approximately equal. It also turns out that the main parameter governing the stratification is the volume fraction φ of the colloid particles (the micelles) in the solution. All factors causing reduction of φ (increase of temperature or electrolyte concentration, or decrease of surfactant concentration) lead to depression or total disappearance of the step-wise transitions and to acceleration of the film drainage. Alternatively, all factors leading to increase of φ favor the formation of stable colloid crystal structures inside the film. This concept allows a unified interpretation of the observed effects, both with nonionic and with anionic surfactant solutions.

The practical importance of the studied effects is the following. By investigating the effect of different additives on the stratification one can predict their influence on stability of the respective microheterogeneous systems. When stable dispersions are needed one can utilize the property of nonionic surfactants to form stable, thick films containing several micellar layers at room temperatures (cf. Figs. 5 and 6). For example, this effect can find an application in enhanced tertiary oil recovery [19]. In other cases the opposite effect, that of destruction of the microheterogeneous system and phase separation might be needed. This aim can be achieved by enhancing some factor that reduces the effective micellar volume fraction and depresses the stratification.

Acknowlegements

This work was supported in part by the National Science Foundation, the U.S. Department of Energy, and the Bulgarian Committee for Science and Education.

References

1. Becker P (1967) In: Shick N (ed) Nonionic surfactants. Marcel Dekker, New York, pp 478–515
2. Shinoda K, Nakagava T, Tamamashi B, Isemura T (1963) In: Colloidal surfactants — some physical properties. Academic Press, New York, pp 118–200
3. Nikolov AD, Wasan DT, Kralchevsky PA, Ivanov IB (1988) In: Ise N, Sogami I (eds) Ordering and organisation in ionic solutions. World Scientific, London, pp 302–314
4. Johnnot ES (1906) Phil Mag 70:1339–1341
5. Bruil HG, Lyklema J (1971) Nature 232:19–21
6. Nikolov AD, Wasan DT (1989) J Colloid Interface Sci 133:1–12
7. Nikolov AD, Kralchevsky PA, Ivanov IB, Wasan DT (1989) J Colloid Interface Sci 133:13–22
8. Mitchell DJ, Tiddy GJT, Waring L, Bostock T, Mc Donald MP (1983) J Chem Soc Faraday Trans I 79:975–992
9. Degiorgio V, Corti M (1984) In: Mittal KL, Lindman B (eds) Surfactants in solution, vol. 1. Plenum Press, New York, pp 471–502
10. Rosch M (1967) In: Shick N (ed) Nonionic surfactants. Marcel Dekker, New York, pp 753–796
11. Kralchevsky PA, Nikolov AD, Wasan DT, Ivanov IB (1990) Langmuir — in press
12. Sigter D, Mysels KJ (1955) J Phys Chem 59:45–68
13. Nakagava T, Kuriyama K, Inone H (1959) Symposia on colloid chemistry (Chem Soc Japan) 12th Symposium: pp 29–37
14. Napper DH (1977) In: Kerker M, Zettlemoyer A, Rowell R (eds) Colloid and interface science, Vol. 1. Academic Press, New York, pp 413–430
15. Woods DR, Diamandopulos E (1988) In: Wasan DT, Ginn ME, Shah DO (eds) Surfactants in chemical process engineering. Marcel Dekker, New York, pp 369–539
16. Becher P (1962) J Colloid Sci 17:pp 325–351
17. Lange H (1967) In: Shick N (ed) Nonionic surfactants. Marcel Dekker, New York, pp 224–253
18. Nakagava T, Kuriyama K, Inone H (1960) J Colloid Sci 15:268–277
19. Wasan DT, Nikolov AD, Huang DD, Edwards DA (1988) In: Smith (ed) Surfactant-based mobility control: Progress in miscible flood enhanced oil recovery. ACS Symposium series 373, Washington DC, pp 312–324
20. van Megen W, Snook I (1984) Adv Colloid Interface Sci 21: pp 119–228

Authors' address:

Professor I. B. Ivanov, D. Sc.,
University of Sofia,
Faculty of Chemistry
1 Anton Ivanov Blvd.
Sofia 1126, Bulgaria

The structure of surfaces of surfactant solutions above the critical micelle concentration

E. M. Lee, E. A. Simister, R. K. Thomas and J. Penfold*)

Physical Chemistry Laboratory, South Parks Road, Oxford, England

Abstract: Specular reflection of neutrons from surfactant solutions above their critical micelle concentrations indicates that layer structures are formed in the vicinity of the surface. The mean spacing of the structures varies with chainlength in a way that suggests the presence of surfactant bilayers.

Key words: Neutron reflection; adsorption; surface multilayers

1. Introduction to neutron reflection

Neutrons are specularly reflected from the interface between two media if the neutron refractive index (a quantity which is simply related to composition) of the two are different. Total external reflection occurs at the interface between a less dense and a more dense medium below a certain incident angle known as the critical angle, which also depends on the compositions of the two phases. This angle is usually small, typically less than a degree. For incident angles greater than the critical angle, the reflectivity decays sharply in a way which depends on the distribution of scattering matter in the surface region, the scattering-length density profile. The calculation of specular reflectivity profiles can be done exactly for any model scattering-length density profile using the optical matrix method for light polarised perpendicular to the plane of reflection [1]. Since the scattering-length density is directly determined by the composition profile the match between experimental and observed profiles provides a means of assessing the validity of any structural model of the interface. Throughout this paper we use the optical matrix method. Fuller details are given in [1, 2].

A particularly powerful tool in elucidating surface structure using neutron reflection is isotopic substitution. It so happens that the scattering lengths of the proton and the deuteron are of opposite sign. Therefore, by suitable mixing of protonated and deuterated materials, null scattering solvents or solutes may be produced. By judicious use of isotopic substitution the sensitivity of the reflected signal to a certain portion of the surface structure, for example, the adsorbed layer, adsorbed counterions, or solvent, may be enhanced. As in small-angle neutron scattering (SANS) this has the effect of simplifying the structural analysis and increasing the spatial resolution of the experiment [3].

The versatility of this technique has been demonstrated by experiments on surfactants and polymers adsorbed at the air/solution interface [2–4] and at the quartz/water interface [5, 6].

2. Adsorption of surfactant at the surface of solutions

Adsorption at the surface of a surfactant solution is manifested by a fall in the surface tension γ. The decrease in the surface tension is related to the amount adsorbed by the Gibbs equation

$$d\gamma/d\ln a = -\Gamma RT , \quad (1)$$

where a is the activity of the surfactant in the solution, and Γ is the surface excess measured with respect to the solvent. Measurement of surface tension in conjunction with the Gibbs equation is commonly used to follow the adsorption of a surfactant

*) Rutherford Appleton Laboratory, Chilton, Didcot, Oxon, England.

up to the critical micelle concentration (CMC). As with the other main method for determining adsorption at the air/solution interface — radiotracer measurement — the surface-tension method gives no structural information about the adsorbed layer. At the CMC there is a sharp change in the variation of γ with $\ln a$, and above the CMC both γ and $\ln a$ change much more slowly with concentration. When both $d\gamma$ and $d\ln a$ are close to zero it becomes very difficult to apply the Gibbs equation to determine the surface excess of surfactant. Cutler et al. [7] have attempted to apply the Gibbs equation to data for sodium dodecyl sulphate (SDS) and found that the surface excess of surfactant continued to increase with concentration above the CMC.

3. Experimental details

The experiments were all done on the reflectometer CRISP at ISIS which has been fully described elsewhere [8]. The reflectivity was measured over a limited range of momentum transfer κ (= $4\pi \sin\theta$ is the scattering angle and λ is the wavelength of the neutrons), from 0.05—0.4 Å^{-1}, and calibrated by reference to D_2O. However, because some of the features observed in the present work were unusual, we also measured a number of complete profiles down to values of the momentum transfer below the critical angle, which for D_2O is at about 0.017 Å^{-1}.

The samples were contained in Teflon troughs which had been cleaned by prolonged soaking in heptane to remove organic materials, followed by soaking in concentrated HNO_3 containing 4% HF. The troughs were rinsed in clean water (Elga Elgastat) before use. The surface of the liquid was 70 × 250 mm, sufficiently large so that only about one-third of the liquid was illuminated by the beam and there were no edge effects. The troughs were enclosed in airtight containers to prevent evaporation. The windows of the containers were quartz (Suprasil) which is transparent to neutrons. All the surfactants were purified immediately before the experiment, either by recrystallization in the case of the CTABs [3] or by Soxhlet extraction with heptane in the case of SDS and LiDS.

4. Specular reflection from surfactant solutions above the CMC

Figure 1 compares the reflectivity profile for a solution of protonated $C_{10}TAB$ in D_2O at a concentration of 0.1 M with the reflectivity profile for pure D_2O. The CMC of $C_{10}TAB$ is at 0.065 M [9] so that the amount of surfactant present as micelles is 0.035 M. The reflectivity of the solution is unexpectedly enhanced compared to that of D_2O, as can be seen in Fig. 1a. Since the ordinate in the figure is on a logarithmic scale, the magnitude of the effect is particularly striking. In Fig. 1b we plot the ratio of the reflectivity of the surfactant solution to the profile calculated for the D_2O surface with no surface roughness (the reflectivity of D_2O is somewhat lower than the ideal surface, which would make the enhancement even greater). The ratio plot shows that at a momentum transfer of about 0.16 Å^{-1} the reflectivity is almost doubled. Since the scattering length of the D_2O has been lowered by the addition of surfactant, by about 3%, the reflectivity should also have been lowered. In-

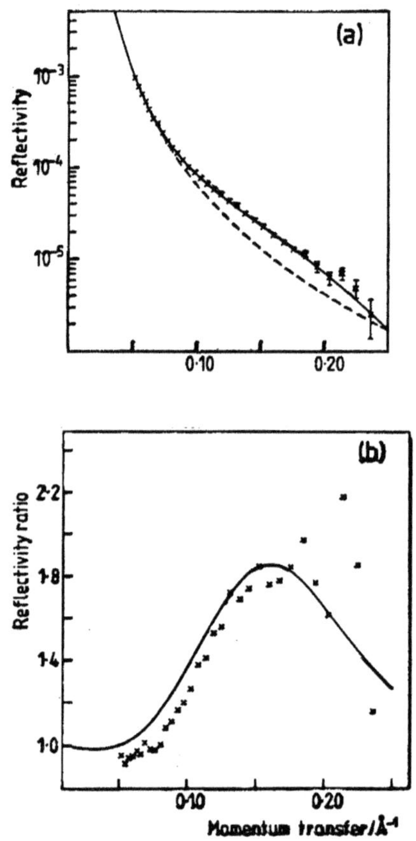

Fig. 1a. Observed (×) and calculated specular reflectivity profiles of a 0.1-M solution of fully protonated $C_{10}TAB$ in D_2O. The model for the calculated profile (solid line) is the three-layer model discussed in the text, and for comparison, the broken line is the calculated reflectivity for a simple D_2O/air interface; *b*. Ratio of the observed reflectivity of the 0.1-M $C_{10}TAB$ solution to that of the simple D_2O/air interface. The continuous line is that for the same three-layer model as in *a*

deed, at concentrations just below the CMC the reflectivity of the surfactant solution was found to be lower than that of D_2O. Something unusual must, therefore, be happening at the surface to cause the enhancement.

Similar measurements on CTABs of different chain lengths showed that enhancement of the reflectivity is related more to the fraction of surfactant present as micelles than to the CMC itself. For example, the CMC of C_{14}TAB is 3.5×10^{-3} M in comparison with 0.065 M for C_{10}TAB [9]. At a concentration of double its CMC C_{14}TAB only enhances the reflectivity of D_2O by a small amount. However, at a micellar concentration the same as used for Fig. 1, i.e. at the CMC + 0.035 M, there is a large enhancement. The enhancement is shown in Fig. 2 for C_{10}, C_{12}, C_{14} and C_{16}TAB. The extent of the enhancement is similar in all cases but the value of κ at which maximum occurs is different, being 0.16, 0.135, 0.115 and 0.100 $Å^{-1}$ respectively. As for C_{10}TAB the reflectivity of solutions below the CMC was lower than that of D_2O.

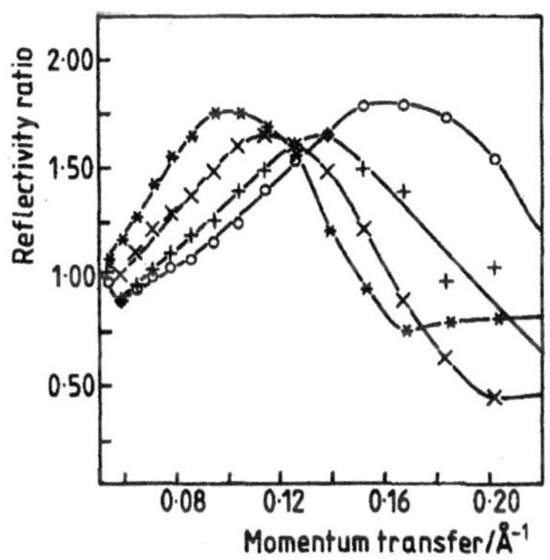

Fig. 2. The observed ratios of the reflectivities of solutions of CTABs to the calculated reflectivity of perfectly smooth D_2O. The concentrations were 0.1 M for C_{10}TAB (o), 0.051 M for C_{12}TAB (+) and 0.0385 M for C_{14}TAB (×), and 0.036 M for C_{16}TAB (*). The solid lines are to guide the eye

This phenomenon is not restricted to cationic surfactants. It has been observed in both SDS and LiDS solutions at about double the CMC (0.008 M for SDS), although the magnitude of the effect is difficult to reproduce accurately. In this case, the reflectivity enhancement is sensitive to pH, to the presence of the hydrolysis impurity dodecanol, and to ionic strength. Other anionic surfactants showing the effect are sodium octyl and decyl sulphate, the fluorocarbon surfactant caesium perfluorooctanoate and sodium decanoate. In Fig. 3 we show the reflectivity profiles of a) protonated sodium decanoate in D_2O which shows the usual supra CMC enhancement of reflectivity, and b) deuterated surfactant in null reflecting water both above and below the CMC. The reflectivity in contrast a) is most sensitive to the distribution of solvent in the interfacial region, whilst in contrast b) it is most sensitive to the distribution of surfactant.

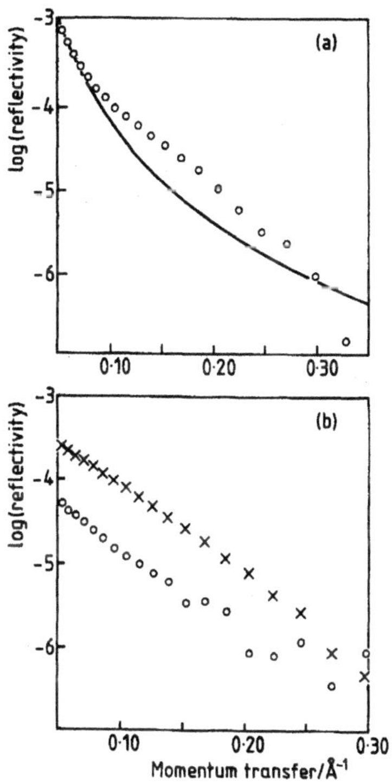

Fig. 3a. The observed reflectivity profile of a 0.024 M solution of protonated Na decanoate in D_2O (CMC = 9.4×10^{-2} M). The solid line, by comparison, is the calculated reflectivity of a simple D_2O/air interface; *b*). The observed reflectivity profiles of deuterated Na decanoate on null reflecting water at bulk concentrations of (o) 7×10^{-2} M and (×) 2.4×10^{-1} M

Non-ionic surfactants also show the effect; for example, we have observed a large reflectivity enhancement in a 0.068 M solution of protonated $C_{12}E_6$ in D_2O (CMC 8.7×10^{-5} M).

5. Discussion

The reflectivity ratio shown in Figs. 1 and 2 is closely related to the form factor of the inhomogeneity at the interface [10]. It was shown in [3] that for simple structures involving no more than three layers a reflectivity enhancement can be brought about only if one of the layers has a scattering-length density greater than that of D_2O. Indeed, a surface structure consisting of a monolayer of surfactant, a layer of D_2O and counterions with scattering-length density about 10% greater than D_2O, and a layer of micelles was found to reproduce the $C_{10}TAB$ data shown in Fig. 1 with great accuracy. The continuous lines in Figs. 1a and b are the result of just such a fit. However, there are three problems with this interpretation. Firstly, it is difficult to understand how the density of such an incompressible fluid as water could increase by such a large factor. Only in the most strongly hydrated salts (for example, involving Li^+ of F^- ions) is the molar volume of water reduced on this scale. Secondly, the variation of the maximum in the reflectivity enhancement shown in Fig. 2 shows that the position of the maximum depends strongly on the chainlength. Unless the thickness of the abnormal water layer correlates perfectly with the chain length of the surfactant, and there seems little reason for it to do so, the model used to fit the data of Fig. 1 cannot be extended to fit the profiles of $C_{12}TAB$, $C_{14}TAB$ and $C_{16}TAB$ in Fig. 2. Finally, for models of the type used to fit the data in Fig. 1, the amount adsorbed Γ is expected to be in the range 20—50% higher than the monolayer just below the CMC. The contrast most sensitive to the total amount of surfactant in the interfacial region is that of deuterated surfactant in null reflecting water. Comparing the reflectivities shown in Fig. 3b for deuterated sodium decanoate below and above the CMC, we immediately see a large enhancement of signal upon passing above the CMC. Fitting the below-CMC data using a uniform density monolayer model gives a surface excess of 2.56×10^{-10} Mol cm^{-2}. Consideration of the level of the reflected signal for the solution above the CMC gives a surface excess of 5.75×10^{-10} Mol cm^{-2}, over twice the amount present below the CMC. This increase is not compatible with the three-layer model.

The amount adsorbed and the strong correlation of the maximum in the reflectivity ratio with chain length suggests that a multilayer structure is occurring, Bragg scattering from layers normal to the surface being responsible for the maximum. Consideration of the reflectivity ratio profiles shown in Fig. 2 leads us to the following qualitative conclusions about the nature of the multilayer structure:

i) Table 1 shows the spacings perpendicular to the interface obtained from the positions of the reflection ratio maxima and the estimated extended chain lengths for most of the surfactants mentioned above [11]. It is clear that the spacing correlate closely with the expected bilayer thickness. It therefore seems that the surface structure may involve a series of bilayers.

ii) The overall amount adsorbed is a little over twice that of a monolayer. In order to make this observation consistent with the observed spacings normal to the surface, either the surface structure is that of a series of highly defective bilayers or of layers of discrete aggregates of similar thickness to bilayers, for example, discoid micelles.

iii) The Bragg peak is very broad, suggesting that there is considerable variation in the spacing between successive layers.

Table 1. Comparing the fully extended chain length of some surfactants [11] with the perpendicular spacing obtained from neutron reflection measurements on the assumption of a multilayer structure (see text)

Surfactant	No. of carbon atoms	Extended chain length (Å)	Experimental spacing (Å)
Na octyl sulphate	8	11.5	33
$C_{10}TAB$	10	14.1	40
Na decanoate	10	14.1	40
$C_{12}TAB$	12	16.7	46
SDS	12	16.7	44
$C_{14}TAB$	14	19.3	55
$C_{16}TAB$	16	21.9	63

The most probable physical model is that of domains consisting of a series of bilayers beneath a surface monolayer. Within each domain there is a gradual transition from ordered bilayer to bulk micellar solution. This might account for the variation in spacing between successive layers required to explain the large width of the diffraction peak.

References

1. Lekner J (1987) The Thery of Reflection, Martinus Nijhoff, Dordrecht
2. Bradley JE, Lee EM, Thomas RK, Willatt AJ, Penfold J, Ward RC, Gregory DP, Waschkowski W (1988) Langmuir 4:821
3. Lee EM, Thomas RK, Penfold J, Ward RC (1989) J Phys Chem 93:381
4. Rennie AR, Crawford RJ, Lee EM, Thomas RK, Crowley TL, Roberts S, Qureshi MS, Richards RW (1989) Macromolecules 22:3466
5. Lee EM, Thomas RK, Rennie AR, Penfold J, Staples EJ, Cummins PG (1989) Chem Phys Lett 162:196
6. Rennie AR, Lee EM, Thomas RK, Simister EA (1990) Langmuir 6:1031
7. Cutler SG, Meares P, Hall DG (1978) J Chem Soc Far I 74:1758
8. Penfold J, Ward RC, Williams WG (1987) J Phys E 20:1411
9. Mukerjee P, Mysels KJ (1971) Critical Micelle Concentrations of Aqueous Surfactant Systems, U.S. Department of Commerce
10. Crowley TL (1982) D Phil Thesis, University of Oxford
11. Tanford CJ (1972) J Phys Chem 76:3020

Authors' address:

E. M. Lee
Physical Chemistry Laboratory
South Parks Road
Oxford OX1 3QZ, England

Lipid phase transition in monolayers [polymorphism]

K. S. Birdi**), R. Sanchez**) and N. Krog*)

*) Grindsted Products, Aarhus, Denmark
**) Fysisk-Kemisk Institut, Technical University, Lyngby, Denmark

Abstract: Spread monolayers of different kinds of lipids, e.g., fatty alcohols, fatty acids, and alkyl [mono- & di-] glycerides at air -water interface, have been analyzed regarding the thermodynamics of phase transition (melting of monolayer). — There were found characteristic differences which allowed to classify the spread lipid monolayer melting at the air-water interface into four distinct groups: A) fatty acid monolayers melted at the same temperature as found from calorimetry; B) fatty alcohol monolayers showed to distinct melting points: one transition-point temperature was the same as the melting point determined by calorimetry method, while the other transition was observed at ca. 26 °C lower than the former; C) mono-glyceride monolayers exhibited only one transition temperature, but at ca. 26 °C below their calorimetry melting point; and D) di-alkyl lecithin monolayers exhibited transition temperatures which were very different than the data for other lipid monolayers.

Key words: Monolayers; fatty alcohols; fatty acids; alkyl [mono- & di-] glycerides; air-water interface; phase transition thermodynamics

Introduction

The spread monolayers of lipids at air-water (or oil-water) interface are known to exhibit very characteristic phase transition at their critical temperature T_{cr} [1—3].

The lipid phase transition temperature, as determined from these data, has been the subject of much investigation since the monolayers are found to be very suitable model membrane systems.

The lipid monolayers of long chain alcohols, fatty acids, and lecithins are analyzed here since both the data and the analyses as reported in literature were found to not be consistent with theory. The lipid phase transition temperature and enthalpy, as determined from monolayer studies, is compared with the calorimetric data.

However, before describing the monolayer data it is necessary to discuss the lipid polymorphism (and lyotropism).

Phospholipid Polymorphism

In earlier studies during the mid- and late 1960s, the polymorphic nature of phospholipids, i.e., their lyotropic and thermotropic phase properties were investigated by x-ray diffraction techniques [4—6]. Phase diagrams constructed as a function of temperature (thermotropism) and water content (lyotropism) clearly indicated that the dominant phase is to be considered as equivalent to the biological membrane, i.e., the bilayer phase was the smetic phase. It was also found that this class of lipids, i.e., two-alkyl chain amphiphile, exhibited other phases which could also be expected in the membrane structures.

This observed transition was, of course, the phase transition from their solid or gel state (solid in three-dimensions) to the fluid or liquid crystalline state (fluid in the planar of the lipid bilayer, but having some solid-like characteristics in the plane normal to the bilayer [7].

This transition, as studied by x-ray diffraction, shows a change in distance between alkyl chains (carbon-carbon) from 4.2 Å (= 0.42 nm) to 4.6 Å as the membrane goes from the solid to the liquid phase. Assuming hexagonal packing, this transition gives an area/molecule change of 30.55 $Å^2$ [= 3055 nm^2] to 37 Å (one can assume for phospholipids that the magnitude of area per alkyl chain is the same for each chain).

The molecular dynamics (CNDO/2) investigations on the confirmation of the α-chain of lecithin was found to indicate a strong preference for a gauche-gauche arrangement about the phospho-diester group [8]. Three levels of water-binding energies were estimated. The highest binding of water was to the unesterified phosphate oxygens. By assuming complete incorporation of 9 water molecules into a lipid structure, a plausible lecithin-water geometry was deduced for a liquid crystalline system.

In a recent study the data on transitions and molecular packing in highly purified DPL — water phases was described [9]. The purity criteria is important when considering that the complex phase diagrams of these systems are very sensitive to even minute impurities. A new procedure was used in order to purify DPL. A sophisticated method of solvent crystallization was used which gave a large supply of single crystals of purity >99.94%. The single crystals were studied by x-ray crystallography. The thermal analyses of DPL-water systems is given in Table 1. As already observed, the effect of water on DPL phases is striking and complex. The adsorbed water in DPL can only be removed upon heating to 200 °C. Thus, anhydrous DPL can only be obtained after prolonged exposure to vacuum. When excess water is present, as in phases 10, 11, and 12 (Table 1), the polar groups of different bilayers do not interact appreciably and in each layer the charged groups will pack in contact within a planar array. The forces between polar groups across each bilayer will tend to compress the bilayer. This will give rise to the reduction in bilayer thickness above the main transition, in phase 12, when the forces between the alkyl-chains are weak.

Phase transition temperature as determined by different monolayer spreading methods

In the current literature many procedures other than the Langmuir film balance method are used to form monolayers at the air-water interface. These different methods are described below.

π vs A isotherms: These isotherms are obtained by the Langmuir film balance, as described elsewhere [3]. These isotherms will, therefore, be expected to resemble the three-dimensional P vs V isotherms.

π vs surface concentration (C_s) isotherms: These isotherms are obtained by applying an increasing amount of the film-forming substance (such as a solution or in liquid or solid form) to a surface of constant area (generally 100 cm^2). These π vs C_s isotherms are thus different than those obtained as π vs A, since in the former the surface concentration C_s is varied.

π_{eq} (equilibrium surface pressure): The value of the equilibrium surface pressure π_{eq} is obtained when a bulk solid or liquid lipid is brought into equilibrium with the monolayer film at the surface of a liquid (or the oil-water interface) [3, 10, 11]. The magnitude of π_{eq} was only found to be the same as the π_{col} when the bulk phase is in liquid form [3].

Accordingly, the lipid phase transition data as obtained from these different spreading methods will be expected to be different under varying experimental conditions [3].

Thermodynamics of lipid phase transition (from π_{eq} vs T)

[Π_{eq}] of different lipid monolayers: In the study of phase transitions the Clausius-Clapeyron equation is the most fundamental thermodynamic relationship. In the monolayers the heat of spreading Q_s of an insoluble film-forming is related to the temperature coefficient of its π_{eq} as follows:

$$d\pi_{eq}/dT = Q_s/T\,[A_{ex} - A_{co}]\,, \tag{1}$$

where A_{ex} is molar area of spread monolayer and A_{co} (which is negligible to A_{ex}) is the molar area of solid or liquid bulk phase, placed at the interface. This formula has been used by various investigators [12—15].

The derivation of the Clausius-Clapeyron equation can be carried out under more explicit grounds [3, 16]. The interfacial region denoted by s is a mixture of n_w moles of subphase, i.e., water (w) and n_i moles of lipid (i), separated mathematically from the adjacent bulk phases by boundaries drawn as dotted lines in Fig. 1.

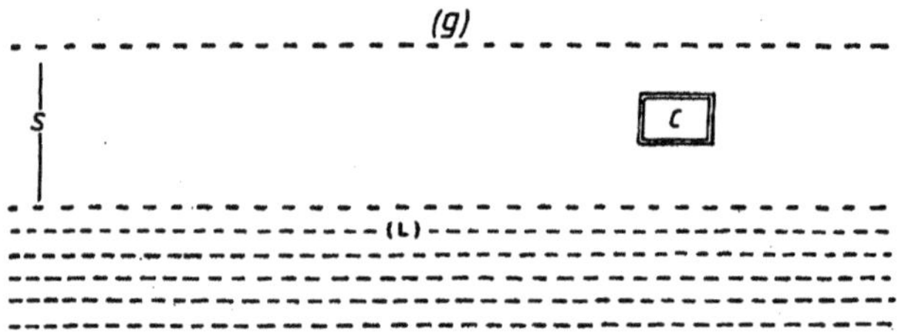

Fig. 1. Mathematical separation of interfacial region s from the gas phase g and the liquid phase L. Bulk crystal phase is c [3]

The thermodynamic relations will be independent of the location of the boundaries as long as they enclose the entire region where deviations from the bulk phase properties are to be expected [2, 3, 17].

At equilibrium the intensive quantities will be uniform throughout the system. If the temperature T is changed by dT, then the equilibrium will be re-established and the changes in the intensive properties of the system will be related by the Gibbs-Duhem equation:

$$S^s dT - V^s dP + A\,d\tau_{eq} + n^s_w d\,\mu_i + n_i d\mu_i = 0 \quad (2)$$

$$d\mu_w = -\bar{S}^L_w dT + \bar{V}^L_w dP \quad (3)$$

$$d\mu_w = -\bar{S}^g_w dT + \bar{V}^g_w dP \quad (4)$$

$$d\mu_i = -\bar{S}^c_i dT + \bar{V}^c_i dP\,, \quad (5)$$

where S^s and V^s are the total entropy and volume of this interface, respectively, of area A, τ_{eq} is the equilibrium surface tension, P is the vapor pressure of w, and μ_w and μ_i are the chemical potentials of w and i. The entire system is composed of: vapor g, liquid phase L, crystal c, and interface s.

In the above equations the bars denote the partial molar quantities. By combining these equations and neglecting the pressure changes one obtains:

$$d\tau_{eq}/dT = -[S^s - n^s_w \bar{S}^L{}_w - n^c_i \bar{S}^c_i]/A\,, \quad (6)$$

which is the Clausius-Clapeyron equation relating the temperature coefficient of the surface tension τ of the film-covered surface. The enthalpy change can be written as

$$d\tau_{eq}/dT = -[H^s - n^s_w \bar{H}^L_w - n^s_i \bar{H}^c_i]/TA \quad (7)$$

$$= \blacktriangle H_m/TA\,. \quad (8)$$

The numerator in above equation, $\blacktriangle H$, is the enthalpy of formation of a film-covered interface from its bulk constituents, i.e., it is the enthalpy change involved in bringing n^s_w moles of substrate from the bulk phase L and n^s_i moles of film-forming molecules from its bulk phase (crystal c, or liquid) into the interface s (Fig. 2).

From similar considerations the enthalpy of formation of a pure substrate interface is given as

$$d\tau_0/dT = -[H^s_0 - n^s_0 \bar{H}^L_w]/TA\,, \quad (9)$$

where τ_0 and H^s_0 are the surface tension and the enthalpy of a pure liquid interface, respectively. Further, if

$$\pi_{eq} = \tau_0 - \tau_{eq}\,, \quad (10)$$

then one obtains

$$d\pi_{eq}/dT = d\tau_0/dT - d\tau_{eq}/dT\,. \quad (11)$$

From these equations one can derive

$$TA[d\pi_{eq}/dT] = \{[H^s - H^s_0] - [n^s_w - n^s_0] \times \bar{H}^L_w - n^s_i \bar{H}^c_i\}\,. \quad (12)$$

This is the enthalpy of formation of a film-covered interface of area A minus the enthalpy of equivalent area of pure substrate. Thus, the quantity Q_s actually refers to a two-step isothermal reversible process:

1) n^s_w moles of w and n^s_i moles of i are transferred from their bulk phases L and c to the interface s, and mixed; and 2) n^s_0 moles of w are transferred from a clean interface s to L. The heat of fusion $\blacktriangle H_f$ of the crystal can be derived as

$$\blacktriangle H_f = [\bar{H}^l_i - \bar{H}^c_i] \quad (13)$$

$$= \{[H^s_l - H^s_c] - \bar{H}^L_w[n^s_{w,l} - n^s_{w,c}] + TA_m[[d\pi^c_{eq}/dT] - [d\pi^l_{eq}/dT]]\}/n^s_i\,. \quad (14)$$

If the films in equilibrium with lipid crystals and liquid at the melting point T_m are identical, then $H^s_l = H^s_c$ and $n^{l,s}_w = n^{c,s}_w$, from which one obtains

$$\blacktriangle H_f = T_m/A_m[[d\pi^c_{eq}/dT - [d\pi^l_{eq}/dT]]\,, \quad (15)$$

where A_m is the molecular area at π_{eq} at T_m, and $A_i = n^s_i/A_m$. The magnitude of A_m can *only* be

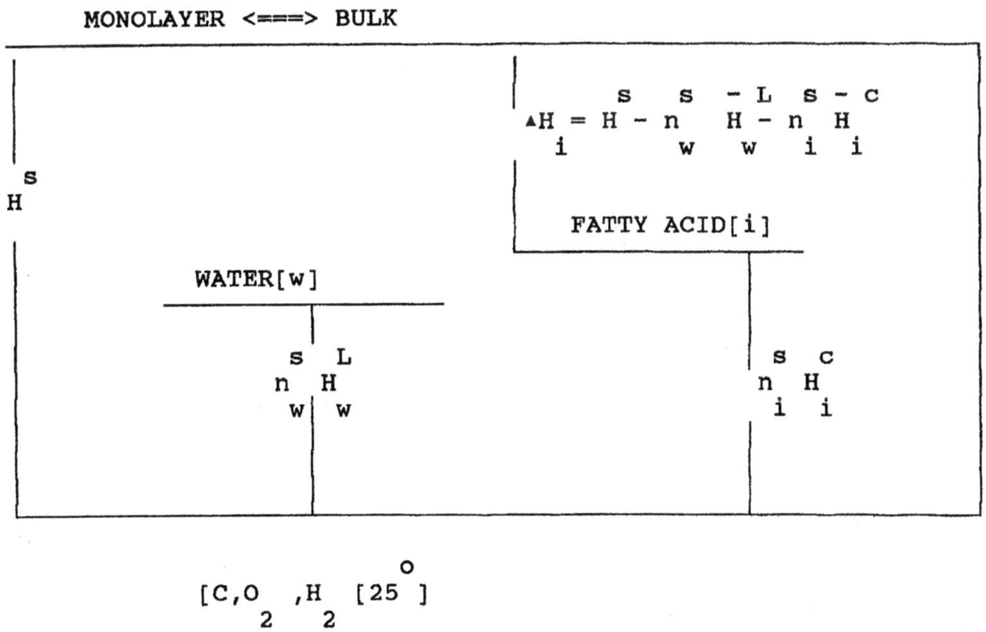

Fig. 2. Monolayer formation from the elements of monolayer and the bulk phase [3]

estimated from π vs A isotherms [3]. This relation has been derived rigorously when the lipid crystal and liquid are in equilibrium with identical films. The magnitude of $\blacktriangle H_f$ can also be determined by independent calorimetric method, which allows one to verify the relation in Eq. (15).

The formation of the film-covered interfaces from the bulk phase constituents always involves an absorption of heat. The spreading process from the crystalline lipid is always an endothermic process, but spreading from molten or liquid crystalline lipid is exothermic. Further, the magnitude of spreading enthalpy was reported to be ca. half of $\blacktriangle H_f$ for trilaurin, while for the case of di-C_{12}-PE monolayers these two quantities were almost equal [18]. These comparative data allow one to conclude that fusion of trilaurin gives rise to a three-dimensionally disordered melt, while chain-melting in the case of PE occurs in the two-dimensional bilayer. This is supported by the finding that the heat of chain-melting in mesomorphic structures are smaller by a factor ca. two than those of ordinary fusion processes [19].

Results and Discussion

Phase transition temperature and the enthalpy of lipid monolayers (spread at air-water interface)

The equilibrium surface pressure (π_{eq}) data of different lipids, e.g., fatty acids, fatty alcohols, di-alkyl and trialkyl lipids has not been critically analyzed in the current literature [3].

Since the data as reported in the current literature on π_{eq} vs T (temperature) for various lipids are not satisfactorily analyzed, these data are described separately for different types of lipids:

1) Fatty acid monolayers

The π_{eq} vs T (temperature) were analyzed very extensively in a few recent studies [3, 10, 11]. These investigations clearly showed that:

- T_{tr} of the monolayer was the same as the bulk melting point T_{mp} (Fig. 3).
- The magnitudes of $\blacktriangle H_f$ were the same for monolayers as found for the bulk lipid by calorimetry of the dry lipid.

2) Fatty alcohol monolayers

Although the π_{eq} vs T plots were the same, in general, as those found for the fatty acids, there was observed an additional break in the plots which was at a lower temperature (ca. 26°) than the T_{mp} (Fig. 4)

— Near the T_{mp} the monolayers showed strange plots, and a weak break at the bulk melting point.

— $\blacktriangle H_f$, as calculated from the monolayer data, were dependent on n_C (Birdi and Sanchez, to be published (Colloid & Polymer Sci.)).

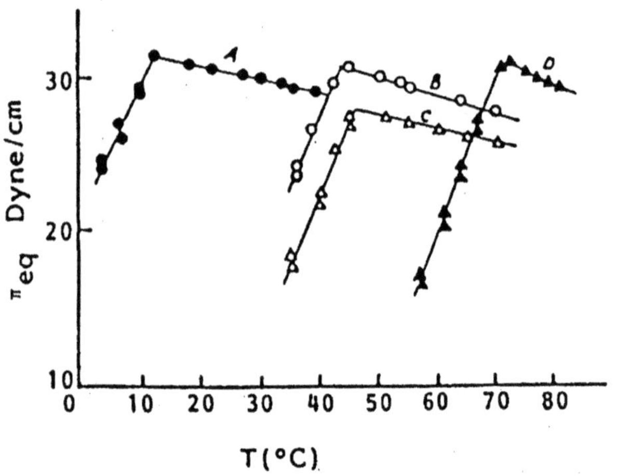

Fig. 3. π_{eq} vs T isotherms of fatty acid monolayer: (A) oleic acid, (B) elaidic acid, (C) 9-stearolic acid, and (D) stearic acid

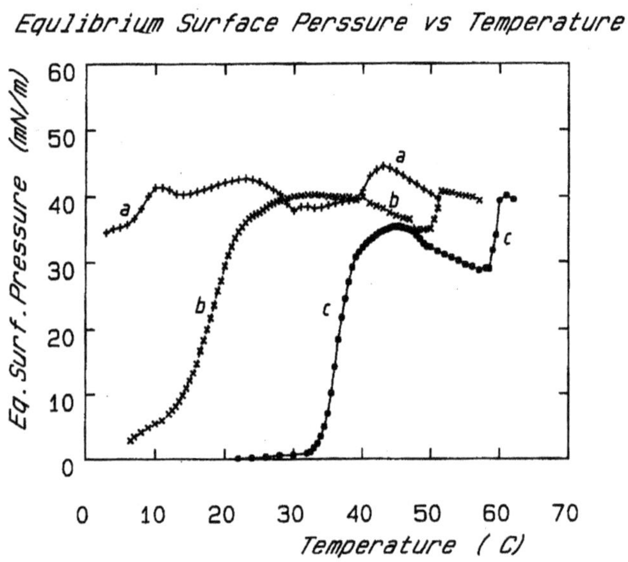

Fig. 4. π_{eq} vs T isotherms of n-alcohol monolayers: a) tetradecanol; b) hexadecanol and c) octadecanol

π_{eq} vs T data for other n-alkoxy propanols (C_{16}-OC_3H_6OH, C_{18}-OC_3H_6OH, C_{20}-C_3H_6OH, C_{22}-OC_3H_6OH) and alkoxy butanols (C_{18}-OC_4H_8OH) π_{eq} vs T plots did not show any breaks [20].

3) Mono-alkyl chain glycerides

π_{eq} vs T isotherms of 1-glycerol-monopalmitate (1-GMP) and 2-glycerol monopalmitate (2-GMP) gave the following results (Fig. 5):

— $T_{tr} < T_{mp}$: It was reported that the break in π_{eq} vs T corresponded to the phase change when 5—10% water is bound in monoglycerides [21].
— $\blacktriangle H_f$ was greater for 1-MP than for 2-MP ([16]; Sanchez and Birdi, to be published (Colloid & Polymer Sci.)).

4) Di-alkyl chain lipids: The di-alkyl lipids can be classified into two sections: i) lecithins, ii) other lipids.

i) Di-glyceride (and tri-glyceride) monolayers (Fig. 6):

— The π_{eq} vs T isotherms show breaks at the fusion temperature. The β-form of tri-laurin melts at 46°C, and the break in isotherm is observed at the same temperature. The same correlation is observed for 1,3 dilaurin [β-form] with melting temperature at 57°C. The di-C_{12}-PE shows a break in π_{eq} vs T = 63°C, which is identical with chain-melting process in the bulk phase.

Lecithins (DML; DPL) [3]:

— The π_{eq} vs T isotherms were completely different from all of those of the lipids as mentioned under 1)—3).
— The magnitudes of $\blacktriangle H_f$, as determined from π_{eq} vs T, were extremely high (2500 kJ/mole) as compared to the calorimeter data (ca. 80 kJ/mole) [2, 3].

ii) Other Lipids: π_{eq} vs T isotherms have been reported for various lipids, e.g., lauryl (C_{12}) glycerides (1,3 dilaurin, trilaurin), phosphatidylethanolamines (PE) (di-C_4-PE, di-C_{12}-PE, di-elaidoyl-PE), lecithins (DML, di-C_{12}-lecithin, dioleoyl lecithin), and cholesterol [18]. In the case of glycerides and PE, the magnitude of $d\pi_{eq}/dT$ was found to be positive below the temperature where monolayer transition took place. It was also observed that longer chain-length lipids spread at higher temperatures. Lecithins were reported to spread only when $T > T_{cr}$ where T_{cr} is the crystal (or gel) to liquid-crystal transition temperature at maximum hydration. For example, DPL was found to give negligible π_{eq} at 22°C even after 24 h. The cholesterol π_{eq} data were completely different from all the other lipids. There was found no change in π_{eq} (= 0) with temperature from 25°—50°C.

It has been asserted that, in principle, π_{eq} measurements as a function of temperature indicate the existence of all phase transitions, and thus provide a thermodynamic basis for calculating the tran-

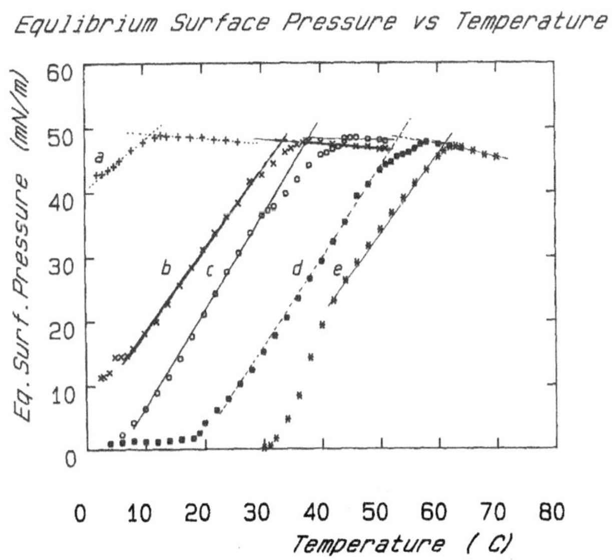

Fig. 5. π_{eq} vs T isotherms of various mono-glyceride monolayers: a) 1-GMO (mono-oleate); b) 1-GME (mono-elaidate): c) 2-GMP (mono-palmitate); d) 1-GMP (mono-palmitate); and e) 1-GMS (mono-stearate)

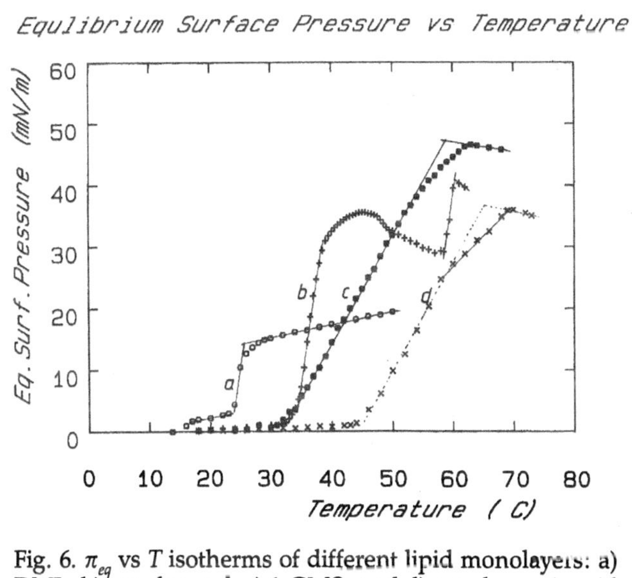

Fig. 6. π_{eq} vs T isotherms of different lipid monolayers: a) DML; b) octadecanol; c) 1-GMS; and d) octadecanoic acid

sition energies. Therefore, it was surprising to find that the π_{eq} vs T plots of DML or DPL or DSL were not in accord with this statement. The same is valid for the data of long chain alcohols.

Especially, the DML data were analyzed in further detail [22] at a higher sensitivity. The coexistence line for the equilibrium of surface film with bulk DML, either gel or liquid crystal, was estimated. The magnitude of T_c was estimated to be 23.5 °C for DML. At temperatures below T_c the bulk gel state is in equilibrium with a gaseous surface film. At temperatures above T_c liquid crystal is in equilibrium with a gaseous film which changes with increasing temperature into a liquid expanded film. The supercooled liquid expanded film line was estimated.

The phase which gives the equilibrium between liquid crystal and liquid expanded film was described as one which exhibits a rapid increase in π_{eq} with temperature. This has not been observed with any of the single chain lipids (to our knowledge). However, one should expect that π_{eq} reaches some finite limiting value at high temperature, i.e., $T > T_c$. In spite of these observations, there is no way by which these data can be analyzed as were the single-chain lipids delineated above.

The data reported by Phillips and Hauser [18] clearly supports this conclusion. The heat of fusion from crystal to melt depends upon the particular polymorphic form of the liquid crystal and the magnitude of $(d\pi_{eq}/dT)$ will reflect this state of equilibrium.

In the case of phospholipids, which are found to swell and take up water only when $T > T_{tr}$, and, thus, the chain melting process for the bulk lipid at the air-water (or oil-water) interface is observed only from crystal to hydrated liquid crystal, and $\blacktriangle H_f$ is, therefore, not the simple heat of chain melting.

However, it is the π vs A isotherms of fatty acid, fatty alcohol, and the di-alkyl lecithin data which provide the correct T_{tr}, and $\blacktriangle H_f$ as well (Table 2). From the π_{eq} vs T plots of DML (Fig. 6) one finds that $\blacktriangle H_f$ = ca. 10 kcal/mole (7.25 kcal/mole by calorimetry) [3]. The π vs A isotherms thus provide the correct enthalpy estimation of the phase transition, while the π_{eq} data is not acceptable.

Lipid chain melting of fatty alcohol and monoglyceride monolayers in anhydrous bulk state

The data in Fig. 7 show that in the case of fatty alcohol monolayers the *first* transition is ca. 26 °C lower than the lipid melting point (T_{mp}). The variation of the first $T_{tr,l}$ is given in Fig. 7.

For comparison the T_{tr} of monoglycerides (where only one transition was observed) is also given. These glycerides are also found to melt at ca. 26 °C lower than their melting point T_{mp}.

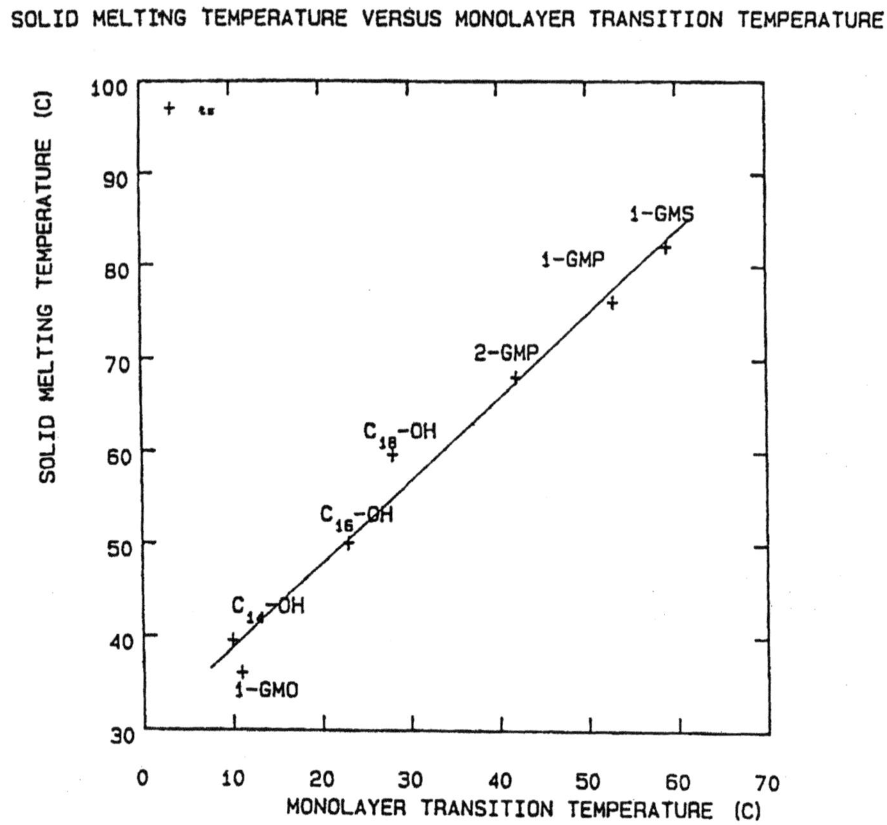

Fig. 7. Plot of melting point of solid (dry) vs monolayer melting temperature of different lipids (see text)

Acknowledgements

RS thanks Academy for Technical Science (A.T.V.) and Grindsted A/S, for a research stipend ("erhuerusforsker") (1986—1988). The excellent technical help by Mrs. J. Klausen and Mrs. G. Hansen is acknowledged.

References

1. Gaines GL (1966) Insoluble Monolayers at Liquid-Gas Interfaces, Interscience Publ, New York
2. Chattoraj DK, Birdi KS (1984) Adsorption & the Gibbs Surface Excess, Plenum Press, New York
3. Birdi KS (1989) Lipid & Biopolymer Monolayers at Liquid Interfaces, Plenum Press, New York
4. Luzzati V (1968) In: Chapman D (ed) Biological Membranes, Academic Press, New York
5. Small DM (1967) J Am Oil Chem Soc 45:3491
6. Reiss-Husson F (1967) 25:363
7. Larsson K (1976) In: Friberg SE (ed) Food Emulsions, Marcel Dekker, New York, chap 2
8. Flaim T, Friberg SE, Plummer PP (1981) J Biol Phys 9:201
9. Albon N (1983) J Chem Phys 78:4676
10. Jalal IM (1978) Ph D thesis, Univ of Wisconsin
11. Sanchez R (1988) thesis ("erhversforsker"), Technical Univ, Lyngby
12. Boyd GE (1958) J Phys Chem 62:537
13. Harkins WD, Nutting G (1939) J Am Chem Soc 61:1702
14. Boyd GE, Schubert J (1957) J Phys Chem 61:1271
15. Brooks JH, Alexander AE (1962) J Phys Chem 66:1851
16. Alexander A, Goodrich FC (1964) J Colloid Interface Sci 19:468
17. Guggenheim EA (1951) Thermodynamics, Interscience, Wiley & Sons, New York
18. Phillips MC, Hauser H (1974) J Colloid Interface Sci 49:31
19. Phillips MC, Williams RM, Chapman D (1969) Chem Phys Lipid 3:234
20. Katti SS, Sansare SD (1970) J Colloid Interface Sci 32:361
21. Lawrence ASC (1969) Mol Cryst Liq Cryst 7:1
22. Gershfeld NL, Tajima K (1979) Nature 279:708

Authors' address:

K. S. Birdi
Fysisk-Kemisk Institut
Technical University
2800 Lyngby, Denmark

Equation of state of mixed monolayers of fatty acid and derivatives of 4-nitroaniline and 2-methyl-4-nitro aniline

R. Sanchez-Morillo[1], S. Rosenkilde[2], A. Holm[2] and K. S. Birdi[1]

[1]) Fysisk-Kemisk Institut, The Technical University of Denmark, Lyngby, Denmark
[2]) Department of General and Organic Chemistry, The H. C. Ørsted Institute, University of Copenhagen, Denmark

Abstract: Surface pressure vs area/molecule (π vs A) isotherms of octadecanoic acid, docosanoic acid and N-octadecanoyl, and N-docosanoyl derivatives of 4-nitroaniline and 2-methyl-4-nitro aniline were measured at 25 °C as a function of composition. — The equation of state for these condensed monolayers was used to analyze the molecular interactions in the mixtures. From these studies it was found that the phase diagram of the mixtures is dependent on the number of carbon atoms in the alkyl chain of fatty acids and on the presence of methyl group in the benzene ring for the abovementioned fatty acid derivatives. The methyl group gave rise to decreased van der Waals interactions which was seen from the significant change in the compressibility in the monolayer isotherms.

Key words: Monolayers; fatty acid; mixed; surface pressure; equation of state

Introduction

The unbranched fatty acids, such as octadecanoic acid (stearic acid = SA) and docosanoic acid (behenic acid = BA), are known to form stable and condensed monolayers (at temperatures below their melting point) at the air-water interface [1, 2]. These fatty acid monolayers are also known to form stable Langmuir-Blodgett [LB] films when transferred to solid substrates [1—3].

The unbranched fatty acid derivatives synthesized here are chosen in the search for amphiphiles possessing possible non-linear optical properties; they include N-octadecanoyl-2-methyl-4-nitro aniline (SAMNA), N-octadecanoyl-4-nitro aniline (SANA), N-docosanoyl-2-methyl-4-nitro aniline (BAMNA), and N-docosanoyl-4-nitro aniline (BANA).

These fatty acid derivatives were found to give very weak LB-film transfer. However, mixed SA + SAMNA, SA + SANA or BA + BAMNA monolayers were found to give stable LB-film assemblies on glass (or quartz) substrates [3].

The purpose of this investigation was to determine the molecular interactions in the mixed fatty acid with some of its derivatives.

The π vs A data of mixed monolayers in the condensed phase region are analyzed with the help of an equation of state in order to determine molecular packing arrangements, as well as interaction forces, as a function of molecular structure.

Materials and methods

Monolayers of synthesized fatty acid anilines octadecanoic acid and docosanoic acid (Sigma, Analytical grade) were spread from their chloroform (Merck, pro analysi.) solutions. The substrate consisted of distilled water that was treated by a Millipore-Q filter system.

Monolayers studies were carried out by using a Langmuir film balance [2] at a temperature of 25 °C. A Sartorius digital micro-balance was used to measure surface pressure (π) by using the Wilhelmy plate method [2]. The precision of this balance was better than +/—0.05 mN/m. The Langmuir trough had dimensions of 55 × 15 × 1 cm, and was constructed of stainless-steel and coated with teflon. The barriers were made of teflon. The Wilhelmy plate was sand-blasted platinum.

Results and disucssion

The π vs A isotherms are known to be very sensitive to the molecular interactions, as determined by the mode of packing in these films [1–9]. The various phase transitions, as determined from such monolayer studies, are shown in Fig. 1.

The liquid-expanded (L_e), liquid-condensed (L_c), and the solid (S) films have been characterized as dependent on the degree of water molecules present in the film [6, 7]. This transition has been ascribed to the sequeezing out of the water molecules until a solid-like film is formed.

The equation of state for the linear part of the isotherm corresponding to the $L_c - S$ or $L_e - L_c$ were analyzed by the following relation [4]:

$$A = \beta - a\pi \,, \tag{1}$$

where the film compressibility is related to the coefficient a.

Mixed films consisting of a derivative containing methyl group in the benzene ring with the corresponding fatty acid, BA + BAMNA and SA + SAMNA, exhibited $L_c - S$ film transitions.

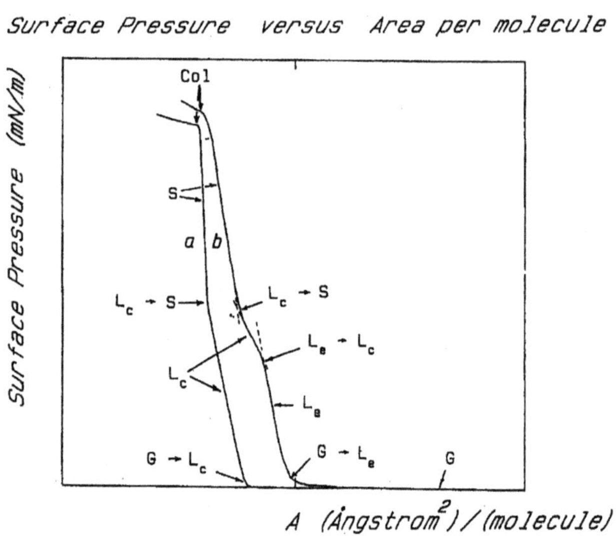

Fig. 1. Surface pressure [π] vs area/molecule [A] for doco- Transition regions [2]: G = gas state, L = Liquid state, L_e = liquid-expanded, L_c = liquid-condensed phase, state, S = solid state, Col = collapse, G-L_e = gas-to-liquid-expanded and $G \rightarrow L_c$ state transition point, $L_e - L_c$ = liquid-expanded to liquid-condensed state transition point, $L_c - S$ = liquid-condensed to solid state transition point

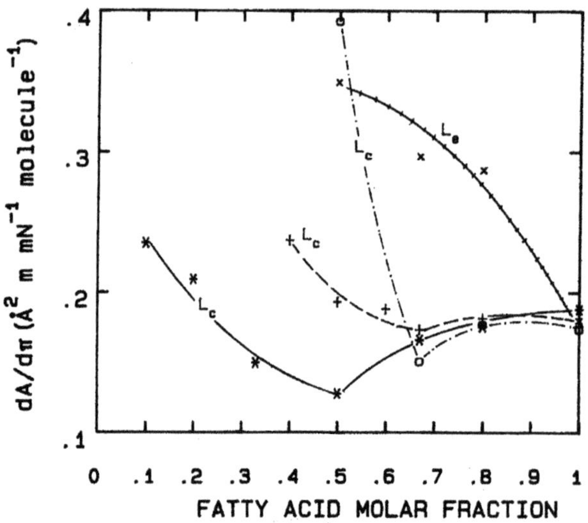

Fig. 2. Variation of compressibility coefficient a with composition (fatty acid molar fraction) for mixtures: (+) = BA + BAMNA, (□) = BA + BANA, and (*) = SA + SANA, (×) = SA + SMNA

The absence of the methyl group in BANA was found to give a more condensed monolayer, as compared to that of BAMNA. The mixed films of BA + BANA π vs A isotherms were found not to differ appreciably from the isotherm for pure BA, when the molar ratio for BA is = <0.5. For mixed films with molar ratio 0.17 and 0.10, the π vs A isotherm show liquid-expanded (L_e) to liquid-condensed (L_c) state transition.

The magnitude of a for the condensed films (i.e., for films where either $L_c - S$ or $L_e - L_c$ transitions were obsered) are given in Fig. 2. These data show that the compressibility is dependent on the composition in the mixed films BA + BANA (minimum at ca. 0.7). In the mixed films of BA + BAMNA the magnitude of a changes to a lesser degree (minimum at ca. 0.7) with composition. In the mixed films of SA + SANA the coefficient a varies with a minimum around mole fraction 0.5. It is also seen that the region within which the L_e or L_c or S phases can co-exist are related to the alkyl chain length, presence or absence of the methyl group and on the composition.

Conclusions

The π vs A isotherms of monolayers of derivatives were found to be very different than the corresponding isotherms for the corresponding fatty acids, e.g., SA and BA.

The difference in the alkyl chain between stearic acid (SA) and behenic acid (BA) was found to have a very strong effect on the compressibility coefficient a of the monolayers of the derivatives.

The methyl group in the MNA moiety of SAMNA and BAMNA was found to give a destabilizing effect. This could be clearly seen from the variation of a with composition in the mixed monolayers. It is shown that molecular interactions in such mixed films in the condensed phase (e.g., liquid-expanded or liquid-condensed or solid) regions can be analyzed with the help of the equation of state, as given by Eq. (1).

Acknowledgements

RSM and SR like to thank the Danish Technical Research Council for generous research support. The authors like to thank Mrs. J. Klausen and Mrs. G. Hansen for their excellent help.

References

1. Gaines GL Jr (1966) Insoluble Monolayers at Liquid-Gas Interfaces. Interscience Publishers, New York
2. Birdi KS (1989) Lipid and Biopolymer Monolayers at Liquid Interfaces. Plenum Press. New York
3. Birdi KS, Sanchez-MR, Rosenkilde S, Holm A, to be published
4. Chattoraj DK, Birdi KS, "Adsorption & the Gibbs Surface Excess", Plenum Press, New York, 1984
5. Matuo H (1978) Bulletin of the Chemical Society of Japan, Vol 51, 690
6. Matuo H (1979) Bulletin of the Chemical Society of Japan, Vol 52(3), 667
7. Langmuir I (1933) J Chem Phys 1:756
8. Adam NK (1941) The Physics & Chemistry of Surfaces 3rd ed, Oxford University Press, London
9. Motomura K et al. (1980) Bull Chem Soc Jpn, 53:2217

Authors' address:

K. S. Birdi
Fysisk-Kemisk Institut
The Technical University of Denmark
Building 206
2800 Lyngby, Denmark

Progress in Colloid & Polymer Science Progr Colloid Polym Sci 82:114—121 (1990)

Equilibrium and dynamic behavior of a system containing a mixture of anionic and nonionic surfactants

F. Mori*), J. C. Lim and C. A. Miller

Department of Chemical Engineering, Rice University, Houston, Texas, USA

Abstract: Mixtures of the pure nonionic surfactant $C_{12}E_3$ and an anionic surfactant of the alkyl ethoxy sulfate type were studied. NaCl brine was added to these mixtures and the phase behavior was determined at 30 °C as a function of surfactant composition and salinity. A large range of compositions was found where the lamellar liquid crystalline phase coexisted with a micellar solution in these oil-free systems. The Phase Inversion Temperature (PIT) was determined as a function of surfactant composition and concentration, salinity, and hardness with n-hexadecane as the hydrocarbon. A new contacting technique was developed to observe with videomicroscopy the dynamic behavior which occurred when aqueous solutions of these surfactants were brought into contact with small drops of n-hexadecane of about 100 μm in diameter. This method more closely simulates the situation during actual detergency processes than that employed previously, and enables the entire solubilization process to be observed, including the final stages. Solubilization of hydrocarbon into a middle-phase microemulsion and excellent hydrocarbon soil removal were observed near the PIT, as in our previous work with nonionic surfactants. Thus, our previous conclusions about the solubilization-emulsification mechanism of hydrocarbon detergency are confirmed for surfactant mixtures similar to those employed in many current products.

Key words: Detergency; phase behavior; solubilization

Introduction

Mixtures of surfactants are the rule in industrial applications. One reason is that commercial processes for making surfactants invariably produce a mixture of individual species. For example, commercial nonionic surfactants of the ethoxylated alcohol type contain compounds with a range of ethylene oxide chain lengths and, usually, hydrocarbon chain lengths as well. In addition, different commercial surfactants are frequently mixed deliberately to improve process performance. For these reasons considerable research attention has been given to surfactant mixtures in recent years as may be seen, for instance, in the collection of papers edited by Scamehorn [1].

*) Present Address: Fuji Photo, Japan.

With the widespread use of fabrics containing polyesters and other synthetic fibers, household detergents have shifted from products whose active compounds were all anionic, to those containing mixtures of anionic and nonionic surfactants. In this paper we describe work on mixtures of a pure nonionic surfactant and a commercial anionic surfactant of the alkyl ethoxy sulfate (AES) type. We discuss equilibrium phase behavior of systems containing these surfactants and water, both with and without added hydrocarbon. We also describe the dynamic behavior which occurs when surfactant-water mixtures are brought into contact with oil in the absence of external agitation. A new contacting technique described below was used for this work. Finally, we show that the maximum removal of a hydrocarbon soil from polyester/cotton fabric occurs near the system Phase Inversion Temperature (PIT), a result which is consistent with but extendes our previous results for pure nonionic surfactants [2] and mixtures of pure nonionic surfactants [3].

Experimental

The pure nonionic surfactant n-dodecyl trioxyethylene monoether ($C_{12}E_3$) was obtained from Nikko Chemical Company, Tokyo, Japan, and was used as received. Neodol 23-3S, supplied by Shell Development Company, Houston, TX, USA is a commercial alkyl ethoxy sulfate made up of various individual compounds whose hydrocarbon chains contain 12 or 13 carbon atoms and whose ethylene oxide chains contain, on the average, three ethoxy groups. The inorganic salts and hydrocarbons used were of reagent grade, and the water used for solution preparation was deionized and double distilled. Radioactively tagged n-hexadecane for the detergency experiments was obtained from Amersham, Arlington Heights, IL, USA and had a radiochemical purity exceeding 97%.

The polarized light screening system described previously [4] was used for determining macroscopic phase behavior. Interfacial tensions were measured with a spinning drop tensiometer.

The soil removal experiments were performed at Shell Development Company by Shell personnel. Experimental details of the radiotracer technique have been described by Benson et al. [5]. The fabric used was 65/35 polyester/cotton, the pH was adjusted to between 7 and 8 using triethanol amine, and washing time was 12 min.

A new contacting technique was developed to observe the dynamic behavior of small, individual drops of oily soils immersed in a large quantity of surfactant solution. With this technique, which differs from that of our previous contacting studies where comparable volumes of oil and surfactant solution were employed [2, 6], the latter stages of solubilization, potentially a significant part of the detergency process, can be observed. Moreover, differential partitioning effects, which are inherent for systems containing mixed soils and/or surfactants and are dependent on the relative amounts of surfactant solution and soil contacted, can be made close to those expected in practical systems.

Samples of the surfactant solutions to be studied were introduced into reactangular optical glass capillaries (Vitrodynamics, Inc., Rockaway, NJ, USA) by capillary action. These capillaries were 50-mm long and 4-mm wide and had an optical path length of 200 μm. The capillaries were filled and then sealed at one end and attached to standard microscope slides with an ultraviolet-light-sensitive polymer adhesive which was cured with an ultraviolet fiber-optic gun. A very thin hypodermic needle (0.21 mm OD, Hamilton Company, Reno, NV, USA) was used to inject a few oil drops into the surfactant solution. This procedure gives good control of the number of drops and drop size. Drops about 100 μm in diameter were injeced in the present study.

Temperature control for the samples was obtained with a Mettler FP-52 controller and FP-5 microscope hot stage. The thermal stage was modified and placed on top of a new X-Y positioner (Nikon, Inc.) so that drops could be observed during and immediately after injection, while the sample was maintained at constant temperature.

Results

Phase behavior in oil-free systems

Figure 1 shows the phase behavior of mixtures of anionic-nonionic surfactant combinations with aqueous NaCl solutions over a wide range of compositions with temperature and surfactant concentration fixed at 30 °C and 5.4 wt%, respectively. With only the anionic surfactant Neodol 23-3S present, a single micellar solution L_1 is seen, according to Fig. 1, except at high salinities where coexistence of L_1 and brine B occurs. Addition of the lipophilic nonionic surfactant $C_{12}E_3$ causes the lamellar liquid crystal L_α to form, with less nonionic needed to produce the transformation at higher salinities. The

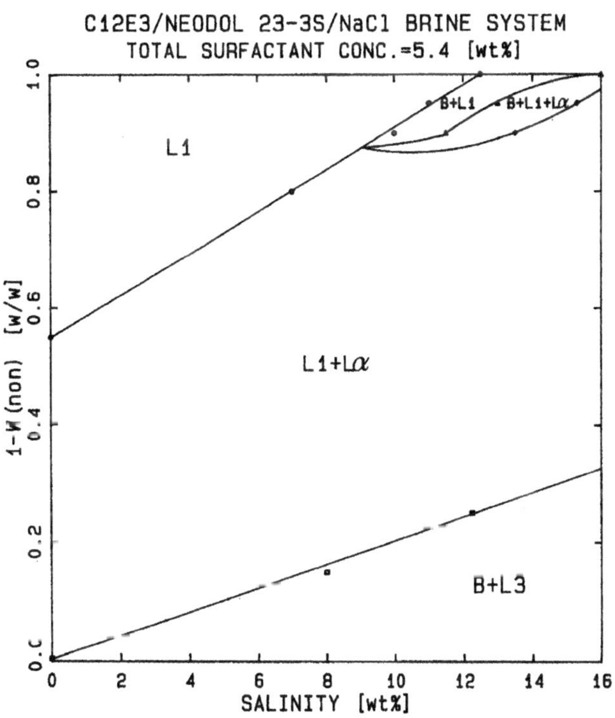

Fig. 1. Phase behavior of mixtures of $C_{12}E_3$, Neodol 23-3S, and NaCl brine with temperature and total surfactant concentration fixed at 30 °C and 5.4 wt%, respectively

latter result is expected since NaCl makes both surfactants less hydrophilic. A three-phase region $B + L_1 + L_\alpha$ is seen in the upper right portion of the diagram.

Since addition of both salt and nonionic surfactant act to make the anionic surfactant solution less hydrophilic, one may ask why the former produces separation into two isotropic phases while the latter produces separation into a lamellar liquid crystal and an isotropic phase. Basically, the added salt has two effects. In the first place, it decreases the electrical repulsion between adjacent ions in a micelle and thus causes a reduction in the area per surfactant molecule. Taken alone, this effect causes the micelles to become anisotropic. If sufficiently large, it produces formation of the lamellar phase. However, salt addition also reduces the electrical double-layer thickness and, hence, the repulsion between micelles. In the absence of nonionic surfactant, the second effect causes phase separation (coacervation) to occur before the area per molecule is reduced sufficiently for the lamellar phase to form. Since nonionic surfactant does not influence electrical double-layer thickness, its principal effect is on micelle shape, and the first new phase to form is the lamellar liquid crystal. Indeed, if sufficient nonionic surfactant is present in a micellar solution, addition of salt produces transformation to the lamellar phase at a relatively low salinity where the electrical repulsion between micelles is still substantial.

The aqueous solution and the liquid crystal coexist over a wide range of compositions, as indicated. While such behavior is common in dilute anionic surfactant systems [4], it is not normally found in dilute systems containing pure nonionic surfactants or mixtures of nonionic surfactants. As we shall see below, optimum removal of hydrocarbon soils occurs for conditions where Fig. 1 would indicate the existence of a $L_1 - L_\alpha$ mixture.

In the absence of salt the pure nonionic surfactant is, in these moderately dilute solutions, near the transition between the L_α phase and the isotropic liquid phase L_3 which exists for such surfactants under rather lipophilic conditions. The liquid crystal disappears upon addition of a little NaCl, according to Fig. 1, and the system enters the two-phase region $B + L_3$. More salt is required for the transition as addition of anionic surfactant makes the surfactant mixture more hydrophilic. The details of the transition are not shown in Fig. 1, but are given in Fig. 2 for another overall surfactant

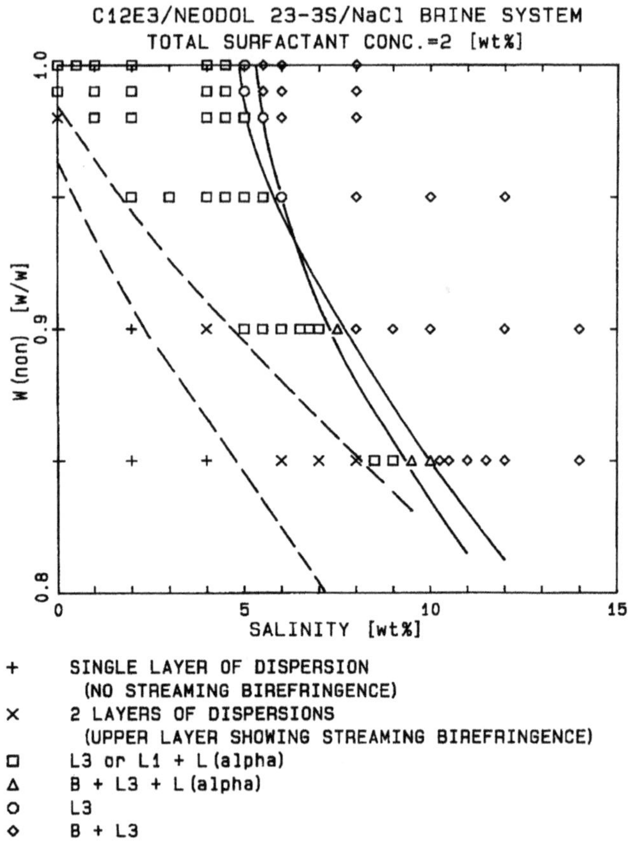

Fig. 2. Partial phase diagram for the same system as in Fig. 1, but with a total surfactant concentration of 2 wt%

concentration. When the surfactant mixture contains more than about 92 wt% of the nonionic, Fig. 2 shows that addition of NaCl produces transformation from $(L_3 + L_\alpha)$ to L_3 to $(B + L_3)$. With more anionic surfactant present the sequence becomes $(L_3 + L_\alpha)$ to $(B + L_3 + L_\alpha)$ to $(B + L_3)$. The manner of transition between the L_1 and L_3 phases is not entirely clear as no single-phase L_α region nor three-phase region with separate L_1, L_3, and L_α layers was seen. It may be that this three-phase region corresponds to the area in Fig. 2 designated as having two layers of dispersions, i.e., the L_α phase may be dispersed in the two immiscible liquid phases.

Phase behavior with added hydrocarbon

Various mixtures of anionic surfactant, nonionic surfactant, and NaCl brine were equilibrated with

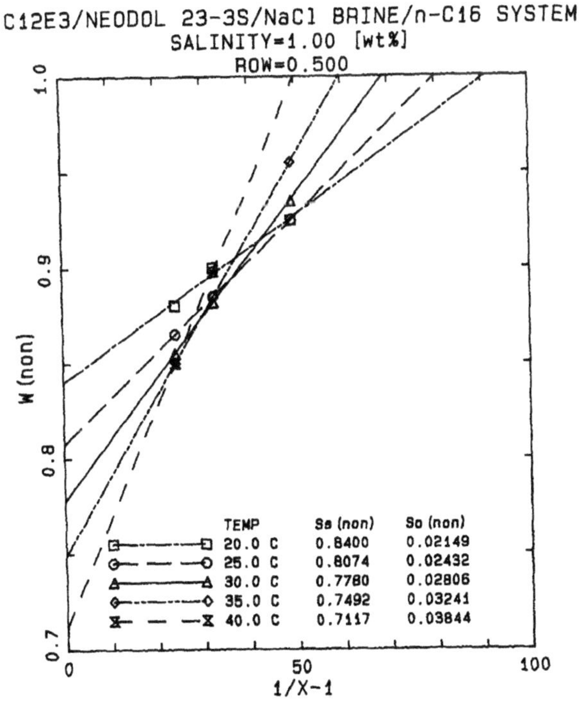

Fig. 3. PIT results for the $C_{12}E_3$/Neodol 23-3S/NaCl brine/n-hexadecane system. The brine contains 1 wt% NaCl

n-hexadecane at several temperatures. The hydrocarbon was 50% by weight of the final mixtures. Surfactant compositions and temperatures where the system was at its Phase Inversion Temperature (PIT), i.e., where a microemulsion containing equal volumes of oil and brine was in equilibrium with excess oil and brine, were recorded. Figure 3 shows the PIT results when NaCl content was 1 wt%. The abscissa of this plot is $(X^{-1} - 1)$, where X is the total surfactant concentration in the system. As Kunieda et al. [7] have shown, the dependence of the nonionic fraction W_n of the overall surfactant mixture on this parameter should be linear at each temperature when partitioning of the nonionic surfactant into the hydrocarbon phase is taken into account. In particular,

$$W_n = S_{sn} + S_{on} R_{ow}[(1 - S_{sn})/(1 - S_{on})] \times (X^{-1} - 1) , \quad (1)$$

where S_{sn} is the weight fraction of nonionic surfactant in the surfactant films of the microemulsion phase, S_{on} is the weight fraction of nonionic surfactant in the excess hydrocarbon phase, and R_{ow} is the weight fraction of oil in the overall mixture.

The plots of Fig. 3 are indeed linear, confirming the applicability of Eq. (1) to this system. The values of S_{sn} and S_{on} obtained from these plots are shown at the bottom of Fig. 3. We see that S_{sn} lies between about 0.70 and 0.85. That is, even with 1 wt% NaCl present, the surfactant films need a high content of nonionic surfactant to achieve the balance between hydrophilic and lipophilic conditions represented by the PIT. Moreover, S_{sn} decreases with increasing temperature, i.e., less nonionic surfactant is required in the surfactant films at higher temperatures for balanced conditions. This result is expected since increasing temperature makes the nonionic surfactant less hydrophilic. We also see that $C_{12}E_3$ content of the oil S_{on} is a few percent by weight and increases with increasing temperature. Direct analysis of the excess oil phase of one sample at 30 °C gave a surfactant content of 2.8 wt% $C_{12}E_3$, in good agreement with the corresponding value in Fig. 3.

Figures 4 and 5 show the effects of NaCl concentration and of hardness at 29 °C, the latter plot being for a salt mixture containing $CaCl_2$ (60 mole %) and $MgCl_2$ (40 mole %). In accordance with common practice, the hardness value given represents the amount of this salt mixture which would be present if the chlorides were converted to carbonates. As the numbers at the bottom of the diagram indicate, S_{sn} decreases with increasing salinity and hardness, i.e., less nonionic surfactant is present at the PIT, because the increased ionic strength makes the anionic surfactant less hydrophilic. Nonionic surfactant content S_{on} of the oil remains constant for each salt, not a surprising result at a fixed temperature. However, S_{on} is significantly lower for the $CaCl_2$-$MgCl_2$ mixture than for NaCl. The value of S_{on} shown for the divalent salt mixture may well be lower than actual nonionic surfactant content of the oil, however, because it was obtained using Eq. (1), which assumes that all anionic surfactant remains in the surfactant films. It is well known that calcium or magnesium salts of anionic surfactants can have appreciable solubility in oil, owing to their two hydrocarbon chains. Inspection of the generalized form of Eq. (1) for two surfactants with different solubilities in an oil phase [8] shows that a higher value of S_{on} would be obtained if solubility of the anionic surfactant in the oil in the form of calcium and magnesium salts were incorporated.

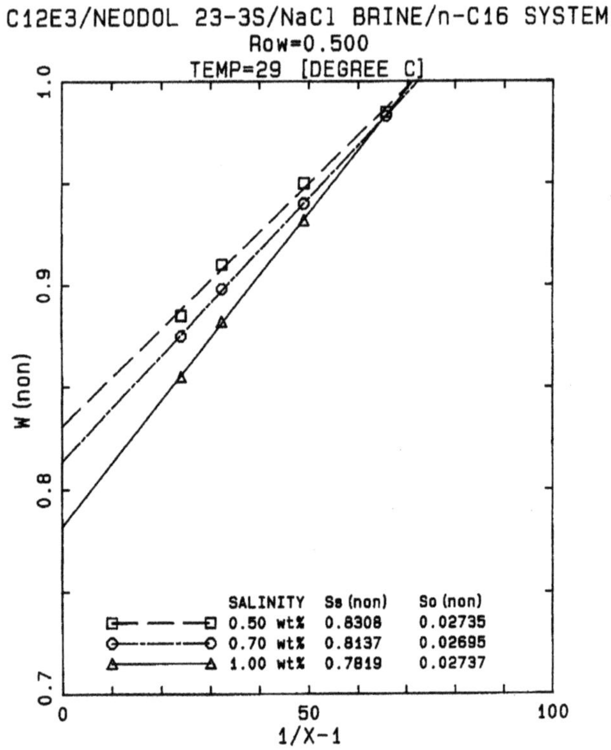

Fig. 4. Effect of NaCl concentration on PIT variation in the same system as in Fig. 3

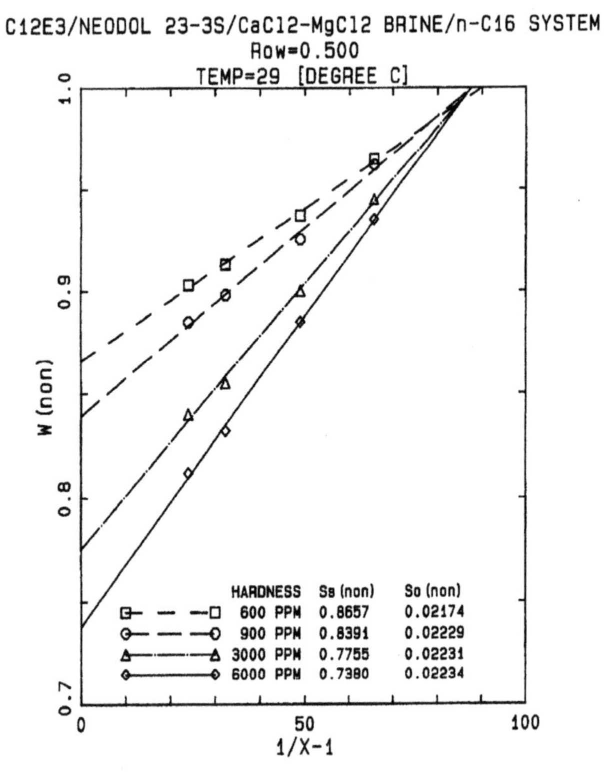

Fig. 5. Effect of hardness (as defined in text) on PIT variation in the same system as in Fig. 3

Interfacial tensions

A surfactant-brine mixture containing 1 wt% NaCl and 2 wt% of a 78/22 nonionic/anionic surfactant mixture was subjected to ultracentrifugation at 30 °C. The initial sample was a dispersion of lamellar liquid crystal in a micellar solution. The transparent liquid after ultracentrifugation was the micellar solution although a few tiny particles of liquid crystal may have been present as well. Analysis performed at Shell Development Company showed that the ratio of anionic to nonionic surfactant was about the same as in the initial sample.

Interfacial tensions were measured between this transparent liquid and n-hexadecane at various temperatures. The results are shown in Fig. 6. Since the oil drop in the spinning drop device is small, very little of the surfactant partitions into the oil. Thus, the PIT of the system should be nearly the same as that of the initial surfactant mixture, which, according to Fig. 3 is about 30 °C. Figure 6 shows that interfacial tension is about 0.01 mN/m at 30 °C. As expected, interfacial tension is greater at lower and higher temperatures.

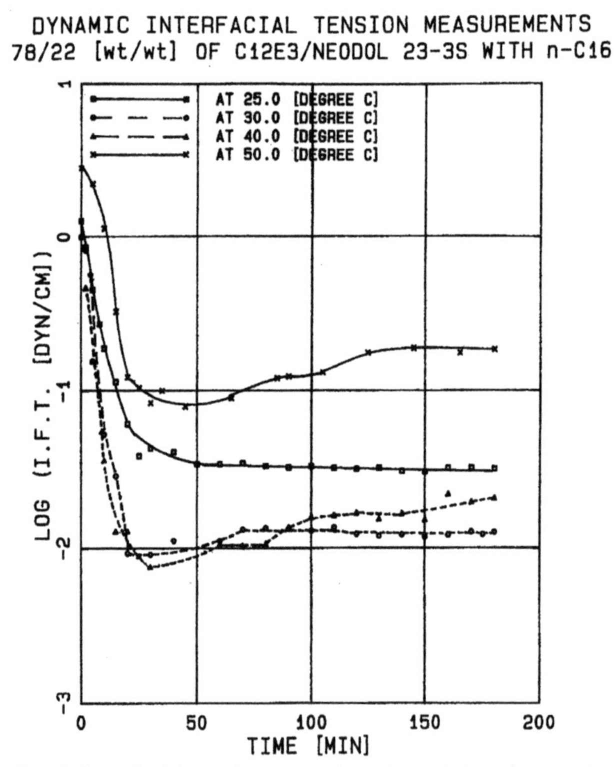

Fig. 6. Interfacial tensions as a function of time for a 78/22 $C_{12}E_3$/Neodol 23-3S mixture and n-hexadecane at several temperatures

Videomicroscopy observations

Oil drop contacting experiments were conducted using the concentrated dispersion of liquid crystalline particles obtained by the ultracentrifugation procedure described in the previous section, i.e., after removal of the transparent liquid. Here, too, the ratio of the two surfactants was found by analysis to be about the same as in the initial mixture, so that the PIT upon injection of a small drop of n-hexadecane was about 30 °C. Drops about 100 μm in diameter were employed.

Neither intermediate phase formation nor spontaneous emulsification was observed during the contacting experiment at 25 °C. Nor could any significant change in drop size be seen over a period of 3 to 4 h, an indication that solubilization is very slow. Figure 7 shows the drop 3.5 h after initial contact.

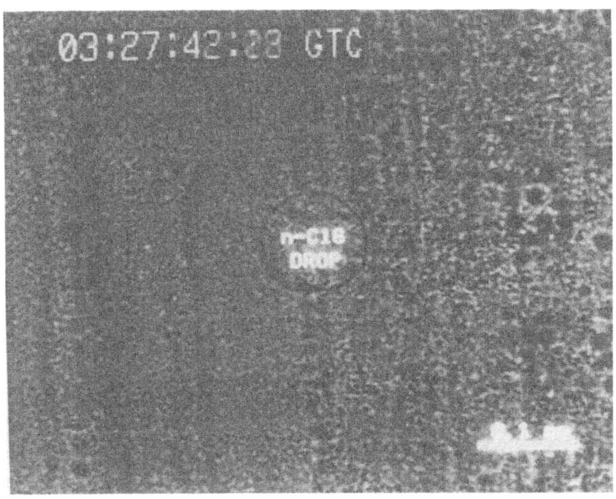

Fig. 7. Video frame showing n-hexadecane drop at 25 °C in concentrated dispersion of liquid crystalline particles 3.5 h after initial contact

At 30 °C, which is near the PIT, an immiscible liquid phase, presumably a middle phase microemulsion formed. By 1 h after initial contact enough microemulsion had formed to completely surround the oil drop (Fig. 8a). Further growth of the microemulsion phase led to disappearance of the oil phase about 10 min later, as seen in Fig. 8b. Subsequently, the microemulsion was itself converted into liquid crystalline particles. Thus, at the final equilibrium state of the system, the same two phases, liquid crystal and micellar solution, were present as initially.

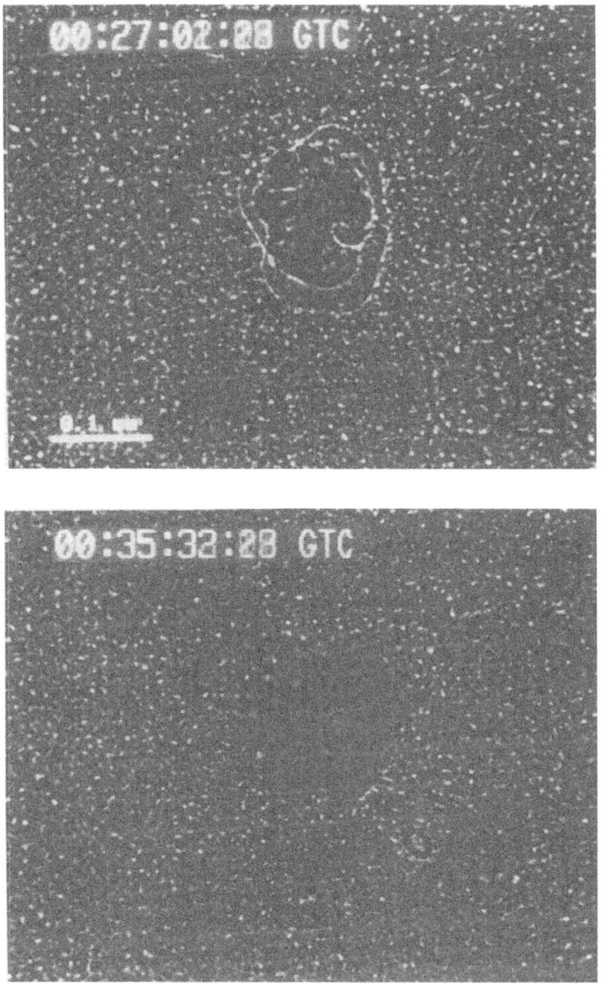

Fig. 8. Video frames for the same system as in Fig. 7 at 30 °C. a) Middle phase microemulsion completely surrounding oil drop about 1 h after initial contact; b) 10 min later, growth of the microemulsion phase and complete solubilization of the hydrocarbon

At 40° and 50 °C the volume of the oil drop increased with time, as would be expected since the oil should take up surfactant and water to form an upper phase (water-in-oil) microemulsion. Figures 9a and 9b show the drop at 40 °C just after injection and after about 80 min; the increase in size is evident. There appeared to be some spontaneous emulsification within the drop, but the presence of the liquid crystalline particles above and below the drop made it difficult to be certain on this point. In

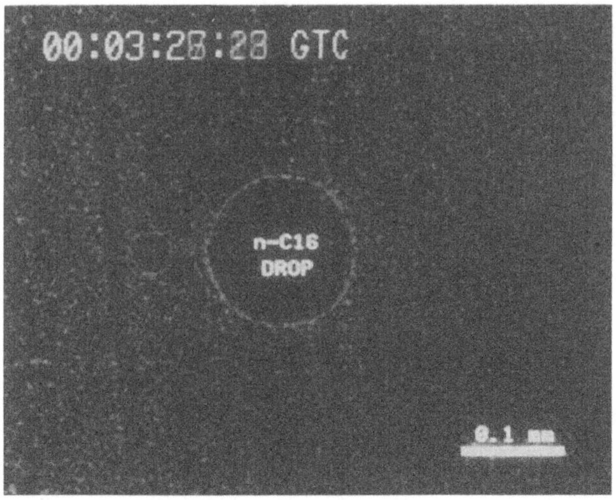

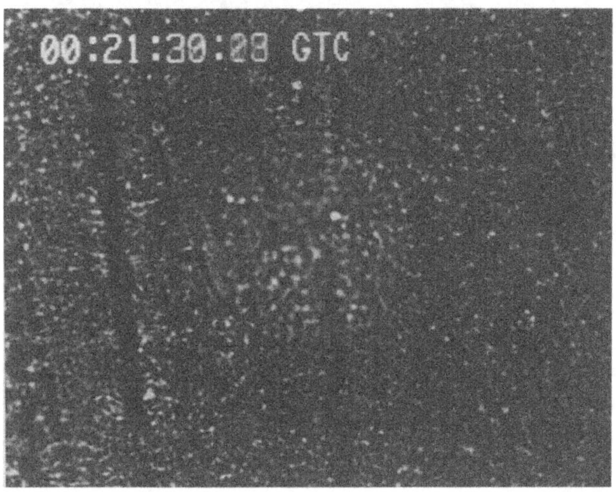

Fig. 9. Video frames for the same system as in Fig. 7 at 40 °C. a) Drop just after injection; b) 80 min later, growth of drop caused by formation of water-in-oil microemulsion

these experiments the drop eventually was converted into liquid crystalline particles in the same manner as seen at 30 °.

Contacting experiments were conducted at the same temperatures with the transparent liquid obtained from ultracentrifugation as described above. Again, little activity was seen at 25 °. At 30 ° solubilization of the oil was much slower than in the experiments reported above with a much higher concentration of surfactant. Both the lamellar liquid crystal and a microemulsion formed as intermediate phases. Above the PIT spontaneous emulsification of water within the oil drop could be clearly seen, a result which is consistent with our observations in pure nonionic surfactant systems using the previous contacting technique [2, 6].

Detergency results

Figure 10 shows the results of detergency experiments in which a mixture of anionic and nonionic surfactants in 1 wt% NaCl was used to wash small pieces of polyester/cotton fabric soiled with n-hexadecane. As the figure indicates, soil removal was greatest at 35 °C but was only slightly less at 30 °C. According to Fig. 3, the PIT of a system with surfactant films having the composition of the initial surfactant mixture (76.5 wt% nonionic surfactant) is about 31 °C. Since the amount of oil present was very small, this PIT is the relevant one according to the same argument as in the preceding sections. Thus, optimum detergency for mixtures of anionic and nonionic surfactants does occur near the appropriate system PIT for hydrocarbon soils, in agreement with our previous results for pure nonionic surfactants and mixtures of nonionic surfactants [2, 3].

Discussion

The oil drop contacting technique developed during this work has enabled the final stages of solubilization of a small drop to be observed. With the previous technique, which employs much larger quantities of oil, the eventual conversion of microemulsion phases to liquid crystalline particles reported above at several temperatures does not occur. Moreover, because only a negligible amount of the nonionic surfactant initially present partitions into the small oil drop, the behavior observed during the contacting experiments should be the same as during the detergency experiments. With the previous technique a large portion of the nonionic surfactant would dissolve in the oil phase and cause a large increase in the PIT.

The present results with mixtures of anionic and nonionic surfactants confirm the generality of our previous conclusions based on experiments with nonionic surfactants and hydrocarbon soils [2, 3, 6]. For temperatures near the PIT, the middle-phase microemulsion forms as an intermediate phase and soil removal is maximized. Solubilization of the soil is greatest under these conditions. In addition, low

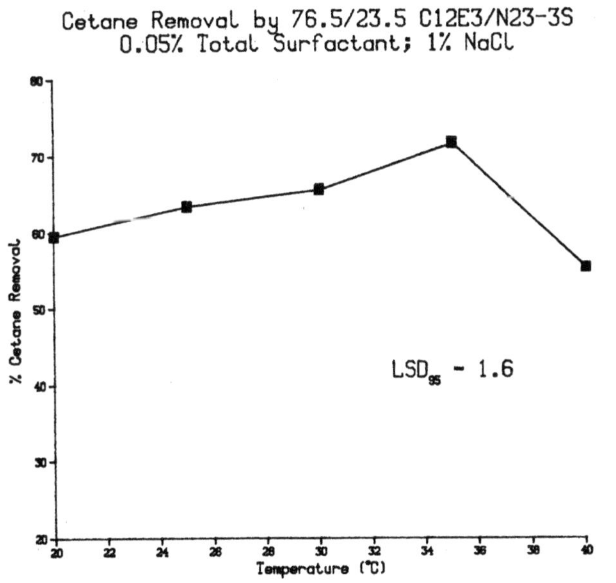

Fig. 10. Removal of n-hexadecane from polyester/cotton fabric as a function of temperature for 76.5/23.5 $C_{12}E_3$/Neodol 23-3S mixture with 1 wt% NaCl

interfacial tensions facilitate emulsification of the microemulsion phase into the washing bath.

Acknowledgement

This work was supported by NSF Grant CBT-8519276 and by the Shell Development Company. Useful discussions with Shell personnel are gratefully acknowledged. Olina Raney, Melissa Robinson, and Jeff Skiba helped with the oil-free phase behavior studies.

References

1. Scamehorn JF (ed) (1986) Phenomena in Mixed Surfactant Systems. ACS Symposium Series 311
2. Raney KH, Benton WJ, Miller CA (1987) J Colloid Interface Sci 117:282
3. Raney KH, Miller CA (1987) J Colloid Interface Sci 119:539
4. Benton WJ, Miller CA (1983) J Phys Chem 87:4941
5. Benson HL, Cox KR, Zweig JE (1985) Household Pers Prod Ind 22:50
6. Benton WJ, Raney KH, Miller CA (1986) J Colloid Interface Sci 110:363
7. Kunieda H, Hanno K, Yamaguchi S, Shinoda K (1985) J Colloid Interface Sci 107:129
8. Kunieda H, Shinoda K (1985) J Colloid Interface Sci 107:107

Authors' address:

C. A. Miller
Department of Chemical Engineering
Rice University
P.O. Box 1892
Houston, Texas 77251, USA

Progress in Colloid & Polymer Science Progr Colloid Polym Sci 82:122—130 (1990)

Effect of hydrogen bonding on water in oil emulsion properties

J. Bus, F. Groeneweg, and F. van Voorst Vader

Unilever Research Laboratorium Vlaardingen, Vlaardingen, The Netherlands

Abstract: The amphiphilic properties of the large majority of edible water in oil (W/O) emulsifiers are due to the presence of hydroxyl groups in their hydrophilic group. These cause such emulsifiers to adsorb at the W/O interface, to stabilize W/O emulsions, and to form inverse micelles in the oil phase. They also cause the formation of hydrogen-bonded complexes with nonaqueous solvents containing proton acceptor groups, e.g., with trigylceride oil. By comparing the critical micelle concentration (CMC) values of a series of such emulsifiers in hydrocarbon oil and in sunflowerseed oil with the infrared spectra of the corresponding solutions, it is shown that the higher CMC values in the latter solvent are due to hydrogen bonding between the emulsifier monomers and the solvent. Formation of such complexes is decreased if the emulsifier structure allows intramolecular hydrogen bonding. — If no hydrogen bonds with the solvent are formed, W/O emulsions are stable when the emulsifier concentration in the oil after preparing the emulsion at least equals the CMC. However, for emulsifiers having a low CMC value a large excess concentration is needed to stabilize W/O emulsions in hydrogen-bonding solvents such as sunflower seed oil.

Key words: Emulsifiers; amphiphilic properties; critical micelle concentration; hydrogen bonding

Introduction

The hydrophilic group of the vast majority of edible emulsifiers contains hydroxy groups. Examples are fatty acid mono- and diglycerides, polyglycerol esters, sugar esters, and sorbitan esters [1]. The hydrogen bonding of these hydroxy groups with water is responsible for their adsorption from the oil phase to the W/O interface and thus for their capability to stabilize W/O emulsions. Within structural limits, mutual hydrogen bonding of emulsifier monomers will cause the formation of inverse micelles in the oil phase. Also, hydrogen bonding with the solvent will occur provided that the oil phase contains proton-accepting groups like ester bonds. For dilute solutions of emulsifier in the oil phase, hydrogen bonding can be described by [2]:

$$K_b = \frac{[M\ ..\ S]}{[M] * [S]}, \tag{1}$$

where $[M\ ..\ S]$ is the activity of the hydrogen-bonded complex,
$[M]$ is the activity of the free emulsifier monomer, and
$[S]$ is the activity of the solvent.

We will assume that the emulsifier activity equals the concentration of its monomer.

All the above reactions are reversible. The values of their equilibrium constants partly characterize the behavior of an emulsifier in a given system. The aim of the present work has been to consider the influence of hydrogen bonding between the emulsifier and the solvent on these constants.

Experimental

Materials

The following emulsifiers were used:

I) α,α'Diglycerol di(bisricinoleate) (pure according to TLC analysis);

II) technical glycerol monooleate from Unichema;
III) purified Hymono 7704, a mixture of glycerol monooleate and glycerol monolinoleate that is liquid at room temperature;
IV) diglycerol monooleate (pure according to TLC analysis);
V) symmetrical triglycerol mono(bisricinoleate) (TLC analysis indicated the presence of a slight amount of triglycerol);
VI) Span 80 (sorbitan monooleate) from Koch Light;
VII) Admul WOL (polyglycerol estrified with polyricinoleic acid) from Unichema;
VIII) Lecithin (Epikuron 100 from Lucas Meyer GmbH.).

The formulae and composition of the emulsifiers are given in Table 1. Compounds I, IV, and V were synthesized by Dr. A. Fröling (Unilever Research Laboratorium, Vlaardingen). Distilled water, squalene ($C_{30}H_{50}$, 98% pure from Merck), sunflowerseed oil, and kerosene (d_4^{20} = 790 kg/m^3) were used without further purification.

High purity solvents are were used in the samples whose infra-red (IR) spectra were measured: *n*-heptane (spectroscopical grade) was used as hydrocarbon solvent, and glycerol trioleate that was pure according to TLC analysis was used as vegetable oil.

Table 1. Structure of emulsifiers

Compound	Name	Structure/Composition
I	*α,α* diglycerol di(bisricinoleate)	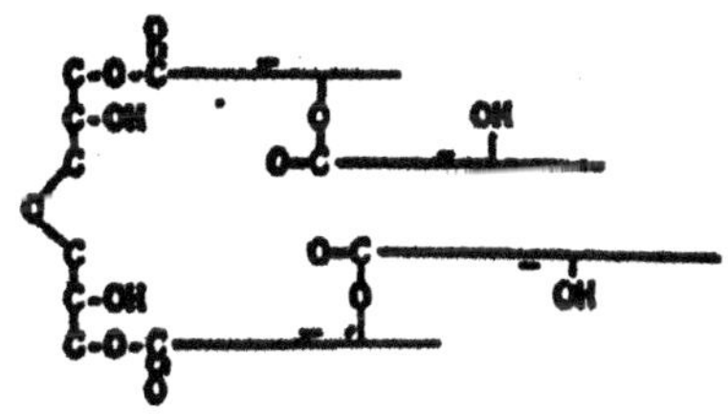
II	Technical glycerol monooleate	contains 50% glycerol monooleate; other comp. mainly di- and triacylgylcerol
III	purified Hymono 7704	69% glycerol monolinolate, 24% glycerol monooleate, no satured compounds
IV	diglycerol monooleate	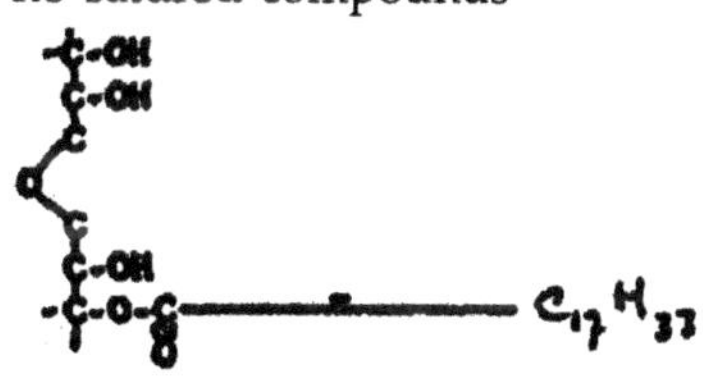
V	triglycerol mono (bisricinoleate)	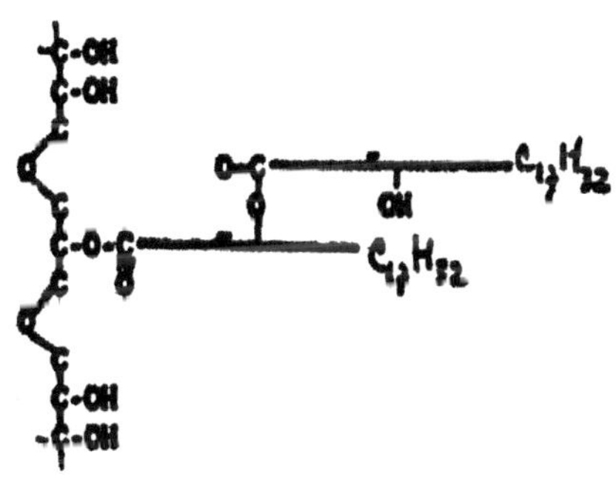
IV	Span 80	sorbitan monooleate 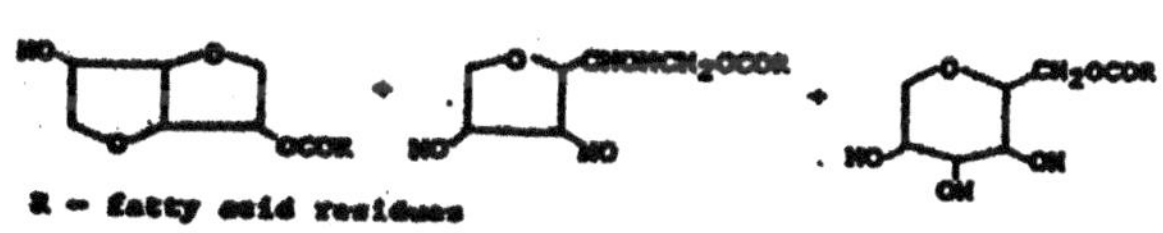

Table 1 *(continued)*

Compound	Name	Structure/Composition
VII	Admul WOL	polyglycerol estrified with polymerized ricinoleic acid Average composition: 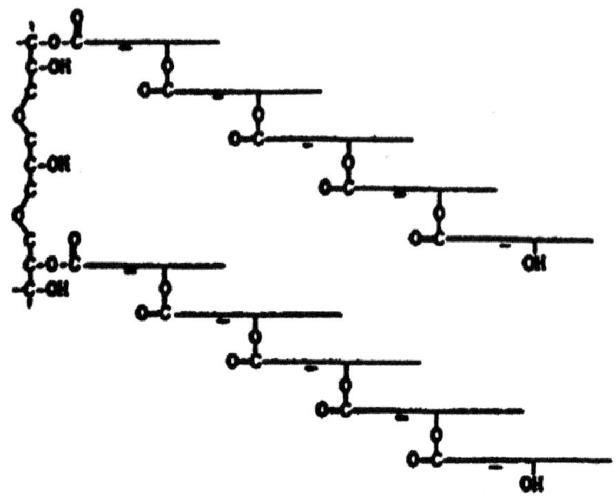
VIII	Epikuron 100 (Lecithin)	20—23% PC 13—16% PE 10—14% PI

Interfacial tension measurements

To determine the presence and the value of the critical micelle concentration (CMC) for inverse micellization interfacial tensions were measured using the Wilhelmy plate method. Because part of the solvents was not very pure and some emulsifiers consisted of technical mixtures, the value of the interfacial tension initially decreased. We used the values obtained after aging the interface for 10 min.

Infrared spectra of the solution

To verify that the associative properties of the hydroxyl-containing emulsifiers (compounds I—VII) are due to hydrogen bonds formed by these polar groups, IR-spectra of their solutions were measured.

Formation of hydrogen bonds by hydroxyl groups causes a considerable change in their stretch vibration bands [2]. Hydrogen bonds between different groups within the sample molecule are designated as intramolecular hydrogen bonds, those formed between two different molecules are designated as intermolecular hydrogen bonds. These two types behave differently on changing the concentration of the OH-containing molecules, which offers a tool for investigating the relative contributions of the free and the associated molecules.

In the case of monoglycerides, only one hydroxyl group is esterified. The following possibilities exist for each of the two remaining OH groups:

1) it may be free, i.e., it does not interact with other groups;

2) the group may form an intramolecular hydrogen bond with the second OH group or with the ester group present in the same molecule;

3) the group may form an intermolecular hydrogen bond with an OH or ester group in another molecule.

The resulting OH stretch absorption bands are shifted to different positions in the IR-spectrum dependent upon the strength of the hydrogen bonds formed.

Because the emulsifier concentration ranges of interest are low in comparison to those used for conventional IR analysis, the IR spectra were measured with a Bruker IFS88 Fourier transform IR spectrometer.

Results and discussion

Interfacial tension measurements

The accuracy of the interfacial tension measurements sufficed to determine the presence and the value of the critical micelle concentration (CMC) for inverse micellization. Though applica-

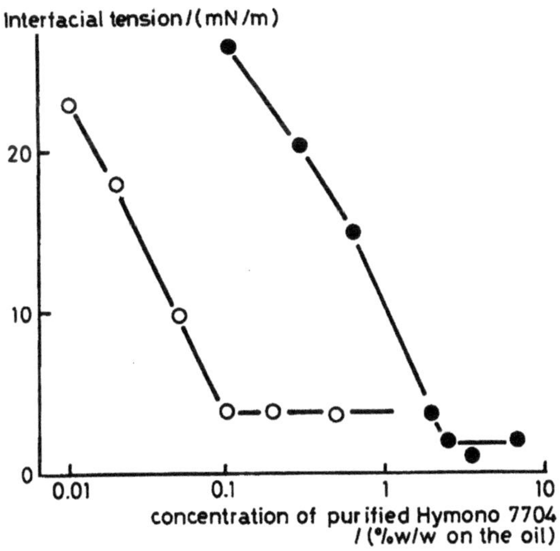

Fig. 1. Interfacial tension as a function of the concentration of purified Hymono 7704 (compound III). o: in kerosene; ●: in sunflowerseed oil

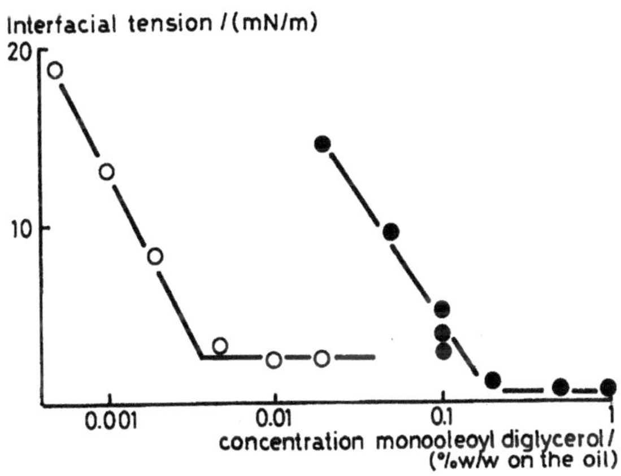

Fig. 2. Interfacial tension as a function of the concentration of diglycerol monooleate (compound IV). o: in kerosene; ●: in sunflowerseed oil

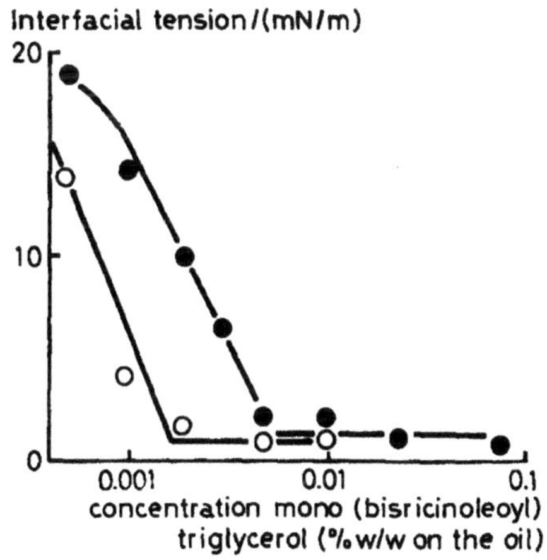

Fig. 3. Interfacial tensions as a function of the concentration of triglycerol mono(bisrinoleate) (compound V). o: in kerosene; ●: in sunflowerseed oil

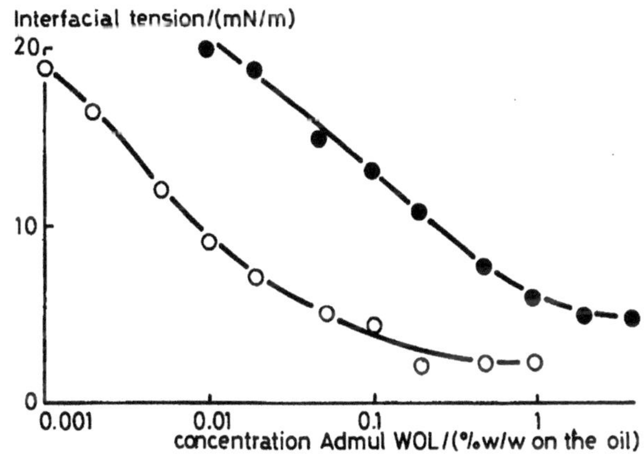

Fig. 4. Interfacial tension as a function of the concentration of Admul WOL (compound VII). o: in kerosene; ●: in sulfowerseed oil

tion of Gibbs' adsorption law to these data leads to reasonable molecular areas of the adsorbed species (40—60 $m^2/10^{20}$), their accuracy is too low to relate differences in area found for various emulsifiers to their molecular structure. Results are given in Figs. 1—4. Plotting the interfacial tension (σ) against the Briggsian logarithm of the w/w concentration (c_w) allows the following classification:

A) Compound I shows little surface activity. No CMC is found in the investigated concentration range;

B) Compounds II, III, IV, V, and VI show sharp breaks in their σ vs $\log c_w$ plots. In analogy to the results obtained for related systems [3, 4], this break can be ascribed to the onset of inverse micellization (CMC). The CMC's of the compounds II and VI dissolved in squalene coincided with those measured in kerosene;

C) On increasing their concentration, compounds VII and VIII show a gradual transition to a plateau value of σ. Both consist of complicated mixtures of components of comparable surface activity, which will promote the formation of mixed aggregates [5].

The "CMC" used in Tables 2—4 corresponds to the concentration at which this plateau is reached.

CMC values in hydrocarbon oils

Both in kerosene and in sunflower oil, the CMC's of the various emulsifiers decrease with increasing polarity (Table 2), which roughly corresponds to the increase in HLB value.

The free enthalpy of micellization [6] equals:

$$G_{mic} = RT \ln x_{cmc} , \quad (2)$$

where R is the gas constant, T is the absolute temperature, x_{cmc} is the mol fraction of emulsifier at the CMC. G_{mic} contains the contributions to the change in free enthalpy due to micellization both of the hydrophilic (phi) and of the hydrophobic (pho) groups of the emulsifier molecule [6]:

$$G_{mic} = G_{phi} + G_{pho} . \quad (3)$$

The latter contribution cancels for compounds III, IV, and VI, so that, for example, $RT \ln(x_{IV}/x_{III}) = -3.34\, RT$ (cf. Table 2) corresponds to the free enthalpy of formation of one additional hydrogen bond. This value is of the order of magnitude expected for hydrogen bonding [2].

Compound VI has a higher CMC than IV, though both contain 3 OH groups. This suggests that the hydrogen bonding between the cyclic sorbitan groups is less effective than that between the linear triglycerol groups.

Comparison of the CMC's of compounds II and III suggests that the CMC of compound II is mainly controlled by its monoglyceride content (cf. Table 1).

Direct evidence on the role of hydrogen bonding in the inverse micellization of these emulsifiers is obtained by comparing the IR spectra of glycerol monooleate solutions in heptane at various concentrations. The spectrum of a 2 mmol/l solution shows the presence of adsorption peaks at 3630 cm^{-1} of the free and at 3600 cm^{-1} of the intramolecularly bonded OH groups; the broad adsorption peak at 3400 cm^{-1} is due to intermolecular hydrogen bonding (Fig. 5a). At 9 mmol/l, the peaks of the OH groups of the unperturbed molecules are very weak; nearly all OH bond intensity is now concentrated in the broad band at 3400 cm^{-1}, indicating that nearly all OH groups have formed inter-

Table 2. CMC's of emulsifiers in various solvents. T = 20°C

Compound	Kerosene or Squalene			Sunflower-seed oil
	n*)	CMC % w/w	G_{cmc}/RT	
I	2	—	—	—
II	—	0,25	—	4.8
III	2	0.1	− 7.56	2.5
IV	3	0.004	−10.90	0.18
V	4	0.0017	−12.37	0.005
VI	3	0.027	− 8.99	0.7
VII	—	(0.2)**)	—	(2)**)
IIX	—	(0.5)**)	—	(0.5)**)

*) n — number of OH-groups in the emulsifier molecule.
**) concentration above which $d\sigma/d \ln c_w \sim 0$.

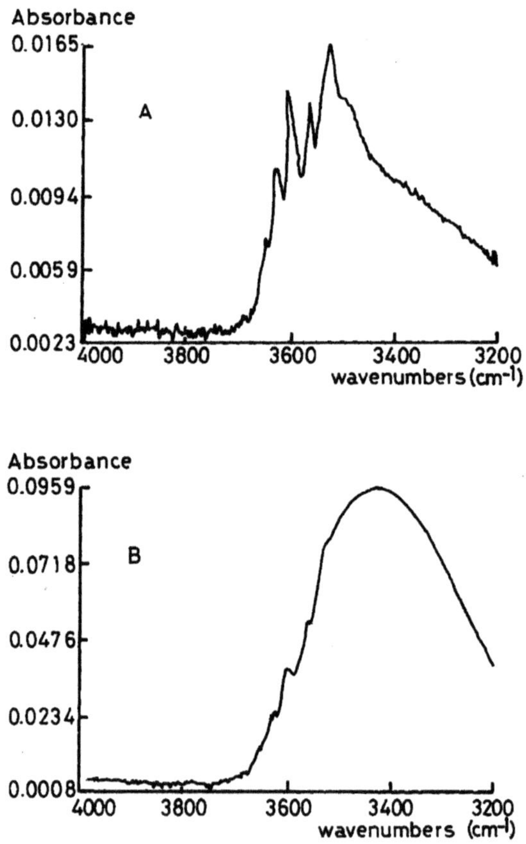

Fig. 5. Infrared spectra of glycerol monooleate in n-heptane at a) 0.002 mol/l; b) 0.009 mol/l

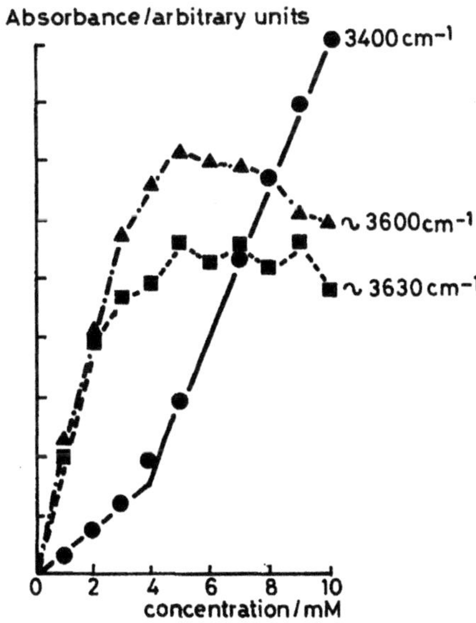

Fig. 6. Relative absorbance maxima as a function of the concentration of glycerol monooleate in *n*-heptane. ▲: ~3600 cm^{-1}; ■: ~3630 cm^{-1}; ●: 3400 cm^{-1}

molecular hydrogen bonds (Fig. 5b). The relative change in intensity of the various absorption bands with monooleate concentration is shown in Fig. 6. The intensities of the bonds at 3630 and at 3600 cm^{-1} increase up to 3—4 mmol/l and then level off or decrease. The increasing intensity of the band associated with intermolecular hydrogen bonded OH groups at 3400 cm^{-1} shows a discontinuity near the same concentration, where its slope increases by a factor 3. The monoglyceride concentration at this point of discontinuity is close to the CMC found for monoglyceride samples from interfacial tension data (2.1 mmol/l). The difference between both CMC values might be caused by solubilization of water in the oil samples used in the latter measurements.

Thus, micelle formation of OH-containing emulsifiers in hydrocarbon systems can approximately be described by hydrogen bonding between the hydrophilic groups of the emulsifiers.

CMC values in vegetable oil

Table 2 shows that the CMC's in vegetable oil are, for all OH-containing emulsifiers, larger than those in a hydrocarbon solvent. Because the CMC values in hydrocarbons are not sensitive to the molecular weight of the hydrocarbon, the following assumptions may be allowed:

1) No association occurs between solvent and emulsifier in hydrocarbon oils. Up to the CMC, the thermodynamical activity of the emulsifier then equals its monomer concentration, which equals the total emulsifier concentration;

2) The values of G_{cmc} for emulsifiers in hydrocarbon and in vegetable oil are equal. It follows that the CMC's found in these solvents for the same emulsifier correspond to the same thermodynamical activity;

3) Up to the CMC, the thermodynamical activities of emulsifiers in vegetable oil deviate from their concentration due to the formation of hydrogen-bonded complexes between the solvent and the emulsifier, as described by Eq. (1).

The association constant between solvent and emulsifier K_b (cf. Eq. (1)) will now be evaluated on the basis of these assumptions. Assumption 1 implies that in hydrocarbons the monomer activity at the CMC equals the CMC itself:

$$[M_h] = [CMC_h] \; . \tag{4}$$

In vegetable oil, assumption 3 implies:

$$[CMC_{vo}] = [M_{vo}] + [M..S] \tag{5}$$

(for the notation see sub Eq. (1)).

According to assumption 2, $[M_h] = [M_{vo}]$. Therefore, combination of Eqs. (1), (4), and (5) yields:

$$K_b = \{[CMC_{vo}]/[CMC_h] - 1\}/[S] \; . \tag{6}$$

The calculated values of K_b are collected in Table 3. For sunflowerseed oil, we use $[S] = 1$ mol/l (for pure glycerol trioleate, $[S] = 0.97$ mol/l). Table 3 indicates that:

Table 3. Hydrogen bonding constants for various emulsifier/sunflowerseed oil systems

Compound	K_b (l/mol)
II	18.2
III	24.0
IV	44
V	1.9
VI	24.9
VII	9
IIX	0

- K_b strongly increases with increasing polarity of the emulsifier relative to the solvent;
- IR spectra indicate that strong intramolecular hydrogen bonding occurs in compound V. This may explain its low degree of hydrogen bonding with triacylglycerols. The rather low value of K_b for Admul WOL (compound VII) may be due to the same mechanism;
- No hydrogen bonding is found between lecithin and triacylglycerols.

An independent check of the validity of the above reasoning was again performed by IR analysis. First, the spectra of a series of solutions of glycerol monooleate in glycerol trioleate with concentrations ranging from 0.002 to 0.060 mol/l were compared. An example is given in Fig. 7. Over the whole concentration range, the form of these spectra changes very little. The most intensive peak is at 3541 cm^{-1}, which is the position of an OH group bonded to a carbonyl group. The lack of concentration effect on the form of the spectrum indicates that the type of hydrogen bonded interaction is virtually independent of the monoglyceride concentration. Apparently, the glycerol trioleate solvent offers such a surplus of ester carbonyl groups to the solute that nearly all its OH groups form an OH...O=C bond. No spectra at or above the CMC could be obtained due to strong absorption.

Assuming the validity of Eq. (1), the value of the association constant K_b in this system was estimated from a series of spectra in which both glycerol trioleate and glycerol monooleate were dissolved in heptane. The glycerol monooleate concentration was kept constant at 2 mmol/l, while the glycerol trioleate concentrations was varied from 0 to 100 mmol/l. The concentration of free monomer $[M]$ was estimated from the absorbance at 3600 cm^{-1} and the concentrations of trioleate $[S]$ and of the $M..S$ complex were then calculated by subtraction. While the accuracy of this method still needs further improvement, preliminary results suggest that $K_b \sim 20$ l/mol with a standard deviation of ± 10 l/mol; this is in fair agreement with the data of Table 3 for monoglyceride systems. We conclude that the strong increase in CMC value of OH-containing emulsifiers in sunflower seed oil relative to their CMC in hydrocarbon oil is due to the formation of hydrogen-bonded complexes with the former solvent.

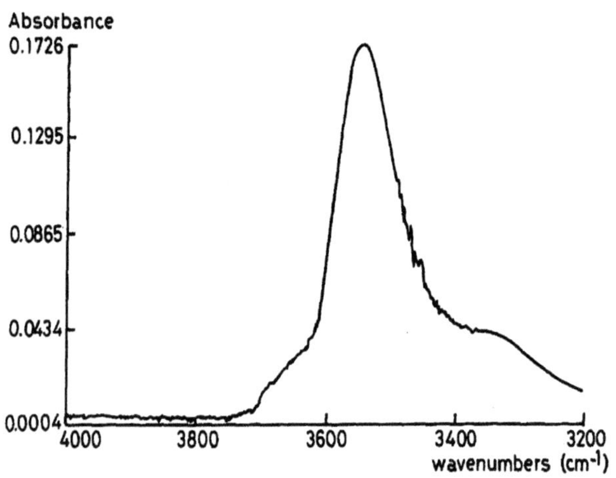

Fig. 7. Infrared spectrum of 0.01 mol/l glycerol monooleate in glycerol trioleate

Emulsion stabilization

According to Kruglyakov et al. [4], W/O emulsions are stable when the emulsifier concentration in the continuous phase after preparing the emulsion exceeds that concentration at which Newton black films between droplets of the emulsion become stable; this concentration is close to the CMC of the emulsifier. If the initial droplet size of the emulsion is so small that emulsifier adsorption at the W/O interface causes the emulsifier concentration in the oil phase to drop below the CMC, rapid coalescence of droplets will occur until the W/O interfacial area is sufficiently reduced to raise this concentration again to the CMC. When all droplets in the emulsion have the same volume, it follows that an emulsion containing a certain volume fraction φ of water droplets with diameter R becomes stable when:

$$c_w(S) \geqslant c_w(\mathrm{CMC}) + 0.3 * \varphi * \Gamma * MW / \{(1 - \varphi) * R * d_s\} , \quad (7)$$

where: $c_w(S)$ is the emulsifier concentration in oil phase at stabilization in % w/w; $c_w(\mathrm{CMC})$ is the critical micelle concentration in the oil phase in % w/w; Γ is the emulsifier adsorption at the W/O interface (mol/m^2); MW is the molecular mass of emulsifier in (g); D is the droplet diameter (m), d_s is the density of oil phase (g/l).

Rearrangement of Eq. (6) allows to predict the droplet size at which the emulsion is stable as a function of its emulsifier content:

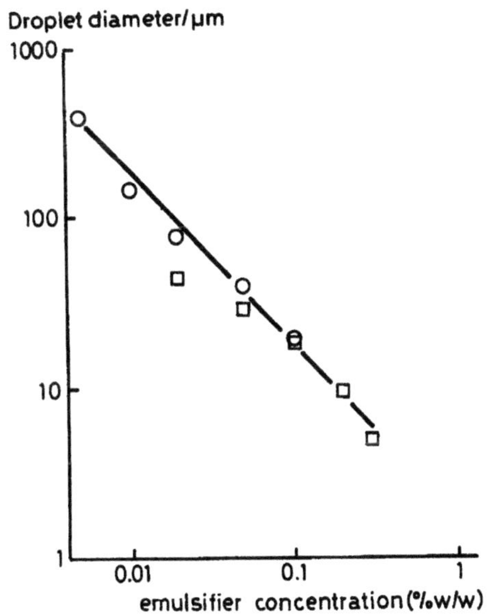

Fig. 8. Mean droplet diameter ($\bar{D}$ 3,3) of 50% (w/w) water in kerosene emulsions as a function of the concentration of emulsifiers in kerosene. □: diglycerol monooleate; ○: triglycerol monobisricinoleate

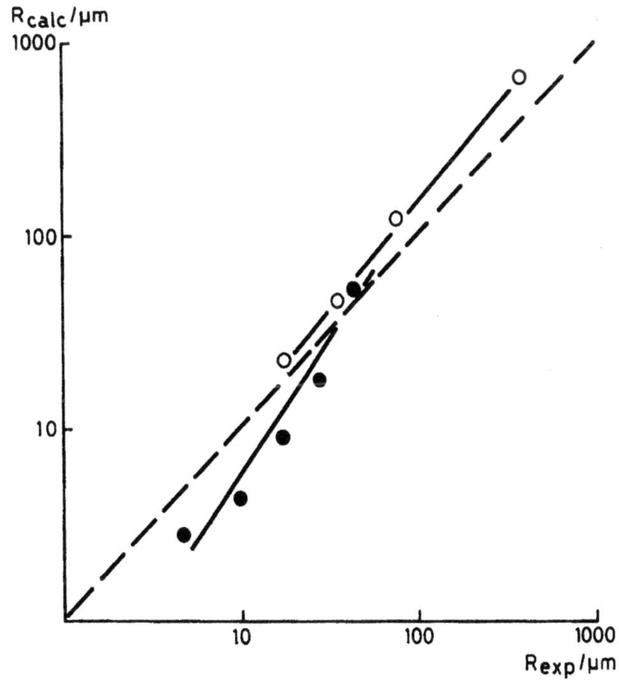

Fig. 9. Mean droplet radius R_{exp} in water/kerosene emulsions. ●: diglycerol monooleate (compound IV) ○: triglycerol mono(bisricinoleate) (compound V)

$$R = 0.3 * \varphi * MW/\{(c_w - c_w(\mathrm{CMC})) * (1 - \varphi) * d_s\} \ . \tag{8}$$

Emulsions of water in kerosene (50% w/w) and of water in sunflowerseed oil were prepared whose oil phase contained various concentrations of the emulsifiers enumerated above. The emulsion in kerosene and some of those in sunflowerseed oil were prepared using an Ultraturrax. Most of the emulsions in sunflowerseed oil were, however, prepared using a wire stirrer at 1200 rev/min for 10 min. For compounds IV and V, the droplet size after emulsification in kerosene was mesured microscopically (Fig. 8), allowing a direct test of Eq. (8).

The values of the adsorption, Γ_{IV} = 3.3 μmol/m^2 and Γ_V = 4.66 μmol/m^2, were obtained from Figs. 2 and 3. In Fig. 9, the theoretical droplet size calculated from Eq. (8) is compared to its experimental value. The agreement is satisfactory.

Table 4 surveys whether Eq. (7) also applies for the other emulsifiers and for sunflowerseed oil. In a number of cases, both coarse and fine 50% (w/w) W/O emulsions were prepared. The droplet size of the coarse emulsions was around 50 μm. Equation (7) indicates that for this droplet size, $c_w - c_w(\mathrm{CMC}) \approx 0.02\%$ (w/w), whereas for a fine emulsion with droplet size of 3 μm, $c_w - c_w(\mathrm{CMC}) \approx 0.3\%$ (w/w). The results obtained in kerosene for compounds II—VII conform fairly well with the stabilizing mechanism proposed above. However, in sunflowerseed oil only the results for compounds III and VII agree with Eq. (7). The concentrations of the other emulsifiers needed to stabilize W/O emulsions in sunflowerseed oil are clearly higher than predicted by Eq. (7). This deviating behavior needs further investigation. Also, the results obtained with lecithin do not agree well with our theory; the stabilizing action of this emulsifier must also be further considered.

Conclusions

The present investigation leads to a fairly coherent picture of the relation between the chemical structure of OH-group containing emulsifiers and their stabilizing action on W/O emulsions. The inverse CMC of these emulsifiers in the oil phase constitutes a key parameter control-

Table 4. Effect of emulsifier concentration on droplet size of W/O emulsions

Compound	Kerosene				Sunflowerseed oil		
	CMC	c_w needed for stabilizing emulsion:		Eq. (7) valid?	CMC	c_w needed for stabilizing emulsion:	Eq. (7) valid?
		coarse	fine			fine	
	(% w/w)	(% w/w)	(% w/w)	yes/no	(% w/w)	(% w/w)	yes/no
I	>1	—	—	—	>9	unstable	—
II	0.25	0.50	0.50	Y	4.8	8	N
III	0.1	0.20	0.50	Y	2.5	2.5	Y
IV	0.004	0.05	0.3	Y	0.18	5	N
V	0.0017	0.05	~0.4	Y	0.005	unstable	N
VI	0.027	0.45	0.45	Y	0.7	10	N
VII	(0.2)	0.1	1	Y?	(2)	0.5	Y
IIX	(0.5)	0.01	≫1	?	(0.5)	1 for 40% W/O unstable at 50% W/O	?

ling emulsion behavior; its value is mainly determined by the hydrogen bonding between the hydrophilic groups of the emulsifier. It has been found elsewhere [6, 7] that the size of the hydrophobic group also influences the CMC value, but this effect is usually smaller than the effect of the structure of the hydrophilic group. Also, the influence of the solvent polarity on the CMC can be understood by considering the extent of hydrogen bonding between solvent and emulsifier. The stabilization of emulsions in kerosene conforms with the condition as proposed by Kruglyakov et al. [4]: the emulsions are stable when the initial emulsifier concentration is sufficient to provide saturation adsorption at the droplet interfaces while maintaining the bulk concentration in the oil close to the CMC.

This simple analysis does, however, not generally apply to the stabilization of W/O emulsions in sunflower seed oil, where an excess of emulsifier is needed, especially for compounds whose monomer activity at the CMC is low. The peculiar properties of such systems need further investigation.

References

1. Off J Eur Comm (1974) No L 189/5–6
2. Pimentel GC, McClellan AL (1960) In: Freeman & Cy (ed) The Hydrogen Bond, chapter 7. San Francisco
3. Debye P, Prins W (1958) J Colloid Sci 13:86
4. Kruglyakov PM, Korezkii AF, Sokolovskaya IP, Rovin YuG, Mikina TV (1976) Isv Sib Otd Ak Nauk SSSR Ser Khim 1:19
5. Hall DG (1972) J Chem Soc Faraday Trans II 68:668
6. Fjedorov AB, Zajchenko LP, Abramzon AA, Proskurjakov VA (1984) Zh Prikl Khim:644
7. Zwierzykowski W, Orzecka B, Spich K (1976) Tenside Det 13:313

Authors' address:

Dr. F. van Voorst Vader
Physical Chemistry Section
Unilever Research Laboratorium
P.O. Box 114
3130 AC Vlaardingen, The Netherlands

Progress in Colloid & Polymer Science Progr Colloid Polym Sci 82:131—139 (1990)

Water-in-crude oil emulsions. Formation, characterization, and destabilization

J. Sjöblom, O. Urdahl, H. Høiland, A. A. Christy, and E. J. Johansen[1])

Department of Chemistry, University of Bergen, Norway
[1]) Statoil A/S, Prolab, Stavanger, Norway

Abstract: A survey of properties of water-in-crude oil emulsions based on crude oils from the Norwegian Continental Shelf is given. Special emphasis is given to the characterization and the separation of the interfacially active fraction in the crude oil. For this purpose a consecutive adsorption method was developed. Among the interfacially active constituents isolated palmitic acid was identified. — After adsorption *all* crude oils are incapable of stabilizing water-in-crude oil emulsions. Model emulsions stabilized by the adsorbed fraction show that the original crude oil emulsion stability can be reproduced depending on the stabilization mechanisms. A similar destabilization sequence for model and authentic crude oil emulsions can be obtained when medium-chain alcohols and fatty amines are used as destabilizers. Fundamentally different destabilization mechanisms for these species are proposed.

Key words. W/O emulsions, North Sea crude oils, stability, destabilization, surface active fraction, modelsystems

Introduction

Water-in-crude oil emulsions constitute a theoretical, as well as a practical problem for the chemist. The emulsions represent oil-continuous emulsions and, hence, a lot of fundamental questions concerning stability mechanisms and destabilization processes are still unsolved. Practically, the co-production of crude oil and water offshore gives rise to a sophisticated separation problem where mechanical energy input and chemicals should be combined at the platform. In the future, the water-in crude oil emulsion problem will also be encountered in the transport phase of crude oil from offshore to onshore conditions in multiphase flow systems transporting simultaneously, crude oil, water, and gas.

Obviously, questions like the predictability of emulsion stability together with an extensive readiness for mechanical and chemical destabilization will be focussed in the near future on the Norwegian Continental Shelf. In the predictability of emulsion stability it is of crucial importance to be capable of separating the surface-active fractions in the crude that are responsible for the stabilization. After this the physical chemistry of this separated fraction should be investigated with regard to film formation and emulsion stabilization.

When discussing the equally important impact of chemical destabilization a proper understanding of the stability mechanisms must be the basis. The interaction between the destabilizer and the interfacially active substances in the crude oil will dictate the destabilization sequence of the colloidal system. In order to be able to direct these interactions into a preferred direction an extensive mapping of the solution chemical properties of the stabilizing agents is requirred.

In this paper we give a survey of physicochemical properties of crude oils from the

Norwegian Continental Shelf, together with their readiness to form emulsions. The emulsions are characterized by means of droplet sizes and deviations in these for different emulsification conditions. In order to penetrate into the fundamental chemistry behind the stabilization of the emulsions we have developed a consecutive adsorption procedure of the surface active components in the crude oil onto solid silicabased surfaces. We also present some physico-chemical properties of this adsorbate together with its stabilizing ability of the oil-continuous emulsions. Finally, a comparison between authentic crude oil emulsions and these model emulsions is undertaken.

Experimental

Geographically, the crude oil samples cover the largest possible area from the Edda field in the south to the Snöhvit field in the north on the Norwegian Continental Shelf.

In determining the analytical and the physico-chemical properties of the crude oils, we have adopted procedures described before [1]. The emulsification procedures and the composition of the aqueous phase (when it is not distilled water) has also been reported before [1].

Chemicals

The purity of the chemicals was reported earlier [2].

Separation of the surface active fraction

Preparative silica from Waters Associates, Millipore Corp., was used as adsorbent in the separation process. 10 ml of the crude oil was diluted to 100 ml with hexane to decrease the viscosity. 5 gram of the solid adsorbent was added and the mixture was stirred in order to disperse the solids. The suspension was then allowed to sediment and was afterward filtered to remove the solids. A consecutive adsorption was undertaken on the residual oil, but during this time the adsorbent was added in a quantitiy sufficient to remove the black color of the oil. After the second adsorption n-hexane was evaporated, and the remaining crude oil was used for preparing W/O emulsions. The result of this emulsification was that *no* stable emulsions could form.

Model system

The purpose of the model system is to show that the adsorbed fraction is capable of stabilizing water-in-oil emulsions.

The adsorbed fraction was dissolved from the solid silica using a mixture of 7% methanol in dichloromethane. The solvent was later evaporated and the adsorbate was dissolved in decane. Together with water (or formation water) and dodecane in equal proportions this decane solution was emulsified under similar conditions as was the authentic crude oil emulsions reported on previously [1, 2]. For the sake of comparison with the authentic crude oil systems, we have kept the amount of emulsifier equal in both cases.

Apparatus

FTIR (Fourier Transform IR) spectra of the adsorbed material were obtained by using the diffuse reflectance technique. The analysis was performed with a Perkin Elmer 1720 FTIR Spectrometer equipped with a DTGS detector and diffuse reflectance accessory. The samples were deposited from carbon tetrachloride solution on firmly packed KBr powder [3] and the solvent was evaporated by warming the sample cup at 60 °C. The spectra were recorded from 4000 to 600 cm^{-1} using 20 scans and a resolution of 4 cm^{-1}. The obtained spectra were then ratioed with reference spectra of KBr. The remaining diffuse reflectance spectra of the adsorbed material were then transferred into Kubelka-Munk [4–5] format for comparative analysis.

A Hewlett-Packard MSD 5970 gaschromatograph linked to a masspectrometer was also used for identification of species in the adsorbed fraction.

Results

Table 1 gives a survey of analytical chemical properties of the crude oils from our research matrix. The table displays that the content of aromatic compounds ranges from 10% to 50% (w/w) and satur-

Table 1. Analytical chemical properties of appropriate representative of crude oils from the Norwegian Continental Shelf

Crude oils	Aromatics (%, w/w)	Saturates (%, w/w)	Polar compounds	Waxes (%, w/w)	Asphaltenes (%, w/w)
A	11.2	84.7	4.0	3.9	0.2
B	17.4	78.2	4.4	4.3	0.7
C	36.4	50.7	12.9	4.5	0.6
D	28.5	66.3	5.2	10.1	0.4
E	36.0	59.1	4.9	14.5	0.4
F	29.2	68.2	2.6	2.9	0.0
G	36.1	58.2	6.0	8.6	1.5
H	47.7	46.0	6.2	3.6	0.8
I	32.9	61.0	6.1	4.6	0.2
J	40.6	53.0	6.4	2.4	0.4

Table 2. Physicochemical properties at 20° and 40 °C

Crude oils	Density (g/cm³)	Viscosity (mPa s) (at 100 s^{-1})	Surface tension σ (mN/m) 20 °C	Surface tension (mN/m) 40 °C	Interfacial tension $\sigma_{0/w}$ (mN/m) 20 °C	Interfacial tension 40 °C
A	0.828	6.5	26.3	25.3	23.0	23.8
B	0.835	9.5	27.4	25.2	23.8	21.2
C	0.876	16.6	28.5	27.0	23.7	23.6
D	0.858	41.2	30.0	27.7	33.5	27.2
E	0.838	34.2	29.9	25.2	36.3	27.3
F	0.793	2.2	24.3	23.3	25.3	23.4
G	0.873	24.4	29.0	27.4	27.5	27.8
H	0.887	17.9	29.0	27.7	27.0	27.0
I	0.828	5.8	25.9	24.9	23.9	26.1
J	0.874	15.6	28.2	27.0	26.3	25.7

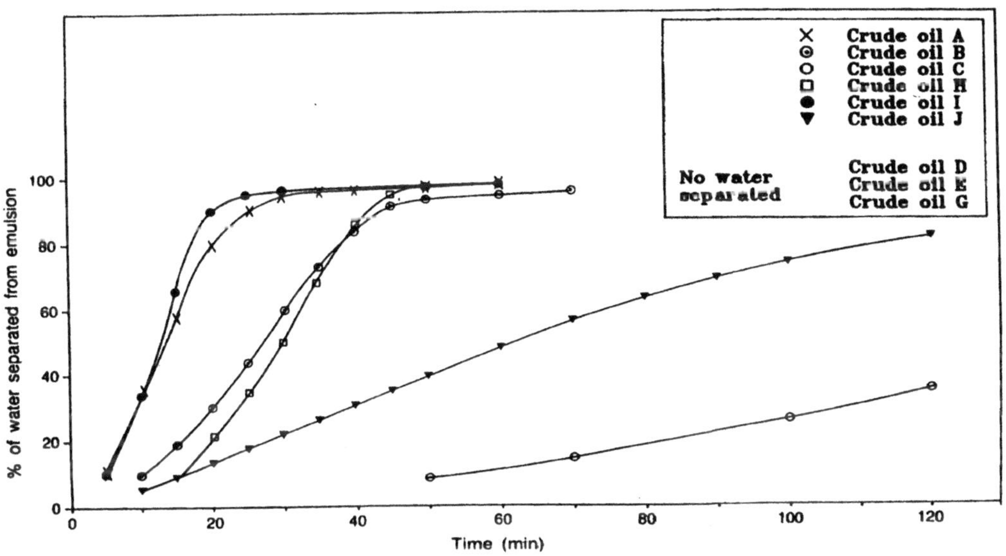

Fig. 1. Separation of formation water from w/o emulsions based on different crude oils from the Norwegian Continental Shelf. Separation was undertaken at room temperature. The content of synthetic formation water was 50% by volume

ated hydrocarbons from 50% to 85%. Non-specified polar compounds are between 4% and 13%. Waxes precipitated in acetone vary from 2% to 15%. The amount of asphaltenes is low, i.e., in the range 0% to 1.5%.

The physico-chemical properties in Table 2 at 20° and 40 °C show that the densities vary between 0.79 and 0.89 g cm^{-3}, the viscosities (as measured at 100 s^{-1}) vary between 2 and 41 mPa s, the surface tensions vary between 24 and 30 mNm^{-1}, and the interfacial tensions (towards synthetic formation water) vary between 23 and 36 mNm^{-1}.

The tables reveal, as expected, a correlation between a high content of wax particles and a high viscosity. In two specific cases the interfacial tension exceeds the surface tension, this is also the case for high amounts of waxes.

The ability of each category of crude oil to form stable emulsions is shown in the following figures. Obviously, one can, based on a visual inspection as in Fig. 1, classify the emulsions as unstable, stable, and very stable on a time scale of minutes, hours, and weeks (or months), respectively.

Figure 1 is based on the separation time for 20% (v/v) of the emulsified aqueous phase. Other independent methods have given a similar stability sequence [1].

When characterizing the emulsion state we have chosen to visualize the droplet size and changes herein when outer conditions are changed, and

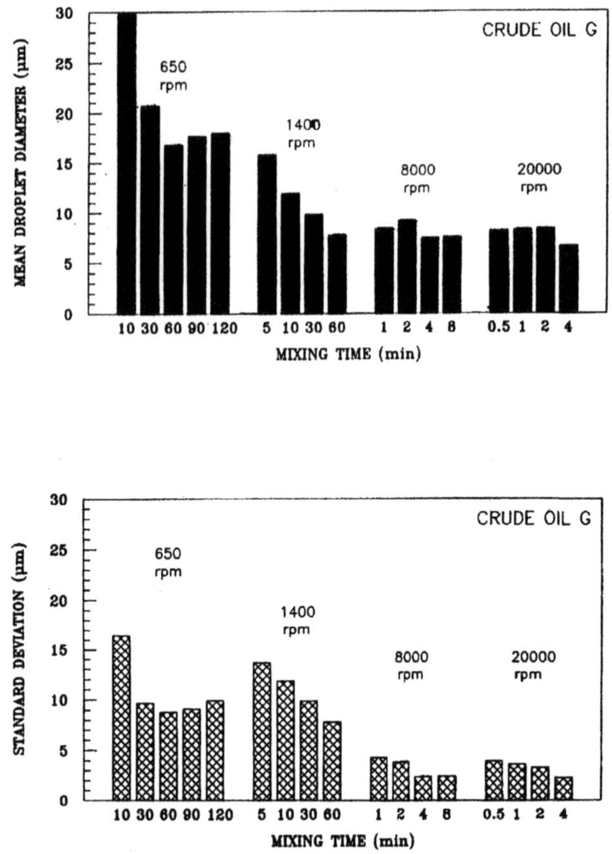

Fig. 2. Mean droplet sizes of water-in-crude-oil emulsions based on crude oil G for different preparation conditions. The aqueous phase was 50% by volume

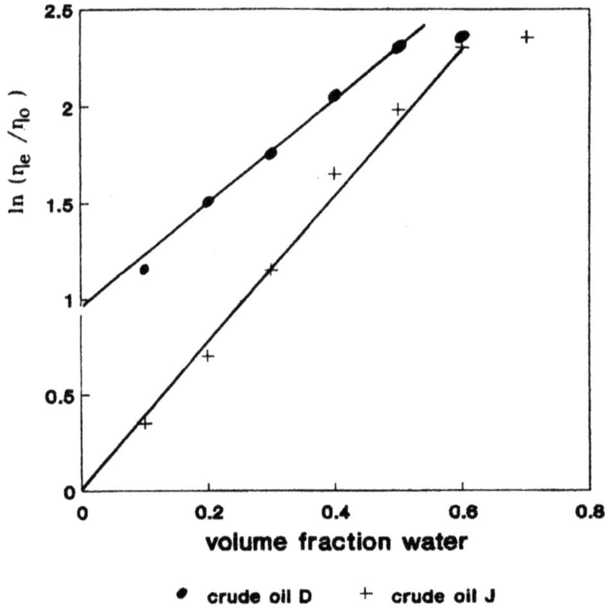

Fig. 3. The logarithm of the ratio of the viscosity of the crude oil emulsion to that of the pure crude oil as a function of the content of water in the emulsion

also macroscopic viscosities. Figure 2 shows how different mixing times and emulsification rates influence the droplet sizes and their standard deviations in the emulsion. Obviously, higher mixing rates will reduce the mean droplet size while a prolongation of the emulsification time will mainly give rise to more equidimensional droplets.

Figure 3 shows that there seems to be a linear relation between the logarithm of the ratio between the viscosity of the emulsion and the crude oil (measured at 100 s^{-1}), and the volume fraction of the water content in the emulsion. Obviously, a Richardson type equation [32], i.e.,

$$\eta_e/\eta_0 = e^{n\Theta+k} \quad (1)$$

will apply in most cases. However, at low shear rates many of the crude oil emulsions show non-Newtonian behavior.

In order to separate the interfacially active components from the crude oils we have practiced different adsorption methods, as described in the Experimental section. We have characterized the adsorbate by means of a mass-spectrometer equipped with a gaschromatograph, and FT-IR, respectively. Figure 4 gives a GC-MS result for a crude oil (denoted *D*). The fragments observed show that the sample contains, for example, $CH_3(CH_2)_{14}COO$ (TMS) with M = 328. The result from the FT-IR characterization is given in Fig. 5. The figure shows a typical spectrum of adsorbate samples.

The stability sequence obtained for the emulsions based on decane/formation water and the adsorbed fraction as stabilizer is given in Fig. 6. As a reference we have inserted the stability curve of the original crude oil emulsions. When comparing the original W/O emulsions based on authentic crude oils and the model emulsions it is, of course, important that the experimental conditions are identical. In our experiments we have consequently used the same ratios between the stabilizing fraction, dispersed phase (water/synthetic formation water), and the oil phase, respectively. Obviously, a lower, a similar, or an enhanced stability can be obtained when comparing the model emulsions with the authentic crude oil emulsions.

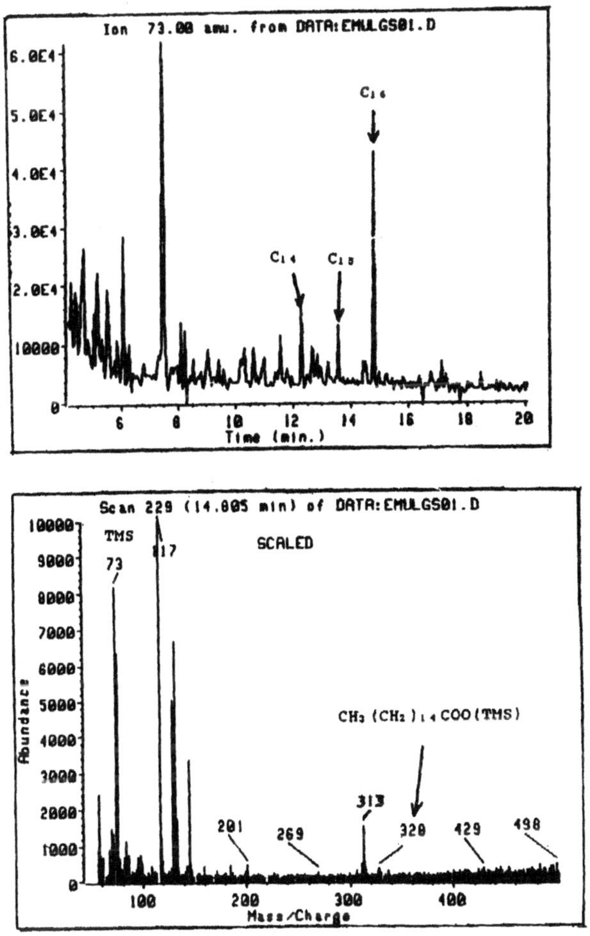

Fig. 4. A combined gaschromatogram and masspectrogram of crude oil *D*. The peak at 328 corresponds to $CH_3(CH_2)COO$ (TMS)

Fig. 5. A combination of FTIR spectra of adsorbed interfacially active fractions from crude oil *D*. The upper curve refers to the result of the first adsorption, while the lower curve refers to the second adsorption from the same crude oil. The spectrum is characteristic for adsorbates from all crude oils investigated

The assignments are:

3200—3300 cm^{-1}, OH stretching;
2850—2950 cm^{-1}, CH stretching;
1600—1650 cm^{-1}, aromatic C=C or asymmetric —COO stretching
1430—1460 cm^{-1}, asymmetric CH_2 or CH_2CH_3
1340—1360 cm^{-1}, symmetric CH_2CH_3 bending

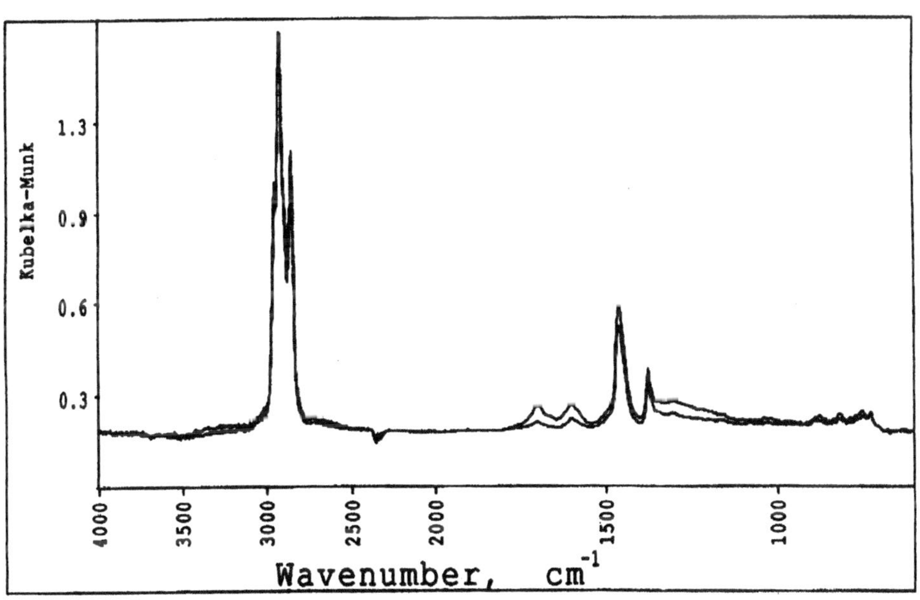

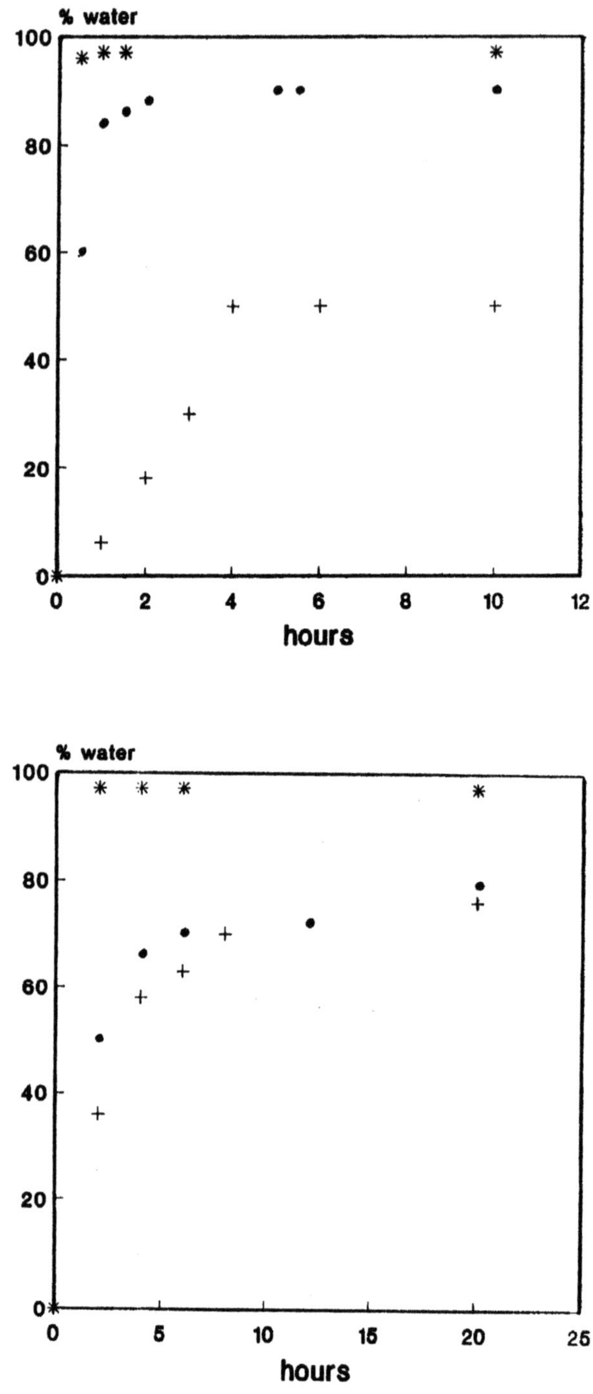

Fig. 6. Separation of water (or synthetic formation water) from water-in-oil emulsions as a function of time. (●) denotes the authentic crude oil, (∗) the residual crude oil after adsorption, and (+) a hydrocarbon/water model system stabilized by the adsorbed interfacially active fraction. The crude oils investigated are *Y* (upper curve) and *B* (lower curve)

Table 3. Effect of different chemical additives on crude oil emulsion stability. The separation of water from the emulsion was measured after 30 min and 24 h, respectively (% = % by volume)

Chemical additive (1.5%)	Water separation (%)			
	Separation time 30 min Crude oil		Separation time 24 h Crude oil	
	D	*G*	*D*	*G*
Ethanol	0		0	
Butanol	50	80	85	95
Pentanol	36		80	
Hexanol	17		17	
Decanol	0		0	
Benzyl alcohol	90	90	100	90
Propylamine	0		0	
Pentylamine	40		40	
Hexylamine	85	80	90	80
Octylamine	85		95	
Decylamine	95		100	
Formic acid	10		10	
Butanoic acid	35		35	
Pentanoic acid	0		0	
Hexanoic acid	0		0	
Heptanoic acid	0		0	
Octanoic acid	0		0	

Finally, Tables 3 and 4 give the destabilization sequence when chemical additives are used. In some cases a distinct similarity is obtained when comparing the authentic crude oil system and the model systems.

Discussion

With the intention of creating fundamental understanding of the stability mechanism behind the crude oil emulsions much effort has been directed towards the chemistry of the chemical constituents in the crude oils. When analyzing the influence of, for example, asphaltenes, waxes, porphyrines, and other nonspecified polar compounds it is very striking how rough the separation techniques are. Asphaltenes are precipitated in hexane or pentane, waxes are precipitated in acetone or at low

Table 4. Effect of different chemical additives on model oil emulsion stability. The separation of water was measured after 20 min, 1 h and 24 h, respectively. (% = % by volume)

Separation of distilled water (%)

Chemical additive (1.5%)	Separation time:								
	20 min			60 min			24 h		
	A	*D*	*X*	*A*	*D*	*X*	*A*	*D*	*X*
Ethanol	15	0	0	50	10	0	90	20	0
Butanol	80	5	0	80	20	15	80	40	55
Pentanol	60	5	0	70	20	20	70	35	35
Hexanol	70	5	0	70	20	20	70	35	50
Benzyl alcohol	90	0	40	90	5	40	90	50	70
Hexylamine	90	50	60	90	50	70	90	50	70
Octylamine	100	40	90	100	50	90	100	80	90
Decylamine	100	0	100	100	20	100	100	100	100

Separation of synthetic formation water (%)

Chemical additive (1.5%)	Separation time:								
	20 min			60 min			24 h		
	A	*D*	*X*	*A*	*D*	*X*	*A*	*D*	*X*
Ethanol	15	0	0	30	0	0	90	0	0
Butanol	60	0	5	60	0	5	60	0	30
Pentanol	60	0	0	60	0	0	60	0	30
Hexanol	65	0	0	65	0	0	65	0	5
Benzyl alcohol	65	40	45	65	40	50	65	40	65
Hexylamine	70	15	20	70	20	20	70	20	30
Octylamine	70	60	0	75	100	0	75	100	55
Decylamine	80	100	15	80	100	45	80	100	80

temperatures, for instance. As a consequence it will be rather difficult to obtain unambiguous experimental results from stabilization/destabilization tests with these operationally defined fractions. In a previous study on crude oil emulsions from the Norwegian Continental Shelf we found a weak correlation between the analytical content of the different constituents and the final emulsion stability [1]. Our explanation is inadequate separation techniques.

In this paper we have approached the separation problem from another direction. We have, from diluted solutions of crude oils, adsorbed active fractions onto solid surfaces. For the sake of internal comparison between the different crude oils used in our test matrix, we started with only one adsorbent, i.e., silica. With this polar surface, we simulate conditions in the water-in-crude oil emulsion, and are capable of removing the material which tends to locate at the interface between the polar and nonpolar phase. In this way we undertake a physico-chemical separation, since we isolate the interfacially active fraction from the crude oils. GC-MS analysis (Fig. 4) reveals that this fraction contain among other constituents, also fatty acids; especially, palmitic acid was identified. For a comprehension of the physico-chemical properties of the adsorbed fraction it is of crucial importance to study its emulsifying properties and, conversely, to show the declining stability properties of the emulsions based on the crude oils from which we have adsorbed the active fractions.

Figure 6 shows very clearly that the residual crude oil is, after adsorption, *completely incapable* of forming stable emulsions. The demulsification is instanteneous and complete, regardless of which crude oil is investigated.

Secondly, we have used the adsorbed fraction in model systems. In these we use a synthetic hydrocarbon and distilled (or alternatively formation) water and the adsorbed fraction as stabilizer. In this way we can, as shown in Fig. 6 and Tables 3—4, compare the stability of the authentic crude oil emulsions and the model emulsions, respectively. It is obvious that the model systems are capable of forming stable emulsions. These emulsions can have a stability comparable with the authentic crude oils emulsions. Obviously the separation technique adopted can remove a fraction responsible for the stabilization of some of the crude oil based emulsions.

In order to create an understanding of the destabilization mechanisms, we have investigated the influence of well-defined chemical additives on the authentic, as well as on the model systems. In our previous study we found that two categories are especially efficient in breaking the crude oil emulsions. These additives were medium-chain alcohols (for instance 1-butanol and benzylalcohol) and fatty amines, respectively. In the present study we have a similar conclusion for the model emulsions stabilized by the adsorbed fraction. As seen from Table 4 the model emulsions can be fairly easily destabilized with fatty amines while 1-butanol and benzylalcohol, although the best of the hydroxylic compounds, are less efficient. It must be pointed out that there can be distinct differences between the authentic and the model systems, respectively. The reason for this lies in the separation technique. With the adsorption procedure adopted in this study we have been able to remove rather similar interfacially active fractions according to characterization by means of FT-IR and GC-MS. In cases where the water-in-crude-oil emulsion is further stabilized by particles (waxes), which is the case for many of the crude oils in the Table, a complete agreement between model and authentic systems cannot be achieved. This is most conspicuous for crude oil *D*, which is characterized by the highest wax content. Furthermore, the model emulsions will fail to reproduce the full influence of an enhanced viscosity on the emulsion stability. The viscosity of the model emulsion is at least one order of magnitude lower than the corresponding crude oil emulsion. However, it is important to point out that when the interfacially active fraction is removed by means of adsorption the residual crude oil is completely incapable of forming stable emulsions. Hence, although other mechanisms can be and are involved in the stabilization, they are *not* themselves capable of giving rise to stable water-in-crude oil emulsions.

Qualitative destabilization mechanisms for the polar additives used in this study have been proposed. A crucial parameter for alcohols to act as destabilizers seems to be their ability to dissolve into the different regions in the emulsified system („pseudophases"). Hence, medium-chain alcohols (1-butanol, 1-pentanol and benzylalcohol in the table) readily dissolves in the aqueous, the interfacial, and in the bulk (oil) phase. The partitioning of the molecules between the pseudophases will create disadvantageous conditions for the stability of the water-in-crude-oil emulsions [2, 6]. The interfacial region will undergo changes towards less rigidity and structuring. One can expect this region to possess higher dynamics and considerable interfacial fluctuations in the presence of a medium-chain alcohol. The situation will be somewhat reminiscent of that of the so-called microemulsions, where a medium chain alcohol is added to a solution consisting of water, an ionic surfactant, and a hydrocarbon. In this case the alcohol will break down the strict micellar geometry and create other, highly dynamic solution conditions [7—18]. It should be noted that the parallel between microemulsions and true emulsions is not completely valid since the former represents *thermodynamically stable solutions* while the emulsions are a dispersed state. Otherwise, droplet sizes and interfacial conditions also seem to deviate. A significant deviation seems to be the importance of interfacial multilayer constellations as a criterion for emulsion stability; this has been proved very convincingly by Friberg [19—22]. The components in crude oils provide many possibilities to create multilayers at the interface oil/water. It is well-known that aromatic molecules like the asphaltenes will stack into a sandwich-like structure when they associate [23]. Further, it is also known that fatty acids will, depending on salinity and pH, associate into lamellar liquid crystalline phases [24—27]. Hence it cannot be any surprise to find multilayers at the interface in the water-in-crude oil emulsions. It is, therefore, interesting to point out that medium chain alcohols like 1-butanol have a characteristic tendency to reduce the existence regime of the lamellar liquid crystalline *D* phase as viewed by means of phase diagrams [28—30]. It seems likely that a similar mechanism will influence the stability of the emulsions reported on in this paper.

In order to understand the effect of fatty amines on water-in-crude oil emulsions, it is essential to point out that there is a completely different concentration behavior from that of the medium chain alcohols. After about 2% of the amine the separation of water stagnates. This could be attributed to a very specific interaction between the amine added and naturally occurring constituents in the interfacial film. Since the adsorbed and stabilizing fraction has been proved to contain fatty acids (palmitic acid) the interaction can be located to these components and the fatty amines. The complexation between the fatty acid and the fatty amine will result in polar (charged) complexes. These charged sites will create a lateral electrostatic repulsion and, hence, make impossible a denser molecular packing in the interfacial film. With these hydrophilic groups present the film will no longer provide stabilization of the water-in-crude-oil emulsion towards coalescense. The complexation of fatty acids and amines in nonpolar media, together with the interfacial activity of the complexes is under investigation and will be reported on in a future paper [31].

Acknowledgement

This project has been financed by Norsk Hydro A/S, Saga Petroleum A/S and Statoil A/S. Financial support from Statoil A/S (Stavanger) and the Nansen Foundation for the FTIR Spectrometer is also gratefully acknowledged. Dr. Ronald Scheline is thanked for the skillful uptake of the gaschromatograms and masspectrograms.

References

1. Johansen EJ, Skjärvö IM, Lund T, Sjöblom J, Söderlund H, Boström G (1988) Colloids Surfaces 34:353
2. Sjöblom J, Söderlund H, Lindblad S, Johansen EJ, Skjärvö IM (1990) Colloid Polym Sci 268:389
3. Christy AA, Dahl B, Kvalheim OM (1989) Fuel 68:430
4. Kubelka P, Munk FZ (1931) Tech Phys 11a:593
5. Christy AA, Tvedt JE, Karstang TV, Velapoldi RA (1988) Rev Sci Instrum 59:423
6. Sjöblom J, Li Mingyuan, Höiland H, Johansen EJ (1990) Colloid Surfaces 46:127
7. Danielsson I, Lindman B (1981) Colloids Surfaces 3:391
8. Hoar TP, Schulman JH (1943) Nature (London) 152:102
9. Zulauf M, Eicke HF (1979) J Phys Chem 83:480
10. Cazabat AM, Langevin D, Pouchelo A (1986) J Colloid Interface Sci 73:1
11. Cebula DJ, Ottewill RH, Ralston J (1981) J Chem Soc, Faraday Trans I 77:2585
12. Cole RH, Delbos G, Winsor IV P, Bosaer TK, Moreau JM (1985) J Phys Chem 89:3338
13. Lindman B, Stilbs P in Friberg S, Borhorel P (eds) (1987) Microemulsions. CRC Press
14. Lindman B, Stilbs P, Moseley E (1981) J Colloid Interface Sci 83:569
15. Wärnheim T, Sjöblom E, Henriksson U, Stilbs P (1981) J Phys Chem 88:136
16. Biais J, Ödberg L, Stenius P (1982) J Colloid Interface Sci 86:350
17. de Gennes PG, Taupin C (1982) J Phys Chem 86:2294
18. Binks BP, Meunier J, Abillon O, Langevin D (1989) Langmuir 5:415
19. Friberg S, Jansson PO, Cederberg E (1976) J Colloid Interface Sci 55:614
20. Friberg S (1971) J Colloid Interface Sci 37:291
21. Friberg S, Mandell L, Larsson M (1969) J Colloid Interface Sci 29:155
22. Friberg S, Solans C (1986) Langmuir 2:121
23. Siffert B, Bourgeois I, Papierer E (1984) Fuel 63:834
24. Friberg S, Mandell L, Ekwall P (1966) Acta Chem Scand 20:2632
25. Ekwall P, Mandell L (1968) Kolloid Z Z Polym 233:938
26. Ekwall P (1969) J Colloid Interface Sci 29:16
27. Ekwall P, Mandell L, Fontell K (1969) Mol Cryst Liq Cryst 8:157
28. Ekwall P In Brown GH (eds) (1975) Advances in Liquid Crystals 1:1. Academic Press, New York
29. Sjöblom J (1982) Thesis, Åbo Akademi, Åbo, Finland
30. Wärnheim T (1986) Thesis, The Royal Institute of Technology, Stockholm, Sweden
31. Saeten JO, Sjöblom J, Gestblom B, J Phys Chem. In press
32. Richardson EG (1933) Kolloid Z 65:32

Authors' address.

J. Sjöblom
Department of Chemistry
University of Bergen
5014 Bergen-U., Norway

Progress in Colloid & Polymer Science Progr Colloid Polym Sci 82:140—154 (1990)

Gas-blocking foams in porous media: effects of oil and surfactant hydrophobe carbon number

T. Meling and J. E. Hanssen

Institute of Petroleum Engineering, Rogaland University Centre, Stavanger, Norway

Abstract: An experimental study of the ability of foams to reduce gas mobility in a porous medium is reported. Single carbon number fractions of a surfactant and pure n-alkanes were used as model liquids, and a glass-bead pack was used as a model porous medium. In this simple, idealized model system of a gas-blocking foam in an oil reservoir, clear trends in the mobility reduction, as measured in a gas-blockage test, with varying combinations of oil and surfactant hydrophobe carbon number were observed. These trends could not be reproduced by foam stability as measured in conventional bulk tests, and were not explainable by several potential controlling factors that have been sought to explain similar data before. However, the change in interfacial tension between gas and the aqueous phase upon equilibration with oil appears to be a good predictor of gas-blockage efficiency for the systems studied.

Key words: Foams; porous media; gas blockage; oil sensitivity; correlation

Introduction

Background

The presence of a foam in a hydrocarbon reservoir, which for typical foams and pore sizes exists in the form of elongated lamellae (as illustrated in Fig. 1) can have a profound influence on the mobility of reservoir fluids. In particular, the mobility of gas can be reduced by several orders of magnitude.

Holm [1] showed that fluids flow by different mechanisms in a sandpack filled with foam: liquid flows as a continuous phase while gas flows by breaking and re-forming lamellae. He ascribed the apparent gas blockage to the latter process, which is very slow if the number and strength of liquid films is high. The potential value of foams in controlling reservoir fluid mobilities has led to extensive research that has shown foams to have an exceedingly complex flow behavior in porous media. Reviews on the subject are scarce; the reader is referred to a recent paper by Radke et al. [2] which discusses previous work and attempts to generalize this with new measurements of foam flow in sandstone.

When a foam exists in pores which also contain oil its efficiency in reducing fluid mobilities can be reduced. This oil-sensitivity of the foam, however, varies considerably between different surfactants. No clear trends have yet been presented for the relations between the mobility-reducing efficiency of foams in porous media and surfactant and oil properties. For applications which require maximum gas blockage, such as the authors' own interest in improving production from thin oil zones [3], high oil tolerance is, of course, essential. For other applications, a somewhat oil-sensitive foam may be desirable in order to optimize displacement efficiency.

Scope

The scope of the present work was to examine a model system with a small number of variables, and attempt to characterize some of the factors controlling the ability of foams to reduce gas mobility and the influence of oil on gas blockage efficiency. A secondary objective was to evaluate the predictive power of bulk tests for screening purposes and, if

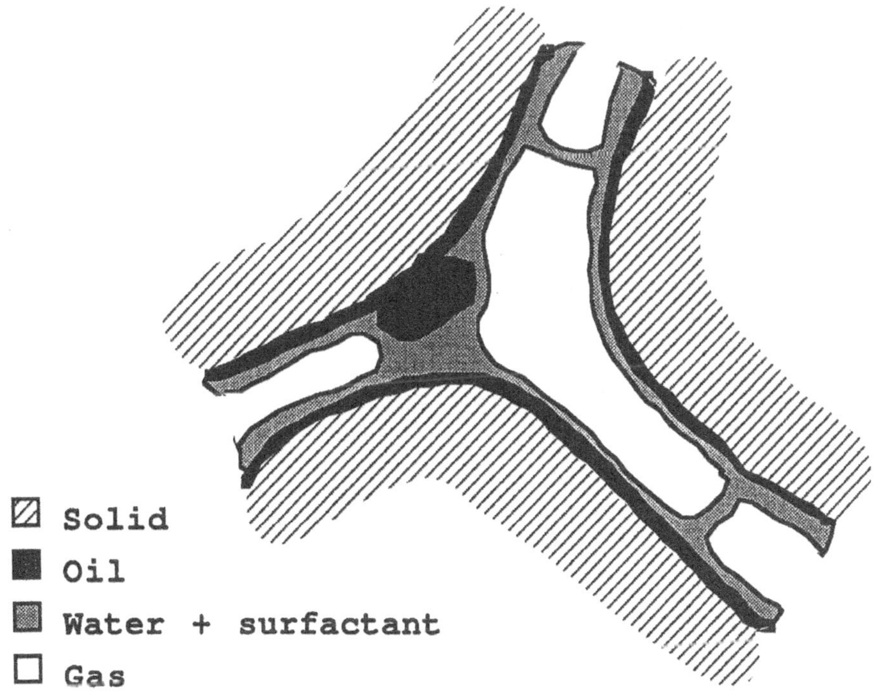

Fig. 1. Conceptual sketch of foam in porous medium with residual oil present. A lamella is the gas "bubble" trapped by surfactant films and the assumed water-wettable pore walls. From film has gas on both sides, pseudoemulsion film separates gas from oil. Drawing based on micromodel observations

unsatisfactory, to search for better prediction criteria. Varying only the combination of pure n-alkanes with foams made from single carbon number fractions of a surfactant in a well-characterized, homogeneous porous medium, the effects of oil and surfactant hydrophobe carbon number on foam gas-blocking ability, bulk foam stability and some physico-chemical properties were measured.

Experimental

Materials

The surfactant used was Hostapur SAS, a secondary alkyl sulfonate (Na salt) with the SO_3^- group statistically distributed along the hydrocarbon chain. The commercial product SAS 60 contains 58% C_{13}—C_{15} and 39% C_{16}—C_{17} alkyl sulfonates and has an average molecular weight of 328 [4]. Fractions of single carbon number C_{11} through C_{17} alkyl sulfonates were previously made available for our research by Hoechst. The C_{13}, C_{15}, and C_{17} SAS fractions and SAS 60 were used in the present study. The surfactant concentration was 2.0% (w/w) active material except for the documented gas-blockage tests in the absence of oil where 0.50% was employed. Distilled water was used as solvent. The oils used were n-octane, n-dodecane, and n-hexadecane, all of 99% purity, with a nonadsorbing dye (Sudan blue) added at 0.1% (w/w) to enable visual saturation observations. For bulk foam tests, compressed air was used as the gas phase; for gas-blockage tests, bottled nitrogen of 99.5% purity was used without further treatment.

Bulk foam tests

Bulk foams were created by bubbling air at 200 ml/min through a 25 μm glass sinter into 25 ml of surfactant solution in a 250 ml calibrated glass cylinder, at room temperature. Volumes of foam and drained liquid were measured as functions of time, and foam texture was observed.

Gas-blockage tests

A model porous medium was made by packing spherical glass particles (Jencons Scientific, UK) of 80—105-μm diameter in glass columns (Omnifit, UK) 1-m long and 10 mm in inner diameter, rated at 40 bar. The bead pack was mounted in a constant pressure fluid displacement apparatus, illustrated in Fig. 2, with maximum operating conditions of 20 bar and 100 °C. All experiments reported in this paper were performed at room temperature. Constant back pressure within ±0.2 bar during flow was maintained at a set value between 6.5 and 7.5 bar in all experiments, and measured to ±10 mbar with a Schaevitz P-721 transducer. Differential pressure

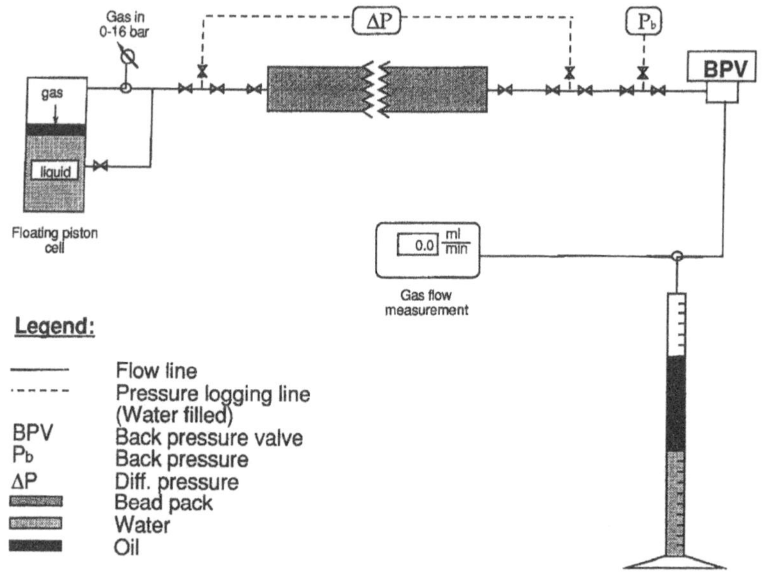

Fig. 2. Experimental apparatus used, simplified. Heater bath for operation at elevated temperatures not shown. Actually contains two columns with sets of pressure sensors, BPR, and flow measurement for increased efficiency

was measured to ±2 mbar using a Honeywell Smart differential transmitter. Liquid production was measured in a byrette and gas-flow rates were measured at atmospheric pressure using a bubble flow meter. Saturations were measured by weighing the column and by material balance.

The gas-blockage test method has been described elsewhere [5]; it briefly consists of generating foam in the porous medium by displacement of surfactant solution with nitrogen gas at a constant differential pressure (target value 1.5 bar), and subsequently testing the foam for gas-blockage efficiency by measuring stabilizing gas flow rates as a function of step-wise increasing gas-pressure gradients. In order to perform the gas-blockage test with a repeatable saturation of oil in the porous medium, the bead pack was evacuated and oil introduced by forced imbibition. The pack was then flooeded by water to residual oil saturation, and the water was displaced by surfactant solution.

Physicochemical properties

Phase to be equilibrated were left standing for 1 week with occasional agitation. Solubilization of oil into the aqueous phase was measured by gas chromatography (HP 5880A instrument, 20% SP 2100/0.1% Carbowax 1500 column, FID) using 1.00% (w/w) methanol as internal standard. Surfactant and oil bulk viscosities were measured on a Contraves LS-30 rheometer, and densities measured on a Paar DMA 46 density meter. Bulk interfacial tensions (air/water, air/oil, oil/water) were measured by the du Nüoy ring method using a Krüss tensiometer. Relative wetting of a glass surface by surfactant solution was measured by micropipetting 50 μl droplets onto a glass plate cleaned with detergent, rinsed with acetone and dried, and measuring droplet areas after a period of 5 min.

Results

Bulk foam stability

Measured bulk foam volume for the commercial SAS blend at regular intervals is shown by Fig. 3. The presence of oil generally enhances bulk stability. This was seen for all SAS fractions and is not uncommon among other foams studied [6]. A positive effect of oil was also seen on liquid drainage for all but two systems, drainage half-life values being 6 to 7 min without oil, but up to 11 minutes with oil. The different degree of enhancement by each oil on foams made from three SAS fractions is seen in Fig. 4; dodecane enhances bulk stability the least, and C_{13} SAS gives the least stable foam. Complete data are given elsewhere [7].

Gas-blockage results

The glass beads used in this work have been characterized for pore-size distribution by mercury porosimetry [8]. A Gaussian distribution with a mean throat radius of 20 μm was found. In this

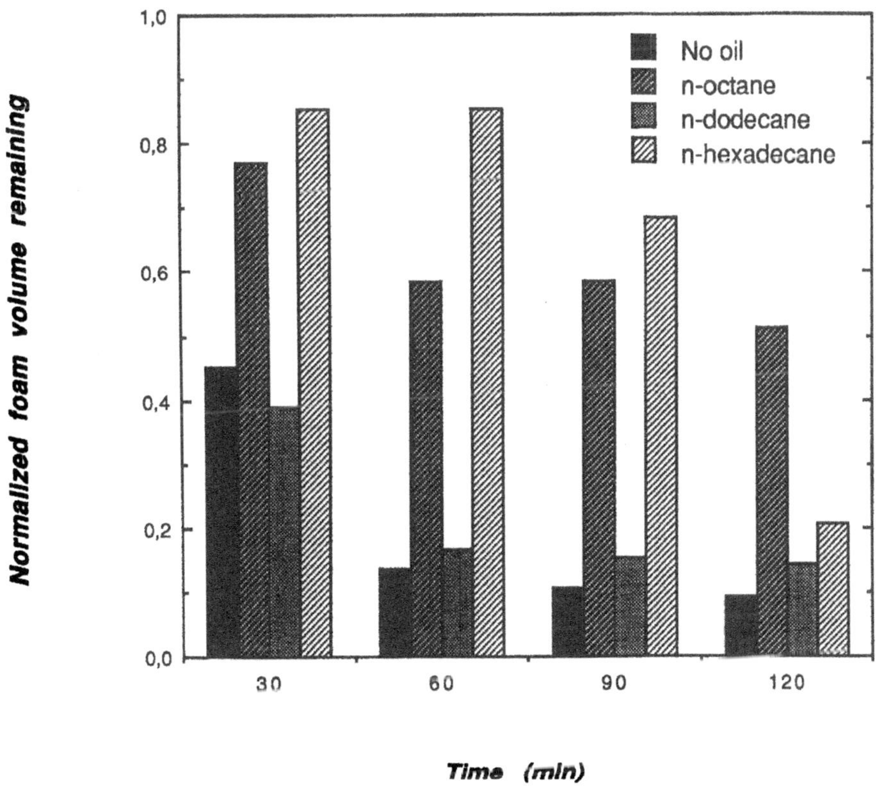

Fig. 3. Bulk stability of 2.0% (w/w, active) Hostapur SAS 60 foam including the effect of different oils on foam volume

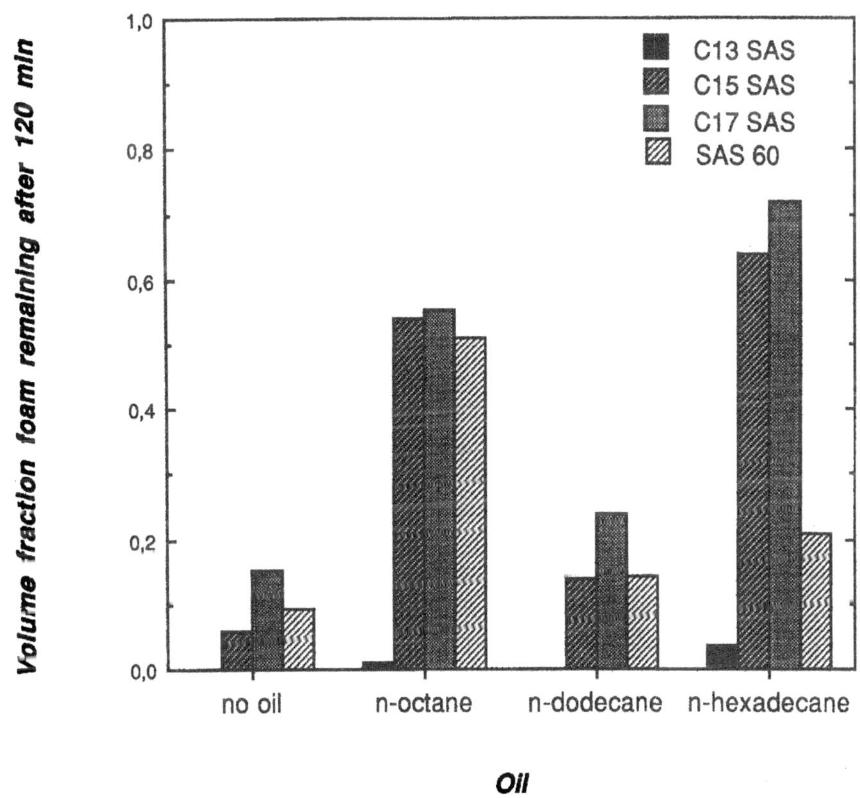

Fig. 4. Bulk stabilaity of 2.0% (w/w, active) SAS 60 and single carbon number Hostapur SAS-fraction foams as observed after 120 min of foam drainage. Effect of different oils

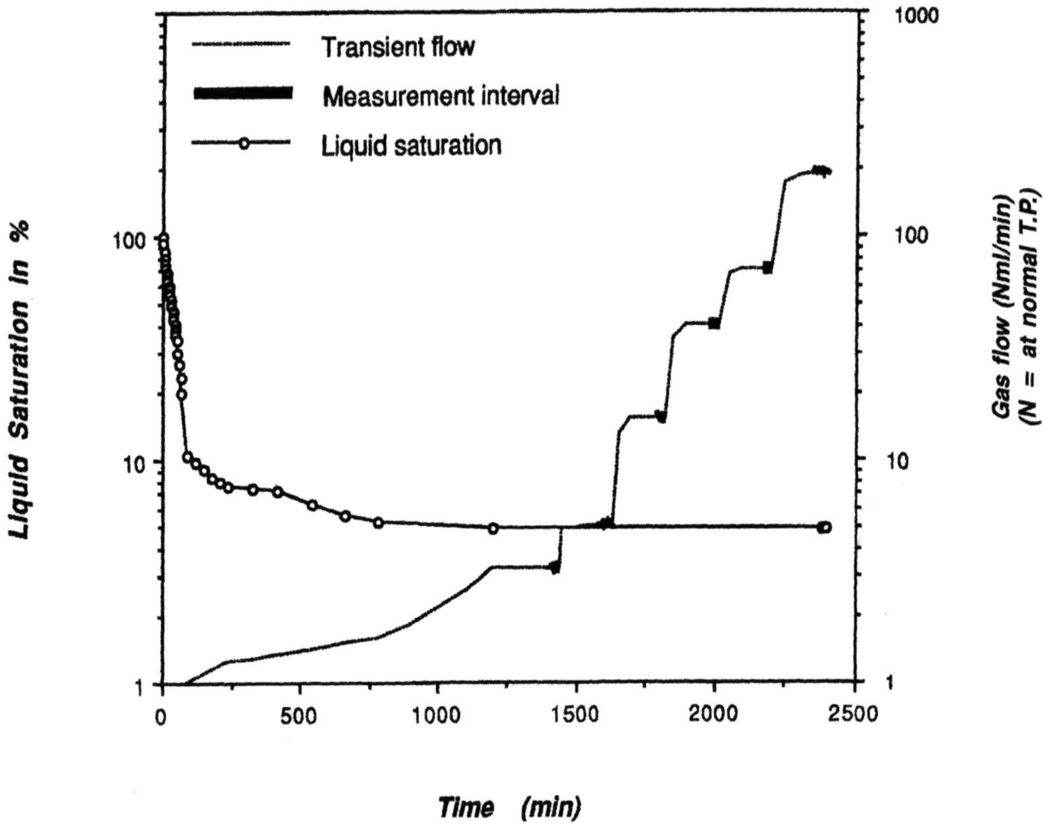

Fig. 5. Measured production data in a gas-blockage test. The vertical portions of the gas-flow curve indicate times when differential pressure was increased, typically in 1-bar increments from a foam-generation pressure of 1.5 bar. Lines are traces

study, a porosity of 0.38 ± 0.01 and an absolute permeability of 7.5 ± 0.2 μm^2 (7500 ± 200 milliDarcy, mD) was obtained in dry packings which were always done using new beads for each experiment. Thus, we state that the model porous medium used gives a mean pore size and permeability that is reasonably representative of some high-permeability hydrocarbon reservoirs of interest for gas-blockage application of foams. Similarly, the test procedure used is believed to be mechanistically relevant to the mode of displacement likely to occur during *in situ* foam generation in such reservoirs [3].

Water flooding and subsequent surfactant flooding were observed to produce a generally homogeneously distributed residual oil saturation between 0.03 and 0.07 (fractions of pore volume) in all experiments. It was verified that no measurable amounts of oil were produced during foam generation and gas-blockage testing, although oil could be produced by continued flooding with several pore volumes of surfactant because of solubilization. The aqueous-phase saturation remained constant during gas-blockage tests at values ranging from 0.04 to 0.37; it was lowest (0.04 to 0.06) for the oil-free experiments, but for the oil-containing systems, could not be correlated in any systematic way. Foam quality consequently varied between 0.61 and 0.97 and also was uncorrelatable to gas-blockage efficiency or to surfactant and oil carbon number, except being high for the oil-free systems.

Measured gas flow and liquid saturation data in a typical experiment is shown by Fig. 5. Foam generation is considered complete when liquid saturation has reached a constant level. Gas-flow rate is recorded when stable. The pressure drop across the foam-filled column is then increased by raising the front pressure, and a new gas-flow rate is allowed to stabilize and is subsequently measured. A waiting period of typically 2.5 h was need-

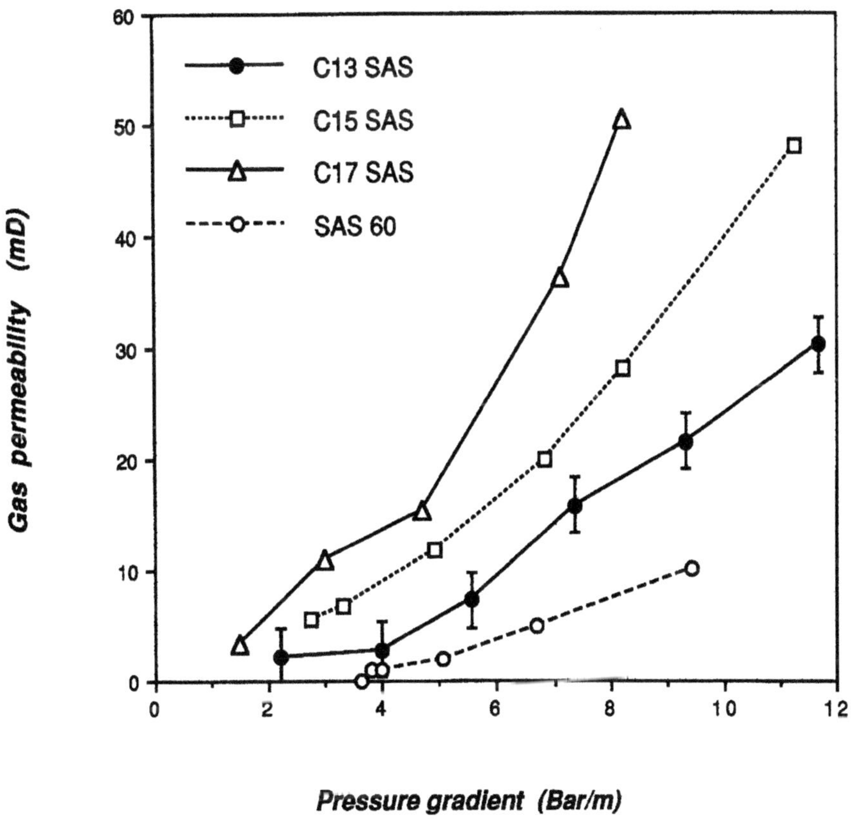

Fig. 6. Gas-blockage efficiency curves obtained for 0.5% (w/w, active) Hostapur SAS foams at zero oil saturation. Estimated error bars corresponding to ±10% of mean gas permeability value are indicated. Lines are interpolated

ed after each pressure increase for the systems studied. No liquid is displaced from the porous medium during the gas-blockage test; only the gas enters and leaves. By the use of optical tracers, foam and oil have been observed to be evenly distributed throughout the pack [9]. Gas permeability through the foam-filled column is calculated by Darcy's law from the measured gas flow rate, pressure drop, and free gas viscosity.

A set of what we have proposed to call the characteristic gas-blockage curves shown in Fig. 6 for oil-free packs. Gas permeability is very low compared to the absolute value of 7.5 μm^2; it is between zero and 0.05 μm^2, even for high pressure gradients. The rise in permeability with increasing pressure gradients testifies to the non-Newtonian nature of fluid flow through the foam-filled medium, and a threshold or mobilization pressure is clearly seen for SAS 60. There is a significant trend in the surfactants with shorter hydrophobic chains producing the better blockage, while the blend is superior to all. Interestingly, bulk stability gave an opposite ranking of the fractions. Note that we had to reduce surfactant concentration to 0.5% in order to obtain any gas flow at all. At 2.0%, C_{13} SAS foam gave zero permeability for the maximum pressure gradient of 13.7 bar/m. Generation of the SAS 60 foam could not even be completed because of premature blockage when the foam front had reached 35 cm.

Figure 7 shows the impairment of gas-blockage efficiency by SAS 60 when octane is present, compared with an n-alkyl ethoxysulfate foam studied in a previous work [9]. The better performance of Hostapur SAS is evident, but more systems will have to be studied before one can say whether the secondary alkyl chain is a structural feature that gives superior mobility reduction in porous media.

Results from gas-blockage tests of each SAS-fraction foam in the presence of each oil are given by Figs. 8, 9, and 10. The best systems containing oil (SAS 60 with octane, C_{13} and C_{15} SAS with hexadecane) approach the efficiency of the oil-free systems (at 0.5%). The poorest gas-blocking foams (C_{13} and C_{17} SAS with dodecane, C_{17} SAS with hexadecane) produce only modest mobility reduc-

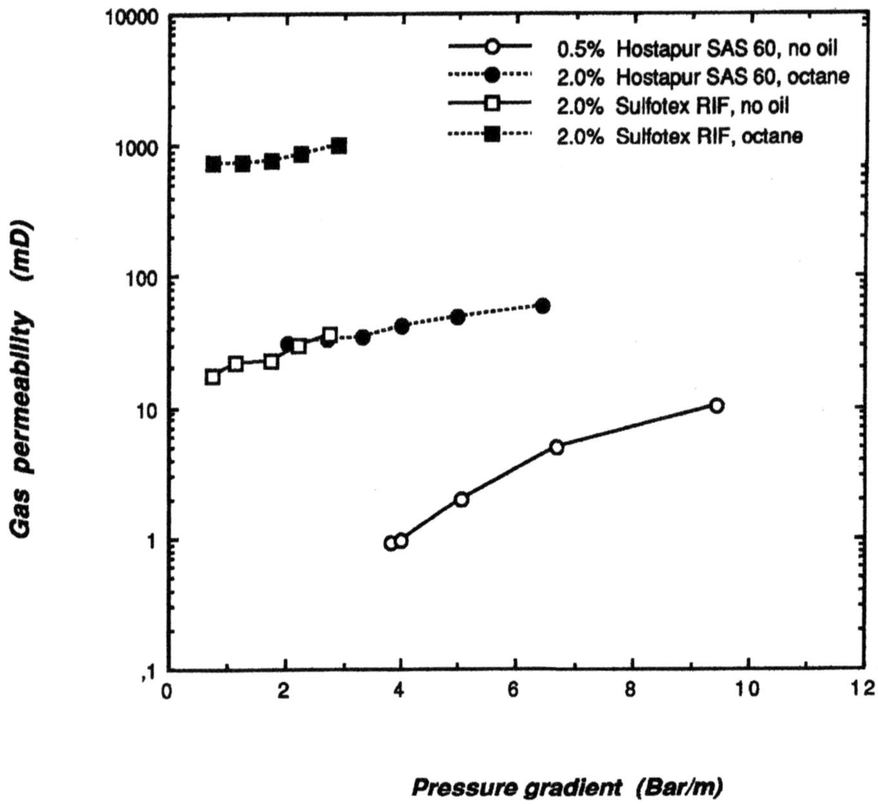

Fig. 7. Gas-blockage efficiency of Hostapur SAS 60 blend and Sulfotex RIF foams with and without n-octane in same poroous medium, with distilled water. (Note that 2.0% SAS 60 produced zero gas permeability in the entire pressure-gradient range for the oil-free system.)

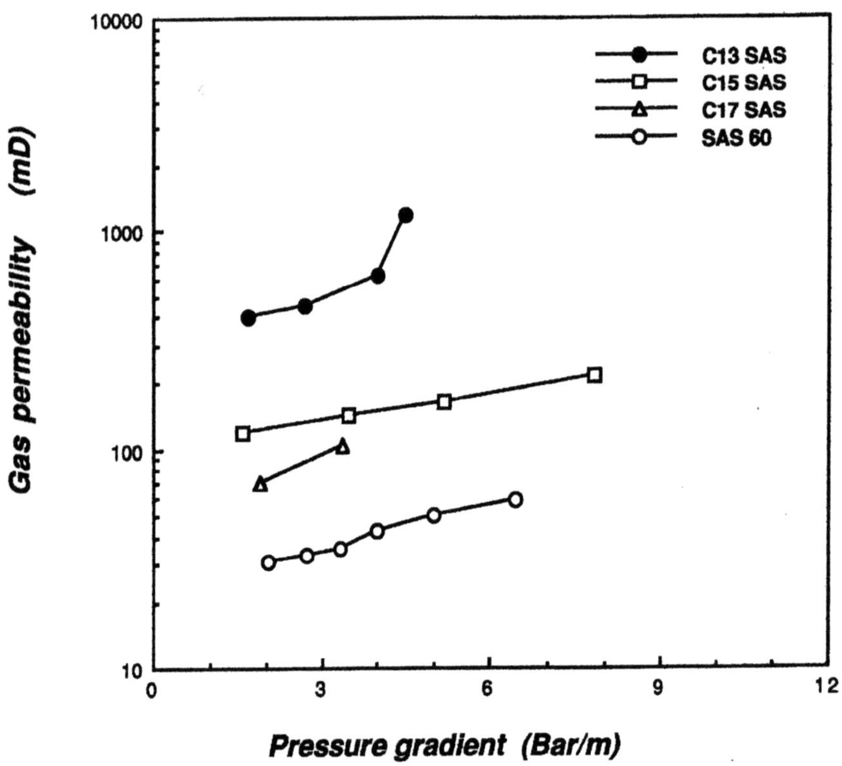

Fig. 8. Gas-blockage efficiency of 2.0% (w/w, active) Hostapur SAS foams in glass bead pack of 7500 mD absolute permeability in the presence of octane. Lines are interpolated

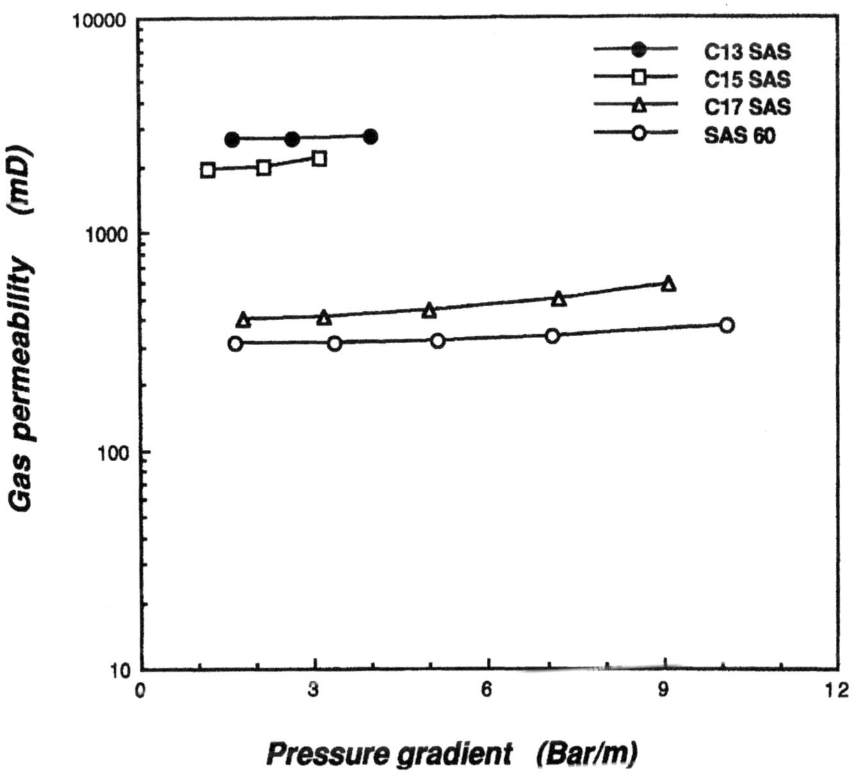

Fig. 9. Gas-blockage efficiency of 2.0% (w/w, active) Hostapur SAS foams in glass bead pack of 7500 mD absolute permeability in the presence of dodecane. Lines are interpolated

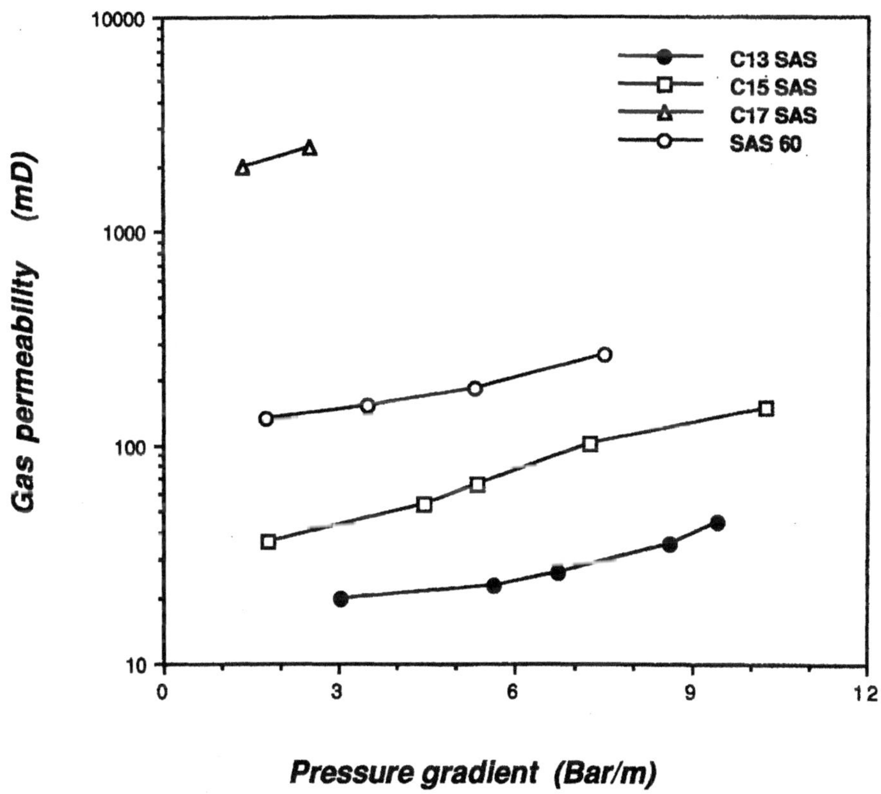

Fig. 10. Gas-blockage efficency of 2.0% (w/w, active) Hostapur SAS foams in glass bead pack of 7500 mD absolute pemeability in the presence of hexadecane. Lines are interpolated

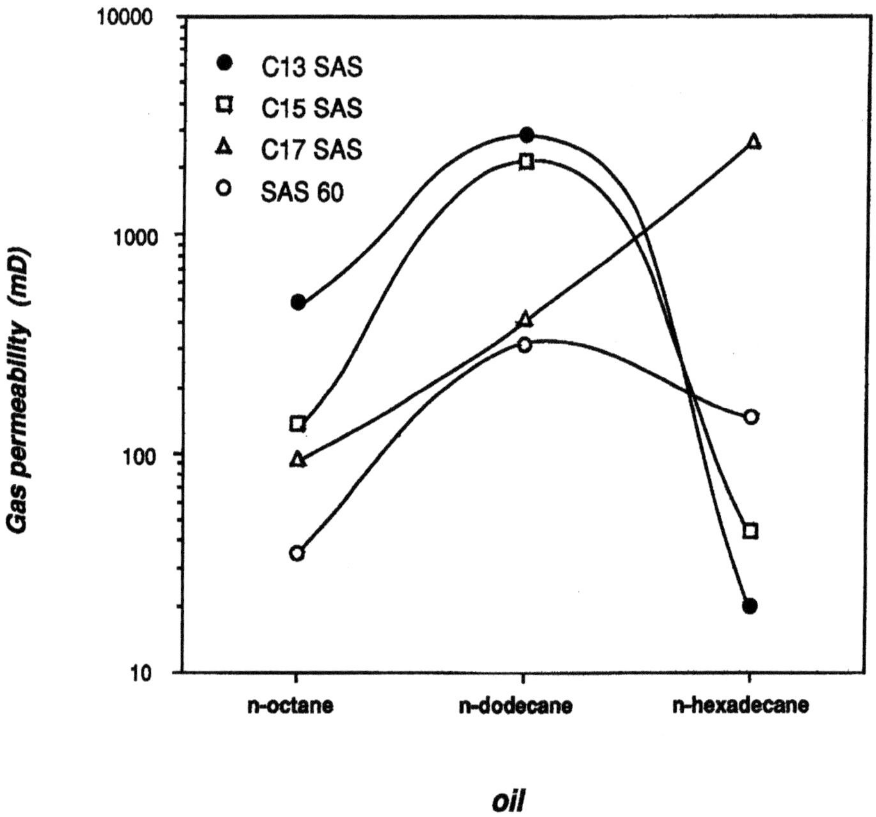

Fig. 11. Gas permeabilities at a pressure gradient of 3 bar/m taken from Figs. 8—10 and plotted as function of oil type for the surfactants used. Lines are eye fit

tion factors of 5 or less compared to the absolute permeabiality. Interesting relative differences are also revealed. Of the oils, dodecane has the strongest negative effect, except for C_{17} SAS foam which is more strongly impaired by hexadecane.

By plotting gas permeabilities at a fixed pressure gradient as a function of oil for each surfactant, a summary of the gas-blockage data is obtained which we propose to call an *oil sensitivity scan.* In Fig. 11, the systematic trends obtained by cross-comparing the previous three figures are shown at a glance. Note, in particular, the shift in the order of rank from $C_{17} > C_{15} > C_{13}$ to $C_{13} > C_{15} > C_{17}$ (the symbol ">" is used to mean "superior to" in terms of gas-blockage efficiency) that apprently occurs for an oil carbon number somewhere between 12 and 16. It is intriguing that the full range of gas-blocking abilities is represented in a narrow model system. If this result is generally applicable, it means that the poorest blockage in practical cases where the oil is given, would be obtained by selecting a surfactant of similar hydrophobe carbon number to that of the oil.

A trend similar to that found above for gas-blockage efficiency has been reported for bulk foams [12], but it should be noted that bulk stability for our system does *not* give the same result consistently, as seen by plotting the bulk stability results in a similar way; Fig. 12 shows that only a part of the systematic trends is repeated. The bulk tests do pick dodecane as a relatively unfavorable oil, but fail to show the particular incompatibility of C_{17} SAS foam with hexadecane. From bulk foam stability, signficant differences in the presence of octane are predicted only between C_{13} SAS and the other three. And the shift in the order of rank of SAS-fraction foams between dodecane and hexadecane is not present in the bulk data.

Physicochemical properties

On equilibration with oil, a microemulsion was formed in the aqueous phase for all systems. The oil phase formed a water-in-oil macroemulsion in all systems studied, which could not be broken by standing for 6 months or by exposure to high temperatures, salt, or ultrasonic agitation.

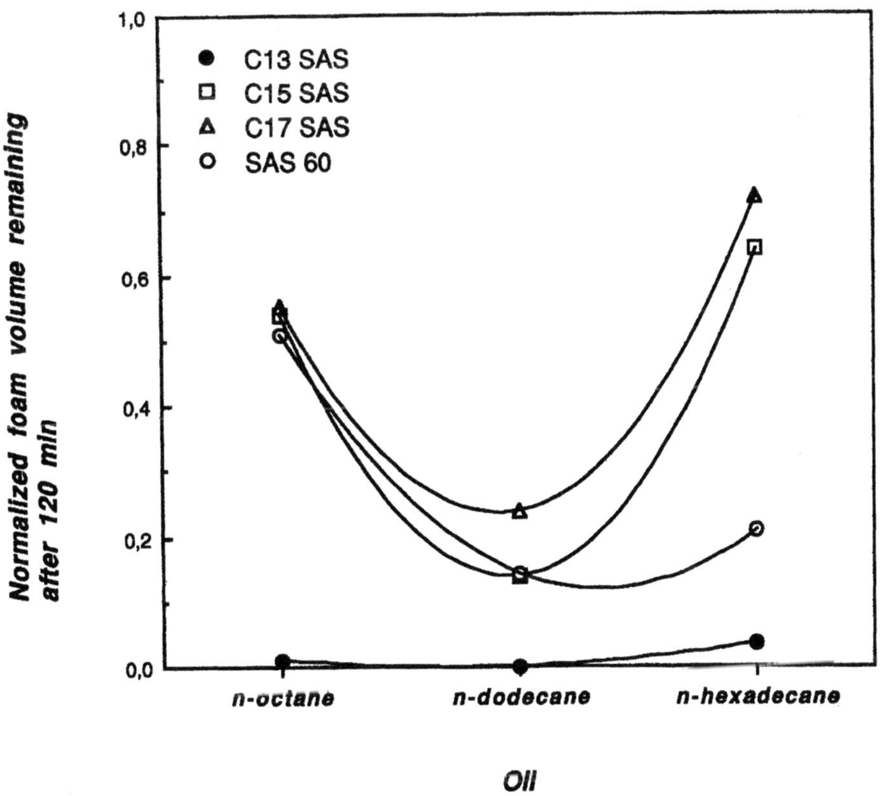

Fig. 12. Results of bulk foam tests after 120 min of drainage, plotted as function of oil type for the surfactants used. Lines are eye fit

Table 1. Bulk viscosity, density and interfacial tension values. All measurements at 22 ± 2°C

Surfactant	Oil	$\gamma_{w/g}$	$\gamma_{O/w}$ *) —mN/m—	$\gamma_{O/g}$	Density	Viscosity
SAS 60	none	31.6			1.000	1.59
	n-C_8H_{18}	30.7	⩽1	26.5		1.12
	n-$C_{12}H_{26}$	32.7	⩽1	27.6		1.12
	n-$C_{16}H_{34}$	31.4	⩽1	30.8		1.11
C_{13}SAS	none	31.4			1.000	1.31
	n-C_8H_{18}	33.5	⩽1	27.2		1.10
	n-$C_{12}H_{26}$	34.1	⩽1	32.5		1.09
	n-$C_{16}H_{34}$	31.0	⩽1	30.8		1.09
C_{15}SAS	none	31.9			1.000	1.31
	n-C_8H_{18}	30.8	⩽1	26.3		1.12
	n-$C_{12}H_{26}$	33.0	⩽1	28.4		1.12
	n-$C_{16}H_{34}$	33.2	⩽1	33.0		1.09
C_{17}SAS	none	30.8			1.000	1.28
	n-C_8H_{18}	30.0	⩽1	23.3		1.11
	n-$C_{12}H_{26}$	32.2	⩽1	26.8		1.09
	n-$C_{16}H_{34}$	34.3	⩽1	33.4		1.07
None	n-C_8H_{18}			22.9	0.703	0.52
	n-$C_{12}H_{26}$			26.7	0.750	1.40
	n-$C_{16}H_{34}$			28.7	0.773	3.30

*) The data obtained ranged from 0.4 to 0.7 mN/m, but since the ring method is not considered reliable below 1 mN/m, they are given as ⩽1 which is sufficient for estimating spreading coefficients.

Table 2. Relative wetting of surfactant solution on solid. Droplets were circular or elliptical. An error of ±10% in droplet area is typical

Surfactant	Concentration % w/w	Area on dry glass mm^2	Area on dry PTFE mm^2
SAS 60	2.0	130	36
C_{11} SAS	2.0	148	21
C_{12} SAS	2.0	134	29
C_{13} SAS	2.0	132	32
C_{14} SAS	2.0	123	27
C_{15} SAS	2.0	112	27
C_{16} SAS	2.0	98	31
C_{17} SAS	2.0	88	27
NaDS	2.0	113	17
3M FC 99	0.1	60	58
dist. water	—	49	12

Table 3. Oil solubilization in aqueous phase

Surfactant	Oil	% w/w oil in microemulsion
C_{13} SAS	n-C_8H_{18}	0.89
	n-$C_{12}H_{26}$	0.68
	n-$C_{16}H_{34}$	0.44
C_{15} SAS	n-C_8H_{18}	1.76
	n-$C_{12}H_{26}$	1.28
	n-$C_{16}H_{34}$	0.92
C_{17} SAS	n-C_8H_{18}	4.10
	n-$C_{12}H_{26}$	2.67
	n-$C_{16}H_{34}$	0.77

Some comments to the results in Table 1 are given below. Densities of the aqueous phases were found to be unaffected by equilibration with oil. For the oil/gas IFT values, a larger error than normal for ring tensiometer values must be reckoned with, due to the problems of measuring for an emulsion. We estimate this error to be ±0.5 mN/m, as opposed to ±0.2 mN/m for the water/gas values. Oil/water IFT values should be measured by a different technique, because they are at the detection limit of the ring method, but within our scope of work the data in Table 1 are sufficient.

Results of relative wetting measurements are given in Table 2, also showing wetting of a hydrophobic surface. NaDS is included as a reference surfactant, and a fluorinated surfactant is included to illustrate a property of these surfactants, which are studied more extensively in our parallel work [6]. Measured oil solubilization in the SAS microemulsions is given in Table 3.

Discussion

Bulk vs. blockage test results

Without oil, reversed surfactant carbon-number effects were observed for the gas-blocking ability and the bulk stability of SAS-fraction foams. The gas-blocking ability of SAS foams was found to be *impaired* by all oils used, while nearly all systems had their bulk foam stability *enhanced* by contact with the same oils. No correspondence was found between the relative oil tolerance of SAS-fraction foams in bulk and in a porous medium, and only a weak correspondence was observed between the relative influence of different n-alkanes on an SAS-fraction foam in bulk and when confined into the pore network. Thus, the observed systematic variations in gas-blocking ability can*not* be satisfactorily explained by differences in foam stability as measurable in our bulk test.

Relative wetting

Assuming largely stationary lamellae, a foam should intuitively be more able to immobilize gas the stronger its liquid films adhere to the solid surface. This "adhesive strength" can be identified with the ability of the film to wet the pore wall. Kanda [10] found a correlation between the gas mobility reduction by foams in glass-bead packs and the relative wetting, measured as the area of constant-volume spread droplets, of the corresponding surfactant solutions on glass. A similar relationship exists in our data for oil-free systems and is shown by Fig. 13: the larger the droplet of surfactant, the better its gas-blocking efficiency. On oil-covered or hydrophobic surfaces no corresponding trend has been found (cf. the data for spreading on PTFE in Table 2).

A number of foam/oil interaction mechanisms that conceivably could explain our observed differences in oil sensitivity of foams have been suggested.

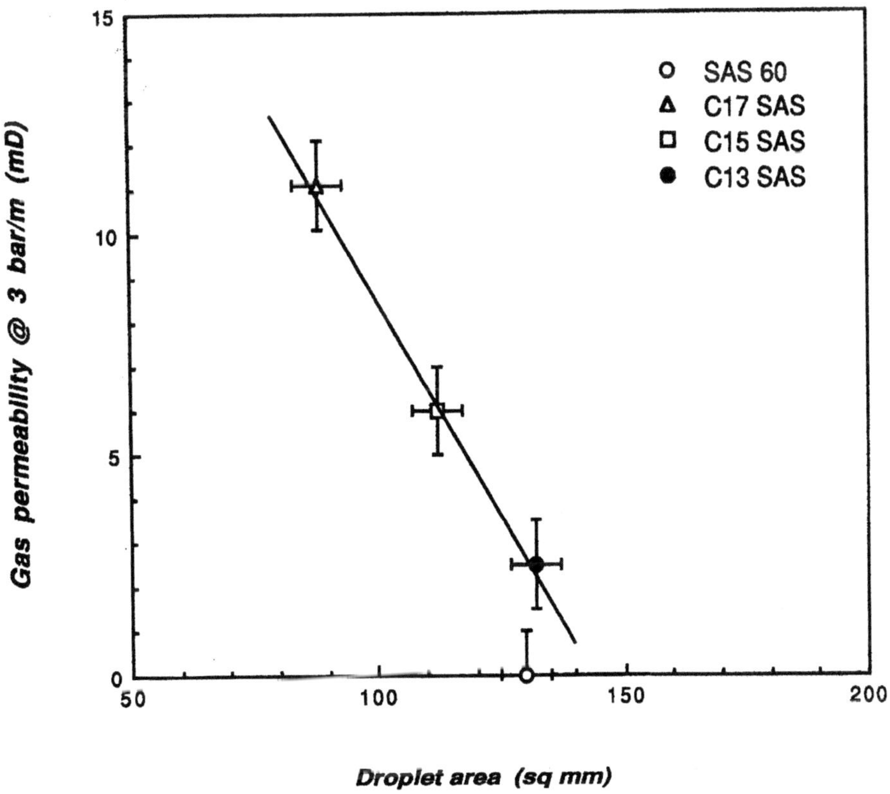

Fig. 13. Gas permeability at 3 bar/m of 0.5% (w/w, active) SAS foams in glass bead packs as a function of the relative wetting ability of corresponding 2.0% (w/w, active) surfactant solutions on clean, dry glass plate. Line is eye fit and correlates SAS fraction data points

Oil spreading

It has been found that, of two equal-viscosity oils, one which spontaneously spreads on a given foam solution and one of which does not, the nonspreading oil allows the foam to have the lower mobility in steady-state foam flow [11]. Higher tolerance for nonspreading oils is also reported in bulk foam stability results [12]. For our systems, however, spreading coefficients for oil-on-water calculated from the IFT data in Table 1 are positive or zero within experimental error in all cases. Spontaneous spreading of oil was also verified by direct observation. Thus, provided that bulk spreading is also representative of pore-level phenomena, the differences in oil spreading on the aqueous phase cannot explain our observed differences in gas-blockage efficiency.

Surface wettability reversal

From the observed similarity in the impairment of gas mobility reduction by foam when oil was present and when the glass beads had been made oil-wet by a silane treatment, Kanda [10] suggested that oil could reverse the wettability of the surface and that this caused the loss of gas blockage. While certainly possible for complex reservoir mineral surfaces in contact with crude oils containing surface-active compounds, it is unlikely that contact with a pure n-alkane in itself could reverse the wettability of a hydrophilic glass surface. It has been reported that silica surfaces can, at most, be modified by contact with oils from strongly water-wet to preferentially water-wet [13]. Recent reports suggests formation of a film of oil between gas in the pore and the adsorbed water layer under spreading oil conditions [14, 15], which to a lamella could appear similar to a wetting oil film. However, oil is spreading in *all* our systems while large variations in oil tolerance are seen. Such "superficially reversed" wettability cannot, therefore, be causing our systematic trends.

Intermicellar attraction and solubilization

Intermicellar attraction has been found to be important in stabilizing foam films to drainage [16].

Oil weakens this attraction and makes the film less ordered when dissolved into the micelles. Thus we might expect more oil to be solubilized in those of our systems which show poorer gas-blocking efficiency, i.e., weaker films. However, the oil solubilization data in Table 3 show only the expected general trends of higher solubility for lighter oil and longer surfactant hydrophobe and are not similar to the clear incompatibilities revealed in Fig. 11.

A high microemulsion viscosity is sometimes taken to indicate the presence of long-range micellar interaction. The somewhat elevated viscosity of our oil-free systems could be due to this, but all viscosities were seen to decrease strongly after equilibration with oil. Since gas blockage approaching that for oil-free systems was achieved also for some oil-containing systems, one would expect a smaller loss of viscosity for these if micellar attraction were a controlling factor.

Conversely, surfactant could be depleted by solution into the oil phase, or used up in stabilizing an emulsion, so that the effective concentration in the foam films, and thus their strength, is reduced. Surfactant loss to oil has, however, generally been found to be slight [17]. In our systems, where surfactant concentration is high and the volume of w/o emulsion less than the volume of aqueous phase in the column during gas blockage tests — and, more importantly, with no correlation between saturations and gas blockage — surfactant loss to oil is not a likely cause of the large spread measured in gas blockage efficiency.

Change in $\gamma_{w/g}$ by equilibration with oil

The measured IFT values in Table 1 appear unsurprising at first, with $\gamma_{w/g}$ for oil-equilibrated systems within 3.5 mN/m of those of the pure surfactant solution. There are somewhat larger and always positive changes for $\gamma_{0/g}$ relative to the pure oils, probably ascribable to emulsion formation. But on closer inspection, there is a systematic variation in the $\gamma_{w/g}$ data which is strikingly similar to the observed trends in foam/oil compatibility. From the graph of $\Delta\gamma_{w/g} = (\gamma_{w/g} - \gamma^0_{w/g})$, where $\gamma^0_{w/g}$ is the surface tension of oil-free SAS, in Fig. 14, we see that the order of rank for gas-blocking ability in the

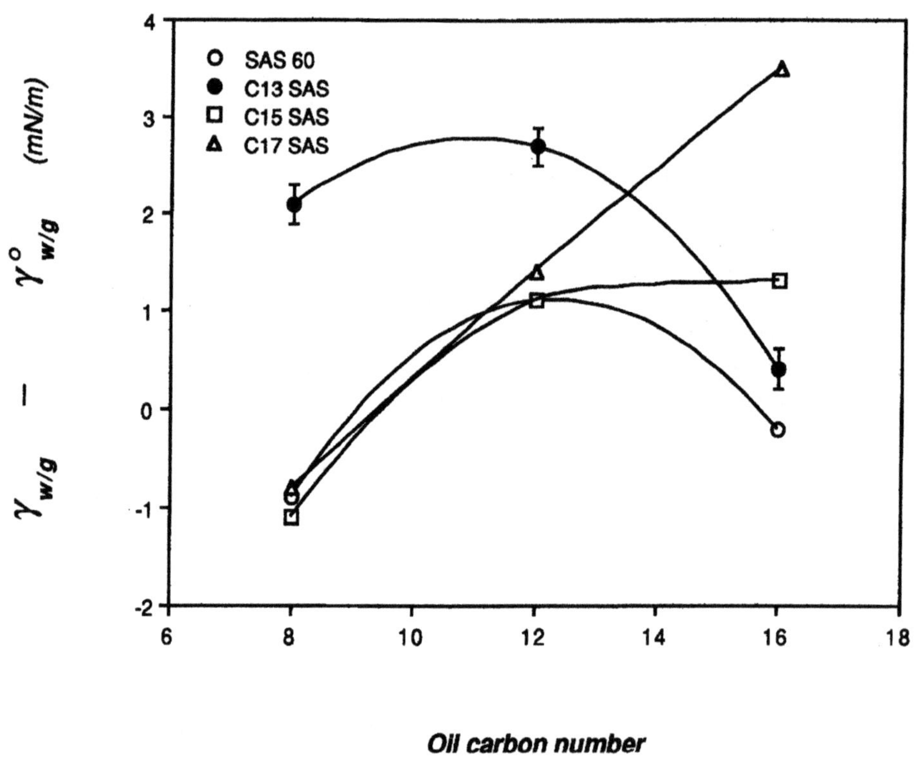

Fig. 14. Change in aqueous phase/gas interfacial tension upon equilibration of surfactant solution with oil as a function of oil type for the various surfactants used. Lines are eye fit

presence of hexadecane is closely matched by the $\Delta\gamma_{w/g}$ curve. The particular incompatibility of C_{17} SAS foam with hexadecane, breaking the general trend that dodecane is the most unfavorable oil, is also reproduced, as are the relative effects of hexadecane and octane for C_{13} SAS and SAS 60 foam. Although the match is not perfect in terms of order of rank — as in the bulk data, only C_{13} SAS is distinguishable from the other foams with octane — we consider the agreement too good to be a coincidence.

A phenomenological interpretation of the result for our model system that the increase in $\gamma_{w/g}$ by equilibration with an oil is a good predictor of foam gas-blocking ability in the presence of that oil, could be that oil adsorbs preferentially to the aqueous/gas interface and weakens it mechanically, so that the film becomes more prone to breaking. This has been suggested in the literature [10], but to our knowledge, experimental evidence of preferential adsorption controlling foam/oil interaction has not been presented before. However, as yet, such an interpretation is only speculative and must be substantiated by interfacial rheology studies; in particular, by measuring film elasticity, which has been shown to be important to foam properties in porous media [18]. For preferential adsorption to occur oil molecules must find the interface microenvironment preferable to the continuous phase of the emulsion and to solubilization in the micelles of the aqueous phase, and this is not intuitively easy to rationalize. Besides, it is not obvious why our gas-blockage test conducted in a porous medium should be a more sensitive indicator of film destabilization than bulk stability measurements.

If, however, the correlation found for this idealized model system should prove to be general, it may greatly simplify procedures in selecting products for foam applications, both where maximum gas blockage is required and where some oil sensitivity is considered desirable. Potentially, the mechanistic implications of this simple way of predicting some of the complex properties of foams in oil-containing porous media could also be signficant.

Conclusions

A model foam/oil system was found useful for studying systematic effects in oil and surfactant hydrophobe carbon number compatibility. Using a gas-blockage test measuring gas flow through a foam-filled glass-bead pack at a stepwise increasing pressure gradient, at constant oil saturation, significant trends in the degree of gas mobility reduction were measurable.

For secondary alkyl sulfonate foams without oil, total gas blockage (zero permeability) was easily obtained and the concentration had to be reduced in order to get measurable gas flow rates. a clear order of rank showing higher gasblockage efficiency for shorter surfactant hydrophobe was measured without oil. For the oil-containing systems, a range of gas mobility reduction from that approaching the oil-free systems to essentially no mobility reduction apart from that ascribable to relative permeability was observed as a function of oil and surfactant hydrophobe carbon number. The observed trends were rather complex, and only for one surfactant-fraction foam was gas-blocking ability found to be monotonically increasing with increasing oil carbon number. For two other surfactant fractions and a commercial blend, minima were observed. The order of rank in surfactant-fraction foam oil tolerance was reversed for the different oils.

The systematic variation in foam efficiency as measured by the gas-blockage test could not be reproduced by bulk foam tests. Rather, bulk foam stability was enhanced by oil while gas blockage was always impaired. Only a weak correspondence between the relative influence of the oils in bulk and in the porous medium was noted. Thus, not even for this well-described, idealized system is foam stability, as measured in bulk tests, a good predictor of gas-blockage efficiency in a model porous medium.

Relative wetting of surfactant on the pore wall, wettability reversal of the surface, oil spreading on surfactant films, and solubilization effects are all unlikely explanations of the measured gas-blockage trends. But the change in gas/water interfacial tension upon equilibration with oil as a function of oil carbon number shows a striking resemblance to the variation in gas-blockage efficiency, and captures many features of the trends observed. If this is a general phenomenon, and a satisfactory mechanistic explanation can be found, this easily measured fluid property could become a useful predictor of foram gas-blockage efficiency.

References

1. Holm LW (1968) Soc Petr Eng J 8:2, pp 359–369
2. Persoff P, Radke CJ, Pruess K, Benson SM, Witherspoon PA (1989) Paper SPE 18781 in Proceedings, Soc

Petr Eng California Regional Meeting, Bakersfield, April, 1989
3. Hanssen JE, Rolfsvåg TA, Corneliussen R, Dalland M (1989) Proceedings, 5th European Symposium on Improved Oil Recovery, Budapest, April 1989, pp 737—746
4. Product information from Hoechst
5. Hanssen JE (1988) 6th Soc Petr Eng/US Dep Energy Symposium on Enhanced Oil Recovery, Tulsa, April 1988. Paper no. SPE/DOE 17362
6. Hanssen JE, Dalland M, Goobewardene AP (1990) Paper SPE/DOE 20193 7th SPE/DOE Symposium on Enhanced Oil Recovery, Tulsa April 1990
7. Meling T (1989) MSc Thesis in Petroleum Engineering, Rogaland University Centre, December 1989
8. Dalland M, Hansen EW, Thorkildsen G (1988) Proceedings, Symposium on Magnetic Resonance and Relaxation, Portoroz, p 310
9. Corneliussen R (1988) MSc thesis in Petroleum Engineering, Rogaland University Centre, June 1988
10. Kanda M (1976) PhD thesis, University of Texas-Austin, 1976
11. Lau HC, O'Brien SM (1988) SPE Reservoir Eng, 3:3, 893—896
12. Kuhlman M (1988) Proceedings, 6th Soc Petr Eng/US Dep Energy Symposium on Enhanced Oil Recovery, Tulsa, April 1988. Paper no. SPE/DOE 17356
13. Brown CE, Neustadter EL (1986) J Canad Petr Tech, pp 100—110
14. Dullien FAL, Chatzis I, Kantzas A (1989) Proceedings, III Int Symp on Enhanced Oil Recovery, Maracaibo, 1989, pp 425—438
15. Danesh A, Krinis D, Henderson GD, Peden JM (1989) Proceedings, 5th European Symposium on Improved Oil Recovery, Budapest: pp 135—144
16. Wasan DT, Nikolov AD, Huang DD, Edwards DA (1988) in Smith DH (ed): "Surfactant-Based Mobility Control, ACS Symp Series 373, Am Chem Soc, Washington: pp 136—163
17. Friedmann F, Jensen JV (1986) Proceedings, Soc Petr Eng 56th Calif Reg Mtg, Oakland, California. Paper no. SPE 15087
18. Wasan DT, Nikolov, A, Huang DD (1986) Langmuir (2): pp 672—677

Authors' address:

J. E. Hanssen, MSc
Institute of Petroleum Engineering
Rogaland University Centre
P. O. Box 2557 Ullandhaug
4004 Stavanger, Norway

The influence of surfactants on coalescence filtration

E. Dahlqvist and F. Setterwall[1])

Asea Brown Boveri corporate research, Västerås, Sweden
[1]) Royal Institute of Technology, Chemical Engineering, Stockholm, Sweden

Abstract: Coalescence filters are used to separate dispersions. We have studied how the efficiency of filters are affected by different filter structure, surface coatings, and additions of surfactants. Filters with parallel fibers bunched like spaghetti showed the same efficiency as fibermats, for the same average pore size. A naked glass surface (negatively charged) showed best performance, an amin-coated was second best, and a non polar was worst, although the difference was not very significant. — An anion tenside with HLB value 9 was compared with a nonion with HLB 9 and a nonion with HLB 14 in fiber filters, but also in thin precoat filters. The nonion tenside with HLB 14 showed the poorest filter efficiency, followed by the nonion with HLB 9, and the anion showing the best result for tenside conc 2.5—100 ppm in water. A correlation between surface tension and filter efficiency can be seen, while interfacial tension shows a correlation only at very high concentrations. Several authors have shown correlations between emulsion stability and surface-film viscosity and elasticity. From our results we conclude that also considering complete filter performance film elasticity and viscosity are of major importance, compared to other properties.

Key words: Coaelescence; filtration; efficiency; surfactants; emulsions

Introduction

Coalescence filters are used in applications where one phase is dispersed in another, e.g., oil in water. The main application for coalescence filters today is to remove water from hydrocarbon fuels. In this article we will concentrate on the removal of oil drops from waters, like produced waters from oil wells or industries like refineries, or to separate phase after extractions.

We will concentrate on how surfactants influence the coalescence efficiency. The importance of filter-media surface properties will also be discussed.

A. Literature review

Not very much is reported in the literature on the influence of surfactants on coalescence filtration of o/w emulsions.

Sareen et al. [1] made a test where they found no direct correlation between surface tension and coalescence efficiency. An oil-soluble surfactant gave a surface tension of 4.5 mN/m, without any decrease in filter efficiency, while a water-soluble surfactant caused a strong reduction although the surface tension was still 20 mN/m.

Most other articles discuss w/o emulsions.

Sareen et al. [1] have reported different critical interfacial tension below which coalescence efficiency is poor, while Hazlett [2] and Bitten [3] showed good coalescence, despite low interfacial tension.

Wasan et al. [4] has shown a correlation between interfacial shear viscosity and coalescer performance. They found no correlation between interfacial tension and coalescence efficiency.

Verdegan [5] showed that in an open coalescer low interfacial tension reduced efficiency, while in a tight coalescer there was no correlation (for w/o emulsions).

Boyd et al. [6] have shown a strong correlation between surface film elasticity, film viscosity and coalescence rate. Malhotra and Wasan [7] has shown that for high interfacial film viscosity this is most important, while for low viscosity, elasticity of the film is of importance. Sabet et al. [8] showed that lecithin-stabilized emulsions (xylen in water) could be destabilized by adding surfactants, decreasing surface film elasticity. Verdegan [9] has reported a correlation between surface tension and coalescence filter efficiency for a number of surfactant systems. Jaisinghani [10] has tested several types of nonionic tensides for w/o systems without getting any influence, except for one of them.

Goubran [11] showed that the molecular weight of a surfactant is important. High molecular weight emulsifier gives much more stabile emulsions than low molecular weight emulsifiers; probably, steric repulsion is the reason.

If we look at fiber properties of the filter media, Madia [12], Chieu [13], and Treybal [14] found that fibers should be wetted by the dispersed phase, while Farrow [15] and Fahim and Akbar [16] say that fibers should not be wetted by the dispersed phase. Voutskij et al. [17] has an opinion in between: the fibers ought to be fairly wetted, but not more than to make it possible for the drops to leave the fibers again after growth. Voutskij [17] and Davies [18] suggest that a mixture of hydrophilic and hydrophobic fibers should be the best. Langdon [19] tested this idea, but he could not find any enhancement.

Jaisinghani [10] suggests that below a critical wettability small drops are released from the fiber surface. A surface that is not wet by the dispersed phase will give a lower pressure drop in the filter.

Bitten [3] and Gudesen [20] found wettability to be of less importance, while a rough surface seemed to be better than a smooth one. Hughes and Foulds [21] treated fibers with amino silanes to get a positive surface potential, which should enhance the removal of negatively charged oil drops.

The aim of our study

Boyd et al. [6] found that the coalescence rate for o/w emulsion stabilized by mixture of polysorbates and sorbitan esthers showed a minimum coalescence rate at HLB 11—14. Below and above this range the coalescence rate increased rapidly to become approx. six times higher at HLB 9 and 17.

We wanted to investigate some o/w systems with additions of nonion-tensides with different HLB values, and to compare one of those with an anion tenside with the same HLB value, to see if these results will hold also for other tensides, and also for the whole filter, except the post separation of coalesced drops or film. We discuss in the present paper why different tensides have different impact on the coalescence efficiency and if there is any difference between polar and unpolar surfaces, dealing with tenside stabilized o/w emulsions compared to tenside free systems.

Experimental method

Three different kind of filters were used. The first is built from smooth, parallel glass fibers (like a bunch of spaghetti), giving pore diameters of approx. 5 or 10 μm, depending on the fiber diameter.

The second type is a glass fiber mat that is compressed to high density, giving an average pore diameter of approx. 5—7 μm. Both of these fiber filters were approx. 30-mm deep.

The third filter type was a precoated filter, coated with a coarse diatomaceous earth particle product (Celite Manville 560, Denver, USA) with a median particle size of 106 μm, giving a median pore size of 22 μm. 200—400 g/m^2 was used, giving a media thickness of approx. 0.75—1 mm.

To study the influence of the pore size, 60% of the finest particles were removed by gravimetric separation in water, giving a median pore size of approx. 60—80 μm.

The amount of drops with a diameter approx. 3—20 μm and <3 μm were measured in the feed solutions and in the filtrate by a gravimetric method. 50-ml samples were separated gravimetrically for approx. 20 min and 20 h (for a density difference of approx. 200 kg/m^3 between oil and water). Thereafter, 10-ml samples were taken from the bottom of the measuring cylinders by pipette and the oil was extracted by freon for an IR-analysis to determine the oil content. The method has been compared to instrumental methods like Malvern laser diffraction, Kratel light blockage, and Coulter Counters drop-volume method [20] and showed good correlation.

The removal efficiency is calculated as

$$(1\text{-filtrate conc/feed conc}) \cdot 100\% \qquad (1)$$

Experiments

Thick filters

First we compared parallel glass fiber filters with glass fiber mat filters (GFM), (see Table 1, (4)) where different surface coatings were tested in some of the experiments (Table 2).

Here we give the results for the best filter in each run and a relative value for the efficiency of the other filters compared to the best one.

Table 1. The effect of tensides on thick coalescence filters

Tenside ppm (water)	Oil %	Surface tension mN/m	(1) Interfacial tension mN/m emulsion	(2) Interfacial tension mN/m dissolved	(3) Removal efficiency 20 μm %	3 μm %	(4) Parallel fibers 5 μm pore	GFM
Berol 026 100 ppm	0.23	33	12	19	69	0	1.1*)	1.0
Dito	0.21	33	12	19	67	40	1.0	1.1
Berol 026 + 25 ppm Berol 822	0.24	—	—	—	66	48	1.1	1.0
No tenside	0.054	60	18	—	98.5	96.4	1.0	1.0
No tenside	0.53	60	18	—	98.3	97.0	1.0	1.0
Tap water/oil		74		20				
Berol 822 100 ppm	0.50	32	13	20	87	79.4	1.0	1.1

*) The best filter is given the efficiency figure 1.0. The other filters have, e.g., 1.1 times higher oil content in the filtrate. The removal efficiency is within ±2% and the relative efficiency between the different filter structures is within ±0.2 units at a 95% confidence level

Table 2. Relative results for Berol 822 (aniontenside) with different fiber surface treatments

Parallel fibers				Glassfiber mat		
5 μm pores	Naked glass 10 μm	Amin-coated 10 μm	Dodekan coated	Naked glass 10 μm	Amin-coated	Dodekan-coated
1.0	1.0	1.9	2.0	1.1	1.6	1.9
This can be compared with typical results from a test with no tenside added:						
1.2	2.6	2.8	4.8	1.2	1.0	2.0

All results are within ±0.2—0.3 units at a 95% confidence level.

Surface tension and interfacial tensions were measured for filtered water, against air, respectively, the same crude oil as that being used in all the tests (1).

For interfacial tension, we also compared the actual test waters with systems where tensides were dissolved in water or oil, and equilibrated with clean oil or water for 10 and 30 min before measurement with the ring method was performed (2).

Here we can see that the watersoluble tenside gave lower value in the synthetic system, while the oil-soluble tenside gave lower value in the real test emulsion.

The removal efficiency for 3—20 μm and <3 μm drops were calculated (3).

Berol 0.26 is a nonylphenol with 4 EO, HLB 9. In the last experiment with Berol 822, calcium dodecylbenzensulfonate, we also tested different surface coatings. The relative result for these different combinations are shown in Table 2.

Surface energies and drop angles at untreated glass surfaces and surfaces treated with amine and oktadecane silanes are given in Table 3 for clean surfaces, respectively, for surfaces immersed for week in o/w emulsions.

For the system without tensides we had an efficiency for the parallel fiber filter (approx. 5 μm pores) of 98.3% concerning 3—20 μm drops, 97% for <3 μm drops and a total filter efficiency of 99.7% concerning also larger drops (total 5300 ppm oil).

Table 3. Surface energies and drop angles for different surfaces

Contact-angle and surface tension at solid surface	Naked glassurface			Amine-treated glassurface			Octadekan-treated glassurface		
	clean Θ	in emulsion		clean Θ	in emulsion		clean Θ	in emulsion	
		Θ	γ_s mN/m		Θ	γ_s mN/m		Θ	γ_s
Waterdrop advance/recede	19/5	40/9.26 [1]	56 [2]	46/34	47/14.27 [1]	51	88/83	63/29.65 [1]	38
Diiodmethane	25/22	25/18	61	37/37	29/29	59	50/50	29/31	59

[1]) The drop has two angles.
[2]) Surface energies where calculated for the surfaces wetted by the dispersion (thin layer left) by the Eq.: $\cos\Theta = -1 + 2(\gamma_s/\gamma_L)^{1/2}$, where $\gamma_L > \gamma_s$; γ_L = surface tension liquid γ_s = surface tension solid.

After the addition of 100 ppm anion tenside in the water phase, calcium dodecyl benzene sulphonate with a HLB value of 9, the efficiency dropped to 81—87% for 3—20 μm drops, 70—79% for <3 μm drops and to 92—97% concerning the total filter efficiency.

After the addition of 100 ppm noniontenside, nonylphenol with a chain of four ethylenoxide units giving HLB value, the efficiency fell down to 67—69% for 3—20 μm drops, (0)—40% for <3 μm drops, and to 67—70% for the total filter efficiency. The span is due to different results in repeated experiments.

For the combination 100 ppm nonion + 25 ppm anion the corresponding results were 66%, 48%, and 70%, or the same as for just 100 ppm nonion!

No significant difference was seen between the parallel fiber filter and the fiber mat filter with approx. the same pore sizes of the channels (see Table 2). We also can see that naked glass fibers gave better or the same result as the amine-treated, and that the unpolar oktadekane treated fibers gave the poorest result.

Before making any conclusions from these results we also want to look at the results achieved from experiments made with thin precoat filters.

Thin precoat filters

First we tested a coarse diatomaceous earth precoat, Celite 560 from Manville, 200 g/m^2, giving a median pore size of approx. 20 μm in an approx. 1-mm deep filter bed.

The same crude oil as before was used and the results without tensides were 90—94% for 3—20 μm drops and 83—93% for <3 μm drops.

The total filter efficiency was not measured.

After the addition of 2.5 ppm of Berol 026 (nonion nonylphenol with 4 EO groups, HLB 9) the removal efficiency dropped to 84—85% for 3—20 μm drops and 53—60% for ς3 μm drops. An addition of 10 ppm of the same tenside gave 60—62% for 3—20 μm drops and 23—35% <3 μm drops.

An addition of 100 ppm gave 17%, respectively, 5% coalescence efficiency. These results can also be seen in Table 3.

With a more hydrophilic tenside, Berolwasc (nonylphenol with 12 EO groups, HLB 14) the filter efficiencies were 51% (3—20 μm) and 25% (<3 μm) for 10 ppm and 23% (3—20 μm) and 3% (<3 μm) for 100 ppm.

To get the coarse fraction of the celite 560 we separated particles by sedimentation in water three times for 2 min each time, removing 60% of the finest particles. The same amount by weight as for the original Celite 560 was used per m^2. We estimated the median pore size to be 60—80 μm.

From these experiments we can see that there is a good correlation between the coalescence filter efficiency and the tenside concentration, and that the effect is strongest for the noniontenside with HLB 14, a little less for the noniontenside with HLB 9, but much less for the anion tenside with the same HLB value, 9.

If we look at the correlation between surface tension and interfacial tension and coalescence efficiency, there seems to be some correlation at least

Table 4. Thin precoat layer

Tenside ppm (water)	Oil %	Surface tension mN/m	Interfacial tension mN/m emulsion	Dito dissolved	Removal efficiency	
					3—20 μm %	<3 μm %
Tapwater		74	20 (w/o)			
No tenside	0.2	60	18	—	92	90
Celite 560						
No tenside	0.2	48	16	—	94	68
Coarse fraction of Celite 560						
Berol 026 (4 EO)		58	14	21	85	57
+2.5 ppm						
+5 ppm		56	16	—	74	50
+10 ppm		52	16	20	61	29
+100 ppm		33	12	19	17	5
Berolwasc (12 EO)						
+5 ppm		48	15	—	51	25
+10 ppm				12		
+100 ppm		38	7	4	23	3

For values higher than approx. 60% coalescence efficiency, we have ±5%, within a 95% confidence interval.

for the surface tension and the efficiency for the tested tensides. For interfacial tension, however, no correlation can be seen except for Berolwasc, and perhaps, to some degree for Berol 822 (anion) at high concentrations.

If we compare fibers with different fibercoatings, the relative efficiency is similar for tests with tensides and without.

The result is slightly better for runs with a thick fiber bed with small pores (5 μm) compared to the coarse, thin bed with larger pores (20 μm) when no tensides were present, but a significantly better result was achieved when tensides were present with the thick, fine filter.

Asphaltenes

To see if asphaltenes has any negative influence, the crude oil with 0.12% asphaltenes was compared with "petroleum ether" which is very pure (density 720 kg/m^3, compared to 820 for the crude oil).

In a test with the coarsest precoat, we got an average coalescence efficiency for the very pure "petroleum ether" of 90.8% for 20 μm drops and 73.6% for 3 μm drops in the thin precoat filter. For the crude oil the corresponding efficiency was 94.7 for 3—20 μm drops and 70.1% for <3 μm drops.

From these results, we cannot say that the asphaltenes have any significant negative influence compared to the pure petroleum ether.

Oil viscosity

A test was performed where the viscosity was varied. The crude oil was mixed with freon to get a density of 980 kg/m^3. The viscosity of this mixture was approx. 2 cSt at running conditions. This product was compared to another heavy oil product (Mobilpac MM) with a density of 980 kg/m^3 and a viscosity of approx. 4000 cSt at running conditions. This product gave a surface tension of 52 mN/m (0.5% oil) compared to 56 mN/m for the crude. The interfacial tension was 17 mN/m for both.

For moderate flow rates, approx. 1—3 m^3/m^2 h, the filter efficiency for a thin precoat filter (with celite 560) was 91.1% for the low viscosity mixture and 90.6% for the high viscosity oil for 3—20 μm drops. For <3 μm drops the corresponding values were 58% and 62%.

From these results we can see that high oil viscosity has little influence on the coalescence efficiency.

Salinity

Attached, depicting an experiment performed with a fiber mat filter shows the results with o/w emulsions, where we first run tap water. After approx. 3 h, 5% NaCl was added, and as can be seen, the filter efficiency increased significantly. The oil content in the filtrate was only half of that before the addition. The same crude oil was used in this test as in the other tests. Of course the increased density difference between the two phases will be one of the reasons for the increased efficiency, but this cannot be the reason, as the density difference is just increased with 15%. The main effect is probably due to a decrease in the electrostatic double-layer thickness, theoretically from approx. 15—30 nm down to approx. 2—3 nm [23].

Principals for drainage and film rupture

According to [23] the dominant factor for the transport of drops from the dispersion to the filter media surface is the mechanical separation primarily taking place in the entrance part of the filter, while diffusion, inertial, and gravitational forces are of less importance.

After reaching a fiber surface, or the continuous oil film at the surface, a film is formed between the drop and the surface. This film must drain before the drop will coalescence.

The drainage rate of the film depends on the capillary pressure and the disjoining pressure. The capillary pressure (P_c) is dependent on the surface tension (τ) and the capillary radius (R_c) according to the equation

$$P_c = 2\tau/R_c \, . \qquad (2)$$

While the disjoining pressure is the sum of electrostatic repulsion, steric repulsion, and attractive van der Waals forces.

When the disjoining pressure is negative, corrugations at the film surface will grow and at the critical thickness (h_{cr}) this can lead to either film rupture or stable conditions, not followed by coalescence.

When there are no tensides present the oscillations in the film can be of high amplitude and h_{cr} is large (approx. 0.2 μm), while the critical thickness will decrease as the tenside concentration is increased, until critical micelle concentration (CMC) is reached.

This is due to higher viscosity in both the continuous phase and the film which reduces the drainage rate, but it is also due to enhanced film elasticity as the surfactant concentration increases. Liquid flowing out of the film area will cause local surfactant concentration gradients. This will cause a back holding pressure opposing drainage (Marangoni-Gibbs effect). At the same time, a concentration gradient will increase between the interface and the bulk, causing the surfactants to move towards the interface by diffusion. In Miller and Neogi [24] damping of the film oscillation due to tensides is theoretically calculated, and they conclude that there will be no damping of the oscillations due to the addition of fatty acids reducing the surface tension from 50 to 5 mN/m.

Surfactants with high solubility in just the film phase will cause lower interfacial mobility and lower drainage rate than if the surfactant is preferably soluble in the drop Malhotra [7].

The drainage rate can be expressed by Reynolds equation:

$$dh/dt = 8 \cdot F \cdot h^3/3 \cdot \pi \cdot R^4 \, , \qquad (3)$$

where h is film thickness, F is the force pressing the surface together, μ is viscosity of the drainage film, and R is contact radius drop/surface.

From the experiments we can see that the addition of surfactants is the single factor influencing the coalescence filter removal efficiency most significantly, indicating that the drainage rate is the factor of greatest importance for the total filter efficiency.

Discussion of why the coalescence efficiency is reduced

Principally lower surface tension should enhance the coalescence rate, but also a splitting-up of large coalesced drops can be expected.

A high repulsion barrier would decrease the coalescence rate. A more water-soluble tenside would give a higher repulsion barrier.

The surface film viscosity will increase by tenside absorption. A low oil viscosity will probably give a lower oil-film viscosity too. A pure anion tenside will give a lower surface-film viscosity than a nonion tenside with a good balance between hydrophilic and hydrophobic forces, due to lower packing density.

A high packing density of tensides at the interfacial surface will give a high surface-film elasticity, and not just high film viscosity.

If we now summarize our results, we can see that the net effect of lowering the surface tension was a reduced coalescence efficiency for the tensides we tested. The poorest result was achieved with a rather water-soluble nonion tenside, indicating that a high repulsion barrier (and surface film elasticity) is negative.

If we look at the surface-film viscosity, no difference was seen between oils with different bulk viscosities (and accompanying surface-film viscosities, at least to some degree). However, nonion tensides will give a relationship also correlated to the filter efficiency results.

A high packing density at the interface will reduce the efficiency, which is also seen as higher concentrations giving an increased negative effect. This will primarily increase the surface film elasticity, but probably also the film viscosity.

Asphaltenes have a low surface activity, and would primarily give an increase in the film viscosity if present. In our tests rather high contents of asphaltenes were present without showing any impact on the filter efficiency. This indicates that surface-film viscosity is not the most important factor for the coalescence filter efficiency. This discussion can be compared with the calculations presented by Malhotra and Wasan [7]. They came to the conclusion that film viscosity is the most important factor concerning films with high viscosity, while surface-film elasticity is the dominant factor for films with low viscosity.

If we compare our results from tests, where tensides with different HLB values were added, we can see a good correlation between these results and previous results presented by Boyd et al. [6]. They found that the coalescence rate for drops was lowest in a region HLB 10—15, while the coalescence rate was several times higher for HLB 9 and 16. In this case the authors tested sorbate esters and used no coalescence filter, while we used filters and nonylphenol with different lengths of ethylenoxide chains, which indicates that there is an actual correlation between HLB values and coalescence rate, although the absolute level differs for different tensides, as the anion tenside gave less negative effect than the nonion did for the same HLB value of 9.

High salinity reduces the electrostatic double layer, and this is correlated to the filter efficiency results we got.

Looking at the filter media properties, we can see that a medium that is wet by the continuous phase seems to be a little better than one not wet by the continuous medium, but the difference is small. Probably, an oil film is formed at the fiber surface, neutralizing most of the initial fiber surface properties.

Most of the coalescence effect is also found to take place in the entrance part of the filter, assuming that the transport of the drops to a surface is essential, and if this is achieved directly the rest of the filler media is not very "active". After the addition of tensidies, however, we found an enhanced effect from the deeper filter media, indicating that the actual coalescence rate is reduced, even after contact with the filter media surface. With finer pores at the entrance we can compensate for this and do not need to have such a deep filter media, as higher shear forces will break the film by the passage in a more narrow channel [23].

If we now try to summarize the results, we can see that the surface tension has been found to have a correlation with the filter efficiency for the tested tensides. The measurement of the interfacial tension shows a correlation only for very high concentrations of tensides. The filter media should be wetted by the dispersed phase. Probably, surface film elasticity is the most important factor for low viscosity films, while surface film viscosity is more important for high viscosity films. We cannot distinctly separate the two effects, but just discuss indirect results.

References

1. Sareen SJ (Jan 1966) M Sc Thesis, Illinois Inst of Tech
2. Hazlett RN (1969) Industrial & Eng Fundamentals 8 (1969a), p 625—632 (1969b) p 633—640
3. Bitten JF (1970) Journal of Colloid and Interface Science 33, p 265—271 and (1971) 37, p 312—317
4. Wasan DT, Sampath K, Aderangi N (1980) AICheEng Series: Effects of interfacial fluid properties on emulsion instability
5. Verdegan B (1988) Ann Meeting AIChE Tech Sess Interfacial Phenomena in Sep Proc, Washington DC Nov 27—Dec 2
6. Boyd J, Parkinson C, Sherman P (1972) J Coll Interface Sci, 41:359
7. Malhotra AK, Wasan DT (1987) Chem Eng Comm v55, 95—128 and (1986) AIChE Symposium Series v 82 n 252, p 5—13
8. Sabet VM, Zourab SH, Said W (1982) Colloids and Surfaces 4:359—366

9. Verdegan (1987) Proc Filtech Conf 302—309
10. Jaisinghani RA (1977) Filtr & Sep July/Aug p 1—3
11. Goubran R, Garti N (1988) J Disp Sci Tech 131—149
12. Madia JR, Furth MS, Beerbower A (1976) Env Sci & Techn 10:1044—1046
13. Chieu JN, Gloyna EF, Schechter RS (1977) v 30:611—620
14. Treybal RE (1951) in McGraw-Hill: Liquid Extraction 280—283
15. Farrow RM (1966) Filtr & Sep Nov/Dec 1966, p 490
16. Fahim MA, Akbar AM (1984) J Environ Sci Health A19(3):299—319
17. Voyutskii SS, Kalyanova KA, Fodiman NM, Panich RM (1958) Jzs Vyssh Ucheb Zaved Khim Technol 2 p 170—179
18. Davies GA (1980) 73rd AIChE Annual Meeting
19. Langdon WM (1962) Report No ARF (Now IITRI) 6169-81-Study of Filter/separator Phenomena
20. Gudesen RC (1965) M Sc Thesis Illinois Inst of Tech
21. Hughes VB, Foulds AW (1986) Proc World Filtr Congress p 11.13—11.25
22. Dahlquist E (1987) Tech Lic Treatise Royal Inst of Tech, Stockholm
23. Dahlquist E, Setterwall F (1987) Proc Filtech Conf p 36—43
24. Miller CA, Neogi P (1985) in Surfactant Science Series vol 17, Marcel Dekker

Authors' address:

E. Dahlquist
Dept KDU
ABB Corporate Research
72128 Vasteras, Sweden

Progress in Colloid & Polymer Science Progr Colloid Polym Sci 82:163—165 (1990)

Langmuir-Blodgett film deposition studied by a surface acoustic wave device

I. Vikholm and H. Helle

Technical Research Centre of Finland, Medical Engineering Laboratory, Tampere, Finland

Abstract: A 77.8 MHz surface acoustic wave (SAW) device has been used to characterize the deposition of monolayers of cadmium arachidate and 1,1'-diarachidoylindigo. The deposition is homogeneous and the surface mass density associated with each deposited monolayer agrees with that calculated from the molecular surface area. The SAW device reveals the necessity of allowing the monolayers to dry before successive deposition.

Key words: Langmuir-Blodgett films; surface acoustic wave device; arachidic acid; 1,1'-diarachidoylindigo

1. Introduction

The Langmuir-Blodgett (LB) technique makes it possible to produce very thin organic films on solid substrates [1]. Amphiphilic molecules are spread from a volatile solvent on the water surface and compressed to form a monomolecular layer. Multilayers can be built up on a solid substrate by successive dipping and raising of the substrate through the monolayer/air interface. LB films are known for their good quality and well-organized structure. However, several factors that affect the film quality are involved in the deposition process. The deposition ratio is generally used to control the transfer of molecules, but may not give sufficient information about the transferred film. It is especially important to follow the transfer procedure when working with non-traditional materials.

The surface acoustic wave (SAW) device is one of the most sensitive microsensors and can be utilized to characterize the deposition of LB films [2]. The SAW device consists of a quartz wafer with sputtered interdigital metal electrodes (Fig. 1). A mechanical Rayleigh wave propagates across the surface between the transmitter and receiver electrodes, when the former is excited by a radio frequency (RF) voltage. The device starts to oscillate at a frequency determined by the interdigital-electrode-spacing and the Rayleigh-wave velocity, when the electrode pair is connected through a RF-amplifier. Deposition of material on the device gives rise to a substantial reduction of the Rayleigh-wave velocity and a corresponding decrease in the resonant frequency [3, 4].

In this paper we report the results of using a 77.8 MHz SAW device to monitor multilayer deposition of cadmium arachidate and an indigo dye.

2. Experimental

Deposition of LB films was performed with a KSV 2200 apparatus. Monolayers were spread from chloroform solutions onto a „Milli-Q" water subphase containing 2.5×10^{-4} M $CdCl_2$ and 5×10^{-5} M NaAc + HAc at pH 5.8 and a temperature of 293 K. Materials studied were arachidic acid (Fluka puriss grade) and an indigo dye derivated with two hydrophobic alkyl chains to allow LB film formation [5, 6]. The structure of 1,1'-diarachidoyl-

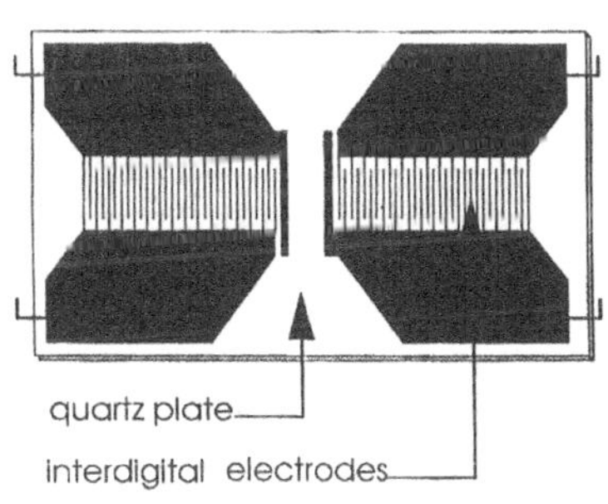

Fig. 1. Top view of the surface acoustic wave device

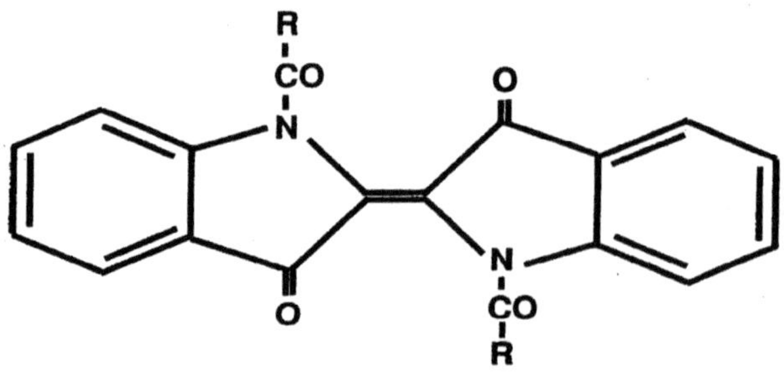

R =$(CH_2)_{18}CH_3$

Fig. 2. The structure of 1,1'-diarachidoylindigo

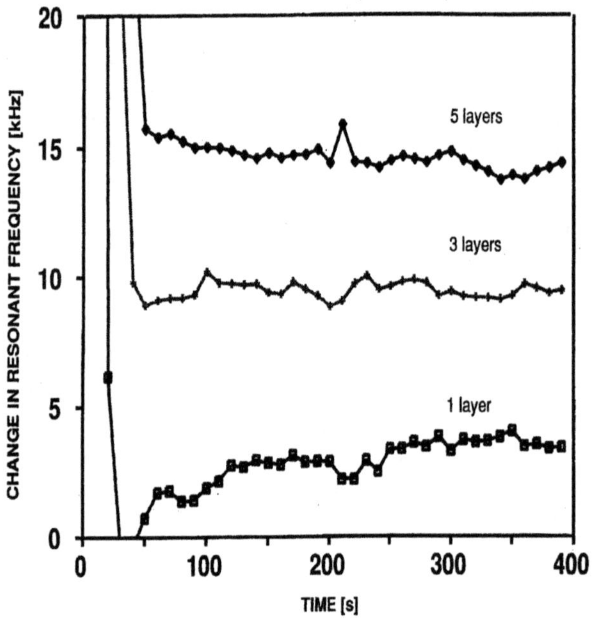

Fig. 3. The change in resonant frequency vs time during deposition of Cd arachidate

indigo, (DAI) is shown in Fig. 2. The 77.8 MHz SAW device was manufactured by Vaisala OY (Helsinki, Finland).

3. Results and discussion

The SAW device was deposited with cadmium arachidate (CdA) at a surface pressure of 30 mN/m and at a deposition speed of 5 mm/min. The resonant frequency was measured when the device had been raised out of the film-water interface. Figure 3 shows the change in resonant frequency, $\delta f = f_0 - f$ vs time for the deposition of 1, 3, and 5 layers. The resonant frequency, f_0 of the device decreases, i.e., the mass deposited onto the device increases with the number of layers deposited. However, the decrease during the first minutes after deposition indicates that subphase is transferred with the layer. The film must be completely dry before deposition of the next layer. Otherwise the last transferred layer is partially lost at the following dipping run [7, 8]. The frequency is stabilized within 1 min and the subsequent layer can be deposited. Figure 4 shows the change in resonant frequency as a function of number of layers; the relationship is linear and indicates a homogeneous deposition of monolayers. A least-squares slope δf of 2.4 kHz/layer obtained from the figure gives a surface mass density, $h\delta$, of 290 ng/cm^2 per layer according to:

$$h\delta = \delta f/(k_1 + k_2)f_0^2 , \quad (1)$$

where k_1 and k_2 are material constants [3]. This is in good agreement with the value obtained from the surface molecular area of the molecule.

The DAI molecule mixed with arachidic acid in a molar ratio of 1:7 gives an isotherm similar to that of CdA and is as stable as a fatty acid monolayer alone [6]. Layers of the mixed film were deposited onto a SAW device, made hydrophobic with five layers of Cd arachidate. The layers were deposited at a surface pressure of 30 mN/m and a deposition speed of 3 mm/min. Figure 5 shows the change in

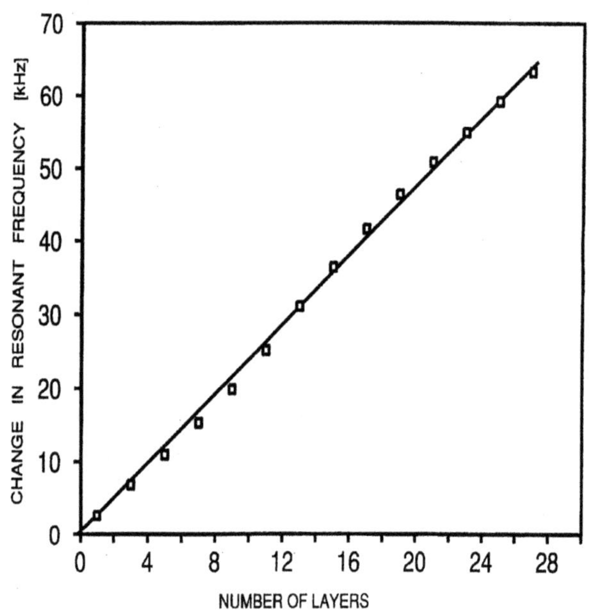

Fig. 4. The change in resonant frequency vs number of layers of Cd arachidate

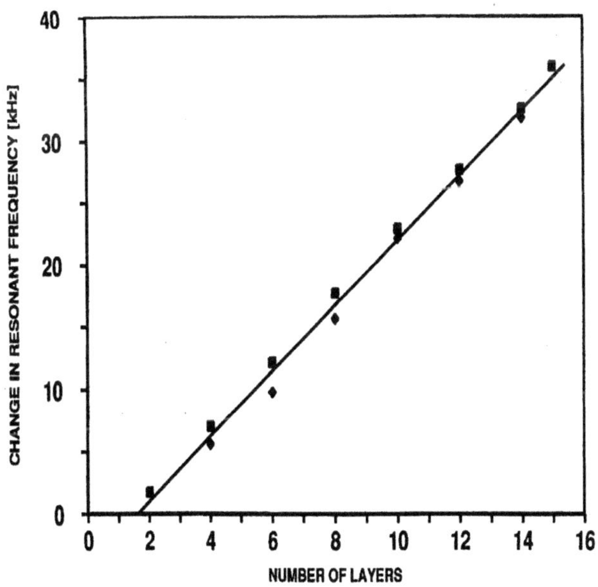

Fig. 5. The change in resonant frequency vs number of mixed monolayers of 1,1'-diarachidoylindigo and Cd arachidate deposited on two separate occasions

resonant frequency vs number of layers deposited onto a SAW device on two separate occasions. The subsequent layers were transferred when the frequency of the device became stable. The relationship is linear and indicates a homogeneous deposition of monolayers. The surface mass density of 310 ng/cm^2 agrees with that calculated from the surface molecular area [6]. However, the existence of a negative intercept suggests that the first layer(s) pack more loosely than the subsequent layers.

4. Conclusion

Monolayers of cadmium arachidate and mixed monolayers of 1,1'-diarachidoylindigo and cadmium arachidate could be homogeneously transferred onto a surface acoustic wave (SAW) device. The SAW-device reveals that subphase is transferred with the layers and they have to dry at least 1 min before successive deposition. In the mixed film the first layer(s) deposited packs more loosely than the subsequent layers.

Acknowledgement

This work was supported by the Finnish Work Environmental Fund and the Academy of Finland.

References

1. Blodgett K (1935) J Am Chem Soc 57:1007—1022
2. Roberts GG, Holcroft B, Barraud H, Richard J (1988) Thin Solid Films 160:53—60
3. Wohltjen H (1984) Sensors and Actuators 5:307—325
4. Venema A, Nieuwkoop E, Vellekoop MJ, Nieuwenhuizen MS, Barendsz AW (1986) Sensors and Actuators 10:47—64
5. Omote Y, Imada S, Matsuzaki R (1979) Bull Chem Soc Jpn 52:3397—3399
6. Vikholm I, Helle H (1989) Thin Solid Films 178: 197—202
7. Srinivasan ML, Higgins BG, Stroeve P, Kowel ST (1988) Thin Solid Films 159:191—205
8. Laschewsky A, Ringsdorf H, Smith G (1985) Thin Solid Films 134:153—172

Authors' address:

Inger Vikholm
Technical Research Centre of Finland
Medical Engineering Laboratory
P.O. Box 316
33101 Tampere, Finland

An improved version of the drop volume tensiometer applied to aqueous/silicone oil systems

P. J. Doyle and B. J. Carroll

Unilever Research Laboratory, Bebington, Merseyside, United Kingdom

Abstract: The drop volume technique for interfacial tension measurement uses relatively small amounts of liquids, but, nevertheless, is capable of high precision. It is, therefore, especially attractive when the substrate phases are limited in quantity. — The introduction of a newly-available micrometer head and alterations to the glass syringe has enabled the traditional technique to be improved on in two ways: first, the effective syringe capacity has been doubled without any loss in accuracy; second, it has been possible to semi-automate the system with considerable time-saving in the calculation of the tension. — The modified technique has been applied to the measurement of the interfacial tension in aqueous/silicone oil systems in situations where the oil, extracted from systems in vivo, had a total volume of about 1 ml. The small density difference in these systems made for large drop volumes, which in certain cases exceeded the capacity of the conventional technique. The modified technique worked well in these systems.

Key words: Drop volume; surface tension; interfacial tension

Introduction

In the measurement of interfacial tension, drop volume tensiometry (DVT) is comparable in precision with the best available techniques, but differs from most of these by requiring only quite small amounts of the liquids concerned. Thus, whereas the DVT uses a few ml of the test liquids, the Wilhelmy Plate method generally uses tens of ml. Both techniques are of comparable precision, measuring tensions to ±0.01 mN/m.

In the DVT, the volume V of the drop which just detaches from a cylindrical supporting tip of radius r is measured (Fig. 1). This drop volume is a function of the volume V' of the largest drop of liquid the tip is capable of supporting before detachment occurs and, also, of the system dimensions with respect to the capillary length, a, defined by

$$a = (2\gamma/\Delta\rho g)^{1/2} , \qquad (1)$$

where $\Delta\rho$ is the operating density difference, g is the acceleration due to gravity, and γ is the interfacial tension. The relationship between the latter quantity and the drop volume V was investigated experimentally by Harkins and Brown [1], who reported an empirical correction term $\varphi(r/V^{1/3})$ for the so-called Tate equation:

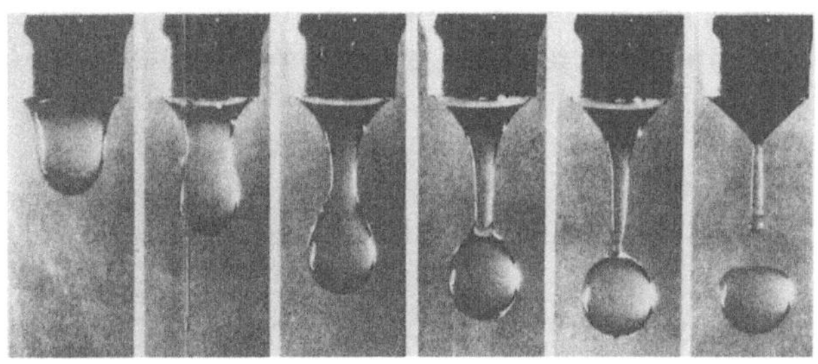

Fig. 1. Detachment of liquid drop from cylindrical tip. A volume V detaches from an initial (equilibrium) volume V'. The volume which detaches is that of the lowermost section (the primary drop) plus that of the middle cylindrical section, which finally breaks off at its topmost point to form the secondary droplet(s)

$$2\pi r\gamma = \Delta\rho g V' , \qquad (2)$$

so that a corresponding expression in V could be used:

$$2\pi r\gamma = \Delta\rho g V/\varphi \; . \qquad (3)$$

However, the Tate expression [1] is itself incorrect, so that the term φ actually embodies corrections both for this fact and for the fact that $V < V'$ [2, 3].

Values of $\varphi(r/V^{1/3})$ over the range $0.3 < r/V^{1/3} < 1.6$ were reported in the original paper [1] and this range has been extended down to zero by Wilkinson [4]. The correction factor over the full range is given in Fig. 2. Note that as the abscissa approaches zero, the Tate expression (2) becomes more nearly exact. This is expected on theoretical grounds [3].

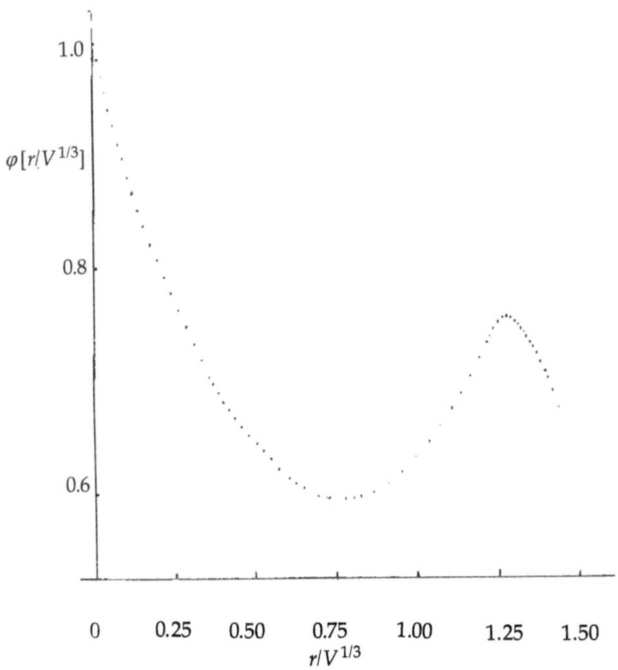

Fig. 2. The correction factor $\varphi(r/V^{1/3})$ from results of Harkins and Brown [1] and Wilkinson [4]

Technique modifications

The modifications introduced in this paper center on the use of a Mitutoyo Digimatic 164 micrometer head in place of the version usually employed.

Delivery system modifications

In its traditional form [5–7], the technique utilizes the "Agla" all-glass syringe (or its equivalent) coupled to a standard vernier micrometer head having a (linear) anvil movement of 25 mm, readable to 0.001 mm (equivalent to an expelled liquid volume of 0.5 ml in steps of 0.0002 ml). Thus, detaching volumes V in excess of 0.25 ml are outside the capacity of the system. The new head used in this work has, unusually, an anvil movement of 50 mm with unchanged readability of 0.001 mm.

By modifying the construction of the syringe piston and changing its mode of attachment to the micrometer head, the enhanced delivery capacity can be fully exploited without actually changing the syringe itself. The modified system is shown in Fig. 3. The glass syringe piston head is re-profiled to allow it to be firmly attached to the micrometer anvil by a spring device (which still allows slight lateral movement). This eliminates a dead space in the syringe due to the compressed length of a helical spring in the assembly. The modified system can thus deliver (without loss of precision) twice the volume formerly possible.

The practical utility of this increased delivery capacity is most manifest when tensiometry of pairs of liquids having a small density differential is needed. The drop volume V can then be comparable with the total syringe capacity, restricting the range of tensions that can be measured. (Although resort can be had to finer dropping tips, tip wetting problems become increasingly troublesome.) Increase in the syringe capacity results in a proportionate increase in the range of tensions accessible (see final section).

Otherwise, in systems where the drop size is much smaller, the effect of the increased syringe capacity is to increase the number of drops obtainable from a single filling. This can sometimes improve the overall precision of the results. The syringe dead space is also relatively (and, in the present case, absolutely) smaller, allowing smaller volumes of the drop-forming liquid to be worked with.

Automation of the system

The new micrometer head, besides having a digital display, can also output data in binary-coded decimal form, which allows it to be interfaced with a microcomputer. Using a Mitutoyo MUX-10 multiplexer, the present system was interfaced with a BBC Master micro programmed to calculate the mean tension from a series of micrometer inputs. Equation (3) was used in conjunction with data on the liquid densities and the tip diameter and values of the factor φ interpolated from tabulations derived from the smoothed data of Fig. 1.

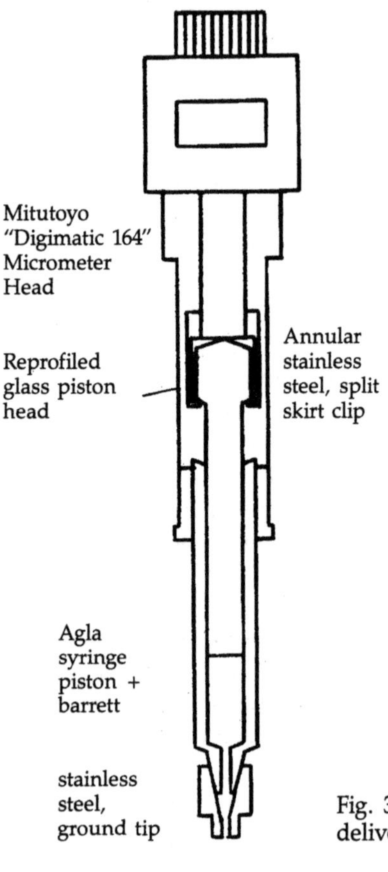

Fig. 3. The modified delivery system

Application to aqueous/silicone oil (SO) systems

Many silicone oils have densities close to that of water, so that aqueous/SO systems provide a good illustration of the features of the improved apparatus. The application involved the measurement of interfacial tension (against water) of an SO which had been implanted in the human eye.

In one surgical treatment of retinal detachment of the eye, an SO is used to replace the vitreous humor in the posterior chamber (Fig. 4). In this system, it is especially important that the aqueous/SO interfacial tension is not lowered by the presence of surface-active material, because of the possibility of forming a stable emulsion, both during the introduction of the oil and subsequently. The interfacial tension has further importance because capillary forces play a dominant role in these systems, where gravity forces are minimized by the small density differential.

The implanted SO, Dimethicone 1000 (Dow Corning, USA) had a density of 0.975 g/ml and was thus the environing phase to the drop in the actual tension measurements. The recovered oil was only obtainable in a total volume of ca. 1.0 ml from a single eye. This necessitated the construction of a containing vessel of appropriate size.

The interfacial tensions against water of the SO as supplied and after its recovery from the eye some months later were measured. Both systems showed comparable values, indicating that surface-active material is probably not being introduced into the system in vivo. The values obtained, ca. 37 mN/m, were, however lower than literature values for similar oils (ca. 42 mN/m, [8]), indicating that purification of the SO as supplied is desirable. Chromatographic techniques are being investigated to this end.

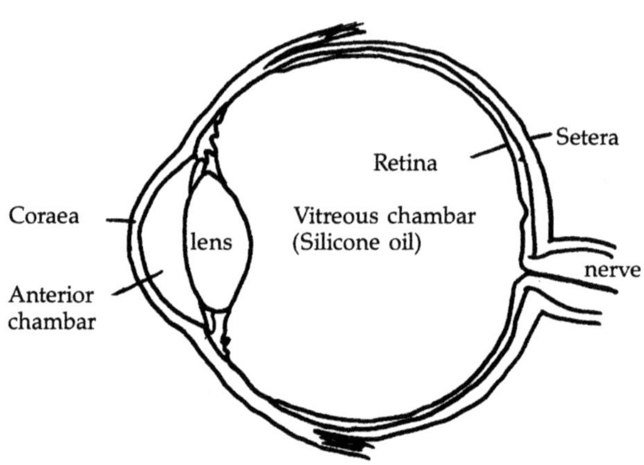

Fig. 4. Section of the human eye showing implant region

Acknowledgements

The SO samples recovered from implants were supplied by Mr. D. Wong of the Royal Liverpool Eye Hospital, U.K. — R. Morgan of this laboratory produced the modified glass syringe piston.

References

1. Harkins WD, Brown FH (1919) J Am Chem Soc 41:499
2. Boucher EA, Evans MJB (1975) Proc Roy Soc A346:349
3. Hartland S, Srinivasan PS (1974) J Colloid Interface Sci 49:320
4. Wilkinson MC (1972) J Colloid Interface Sci 40:14
5. Parreira HC (1965) J Colloid Sci 20:44
6. Aveyard R, Haydon DA (1965) Trans Faraday Soc 61:2255
7. Carroll BJ (1970) Ph D Thesis, Cambridge
8. Kanellopoulos AG, Owen MJ (1972) J Chem Soc Faraday I 67:3127

Authors' address:

Dr. B. J. Carroll, Mr. P. J. Doyle
Unilever Research Laboratory
Quarry Rd East
Bebington, Merseyside, L63 3JW, United Kingdom

Self Assembling System

Progress in Colloid & Polymer Science Progr Colloid Polym Sci 82:173—180 (1990)

Interaction of steroids with a model of stratum corneum lipids

I. Kayali, S. E. Friberg and P. Zuman

Department of Chemistry, Clarkson University Potsdam, New York, USA

Abstract: Lithocholic acid acetate methyl ester, cholesteryl acetate, methyl lithocholate, hydrocortisone, progesterone, and testosterone were added to a model system of stratum corneum lipids and structural changes were investigated using low-angle x-ray diffraction and NMR spectroscopy. — The interactions were dominated by the presence of cholesterol in the model. Without it the three esters gave a radically different relation between water and the structure than the three remaining steroids. With cholesterol present the only difference was less water penetration after addition of the esters. NMR data demonstrated a disordering effect of hydrocortisone contrary to the influence of cholesterol.

Key words: Steroids; stratum corneum lipids; cholesterol; hydrocortisone

Introduction

The relation between steroids and stratum corneum is of decisive importance because of the use of steroids in topical applications that utilize their anti-inflammatory, anti-pruritic, and vasoconstrictive properties [1—3]. It is well known that the extent of percutaneous absorption is determined by many factors [4] including the vehicle [5], the structure of the epidermal barrier [6], and the use of occlusive dressing [7]. Recent contributions have improved the precision and reproducibility of in vivo skin penetration studies [8].

These investigations have meant a pronounced progress in the understanding of percutaneous absorption but a more exact relation between steroids and the stratum corneum lipids is still warranted against the unsolved problem of the surprising non-effect of delipidization on steroid penetration [9].

Earlier research [10] has shown that the partitioning of steroid esters between human stratum corneum and water points to a liquid crystalline liposome as the optimal model for stratum corneum. With this in mind we found useful an investigation of the structural interactions between steroids and our liquid crystalline model for stratum corneum lipids [11, 12].

Experimental

Materials

The materials used for the model lipid were all of the highest available purity, Table 1. Hexadecanoic-d_{31} acid (99 atom % D) was obtained from MSD isotopes; deuterium-depleted water from Sigma; lithocholic acid acetate methyl ester, cholesteryl acetate, methyl lithocholate, hydrocortisone, progesterone, and testosterone from Steraloids, Inc.

Table 1. Composition of model epidermal lipid

Component	Source	Purity	Wt. % in mixture
Free Fatty Acids [a])	Sigma		50
Oleic acid		99%	33.4
Linoleic acid		99%	12.5
Palmitoleic acid		99%	3.6
Palmitic acid		99%	36.8
Myristic acid		99%	3.8
Stearic acid		99%	9.9
Phosphatidyl-ethanolamine (PE)	Avanti polar lipids	99%	13.2
Cholesterol	Fisher	99%	36.8

[a]) Free fatty acids mixture was neutralized to 41% using NaOH with water content of 32% in order to produce a lamellar structure. After the addition of the rest of the lipid, water content was adjusted to required percentages in the range of 32—45% water.

Preparation of samples

The free fatty acids (Table 1) were added together and neutralized to 41 mol%, using sodium hydroxide solution giving a total water content of 32% by weight. The samples were mixed by centrifuging repeatedly through a constriction in a sealed tube. The mixture of acid/soap/water formed a lamellar liquid crystal. The structure was verified by low-angle x-ray diffraction and optical microscopy. Phosphatidylethanolamine, cholesterol and/or the other steroids were then added under nitrogen atmosphere and mixed as above. Water content in the different samples was in the range 32—45% by weight.

Small angle x-ray diffraction

Small-angle x-ray diffraction measurements were obtained using a Kiessig low-angle camera (Richard Seifert, FRG). Nickel-filtered copper radiation was used and the reflections determined by a position-sensitive detector system (Tennelec Model PSD-100, Memphis, USA).

NMR spectroscopy

NMR spectra were obtained at a resonance frequency of 13.71 MHz using a multinuclear pulsed NMR spectrometer (JEOL FX90) operating in the Fourier-transform mode.

Results

The interlayer spacings from the low angle x-ray diffractograms are first described followed by the order parameters changes from the NMR spectra.

Low angle x-ray

The interlayer spacing in the simplified model with cholesterol, Table 1, was increased with water content for all steroids (Fig. 1), but to a varied extent. Methyl lithocholate showed the greatest increase, followed by cholesteryl acetate, lithocholic acid acetate methyl ester, and the hormone group.

With the cholesterol in Table 1 replaced by steroids the results were entirely different. The esters still gave an increase of interlayer distance with added water, but the hormone group now gave a strong reduction in interlayer spacing with increased water content (Fig. 2).

Order parameters

The ^{2}H NMR spectrum for perdeuteriated hexadecanoic acid in unsaturated free fatty acids (41% neutralized)/water (32% by weight) is shown in Fig. 3a. The corresponding spectrum for the composition in Fig. 3a plus cholesterol in amounts according to Table 1 is shown in Fig. 3b, plus methyl lithocholate in Fig. 3c, and plus hydrocortisone in Fig. 3d. The central, sharp signal was approximately 10% of the total signals in Fig. 3a, extremely weak in Fig. 3b, weak in Fig. 3c, and dominant in Fig. 3d.

The signals symmetrically distributed vs the central one are due to quadrupolar splitting [13]. This value $\Delta\nu$ is directly proportional to the order of parameter S for an unoriented lamellar dispersion. One has

$$\Delta\nu = 3/4\chi \mid S \mid ,$$

where χ is the quadrupole coupling constant and S is the order parameter associated with a C—D bond.

In this way the order parameter for each methylene group of hexadecanoic acid chain (Fig. 4) is directly calculated from the spectra in Fig. 3. The direct presentation of order parameters (Fig. 5) does not completely reveal the specific influence, it only focuses attention on the sudden onset of reduction of order parameter for groups farther away from the polar group than the double bonds of the unsaturated chains. A better illustration is obtained by plotting palmitic acid order parameters relative to the ones in its combination with the unsaturated fatty acids (Fig. 6). Addition of cholesterol caused a pronounced ordering of the structure while hydrocortisone displayed a weak disordering effect.

Discussion

The results present information about the change in the order of the stratum corneum model lipids when steroids were added; these changes were distinct. The cholesterol gave a pronounced ordering of the chains, the hydrocortisone gave a disordering while the methyl lithocholate caused no distinct change whatsoever.

It is obvious that this information is not sufficient to explain the differences in the interlayer spacings (Fig. 2). To explain these the total behavior must be discussed.

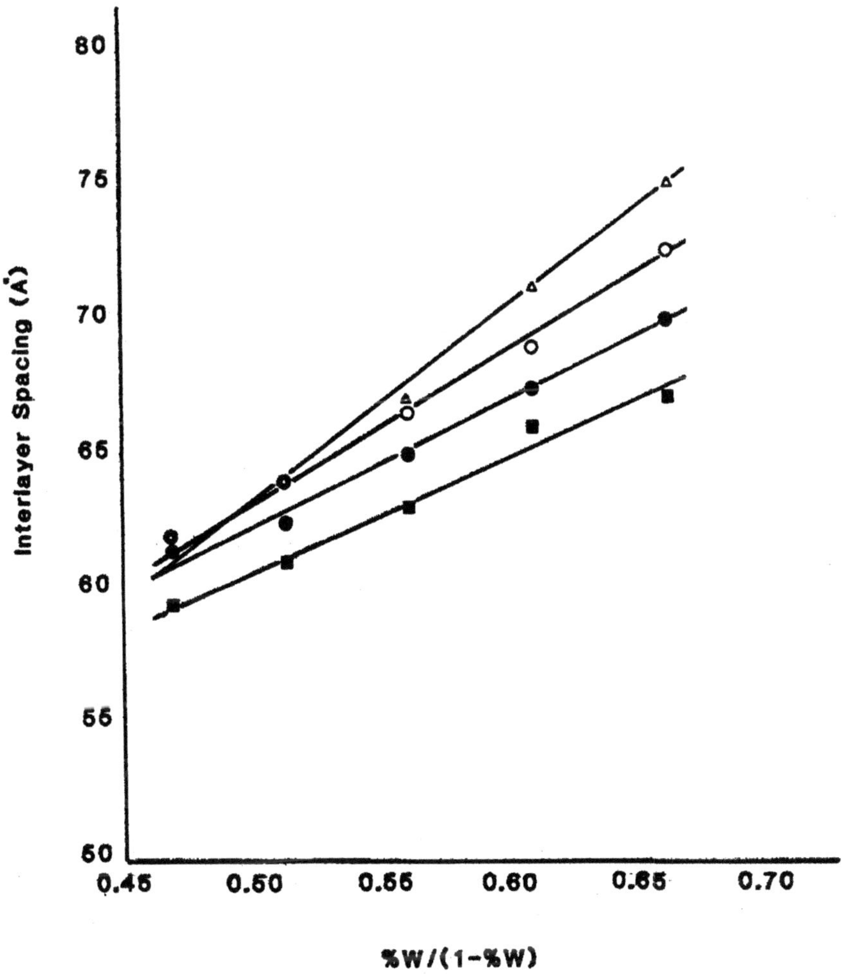

Fig. 1. Interlayer spacings vs water weight ratio. Composition according to Table 1 plus 10% of the steroids counted on the model lipid plus water. Steroid: △ = Methyl lithocholate; ○ = Cholesteryl acetate; ● = Lithocholic acid acetate methyl ester; ■ = Hydrocortisone, Progesterone, and Testosterone

The ordering/disordering as revealed by Fig. 6 is a reflection only of the condition of the palmitic acid located at site B in Fig. 4. It does not take into consideration the transfer of palmitic acid between the two sites. Information about the relative amount of palmitic acid in the two sites is obtained in qualitative form from the NMR spectra, Figs. 3a—d. These are from samples with the minimum water (Figs. 1 and 2). The spectra shows addition of cholesterol and, to a lesser extent, methyl lithocholate to reduce the relative number of acid molecules located at site A, Fig. 4. The addition of hydrocortisone, however, gave the opposite effect in the most pronounced manner — now most of the palmitic acid molecules are located between the methyl group layers.

This information is combined with the results from the x-ray diffraction patterns. These also provide information about transfer between sites A and B in Fig. 4, but now in quantitative manner after simple arithmetical manipulation of the curves in Figs. 1 and 2. They are straight lines obeying the relation

$$d = d_0 + R \cdot \partial d/\partial R \,, \tag{1}$$

in which d is the experimentally determined interlayer spacing, d_0 is its value extrapolated to r = 0, and R is the volume ratio of water to lipid material. With no penetration of water into the structure the increase in interlayer spacing would be proportional to R and Eq. (1) is

$$d = d_0(1 + R) \,. \tag{2}$$

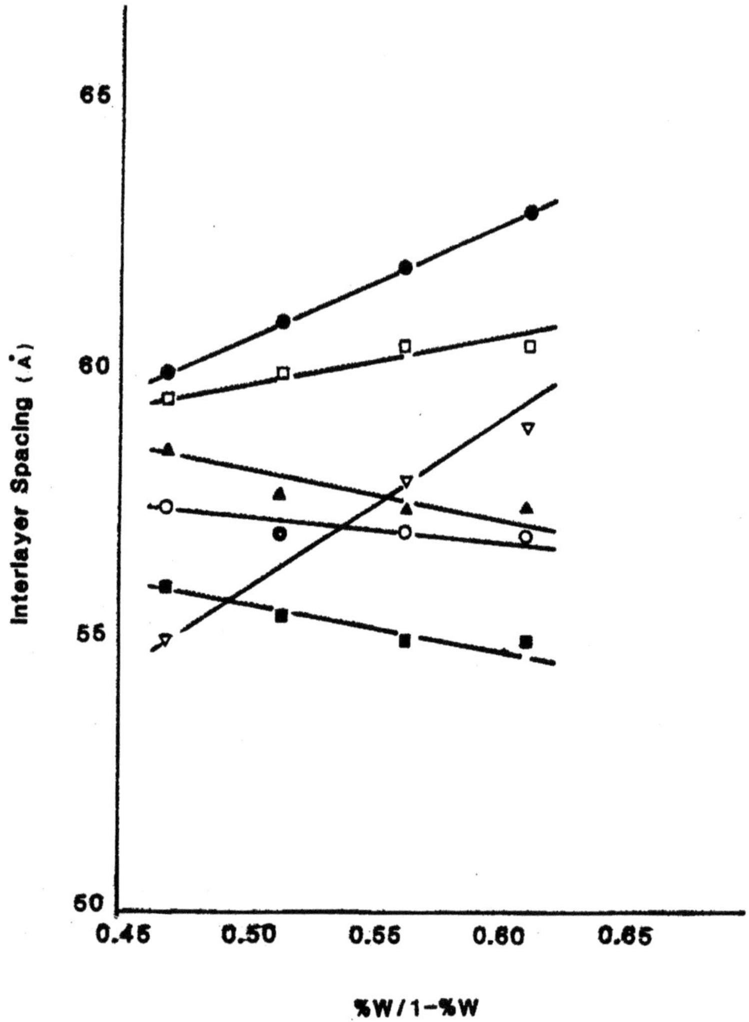

Fig. 2. Interlayer spacing vs water weight ratio. Composition according to Table 1 with the cholesterol substituted by the steroids. Steroid: ● = Choleseryl acetate; □ = Lithocholic acid acetate methyl ester; ▽ = Methyl lithocholate; ▲ = Progesterone; ○ = Hydrocortisone; ■ = Testosterone

A penetration of a fraction a of the water results in a modified formula

$$d = d_0[1 + (1 - a)R] , \quad (3)$$

and a may be calculated

$$a = 1/R(1 + R - d/d_0) .$$

The penetration values are given in Table 2. The values may at first appear confusing; a values less than zero mean an increase of interlayer spacing greater than what corresponds to added volume. In the same manner a penetration in excess of 1.0 seems impossible, it implies a reduction in interlayer spacing with added material. However, the values should be regarded considering the total picture (Fig. 4). A negative a value means that the sum of water penetrating and the lipids transferred from site B to site A (Fig. 4) is less than zero, i.e., the volume of lipids leaving site B and moved to A is greater than the volume of water penetrating into the space between the hydrocarbon chains. In the same manner an a value in excess of 1.0 implies that the volume of water penetrating plus that of lipids moving from A to B is greater than the volume of water added.

The penetration values (Table 2) may now be understood. With cholesterol or one of the esters present (Table 1) addition of water gives an ap-

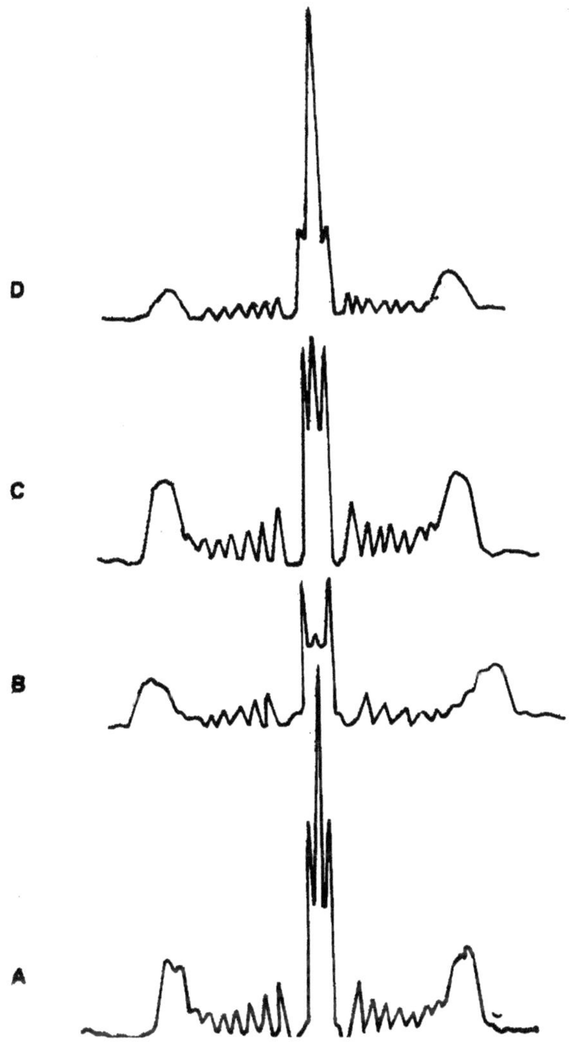

Fig. 3. NMR spectrum of totally deuteriated palmitic acid in a) unsaturated partially saponified fatty acids according to Table 1, 68% plus 32% water; b) unsaturated partially saponified fatty acids plus phosphatidylethanolamine (PE), plus cholesterol according to Table 1, 68% plus 32% water; c) according to b, but with methyl lithocholate replacing cholesterol; d) according to b, but with hydrocortisone replacing cholesterol

parent negative penetration, because the presence of water causes some of the lipid molecules at site B to be transferred to site A. The "negative penetration" of water is in fact a consequence of a coordinated movement resulting in a greater volume of lipids moving from b to A than that of water entering B. With the conditions clarified the results from NMR and x-ray may be explained within one framework.

There is a distinct difference in the behavior of the biologically active steroids and the esters. In a model without cholesterol the former showed a strong action in attracting both lipids and water to penetrate into the space between the lipid hydrocarbon chains when water was added. This phenomenon should be viewed against the fact that at low water content a signficiant part of the lipids are located at site A in the structure (Figs. 2 and 4). The more hydrophobic esters gave lower penetration values with methyl lithocholate giving the lowest.

In the model with cholesterol the hormones had little influence; the water penetration was zero. The cholesterol obviously blocked the interaction between hormones and water. This is in agreement with earlier results [14], which demonstrated cholesterol strongly blocked water penetration and, with the present NMR result, showed cholesterol to cause all but a small fraction of the palmitic acid to be located at site A (Figs. 2 and 4). Methyl lithocholate displayed a similar action and the values in Table 2 show that the combined influence is stronger than that from cholesterol alone.

In summary, the conclusion is that the physiologically active steroids have an entirely different structural influence on the stratum corneum lipids than on the inactive ones; they act as amphiphilic intermediates between water and the more hydrophobic lipids.

Acknowledgement

This research was supported by Dow Chemicals, Midland, Michigan in the Clarkson University CAMP program.

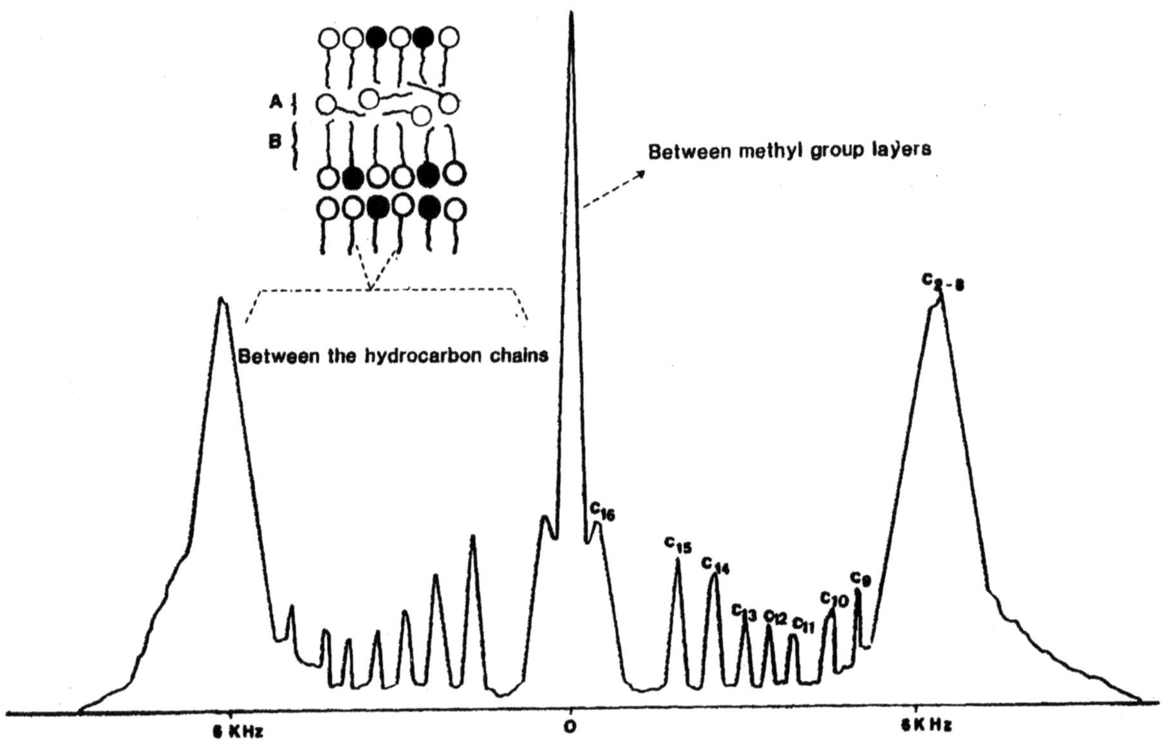

Fig. 4. The NMR spectra of deuteriated palmitic acid in oleic acid in a weight rate, 0.74. The acids were saponified to 41% by mole. Water present to 32% by weight to form a lamellar liquid crystal. The sharp central signal corresponds to acids located at site A. The symmetrically split signals are for moleöcules at site B. The numbers given denote methylene groups counted from the polar group of the acid

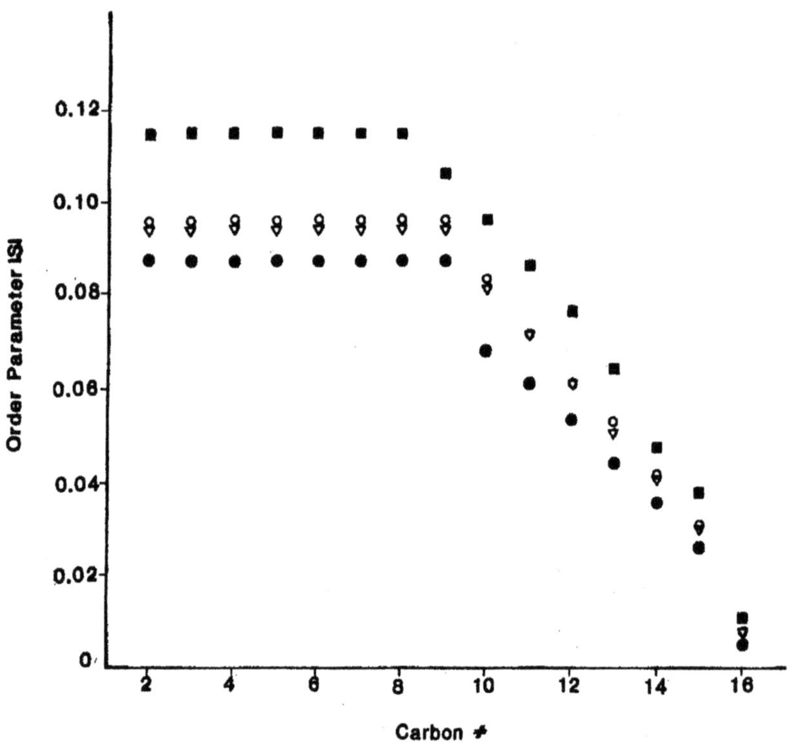

Fig. 5. The order parameter for the methylene groups in palmitic acid according to Figs. 3a–d. ▽ = 3a; ■ = 3b; ○ = 3c; and ● = 3d

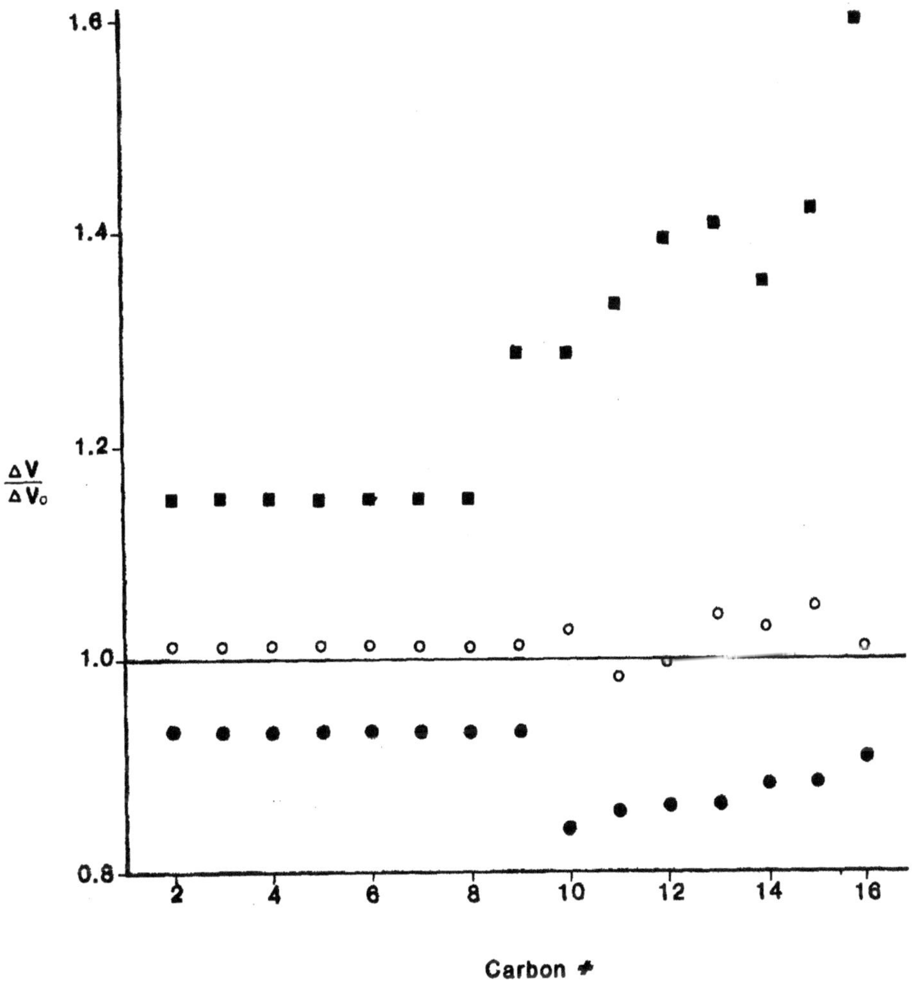

Fig. 6. The ratio of methylene group order parameters of palmitic acid with different additives. Composition: — = 3a/3d; ■ = 3b/3a; ○ = 3c/3a; and ● = 3d/3d

Table 2. Values of interlayer spacing extrapolated to zero water content and the penetration system

FFA/PE/Steroids[a])			FFA/PE/Cholesterol/10% W/W Steroids		
Steroid	Intercept, d_0, A^0	a	Steroid	Intercept, d_0, A^0	a
Lithocholic acid Acetate methyl ester	56.2	0.87	Lithocholic acid Acetate methyl ester	40.0	−0.125
Cholesteryl acetate	50.2	0.58	Cholesteryl acetate	37.0	−0.432
Methyl lithocholate	42.6	0.36	Methyl lithocholate	30.0	−1.250
Hydrocortisone	58.8	1.05	Hydrocortisone	40.0	−0.02
Progesterone	60.7	1.09	Progesterone	40.0	−0.02
Testosterone	59.3	1.12	Testosterone	40.0	−0.02

[a]) Steroids added in the same ratio of cholesterol in the original model lipid, Table 1.

References

1. Viller CA, Engel LC (1961) Mechanism of Action of Steroid Hormones. Pergamon Press, New York
2. Shoppee CW (1964) Chemistry of Steroids. Butterworth, London
3. Pasqualini JR (1977) Receptors and Mechanism of Action of Steroid Hormones. Marcel Dekker, New York
4. Scheuplein RJ, Blank IH, Brauner GJ, MacFarlane DJ (1969) J Invest Dermatol 52:63
5. Gibson JR, Hough JE, Marks P, Webster A (1987) Pharmacol Skin 1:214
6. Kao J, Hall J (1987) J Pharmacol Exp ther 241:482
7. Barry BW, Southwell D, Woodford R (1984) ibid 82:49
8. Guy R, Bucks DAW, McMaster JR, Villaflor DA, Roskos KV, Hinz RS, Maibach HI (1987) Pharmacol Skin 1:70
9. Bucks DAW, Maibach HI, Menczel E, Wester RC (1983) Anh Dermatol Res 275:242
10. Saket MM, James KC, Kellaway IW (1985) Int J Pharmaceutics 27:287
11. Friberg SE, Osborne DW (1985) J Disp Sci Techn 6:485
12. Friberg SE, Suhaimi H, Rhein LK, Goldsmith LB (1988) Colloids § Surfaces 30:1
13. Wernerström H, Lindblom G, Lindman B (1975) Am Chem Soc Symp Ser 9:253
14. Friberg SE, Suhaimi H, Goldsmith LB (1988) J Disp Sci Techn 9:371

Authors' address:

S. E. Friberg
Department of Chemistry
Clarkson University
Potsdam, New York 13699-5810, USA

Progress in Colloid & Polymer Science Progr Colloid Polym Sci 82:181—189 (1990)

Microemulsions as vehicles for lipase catalyzed reactions

K. Holmberg and E. Österberg

Berol Nobel AB, Stenungsund, Sweden

Abstract: Microemulsions have proved to be excellent reaction media for several types of enzymatic lipid transformations. Using a 1,3-specific lipase transesterification, selective hydrolysis and glycerolysis reactions have been performed in good yield with a high degree of regioselectivity. The rate of transesterification was found to be highly dependent on the type of unsaturation of the fatty acid residues; a double bond in the 6-position, as in γ-linolenic acid, inhibited the reaction. — All reactions were made in microemulsion containing 0.5—5% water. Attempts were made to perform glycerolysis in a water-free microemulsion with glycerol as the polar component. The rate of reaction was very low, however, in the absence of water. — The reactions were governed by the choice of surfactant of the microemulsions. Nonionic surfactants were found to be unsuitable for the hydrolysis reactions, giving slow reaction rates and low yields. These surfactants probably form a palisade layer at the interface which prevents the enzyme access to the organic domains in which the substrate is located.

Key words: Lipase; enzyme; microemulsion; triglyceride; monoglyceride

Introduction

It is by now well established that enzymes can be used in reaction systems with very low water content. The choice of the organic solvent has been found to be crucial and the biocatalytic activity varies considerably, depending on the polarity. The logarithm of the partition coefficient between octanol and water, $\log P$, has been used as a parameter to predict the solvent effect [1, 2]. It has been shown that biocatalysis in organic solvents is low in polar solvents having a $\log P < 2$, is moderate in solvents having a $\log P$ between 2 and 4, and is high in apolar solvents having a $\log P > 4$. The correlation between activity and solvent polarity seems to parallel the ability of organic solvents to distort the essential water layer that stabilizes the biocatalysts [1].

Relatively few solvents fulfil the requirement of $\log P > 4$, as can be seen from Table 1. Among the more common solvents only aliphatic hydrocarbons with 7 or more carbon atoms belong to this category. Isooctane, combining a high $\log P$ value with an acceptable volatility, has became the solvent of choice.

Table 1. Log$P$ values of common organic solvents

Solvent	$\log P$
Dimethylsulfoxide	—1.3
Ethanol	—0.24
Acetone	—0.23
Ethylacetate	0.68
Diethylether	0.85
Pentanol	1.3
Chloroform	2.0
Toluene	2.5
Hexane	3.5
Heptane	4.0
Octane	4.5
Dodecane	6.6

In many cases water needs to be incorporated into the reaction mixture in an amount of 0.5—5%. This can be done in the form of an emulsion of w/o type. It may also be done by creating a microemulsion. The main advantages with the microemulsion approach, apart from the inherent stability of the system, are the large interfacial area between oil and water microdomains and the highly dynamic nature of the system. Compared to emulsions and other two-phase systems diffusion of substrate and product in and out of the water pools is extremely rapid in microemulsions.

Experimental

General procedure

All reactions were carried out at 37°C under nitrogen using a magnetic stirrer. The enzyme was used in an amount of 1.5 U/mg substrate.

Microemulsions were made by mixing isooctane, surfactant, aqueous 0.066 M phosphate buffer of pH 7 and substrate. After completed reaction the temperature was raised to 100°C and kept there for 10 min to denature the enzyme. The solvent was evaporated under vacuum.

Hydrolysis

In the hydrolysis reactions 5% phosphate buffer and 8% surfactant were used. The residue obtained after evaporation was dissolved in chloroform. The chloroform solution was extracted four times with 4% aqueous $CaCl_2$ solution. To a 20-ml sample of the chloroform solution was added 0.050 ml 70% perchloric acid in order to induce acyl group migration. After 90 s stirring 50 ml periodic acid was added and analysis for vicinal diol was carried out as described in the literature [3]. In the calculation of total monoglyceride a correction factor of 1.15 suggested in [4] was used.

The combined aqueous phases from the extraction were analyzed for glycerol in the same way, using periodic acid as glycol-splitting agent. No perchloric acid was added and no correction factor was used in the calculation.

Analysis for fatty acid was made by titration of a sample dissolved in 0.1 M KOH in ethanol. Because triglycerides contain small amounts of free fatty acids, analysis must be performed on the starting material and the value of free fatty acids obtained must be subtracted from that of the sample.

The reactions were monitored by gas-liquid chromatography (GLC) analysis of tri-, di-, and monoglycerides, as well as glycerol and fatty acid. TMS derivatives were made and the analysis was performed using literature methods [5]. Due to the large differences in molecular weight, tri-, di-, and monoglycerides were analyzed by a faster GLC program (0—160°—30—350°—6), and monoglycerides, glycerol, and fatty acids by a slower program (2—80°—50—160°—30—350°—6).

The reactions were also monitored by thin-layer chromatography (TLC) on silica gel, using diethyl ether (50 parts), petroleum ether 60—70 (40 parts), and acetone (10 parts) as eluent. The plates were developed by iodine vapor.

Glycerolysis

Analysis of tri-, di- and monoglycerides, as well as glycerol and fatty acid, was made by TLC, as described above under Hydrolysis.

Monoglycerides were also detected using ^{3}H-labelled glycerol which was mixed with inactive glycerol and used as starting material. After completed reaction the monoglycerides were isolated by preparative thin-layer chromatography, using the eluent system described above. The monoglyceride fraction was scraped off and extracted with diethyl ether. The ether solution was evaporated and the monoglycerides were analyzed in a liquid scintillation counter for radioactivity. A standard curve was prepared from the radioactive glycerol used as starting material, and the amount of radioactive monoglyceride formed during the reaction was calculated.

Transesterification

The reaction product was dissolved in diethyl ether and charged on a preparative TLC plate. Elution was made with a combination of hexane, diethyl ether, and formic acid (70:30:0.2). The triglyceride fraction was scraped off, dissolved in diethyl ether, and collected. The triglyceride obtained was hydrolyzed and methylated with $LiOCH_3$. The methyl esters obtained were analyzed by GLC using a Carbowax 1.2 µm FSOT column. Analytical grade, fatty-acid methyl esters were used as references. The amount of tri-, di-, and monoglycerides present in the reaction mixture were monitored by GLC as described above under Hydrolysis.

NMR measurements

Self-diffusion coefficients (D-values) were obtained using the Fourier transform pulsed-gradient spin-echo (FT PGSE) technique with ^{1}H NMR on a Jeol-100 NMR spectrometer. The temperature of the measurements was 37.0 ± 0.5°C. The FT PGSE measurements were performed by varying the duration of the gradient pulse (~1 Gauss/cm) at a constant measuring time interval of 140 ms [6].

Materials

The 1,3-specific lipase was from *Rhizopus delemar* (99 U/mg), obtained from Fluka, Switzerland. The fatty acids and triglycerides, as well as the fatty acid methyl esters, were all from Sigma, USA. Sodium bis(2-ethylhexyl)sulphosuccinate (AOT) was from Sigma and the nonionic surfactants, alcohol ethoxylates, were homologue pure compounds from Nikko Chemicals, Japan. The isooctane used was reagent grade.

Results and discussion

Hydrolysis

Monoglycerides (monoacylglycerols) can be obtained in high yields by lipase catalyzed hydrolysis of the corresponding triglyceride oil in microemulsions of L 2-type [7]. By using a 1,3-specific enzyme the hydrolysis takes place in a fairly regioselective manner, conversion into 2-monoglyceride being completed in 2—3 h at a reaction temperature of 37 °C. Prolonged reaction time results in a decrease in monoglyceride yield, a process that is not due to a lack of regioselectivity of the enzyme, but to an acyl-group migration of 2-monoglyceride to 1-monoglyceride. The latter species are good substrates for the regiospecific enzyme and complete hydrolysis to fatty acids and glycerol will eventually take place (Fig. 1) [7].

Kinetic studies indicated that effects of temperature and pH, as well as kinetic constants, resemble those of aqueous systems. Lipase activity in microemulsions has been found to be dependent on the molar ratio *R* of water to surfactant [7–9]. In most instances, *R*-values between 10 and 15 have been found optimal. The *R*-value governs the size of the water droplets; the higher the value, the larger the droplets. In the AOT-water-isooctane system $R = 11$ was found to correspond to a radius of *approx.* 2 nm [10].

Lipase activity was high in AOT-based systems, both with triglycerides and nitrophenyl alkanoate esters as substrates [7—10]. Nonionic surfactants [11, 12], as well as cetyltrimethylammonium bromide [10], were less effective. AOT also gave excellent enzyme stability [8, 10].

Lipase catalyzed hydrolysis in an emulsion system has also been performed with the enzyme covalently immobilized on silica particles [13]. A comparison was made between lipase attached directly to silane-treated silica and lipase bound via a very long hydrophilic spacer molecule, PEG 1500. As can be seen from Fig. 2, the spacer had a positive effect on enzyme activity. This is probably due to a reduction of the interactions between the protein and the support since both hydrophobic interactions and electrostatic attraction forces will be suppressed by the introduction of the hydrophilic non-charged PEG layer.

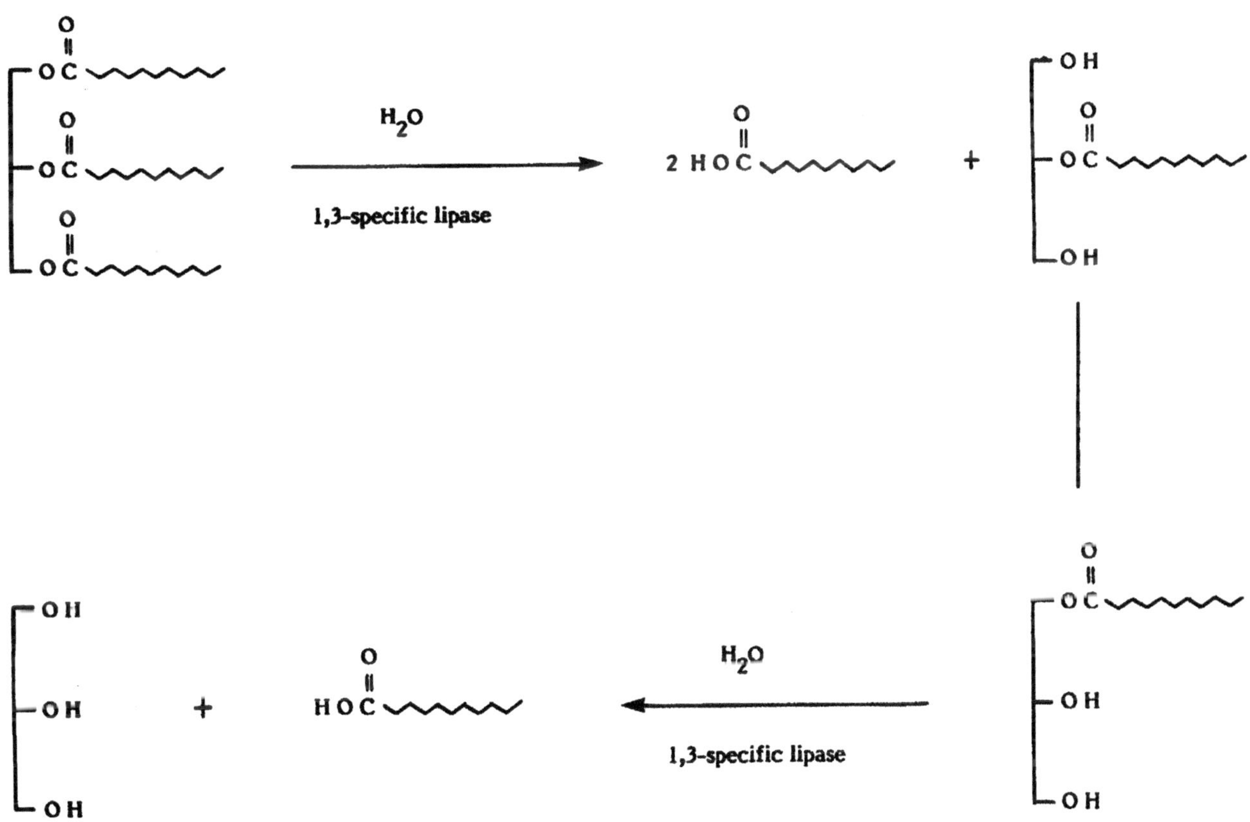

Fig. 1. Enzymatic hydrolysis of a triglyceride in combination with migration of the acyl group from 2- to 1-position

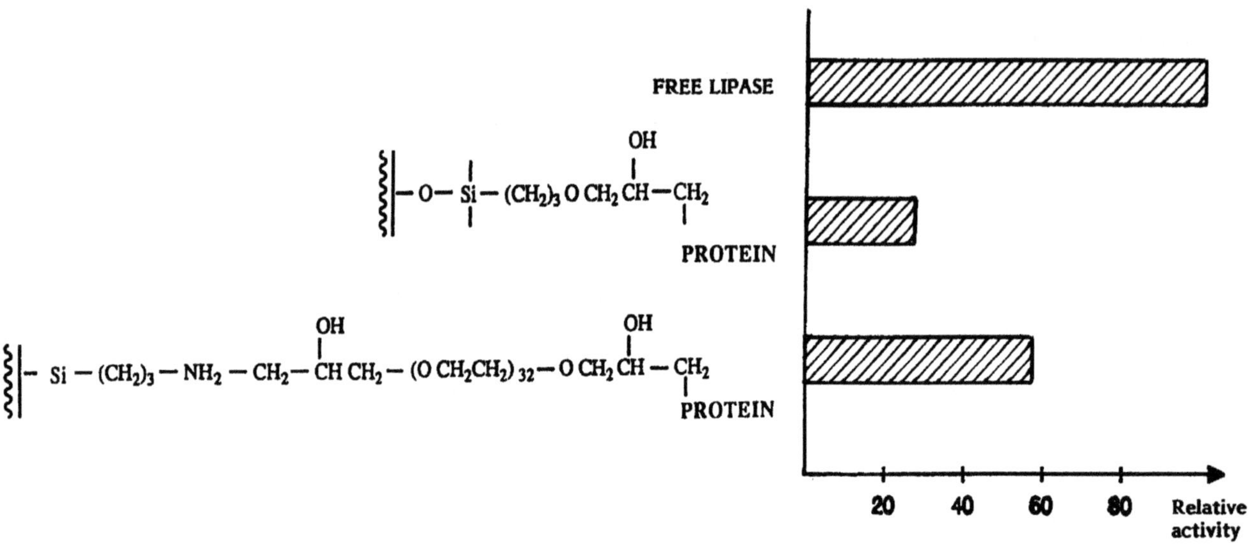

Fig. 2. Activity of free lipase, lipase bound directly to silane-treated silica, and lipase bound via a PEG spacer. Activity was monitored as rate of hydrolysis of triolein in emulsion

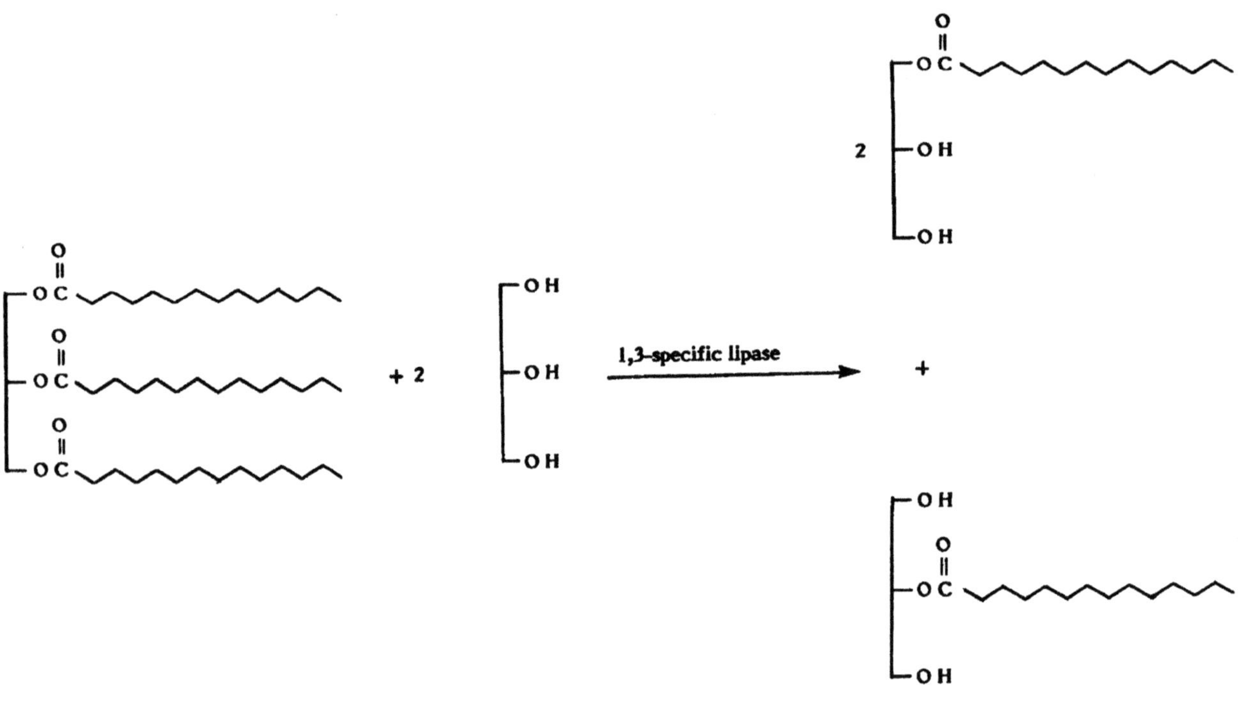

Fig. 3. Enzymatic glycerolysis of a triglyceride

Glycerolysis

Water-free microemulsions can be designed with glycerol as polar component. Such systems are of potential interest for lipid conversions since they open the possibility to perform glycerolysis instead of hydrolysis, provided that glycerol can replace water as activator and stabilizer of the enzyme.

Lipase catalyzed glycerolysis of triglycerides would yield a mixture of 1- and 2-monoglycerides, as is shown in Fig. 3. Attempts to perform this reaction failed, however, since very little enzyme activity was obtained in the completely non-aqueous system [14]. Using a combination of water and glycerol as the polar component of an AOT-based microemulsion led to simultaneous hydrolysis and glycerolysis. Using ^{3}H-labelled material, monoglyceride originating from added glycerol could be distinguished from that originating from the starting triglyceride, as is shown in Fig. 4. It was found that glycerolysis and hydrolysis occur at approximately the same rate; the two reactions probably go through a common intermediate [14]. The work shows that, although water can be substituted by glycerol, both as polar component of the microemulsion and as solvolytic agent in the reaction, it can not be replaced in its role as activator of the enzyme.

Transesterification

Lipase catalyzed transesterification, e.g., acyl group replacement in triglycerides, can be performed in microemulsions of low water content [15—17]. Special attention has been directed toward the production from inexpensive starting materials (such as palm oil [15]) of a triglyceride mixture that corresponds to natural cocoa butter.

A conversion of palm oil into cocoa butter requires a partial replacement of palmitoyl groups by stearoyl groups in 1(3)-position, while leaving the

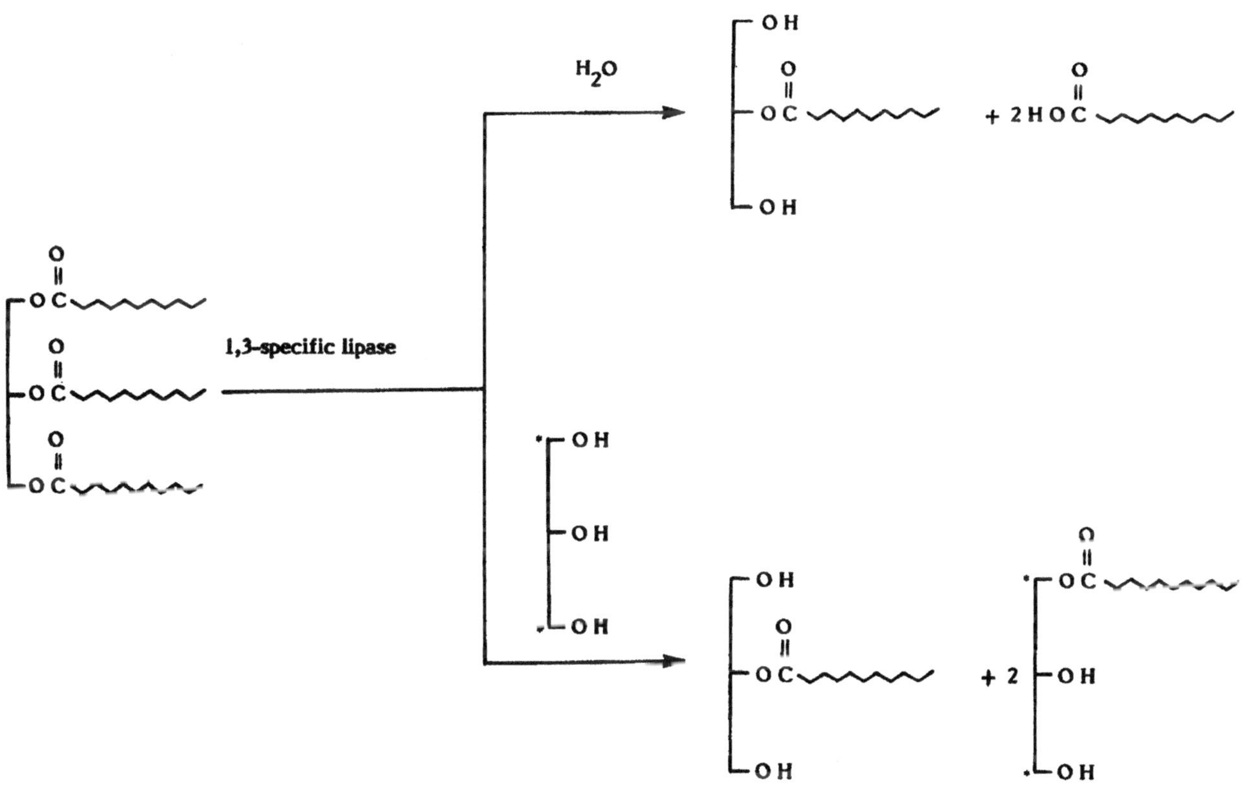

Fig. 4. Enzymatic glycerolysis using ^{3}H-labelled glycerol

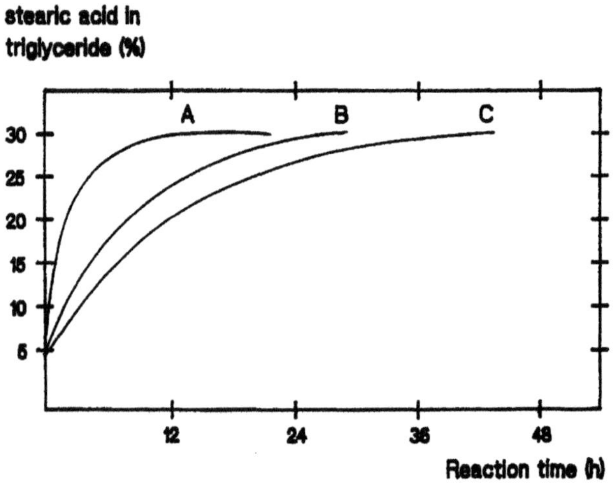

Fig. 5. Reaction rate for lipase catalyzed transesterification of palm oil with stearic acid. Reaction rate is expressed as percentage stearic acid incorporated in the triglyceride fraction vs reaction time. A: microemulsion based on alcohol ethoxylate; B: microemulsion based on AOT; C: hydrated celite dispersed in hydrocarbon

2-position essentially unaffected. Reaction between stearic acid and palm oil was carried out in lipase-containing microemulsions based on either AOT or an alcohol ethoxylate. As a comparison, a reaction in which the lipase was adsorbed on hydrated mineral particles, instead of solubilized in water pools, was performed. Figure 5 shows that the reactions in microemulsions are considerably faster than that in hydrocarbon, presumably due to a larger interfacial area between oil and water domains in the former [15]. The microemulsions remained clear and isotropic during the course of the reaction.

Transesterification was also carried out in microemulsion using lipase immobilized with and without the hydrophilic spacer molecule PEG 1500 [13]. Contrary to the hydrolysis reaction discussed above, the introduction of the hydrophilic spacer had no effect on enzyme activity in the transesterification reaction. Whereas the hydrolysis was performed in water-continuous emulsion, the transesterification was carried out in an almost nonaqueous medium in which PEG is not soluble. Hence, the hydrophilic spacer arm, which in aqueous medium stretches out into the bulk phase, will now adhere to the particles. The positive effect of the spacer arm in terms of reduced interaction between enzyme and particle will, therefore, not appear in the transesterification. The ideal spacer for this reaction would probably be a non-charged hydrophobic molecule instead.

In order to synthesize novel triglycerides with polyunsaturated acyl groups in 1- or 2-position a series of unsaturated triglycerides were transesterified with stearic acid [17]. It was shown that the position of the double bonds, but not the degree of unsaturation, is decisive of the rate or reaction. $\Delta 6$ double bonds which are present in many natural triglycerides prevent reaction.

Unsymmetrical triglycerides containing only one acyl group with $\Delta 6$ unsaturation react regiospecifically with the $\Delta 6$ acyl group facing away from the enzyme.

Interestingly, $\Delta 6$-containing fatty acids, such as γ-linolenic acid, can be incorporated into an unhindered triglyceride through lipase catalyzed transesterification. This seems to indicate that the lipase active site can accommodate a $\Delta 6$ fatty acid, but not the corresponding triglyceride [17]. This is not trivial since it is generally accepted that lipases are insensitive to the structure of the alcohol moiety of the ester. In this case, however, with an extreme steric requirement of the acyl group the alcohol part of the molecule will decide the feasibility of the reaction.

Lipase is believed to be a serine esterase and the enzyme-substrate complex is shown in Fig. 6.

Synthesis

Synthesis of triglycerides from glycerol and fatty acids has been attempted in microemulsions of very low water content [16, 18]. Mono- and diglycerides were formed, but the yield of triglycerides was very low. The reaction reaches an equilibrium position within a few hours, and the distribution between mono- and higher acyl products is sensitive to the amount of water added to the system.

Starting from the diglyceride, triglycerides can be synthesized through lipase catalyzed esterification with a fatty acid in microemulsion [19]. The reaction proceeded in fair yield in a phosphatidylcholine-based system. In an analogous system using phospholipase A2 as catalyst a phospholipid containing $\omega - 3$ fatty acids has been prepared by esterification of lysophosphatidyl choline with the polyunsaturated acids EPA and DHA [20].

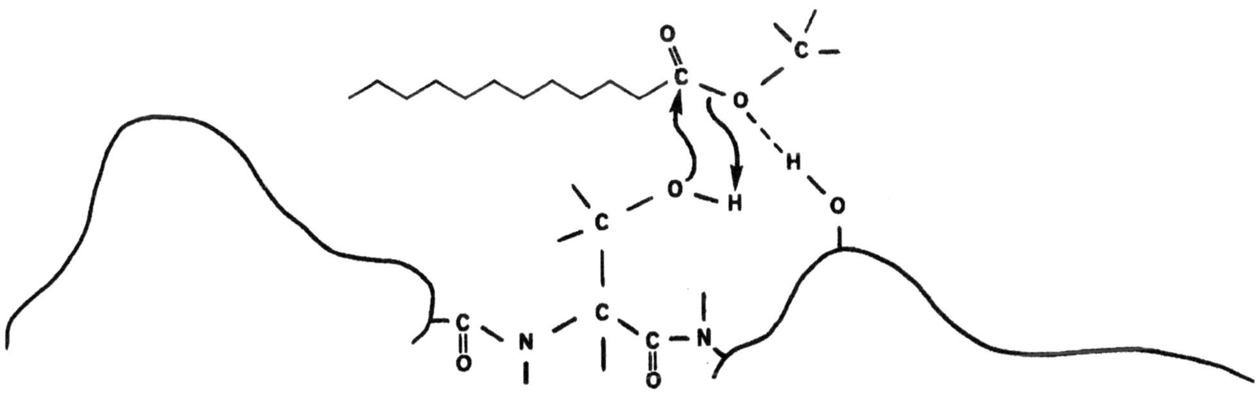

Fig. 6. A schematic illustration of the lipase-substrate complex

Structure of protein-containing reversed micelles

Luisi and coworkers [21] have studied how the size and structure of reverse micelles are modified upon uptake of a large guest molecule, such as a protein. Generally, one is then dealing with three different water pools, i.e., three different R: the initial one, R_0; the one of the protein-containing micelles, R_f; and the one of the unfilled micelles, R_e; (f stands for full and e for empty.) It has been demonstrated that the initial R_0 splits after addition of enzyme into a larger R_f and a smaller R_e. This means that the micelles accommodating the protein grow in size at the expense of the unfilled micelles. The size difference between filled and unfilled micelles vary with the protein, but is in the order of 2:1.

From circular dichroism spectra of trypsin and α-chymotrypsin run at subzero temperature it can be concluded that the conformation in the micelles is very similar to that in aqueous solution. Since there is no temperature-dependence on the cd spectra, the protein in reverse micelles may already be in a frozen conformational state at room temperature [22]. It should be noted, however, that the situation may be somewhat different with lipase which is a more surface active protein with a greater affinity for the water hydrocarbon interface.

The difference in behavior between microemulsions based on AOT and those based on nonionics has, in fact, been attributed to differences in accessibility of the enzyme to the interface. Whereas in the former system the enzyme will have good access to the interfacial region and, consequently, to the hydrophobic triglyceride substrate, in the latter systems the polyethylene glycol chains will prevent the enzyme from reaching the interface [11]. This is illustrated in Fig. 7.

Recent results from our group also indicate that the structure of the palisade layer is decisive of the reaction process [12]. Increasing the amount of water added to a microemulsion based on a nonionic surfactant leads to a decreased rate of lipase catalyzed triglyceride hydrolysis, as can be seen in Fig. 8. NMR self-diffusion measurements were performed on the microemulsion containing from 1%

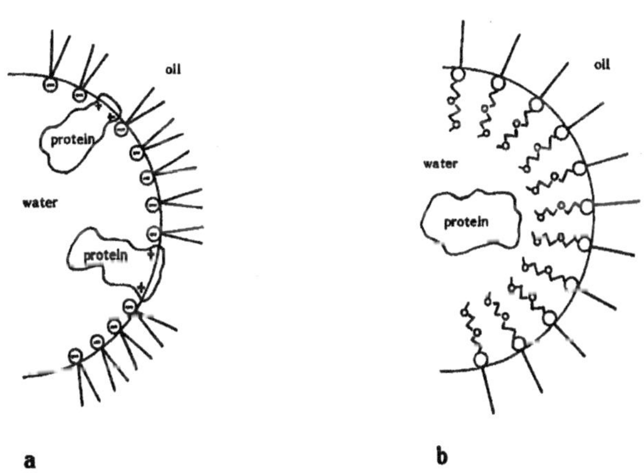

Fig. 7. Schematic picture of the surfactant palisade layer of microemulsion droplets based on AOT (a) and nonionic surfactants (b)

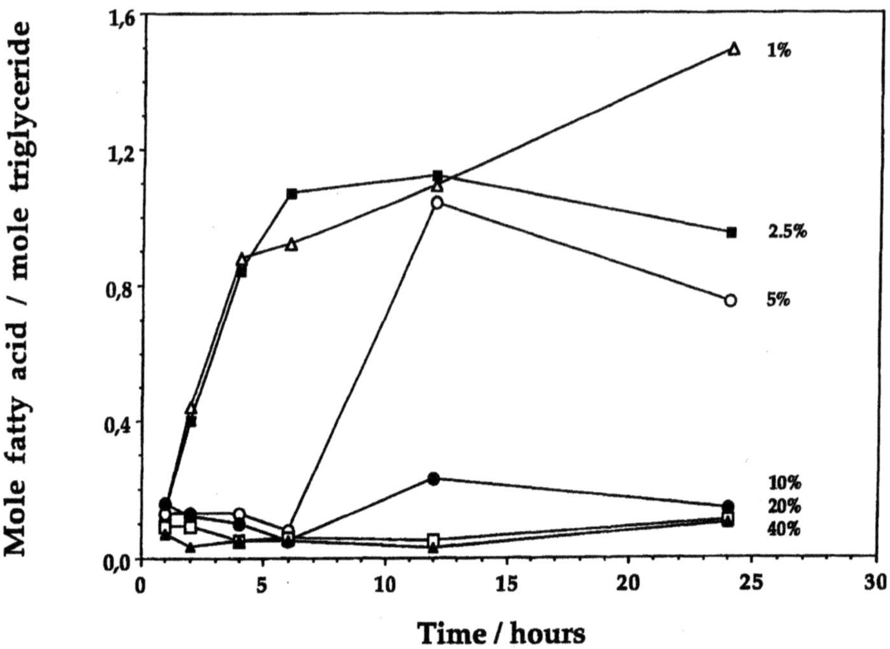

Fig. 8. Effect of water content on the formation of fatty acids from palm oil. Reactions were run at 37°C and the surfactant concentration was kept constant at 17%

up to 40% water while keeping the surfactant concentration constant. As is shown in Table 2, all self-diffusion coefficients, including that of water D_w decrease with increasing water content, although $D_{\text{palm oil}}$ seems to have reached a plateau at about 5% water. From the D_w/D_w^0- and D_H/D_H^0-values it is obvious that, whereas the diffusion of water is highly restricted throughout the series, that of isooctane is of the same magnitude as of the neat solvent. Furthermore, the diffusion constants of the surfactant closely follow those of the water. Evidently, the system is of w/o type microemulsion and the structure changes in the direction of more closed water domains on water addition.

The conclusion that may be drawn from this study is that in the systems with more well-defined water-in-oil spheres, i.e., those of high water content, the hydrolysis of the triglyceride was inhibited compared with the reactions in the more bicontinuous structures. Inhibition may be related to incorporation of surface active reaction products in the palisade layer between water and oil microdomains.

Table 2. Self-diffusion coefficients D (10^{-11} m^2 · s^{-1}) at 37°C for the components of a microemulsion containing 17% C12 EO5, 5% palm oil, 1—40% 0.1 M phosphate buffer pH 8, isooctane constituting the balance. Self-diffusion values for neat components are: D_{water} (D_w^0) 271, $D_{\text{isooctane}}$ (D_H^0) 235, and $D_{\text{palm oil}}$ (as 5% palm oil in isooctane) 60.3

% Water	D_{water}	D_w/D_w^0	$D_{\text{hydrocarbon}}$	D_H/D_H^0	$D_{\text{surfactant}}$	$D_{\text{palm oil}}$
1.05	—	—	168	0.71	36	37
2.09	23	0.085	169	0.72	26	28
5.01	13	0.048	—	—	18	21
10.0	16	0.059	—	—	14	17
19.7	16	0.059	142	0.60	12	20
30.0	8.8	0.032	133	0.57	9.7	20
40.0	6.4	0.024	—	—	7.7	20

References

1. Laane C, Boeren S, Vos K, Veeger C (1987) Biotech Bioeng 30:81
2. Reslow M, Adlercreutz P, Mattiasson B (1987) Appl Microbiol Biotechnol 26:1
3. Pohle WD, Mehlenbacher VC (1950) J Amer Oil Chem Soc 27:54
4. Martin JB (1953) J Amer Chem Soc 75:5482
5. Freedman B, Kwolek WF, Pryde EH (1986) J Amer Oil Chem Soc 63:1370
6. Stilbs P (1987) Progr Nucl Magn Reson Spectrosc 19:1
7. Holmberg K, Österberg E (1988) J Amer Oil Chem Soc 65:1544
8. Han D, Rhee JS (1986) Biotech Bioeng 28:1250
9. Han D, Rhee JS, Lee SB (1987) Biotech Bioeng 30:381
10. Fletcher PDI, Robinson BH, Freedman RB, Oldfield C (1985) J Chem Soc Faraday Trans 1 81:2667
11. Österberg E, Ristoff C, Holmberg K (1988) Tenside 25:293
12. Stark M-B, Skagerlind P, Holmberg K, Carlfors J (1990) Colloid Polymer Sci 268:384
13. Stark M-B, Holmberg K (1989) Biotech Bioeng 34:942
14. Holmberg K, Lassen B, Stark M-B (1989) J Amer Oil Chem Soc 66:1796
15. Holmberg K, Österberg E (1987) Progr Colloid Polymer Sci 74:98
16. Bello M, Thomas D, Legoy MD (1987) Biochem Biophys Res Comm 146:361
17. Österberg E, Blomström A-C, Holmberg K (1989) J Amer Oil Chem Soc 66:1330
18. Fletcher PDI, Freedman RB, Robinson BH, Rees GD, Schomäcker R (1987) Biochim Biophys Acta 912:278
19. Morita S, Narita H, Matoba T, Kito M (1984) J Amer Oil Chem Soc 61:1571
20. Na A, Eriksson C, Eriksson S-G, Österberg E, Holmberg K (1989) J Amer Oil Chem Soc (in press)
21. Zampieri GG, Jäckle H, Luisi PL (1986) J Phys Chem 90:1849
22. Magid L, Walde P, Zampieri GG, Battistel E, Peng Q, Trotta E, Maestro M, Luisi PL (1988) Colloids Surf 30:193

Authors' address:

K. Holmberg
Berol Nobel AB
44485 Stenungsund, Sweden

Kinetics of surfactant-induced leakage and growth of unilamellar vesicles

K. Edwards and M. Almgren

Department of Physical Chemistry, Uppsala University, Uppsala, Sweden

Abstract: The leakage process induced by the addition of nonionic surfactants to small unilamellar lecithin vesicles has been studied. The leakage rate depends strongly on the concentration of the surfactant — to the power 3.4 for $C_{12}E_8$ (octaethylene glycol n-dodecyl monoether) and the power 5 for Triton X-100. This suggests that the process requires the cooperation of several surfactant molecules. The leakage can occur either through a hole or channel created by the surfactants, or by a shuttle mechanism. While the data does not exclude the shuttle mechanism, the fact that different surfactants closely show the same behavior stands against it. The assumption that a hole is formed in the vesicle wall has to be reconciled with the observation that the addition of surfactant does little to enhance the equilibration rate of a labelled lecithin probe between the outer and inner membrane halves. It is argued that a relatively long-lived hole with a high surface coverage of surfactant molecules could explain the observations. — At higher additions of surfactant the vesicles grow into much larger but still unilamellar vesicles, with a relatively sharp size distribution and temperature-independent average final size (600 ± 10 Å at a lipid/surfactant ratio of 1:1, at this and higher surfactant concentrations the structures are mainly open "vesicles"). This transition is very fast, completed within a few minutes at 25 °C, and only a few seconds at 37 °C. At still higher temperatures the final stage of the process proceeds slower; the equilibration of labelled lecithin is fast in this region.

Key words: Vesicle; non-ionic surfactant; leakage; lecithin; growth

Introduction

In previous articles [1, 2], we have described and discussed the influence of nonionic surfactants on small unilamellar vesicles prepared by ultrasonic irradiation from egg lecithin dispersions. Up to a certain concentration — which for both Triton X-100 and $C_{12}E_8$ was about 0.50 mM at a lecithin concentration of 1.2 mM (see Fig. 1) only a slight growth of the vesicles could be observed with either quasielastic light-scattering or Cryo TEM [2]. This growth is not larger than what would be expected from the increase in surface area on the uptake of the surfactant molecules by the membrane. The membrane starts to leak, however, at a rate which increases rapidly with the surfactant concentration. Above a critical concentration the leakage is almost instantaneous, and simultaneously, the vesicles start to grow much more dramatically, forming large but rather uniform structures. Further increase in surfactant concentration gives samples with open vesicles [2]. We believe that in this concentration region vesicles and mixed micelles coexist [1]. At still higher concentrations the vesicles are completely dissolved into much smaller mixed micelles.

The leakage rate depends very strongly on the surfactant concentration. In the search for a mechanism of the leakage process, we have made a quantitative study of this dependence, and discuss several conceivable mechanisms for the leakage process. The leakage measurements were complemented by measurements of the trans-membrane and trans-vesicular movement of an excimer-forming, pyrene-labelled lecithin probe. Earlier investigators [3] have shown that the excimer formation, which monitors the local concentration of the

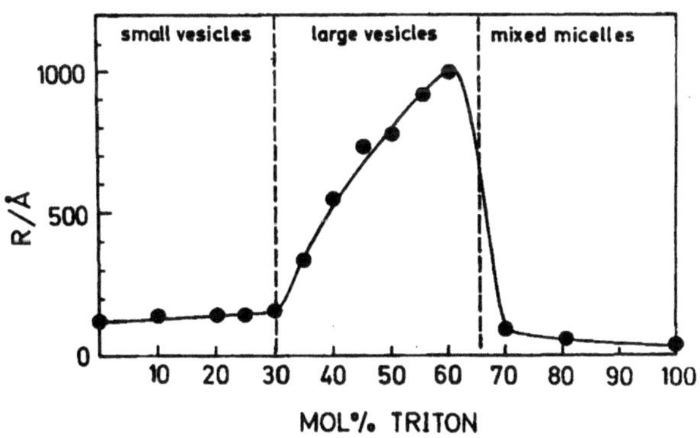

Fig. 1. The hydrodynamic radius (from quasi-elastic light scattering) of sonicated unilamellar vesicles of egg-yolk lecithin (1.2 mM) after addition of the nonionic surfactant Triton X-100. There is evidence for the coexistence of vesicles and mixed micelles in at least part of the middle region. The position of the transition to the mixed micelle region has not been determined exactly

probe, drops in a two-step process after mixing with vesicles lacking the probe. The faster of these is taken to represent the migration of the probe between the outer surfaces of different vesicles, and is followed by the much slower process in which the probe is transported over the membrane (if the transmembrane process were faster only one process would have been observed). Our somewhat surprising finding is that these processes are very little affected by the presence of surfactant in the low-concentration range; the exchange of membrane components between the two sides of the membrane is seemingly totally independent of the transmembrane leakage of the water-soluble dye.

The vesicle growth process in the intermediate range of surfactant concentrations was found to occur rather quickly and lead to a uniform population of mainly unilamellar vesicles [2]. This has been confirmed, and the temperature dependence of the rate determined. Although the final size is not temperature-sensitive, the rate of the growth process shows a complex temperature dependence.

Experimental section

Materials

Egg yolk lecithin of grade 1 was purchased from Lipid Products, Nutfield, UK, and pyrene-lecithin (3-palmitoyl-2-(1-pyrenedecanoyl)-L-a-phosphatidylcholine) from Molecular Probes Inc., Eugene, Oregon, USA. $C_{12}E_8$ (octaethylene glycol n-dodecyl monoether) was bought from Nikko Chemicals, Tokyo, Japan. The products were used without further purification. HEPES (2-[4-(2-hydroxyethylene)-1-piperazinyl]-ethanesulphonic acid) from Merck, FRG, and 6-carboxy-fluorescein from Eastman Kodak Company, Rochester, New York, USA. Sephadex G-50 was obtained from Pharmacia, Uppsala, Sweden.

Preparation of vesicles

The lipid vesicles were prepared by ultrasonic irradiation of samples containing 20—40 mg lecithin in 4 ml of water or 0.1 M 6-carboxy fluorescein. A Soniprep 150 from MSE Scientific Instruments, Crawley, Sussex, England, was used for the sonication. After 40 min sonication the samples were diluted with water of 10 mM HEPES, pH 7.4, to the desired concentration. In all experiments the final lecithin concentration was 1.2 mM.

Vesicle leakage

The vesicle leakage was studied by monitoring the fluorescence increase due to the release, and subsequent dilution of the vesicle-entrapped, self-quenching probe 6-carboxy fluorescein (6-CF) [5]. Vesicles containing 0.1 M 6-CF were prepared by sonication in a water solution of the probe. At this concentration the fluorescence of the probe is very low due to self-quenching. To remove untrapped solutes the sonicated sample was run through a Sephadex G-50 column using 10 mM HEPES, pH 7.4, as the eluent.

The vesicle preparation was diluted with HEPES-buffer to a lecithin concentration of 2.4 mM and transferred to one of the driving syringes in a HI-TECH Rapid Kinetics Accesory, model SFA-11, fast-mixing apparatus from HI-TECH Scientific Limited, Salisbury, Wiltshire, England. The second driving syringe was filled with a solution of $C_{12}E_8$ in HEPES-buffer. The vesicle and surfactant solutions were quickly mixed (<200 ms) and the change in fluorescence followed using a SPEX Fluorolog 2, model F212, spectrometer system comprising a DM1B computer system, from SPEX Industries Inc., Edison, New Jersey, USA. The excitation wavelength was set to 490 nm and the emission to 520 nm. All experiments were performed at 25 °C.

Lipid exchange

The lipid exchange between vescicles was studied by mixing two vesicle populations, one containing only egg lecithin and the other having about 2.5 mol% of the total lipid content exchanged for pyrene-lecithin. Figure 2 shows the emission spectrum for pyrene-lecithin in vesicles of egg yolk lecithin. The lipid exchange was followed as a decrease in the excimer to monomer

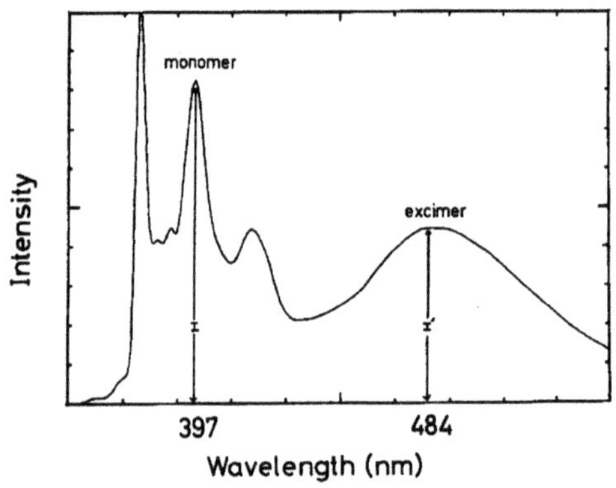

Fig. 2. The fluorescence spectrum of pyrene-labelled lecithin in egg-lecithin vesicles. I'/I represents the ratio of excimer to monomer emission and should be directly proportional to the local pyrene label concentration

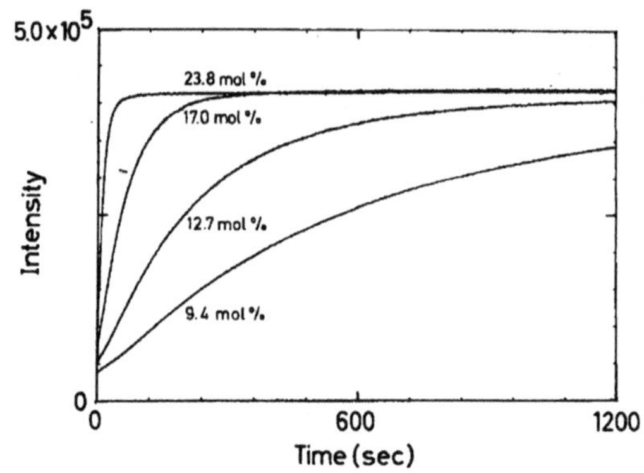

Fig. 3. The leakage of 6-carboxy fluorescein from egg lecithin vesicles after addition of indicated amounts of surfactant $C_{12}E_8$

fluorescence intensity ratio (I'/I) as the probe was evenly distributed between the two different vesicle populations. The mixing was performed using the HI-TECH fast-mixing apparatus and the excimer and monomer fluorescence intensity was recorded using the SPEX-fluorometer. The change in I'/I was followed, both in the absence and presence of $C_{12}E_8$. In the latter case the surfactant was introduced either externally to a premixed vesicle solution containing the two populations or else equilibrated with the probe-free population before the mixing. Measurements were carried out at 25° and 50°C.

Light scattering

To follow the transition from small to large vesicles the light scattered at 90° was measured as a function of time after mixing a vesicle and a surfactant solution. The mixing was again performed using the HI-TECH apparatus and the scattered light was monitored on the SPEX-fluorometer with excitation and emission wavelength set to 350 nm.

Quasi-elastic light-scattering (QELS) was used to determine the size of the aggregates. The method and equipment used have been described previously [2]. QELS measurements were carried out to determine the aggregate size at 25° and 50°C, respectively. The vesicle solution was mixed with surfactant at room temperature and measured at 25°C 20 min after mixing. After heating the sample to 50°C the light scattered was measured again.

Results

Vesicle leakage

At a $C_{12}E_8$ concentration of 29 mol% and higher, complete leakage takes place almost instantaneously. For lower concentrations the leakage rate increases with the added amount of surfactant. Figure 3 shows the fluorescence intensity as a function of time after mixing for different $C_{12}E_8$ concentrations. Similar results were obtained with Triton X-100 as the surfactant.

Lipid exchange

Figure 4 shows the lipid exchange between egg yolk lecithin vesicles at 50°C. The relative fluor-

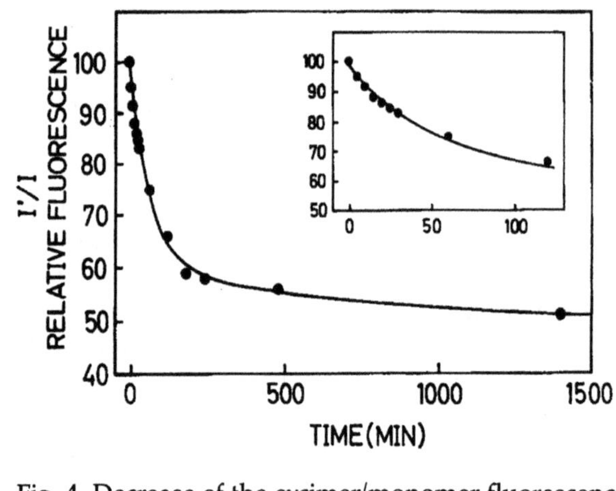

Fig. 4. Decrease of the excimer/monomer fluorescence intensity ratio after mixing of two populations of vesicles, one unlabelled and the other containing 2.5% pyrene-labelled lecithin molecules at 50°C. The initial faster decrease is thought to monitor the exchange of lecithins between the outer halves of the vesicles membranes, whereas the slow tail is due to the transport over the membrane. Virtually identical curves were obtained after addition of 15% $C_{12}E_8$

escence, calculated as the excimer/monomer ratio divided by the excimer/monomer ratio a the start of the experiment, follows a double exponential decay curve. A first fast phase leads to about 30% reduction of the initial intensity ratio and is followed by a slower phase leading to a final intensity ratio of 50% of the initial value. At this temperature $t_{1/2}$ was found to be about 40 min for the first phase and about 7 h for the second.

When 15 mol% $C_{12}E_8$ was present no significant change in the lipid-exchange rate could be observed. Introducing the surfactant by letting one of the vesicle populations contain 30 mol% $C_{12}E_8$ before mixing, or by external addition to a premixed vesicle solution, gave the same result.

The addition of 50 mol% $C_{12}E_8$ however, gave a dramatic effect on the lipid exchange. In these experiments the surfactant was at all times introduced externally, i.e., by mixing with one syringe containing surfactant solution and the other a freshly mixed (<1 min old) vesicle solution. The I'/I intensity ratio — which is proportional to the local concentration of pyrene — dropped quickly during the first second after mixing, as shown in Fig. 5. At 25 °C the intensity dropped to about 45% of the initial value and at 50 °C to about 70% of the initial value during this time.

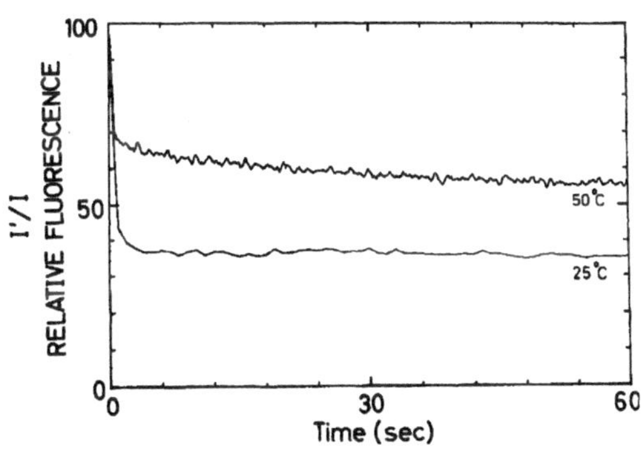

Fig. 5. Lipid exchange after addition of 50% $C_{12}E_8$ at 50° and 25 °C. Note the much shorter timescale than in Fig. 4

The fact that the relative intensity ratio drops below the 50% — which should correspond to the reduction of the local probe concentration when equal amounts of vesicles with and without probe are mixed — has several possible explanations. The surfactant does add some volume to the hydrophobic phase, but not enough to explain the whole effect. The surfactant may also change the mobility and other solvent properties of the hydrophobic phase, and most importantly, induces a formation of mixed micelles which contain both labelled and unlabelled lecithins. The distribution of the labels over small micelles leads to the formation of less excimers, since a good part of the labels will be solitary in their micelles.

Light scattering

Figure 6 shows the light scattered as a function of time after addition of the surfactant to a final concentration of 50 mol%, at a number of temperatures between 16° and 50 °C. Two different pathways can be observed. At temperatures between 16° and 40 °C the aggregate growth seems to follow the same multiexponential curve, and the time needed to reach a constant high-intensity ratio decreases with the experimental temperature. As the temperature exceeds 40 °C an additional factor sets in slowing down the later part of the course. After an initial fast growth, in line with the results at the lower temperatures, the rate slows down unexpectedly. The higher the temperature the longer it takes to reach the final scattering intensity. The final intensity level reached is the same, independent of the experimental temperature. The QELS results indicated a monodisperse preparation of aggregates having a hydrodynamic radii of R = 600 ± 10 Å when the sample was measured at 50 °C, wherease at 25 °C this main population was accompanied by a small population having much smaller radii, corresponding to mixed micelles. This bimodal size distribution will be further investigated.

Discussion

The results for small vesicles at low surfactant concentrations will be discussed first. The division between the regions shown in Fig. 1 pertains to Triton X-100; the turbidity results presented in Fig. 7 indicate similar partitions for $C_{12}E_8$.

The kinetics of 6-CF release

Before discussing the mechanistic aspects of the release measurements, it is in place to consider

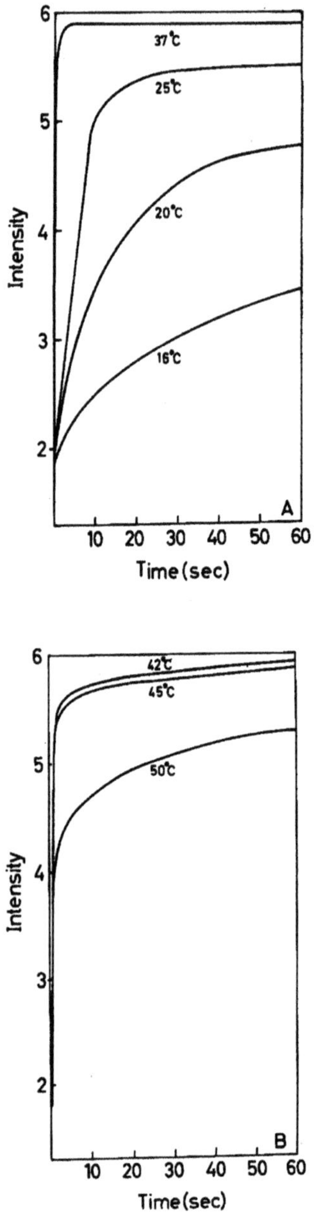

Fig. 6. The increase in size of small unilamellar egg lecithin vesicles (1.2 mM lecithin) after the addition of 1.2 mM $C_{12}E_8$ at a number of temperatures, as monitored by the static light-scattering. The intensity of the scattered light finally rose to the same value at all temperatures (a). The decreasing rate of the final process at the highest temperatures (b) is especially noteworthy

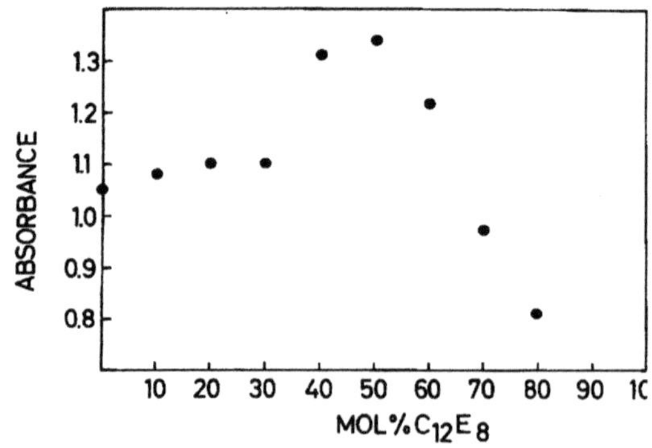

Fig. 7. The turbidity of sonicated egg lecithin (1.2 mM) vesicles after addition of $C_{12}E_8$. The qualitative aspects of the size changes are nearly the same as in Fig. 1

some quantitative aspects of the method as such. The measurements lean on the fluorescence increase that results from the diffusive dilution of the entrapped CF solution when a part of the entrapping wall is removed. A complete removal of the wall should permit diffusion in all directions, and if the mean square displacement is three times the radius, a dilution by a factor of 27 has occurred. Assuming a diffusion coefficient of 5×10^{-10} m^2 s^{-1} and a vesicle radius R_v of 100 Å, this corresponds to a time of about 10^{-6} s. For observed release times in the millisecond range, the dilution is rapid enough to make the release process rate determining.

It is also of interest to estimate how fast a vesicle would be emptied if a channel of radius R_{ch} was opened in the vesicle wall of thickness b. It is easily shown that the inside concentration c — the outside concentration is kept at zero — decays exponentially with a first order rate constant k_l given by

$$k_l = 3D_0R_{ch}^2/(4R_v^3b) \ . \tag{1}$$

Using the same values for the diffusion coefficient and vesicle radius as above, and b = 3 nm, R_{ch} = 1 nm, the release time is obtained as about 1 ms.

As to the mechanism of the surfactant-mediated release, there are two main possibilities:

i) that the surfactants interact directly with the probe, forming some sort of carrier complex that helps to shuttle the dye over the membrane. In this case, one has to consider the equilibrium

$$P + xS = PS_x \ ,$$

and would expect a leakage rate that depends on the surfactant concentration to power x.

ii) that a hole or channel is created by the action of the surfactant molecules. In this case, there are two kinetically distinct alternatives: iia) the channels are long-lived so that a vesicle will be completely emptied once the channel has been created, or IIb) the channels are very transient structures, releasing only a few molecules each time they open. In the former case the vesicles would release their content in distinct bursts, which, in principle, should be possible to observe by measurement of the fluorescence intensity from a very small volume. The kinetics of the process — in normal measurements on a large volume — should be determined by the rate of formation of the channels, which could be a process with a nucleation character, i.e., smaller number x of surfactant molecules come together in the membrane to form a critical nucleus that triggers the formation of the channel. The leakage constant would again be proportional to the surfactant concentration to power x.

If, however, transient channels are involved, the leakage rate would depend on the probability that a channel is open in the vesicle at any given time. The rate constant for leakage would be given by Eq. (1), but multiplied with the probability P_{ch}. We will follow the reasoning of Fromherz [6] when estimating this quantity and its surfactant concentration dependence.

The free-energy cost of opening a channel with a water-exposed hydrophobic area of $A = 2\pi R_{ch} b$ is given by

$$\Delta g_r = \gamma A \ . \tag{2}$$

The interfacial energy γ is related to the surface excess of adsorbed surfactant molecules according to the Gibb's relation

$$d\gamma / d\ln c_s = -kTn_s / A \ . \tag{3}$$

Using a Langmuir-type of isotherm for the adsorption of the surfactants,

$$\frac{n_s}{n_0} = \frac{K_s c_s}{K_s c_s + 1} \ , \tag{4}$$

Eq. (3) may be integrated to yield

$$\gamma = \gamma_0 - \frac{kTn_0}{A} \ln(K_s c_s + 1) \ . \tag{5}$$

The probability of having a channel in the vesicle membrane is given by

$$P_c = BR_v^2 \exp\left(-\frac{A\gamma}{kT}\right) \ , \tag{6}$$

where the preexponential factor is the ratio between the partition functions for the vesicle with and without a channel. The detailed form is not known; there should be a proportionality to the vesicle surface area, however, with the factor R_v^2. Using Eqs. (1, 5, 6) one obtains

$$\ln k_l = \text{const} + n_0 \ln(K_s c_s + 1) \ . \tag{7}$$

In order to utilize Eq. (7) some knowledge of K_s is required. We assume either that saturation has been reached, i.e., that $K_s c_s$ is much larger than unity, in which case the leakage rate would depend on the surfactant concentration to the n_0th order, or that the point $K_s c_s = 1$ is reached just where the vesicle growth sets in.

Figure 8 presents double logarithmic plots of the inverse half-life of the leakage process — proportional to the leakage rate constant — vs the total surfactant concentration — assumed proportional to the concentration in either pseudophase, as in a simple distribution between phases. The regression lines suggest exponents for the dependence on the

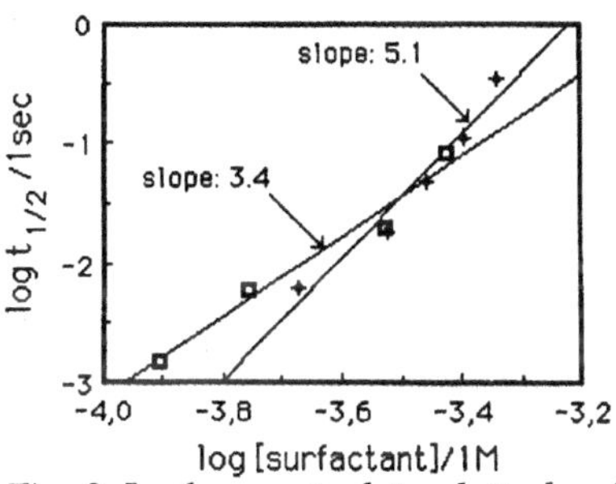

Fig. 8. Leakage rate data plotted as log (inverse halftime) vs log (surfactant concentration) for additions of $C_{12}E_8$ and Triton X-100

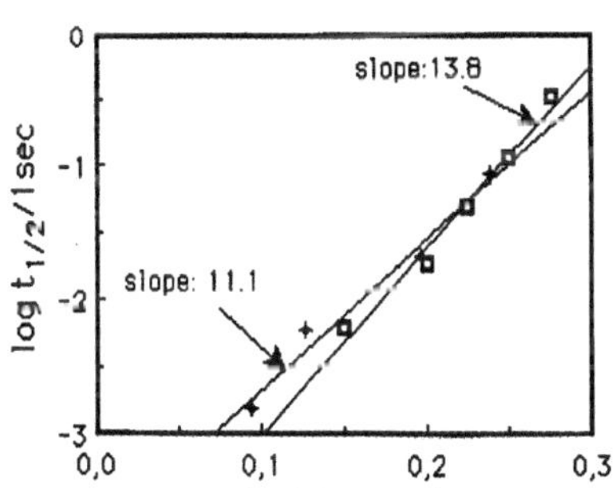

Fig. 9. Leakage rate data from Fig. 8 plotted according to Eq. (7), using $K_s = 1944\ \text{M}^{-1}$

concentration of 3.4 for $C_{12}E_8$ and 5.1 for Triton X-100. These plots are in conformity with the mechanisms i) and iia), and the values of the exponents are reasonable for such mechanisms.

If, however, the mechanism is iib) in its version with $K_s c_s$ much larger than unity, then the exponent should equal the number of surfactant molecules adsorbed on the channel hydrophobic area at saturation; a number of 3—5 is much too small for such an interpretation. The other version of iib) gives the plots shown in Fig. 9. These are about as well linear as the ones in Fig. 8, but much steeper with slopes of 11 and 14 for $C_{12}E_8$ and Triton X-100, respectively. The slopes — and thus the number of adsorbed surfactants at saturation — would increase even more with a higher value chosen for K_s. It is to be expected, however, that the adsorption of these surfactants at a hydrophobic interface should be near saturation when the surfactant concentration is close to the intermediate range, where the mixed micelles start to form.

So far the analysis of the leakage data indicates that the mechanism is either a carrier transport according to scheme i), or due to an induced channel as in iia), or iib) in the version which assumes the adsorption of surfactants on the channel walls to be far from saturation. The measurements on pyrene-labelled lecithins give further information.

Intervesicle and transmembrane transport

The decrease of the excimer to monomer fluorescence ratio after the mixing of vesicle populations with and without labelled lecithin follows precisely the expectations from the accepted mechanisms: in the initial fast process (30 min at 50 °C) the ratio goes down to about two-thirds of the initial value, and is further decreased to about half this value in the final and much slower stage. If the outer vesicle interface is twice as large as the inner — which is reasonable for the small vesicles formed at low surfactant concentrations — then the rapid process should dilute two-thirds of the pyrene labels over twice the initial volume, whereas the remaining one-third would require a longer time since it had to cross over the membranes.

The total insensitivity of these processes to low concentration of surfactants was a surprise: it was expected that a surfactant action that gives a rapid release of entrapped water-soluble dyes should also permit a rapid exchange of lipids between the outer and inner vesicle half membranes. The insensitivity is, of course, in agreement with mechanism i), but it cannot rule out channel mechanisms completely. It requires, however, that if a channel is formed, it must still provide a barrier towards the passage of lecithin lipids. This suggests that a rather rigid structure with a high surface coverage of surfactants is involved. The observation, therefore, tentatively rules out the mechanism iib) with the formation of transient channels, since it is consistent with the leakage data only if a relatively low surfactant coverage is assumed.

It seems, therefore, as if the leakage mechanism either is a carrier mechanism, or depends on the formation of relatively long-lived channels with high surfactant coverage.

The growth of large vesicles

The most striking features of the growth after the addition of surfactant are the rapidity of the process and its strong temperature-dependence. It is also remarkable that the same final scattered intensity is attained, independent of the temperature. QELS measurements at 25° and 50 °C showed that the final size was indeed the same, but with some small structures present at the lower temperature; the scattering intensity alone is not conclusive in this respect since the particles could have reached a size where the dependence of the intensity on the size is weak. The strong nonlinearity of the size-intensity relationship may be important for the interpretation of the time-dependence in that the turnover from the initial fast growth to the slower final process may appear more sudden than it is. In any case, there is certainly a change in the character of the growth process around 37 °C. Below that temperature the process is strongly non-exponential with a rate that increase rapidly with temperature. At 37 °C it takes only a few seconds for the whole change. At higher temperatures the very rapid change seems to occur to a lesser extent, and is followed by a much slower growth up to the final intensity.

From cryo-TEM studies on the lecithin — Triton X-100 system we know that the large structures are big vesicles and not clusters of smaller structures [2]. The light-scattering experiments performed at different temperatures give a good picture of the time needed for the formation of the large vesicles. At 25 °C the transition is accomplished about 2 min after mixing. The lipid-exchange measurements at this temperature show, however, that the excimer/monomer ratio has decreased to about 45% of the initial value less than 1 s after the vesicle and surfactant solution were mixed. A plausible description of this process would be that the surfactants associate rapidly with the vesicles, which then open up and immediately start to fuse together. The excimer formation, determined mainly by the local concentration of pyrene labels, quickly reaches a level close to equilibrium after a few fusion events, whereas the scattering intensity, which is more sensitive to the size, follows the growth longer. For a better understanding of these processes, one must examine the reason why the growth stops at a certain size, which depends on the surfactant concentration but not on the temperature. Related to this is the question of if the final aggregate size is an equilibrium property, or if it is kinetically determined. From dilution experiments we know [1, 2] that vesicles stay at the maximum size when the solution is diluted from the mixed micelle region to a composition beyond that which gives the maximum size. Ollivon et al. [7] reported that size adjustments could occur, both up and (but not below a radius of 305 Å) in their egg lecithin-octyl glucoside system. The most important difference between the systems is the high water-solubility of the octyl glucoside monomer, which has a CMC of about 20 mM, whereas $C_{12}E_8$ has a CMC of only 7×10^{-5} M [8].

The slow rate of formation found at temperatures above 40 °C is a further puzzle. It is interesting to note that the lipid exchange measurements also give a somewhat unexpected result at 50 °C. The excimer/monomer ratio shows, at this temperature, only a decrease with 30% during the first 1 s after mixing, compared to the decrease with 55% during the same period of time at 25 °C, whereas the intensity of scattered light during the first 1 s increases much more at the higher temperature. It seems probable that this change of behavior is connected to the presence of mixed micelles in this surfactant concentration range, as indicated in the QELS measurements. The formation of small micellar aggregates by the nonionic surfactant should be less favorable at higher temperatures. Further investigations are necessary on this point.

Acknowledgements

We are very grateful to Dr. Wyn Brown for help with the QUELS-measurements, and to the Swedish Natural Science Research Council for financial support.

References

1. Nilsson K, Almgren M, Brown W, Jansson M (1987) Mol Cryst Liq Cryst 152:181—203
2. Edwards K, Almgren M, Bellare J, Brown W (1989) Langmuir 5:473—478
3. Galla H-J, Theilen U, Hartmann W (1979) Chem Phys Lipids 23:239—251
4. Almgren M, Swarup S (1982) Chem Phys Lipids 31:13—22
5. Weinstein JN, Yoshikami S, Henkart P, Blumenthal R, hagins WA (1977) Science 195:489—492
6. Fromherz P (1983) Chem Phys Lett 94:259—266
7. Ollivon M, Eidelman O, Blumenthal R, Walker A (1988) Biochemistry 27:489—492
8. Meguro K, Ueno M, Esumi K (1987) In: Schick MJ (ed) Non ionic surfactants, Surfactant Science Series. Marcel Dekker, New York, vol 23 pp 183

Authors' address:

Mats Almgren
Fysikalisk-kemiska institutionen
Uppsala universitet
Box 532
75121 Uppsala, Sweden

Progress in Colloid & Polymer Science Progr Colloid Polym Sci 82:198—210 (1990)

Effects of salts on association and conformational equilibria of macromolecules in solution

L. Piculell and S. Nilsson

Physical Chemistry 1, University of Lund, Chemical Center, Lund, Sweden

Abstract: The physical basis and the theoretical modelling of effects of added electrolytes on the association and conformational equilibria of macromolecules in solution are discussed in the light of our recent detailed studies and analyses of such effects on the coil-to-helix transitions of charged and uncharged polysaccharides (carrageenans and agarose, respectively). Salt effects are divided into 1) generic electrostatic effects (for charged macromolecules), 2) general, ion-specific (but not site-specific) so-called lyotropic effects, which occur also for uncharged macromolecules, and 3) effects of site-specific binding of ions to the macromolecule. Electrostatic salt effects are analyzed within the Poisson-Boltzmann Cell Model (PBCM), and it is found that this model can successfully account for experimentally observed effects on the conformational transitions of carrageenans, which result from changes in the concentrations of the macromolecules and/or of the added salt, and from a variation of the relative proportions of mono- and divalent counterions. Analyses of ion-specific (lyotropic) salt effects on the coil-to-helix transition of agarose suggest that the latter effects result from a change in the interfacial energy contribution to the transition, on addition of salt, due to a non-zero (positive or negative) surface excess of ions at the macromolecular surface which is expected, in general, also for an uncharged macromolecule. The same mechanism also accounts for similar effects of added uncharged cosolutes, such as urea. It is emphasized that water-cosolute interactions may be as important as cosolute-macromolecule interactions in the generation of a non-zero surface excess of the cosolute at the macromolecular surface.

Key words: Salt effect, lyotropic series, conformational, transition, polyelectrolyte, biopolymer

Introduction

Many macromolecules, and especially those of biological origin, have the ability to form well-defined intra- and intermolecular structures, at various levels of complexity, in an aqueous environment. This ability is essential for the biological function of biopolymers and it is also responsible for the ability of many such systems to form physical gels [1, 2] — a property which is widely utilized in numerous applications [3]. Added salts notoriously influence the disorder-order or association processes which are involved in the structure formation. Both salting-in (favoring dissociation) and salting-out (favoring association) effects may be found, depending on the particular system and, also, on the identities of the ions of the added salt [4]. Often enough, changes in the salt concentration or, indeed, the replacement of one type of ion with another one of the same valency may give rise to dramatic effects on the macroscopic properties of a given system at a given temperature. The molecular understanding of these various effects of added salts presents a great challenge to scientists involved in the physical chemistry of macromolecular systems, and the subject has been intensely investigated [4—7].

Our own active interest in this problem was aroused when we began working with gel-forming polysaccharides some 4 years ago. We soon chose to concentrate our studies on algal polysaccharides of the carrageenan/agarose family [8], which are among the best studied, and most widely used systems of this type. On studying these systems, we were immediately confronted with striking effects of the salt concentration and of the valencies and the identities of the ions present on the conformational transitions and the association/gelation behavior of these polysaccharides (in the literature [2, 9—13] as well as in the laboratory). Much of our recent research effort [14—19] has been devoted to the understanding of these salt effects through the combined use of experimental studies and theoretical analyses. In these analyses we have emphasized the use of general models and concepts (which by no means are restricted to gel-forming polysaccharides) and, as it turned out, our case studies led to a number of conclusions regarding salt effects on the conformational transitions of macromolecules which, according to our models, should be quite general. The main purpose of the present unified presentation of these studies is, therefore, to try to extract, from rather detailed analyses of a few particular systems, *general* insights into the nature of, and some non-trivial manifestations of, various salt effects on macromolecular conformational transitions and association processes.

Strictly, the types of processes which we have in mind here are those where the interconversion between some disordered or dissociated state, on the one hand, and some ordered or associated state, on the other, is an equilibrium process which occurs in a single-phase solution. Figure 1 presents, schematically, various types of such reversible conformation and/or association equilibria which may be encountered in macromolecular solutions.

Already at the outset we may note certain features which the processes of Fig. 1 obviously have in common. The first is that all processes when going from left to right in Fig. 1 result in an increase in the number of macromolecular segment-segment contacts or, stated differently, a *decrease in the specific surface area exposed to the solvent*. Thus, in a generalized sense, all the equilibria of Fig. 1 may be said to involve *dissociated* states on the left hand side and *associated* states on the right hand side; from now on we will, therefore, refer to all these equilibria as macromolecular association equilibria. Furthermore, for *charged macromolecules*, each of

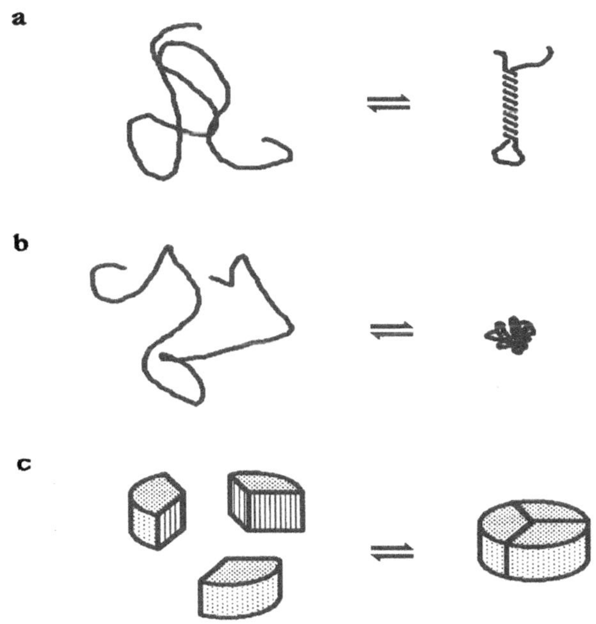

Fig. 1. Schematic illustration of various types of reversible conformation/association equilibria encountered for macromolecules in solution: a) coil helix, b) coil-globule and c) association into supramolecular assembly

the association processes necessarily leads to an *increase in the magnitude of the electrostatic potential* of the macromolecule in the associated state. (We should note, however, that cases when specific binding sites for certain counterions are created in the associated state may represent exceptions to the latter rule.) In our view, these two general features of the processes of Fig. 1 contain the keys to understanding general salt effects, which we will discuss in the following.

II. Phenomenology and classification of salt effects

The processes of Fig. 1 are typically, strongly cooperative in nature, so that a temperature-induced transition occurs over a rather narrow temperature range. A transition in a system of specified composition may thus be characterized by a transition (midpoint) temperature, T_m. Added salts will generally effect T_m, and on phenomenological as well as on mechanistic grounds, we will find reasons to distinguish between three types of salt effects, as presented below.

II.1. Generic electrostatic salt effects

The term "electrostatic salt effects" we will reserve for effects that are specific to macromolecules which possess a net charge. As regards macromolecule-ion interactions, we include under this heading only those which are of the diffuse double-layer type and which depend very weakly on factors other than the signs and the valencies of the ionic charges. As a rule (although there are important exceptions, as will be demonstrated below), these generic electrostatic effects lead to a stabilization of the more highly charged state of the macromolecule as more salt is added. Typically, the transition temperature varies linearly with the logarithm of the salt concentration in systems where the salt concentration is much larger than the concentration of macromolecular charges [5, 6, 11, 20]. Thus, we may write

$$T_m = a + b \log(c_s) , \qquad (1)$$

where c_s is the concentration of added salt and a and b are constants. Accordingly, in such systems changes in the salt concentration may give rise to large effects on T_m (i.e., changes of the order of 10 °C or more) already at salt concentrations below 0.1 M.

II.2. Lyotropic salt effects

Experimentally, it is generally found that also the association equilibria of *uncharged* macromolecules are affected by added salts. Typically, the addition of a salt (or, indeed, any other low-molecular cosolute which does not have a marked tendency to self-assemble in solution) affects the transition temperature in a linear fashion [4], so that

$$T_m = T_m^0 + kc_s , \qquad (2)$$

where T_m^0 is the transition temperature in the absence of added salt. The constant of proportionality, k, may be either positive or negative. Typically, $|k| \leqslant 10\,°\mathrm{CM}^{-1}$ [4], so that large effects on the transition temperature require the addition of salts at concentrations of the order of 1 M or more. It is often found that the contributions from different ions to k are additive, and if different ions or other low-molecular cosolutes are listed in order of increasing destabilizing efficiency, the same (or nearly the same) sequence is commonly found regardless of the particular system [4]. The effects of different ions thus follow the so-called lyotropic series [4, 6]. A typical lyotropic series of anions reads (in order of decreasing value of k):

$$SO_4^{2-} > Cl^- > Br^- > NO_3^- > I^- > SCN^- .$$

Similar series exist for simple cations, but the range of k-values observed for these ions tends to be smaller than that of the anions [4]. We will refer to salt effects of the kind just described, which are found for uncharged as well as for charged macromolecules, as lyotropic salt effects. We note that although the lyotropic effects are *ion* specific, they are still general in the sense that they do not (in contrast to the third category of effects described below) depend critically on the details of the macromolecular surface.

II.3. Effects of site-specific ion binding

When a macromolecule, in either or both of its states, contains specific sites which bind certain ions, the addition of such ions may give rise to large effects on the macromolecular association equilibrium. A thermodynamic treatment of these effects was first presented by Schellman [21]. In the analysis, the macromolecule is considered to have a number of (possibly different) binding sites, and the binding of an ion (or, in general, a ligand) to a given site i is described by a phenomenological binding constant K_i. The macromolecular association reaction will, in general, lead to a change in the number of the sites and/or in their binding constants. The resulting difference in the binding of ligands to the two macromolecular states leads to a change in the chemical potential difference between the associated (A) and the dissociated (D) states, $\Delta\mu_{D\to A}$, which is given by the expression [22]

$$\Delta\mu_{D\to A} = \Delta\mu^0_{D\to A} - RT\{\Sigma_A \ln(1 + K_i c_s) - \Sigma_D \ln(1 + K_j c_s)\} . \qquad (3)$$

Here $\Delta\mu^0_{D\to A}$ is the chemical potential difference in the absence of binding ligands. The sums in Eq. (3) are to be taken over all macromolecular sites i and j exposed in the A and D states, respectively. We should caution that for ionic ligands, Eq. (3) is only valid when electrostatic effects may be neglected.

III. Investigated systems

The associating macromolecules which we will specifically consider here (*viz.* agarose and carrageenans) are linear polysaccharides which undergo coil-to-helix transitions in solution. In this section, we will briefly review some of the properties of these systems. We will also discuss the nature of coil-helix transitions and the application, to such transitions, of the thermodynamic models which we use to describe salt effects.

III.1. Agarose and carrageenans

The polysaccharides of the agarose/carrageenan family are linear galactans composed of repeating disaccharide units which may be more or less sulphated [8]. Many different types of disaccharide residues have been found in polysaccharides from this family, and structural heterogeneity (also within the chains) is, to a varying degree, a characteristic of each extract from the natural source (marine red algae). By proper choice of algal source combined with selective precipitation methods it is, however, possible to obtain different types of polysaccharides with structural purities of *ca.* 80—90% [8]. The regularity of the primary structure may also be improved by chemical methods [23] after extraction. Several gel-forming types of agarose/carrageenans exist, which all are characterized by predominantly regular primary structures. In all known cases, these regular structures have the ability to form helices, and there is strong evidence [2, 24] that the gel junction zones involve the helical forms of the polysaccharides. In some systems, the gel-sol transition is characterized by a pronounced thermal hysteresis, which indicates substantial aggregation of helices into larger aggregates. Thus, upon lowering the temperature of a solution of carrageenan/agarose, the following sequence of transitions may occur [2, 24].

$$\text{coil} \rightarrow \text{double helix} \rightarrow \text{aggregate of helices}\ .$$

It should be emphasized that whether each of the above transitions occurs or not depends not only on the type and concentration of the polysaccharide, but also on the type and concentration of ions present. Thus, not all carrageenans seem to form aggregates of helices, and for those that do aggregate it is sometimes possible to prevent aggregation by adding certain types of ions [12, 13]. Salts, therefore, may have pronounced effects, not only on the coil-helix transition temperature of the system, but also on the gelation properties (optical and mechanical properties) of the helical systems.

In the studies reviewed here, we have concerned ourselves with the best studied gel-forming members of the family, i.e., agarose, kappa-carrageenan, and iota-carrageenan. As remarked by Rees almost two decades ago [25], these systems form an interesting set since the degree of sulphation per repeating disaccharide unit equals zero, one and two for agarose, kappa-carrageenan, and iota-carrageenan, respectively. These systems should, therefore, be well suited for studies of both electrostatic salt effects (and their dependence on the charge density of the polyelectrolyte) and lyotropic salt effects. To avoid the complications of the pronounced ion specificity referred to above (the molecular interpretation of these effects in terms of specific ion binding is still a somewhat controversial issue [12, 13, 26, 27]) we will, in our studies of carrageenans, restrict ourselves to ion forms where such specific effects have not been found and where closely similar effects occur for a large number of ions of the same valency [11].

We have here set out to study the salt effects on the coil-helix *equilibria* of the chosen systems (i.e., the first process of the above transition scheme); from the above said, the question may arise whether these equilibria are at all experimentally accessible in the aggregating kappa-carraageenan and agarose systems. However, it is generally believed that transition temperatures obtained *on cooling* do, in fact, reflect the coil-helix equilibria also in the aggregating systems. (Evidence supporting this view are given by Morris and Norton [2]. Further support is given by the finding [11] that the linear $1/T_m$ vs $\log(c_s)$ plot obtained for kappa carrageenan under non-aggregation conditions [low c_s] extrapolates into the corresponding plot obtained from cooling experiments under aggregating conditions [high c_s].) In particular, this should be true for the (experimentally, quite well-defined) temperature of *onset* of helix formation, T_0. Partly for this reason, we have consistently quoted values of T_0, rather than T_m, as our transition temperatures. (It should also be recognized that T_m bears no particular significance in the case of coil-helix transitions of polydisperse polymers.)

III.2. Coil-helix transitions

A coil-helix transition is not, in general, an all-or-non transition [28]; only when the polymers are short compared to the cooperative length of a helical region should this be a valid approximation. Thus, under conditions when coil and helix conformations coexist, a sufficiently long polymer will generally contain regions of both conformations. In terms of the Zimm-Bragg model [29] of coil-helix transitions, the partition function for the system is determined by the free energy of forming the first helical unit, which is written as $-kT\ln \sigma s$, and the free energy of forming subsequent units, which is given by $-kT\ln s$. Here σ and s are the so-called initiation and propagation parameters of the transition, respectively. Physically, the existence of an initiation parameter is generally attributed [28, 29] to the extra entropic cost of forming the first unit of a helical region ($\sigma \ll 1$, in general). (In the case of double-helix formation, this also involves the entropy cost of bringing together two distant polymer segments.) This extra entropy cost strongly disfavors random mixing of helix and coil units in the chain, thus accounting for the cooperative of the transition. In line with this interpretation, we will in the following assume σ to be constant, for a given polymer (at a given concentration), which is independent of temperature and of the ionic content of the system. (This assumption is supported by experiment also for the present systems. Thus, in excess salt, the sharpness of the transition of iota-carrageenan is virtually independent of the salt concentration [15].) The salt-dependent chemical potential difference $\Delta\mu_{D\to A}$ which we will concern ourselves with in the following (cf. also Eq. (3) above) may, therefore, in the case of a coil-helix transition, be identified with the free energy of the *propagation* step. In terms of the Zimm-Bragg model, we may, therefore, write

$$\Delta\mu_{\text{coil}\to\text{helix}} = -RT\ln s \ . \qquad (4)$$

IV. Electrostatic effects

IV.1. Thermodynamic model

In the thermodynamic treatment of the electrostatic effects, we make the basic assumption that the difference in chemical potential between the two states of an association equilibrium involving a charged macromolecule may be expressed as a sum of an electrostatic and a non-electrostatic contribution,

$$\Delta\mu_{D\to A} = \Delta\mu_{el} + \Delta\mu_{nonel} \ , \qquad (5)$$

where the non-electrostatic contribution, $\Delta\mu_{nonel}$, is *independent* of the concentrations or types of mobile ions present in the system and where $\Delta\mu_{el}$ may be evaluated in some suitable theory. In cases where we are interested in the temperature dependence in $\Delta\mu_{D\to A}$, we make the further assumption that $\Delta\mu_{D\to A}$ may be divided into enthalpic and entropic contributions,

$$\Delta\mu_{D\to A} = \Delta H_{D\to A} - T\Delta S_{D\to A} \ , \qquad (6)$$

where both $\Delta H_{D\to A}$ and $\Delta S_{D\to A}$ are independent of temperature or ionic content of the system.

In our applications of the above thermodynamic model [14, 15], we have evaluated $\Delta\mu_{el}$ within the so-called Poisson-Boltzmann Cell Model (PBCM) [30]. The latter model has been used extensively in other contexts, but only rarely in the calculation of $\Delta\mu_{el}$ of macromolecular association equilibria. In the PBCM, the macromolecule in either state of association is modelled as a sphere, an ellipsoid or a cylinder (or some other suitable object of simple geometry) of uniform surface charge density, centered in a cell of the same geometry, which contains water and mobile ions. In the specific applications to be discussed here, the polyelectrolytes have been modelled as infinite cylindrical rods in both conformations. (For details, we refer the reader to the original works [14, 15]. Note that the sign of $\Delta\mu_{el}$ used here differs from that [14] and [15], since we here consistently consider the process D → A, rather than A → D.) For a system containing only one state of the macromolecule, the cell size is determined by the global concentration of the macromolecule. In the present context, however, we are specifically interested in systems containing two states of association of the macromolecule and, thus, two different cell sizes. The partitioning of water and mobile ions between the two cell types is then determined by the requirement that the chemical potentials of all these mobile species be uniform throughout the system. The electrostatic potential profiles and the cell sizes are determined self-consistently by iterative solutions of the Poisson-Boltzmann equation in each cell, and from these solutions the electrostatic contribution to the chemical potential difference, $\Delta\mu_{el}$, may be computed. (See [14] for details.)

We wish here to draw attention to two features of the PBCM, both of which will be found to be important in our applications. The first is that the full Poisson-Boltzmann equation is solved, i.e., no linearization approximation is introduced. The other is that the cell model treats systems where the concentrations of all species are finite. Electrostatic polyion-polyion interactions are thus treated implicity in the model. These two features are the important differences between the PBCM and other models [6, 31, 32] which have been used previously in the calculation of $\Delta\mu_{el}$ associated with the coil-helix transitions of linear polyelectrolytes.

IV.2. Theory vs experiment

We begin by looking at the effects of increasing the concentration of a single salt on the coil-helix transition of a non-titrating polyelectrolyte. Figure 2 shows the results of such an experiment [15] performed on iota carrageenan, compared with the predictions of the PBCM theory using a fitted value of −17.2 kJ/(mole disaccharide) for $\Delta H_{coil \rightarrow helix}$ [15]. As expected, the stability of the helix conformation increases when salt is added. A more subtle effect is that the sharpness of the transition increases with increasing salt content, as long as the concentration of salt does not greatly exceed the concentration of polyelectrolyte charges. The latter feature has also been reported for DNA [20, 33], where the effect is larger owing to, as we shall see, the higher charge density of the DNA molecule.

Both of the above salt effects are seen to be reproduced quantitatively by the PBCM. In particular, the PBCM offers an explanation for the dependence of the sharpness of the transition on the salt content. According to our analysis, the latter effect is not due to a change in the cooperativity of the transition — the initiation parameter is assumed to be constant — but to a gradual variation in $\Delta\mu_{el}$ (and, thus, in the helix propagation parameter) with the helical content in systems where the concentration of added salt does not greatly exceed the concentration of polyelectrolyte charges. This effect we may understand as follows. The helix is more highly charged than the coil, and, therefore, an increasingly larger fraction of the counterions will accumulate near the polyion as the helical content increases. In the absence of large quantities of excess salt, this will lead to a significant decrease in the activity of the counterions and, thus, to a decreasing stabilization of the helical conformation. The effect may be said to be a result of electrostatic interpolyion interactions, and its theoretical treatment requires that the concentration of the polyion be considered explicitly. The reason why this electrostatic effect seems to have been overlooked in

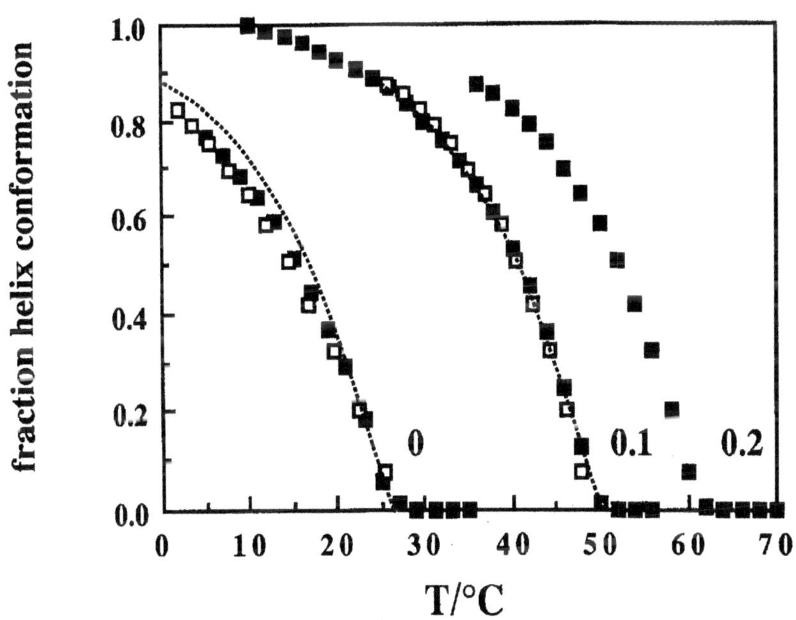

Fig. 2. Effects of increasing salt concentration on the thermally induced coil-helix transition of 49 mM (disaccharide) iota carrageenan (from [15]). The concentration of added NaCl is indicated beside each curve. Solid symbols show experimental data, open symbols show the PBCM predictions with $\Delta H_{coil \rightarrow helix}$ = −17.2 kJ/(mole disaccharide), using the data in the presence of 0.2 M NaCl as reference points. The dashed lines, added to illustrate the increase in sharpness of the transition with the addition of salt, are parallel transpositions of a curve drawn through the experimental data obtained at 0.2 M NaCl

past theoretical studies probably lies in the fact that the theories more commonly used assume infinitely dilute polyion solutions [6, 31, 32]. Needless to say, the effect is not limited to coil-helix transitions in principle; however, its significance also for other types of macromolecular association equilibria remains to be established.

The next case to be examined is the effect of an increase in the concentration of the polyion itself. Obviously, the theoretical treatment of this effect again requires a model using finite polyion concentrations, like the PBCM. Using the same arguments as above, we can understand that an increase in the polyion concentration, in the limit of low concentrations of added salt, will result in an increased electrostatic stabilization of the helix conformation. In the case of iota carrageenan the helix conformation is, in fact, stable even in a salt-free solution when the polyion concentration is sufficiently high. By balancing the concentrations of polyion and added salt, one may thus obtain a series of systems which are all characterized by the same transition temperature. The result of such an experiment [14] is displayed in Fig. 3 in the form of diagram which shows the salt and polyion concentrations of systems that are characterized by a common, specific value of the transition temperature, T_0. By analogy with an isothermal phase diagram, we refer to a diagram of this type as an "isothermal conformational stability diagram".

Since the experiment is isothermal, the basic assumptions of our model require that $\Delta\mu_{el}$ be constant over the entire "conformation boundary" of Fig. 3. The theoretical prediction of such a conformation boundary, with $\Delta\mu_{el}$ calculated for the salt-free systems used as a reference point, is also indicated in Fig. 3. The agreement between theory and experiment is quite satisfactory, giving additional indications that the important contributions to $\Delta\mu_{el}$ deriving from electrostatic interpolyion interactions are well captured by the PBCM.

The last topic of our investigation of electrostatic effects on macromolecular association equilibria concerns the influence of the valency of the counterions. In the specific case of carrageenans, divalent ions are found to be quite potent in stabilizing the helix, which has led to speculations that a different helical conformation might be involved in the case of the divalent ion form [11]. In our investigations of the separate and combined effects of mono- and divalent counterions [14], we again determined isothermal conformational stability diagrams, this time by varying the amounts of added NaCl and $CaCl_2$ to obtain systems which differed in their salt content, but which were all characterized by the same transition temperature, T_0. The results of such experiments for kappa- and iota-carrageenan are shown in Figs. 4a and b, respectively, together with the corresponding theoretical predictions by the PBCM, using the

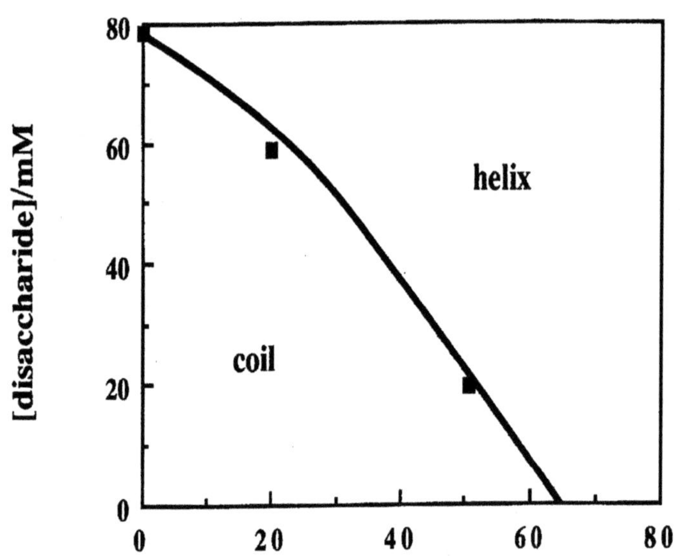

Fig. 3. Polyion and salt concentrations of Na iota-carrageenan/NaCl solutions characterized by a common temperature of onset of helix formation, T_0 = 32.5°C (from [14]). Points show experimental data, the solid line shows the PBCM prediction using the salt-free system as a reference point

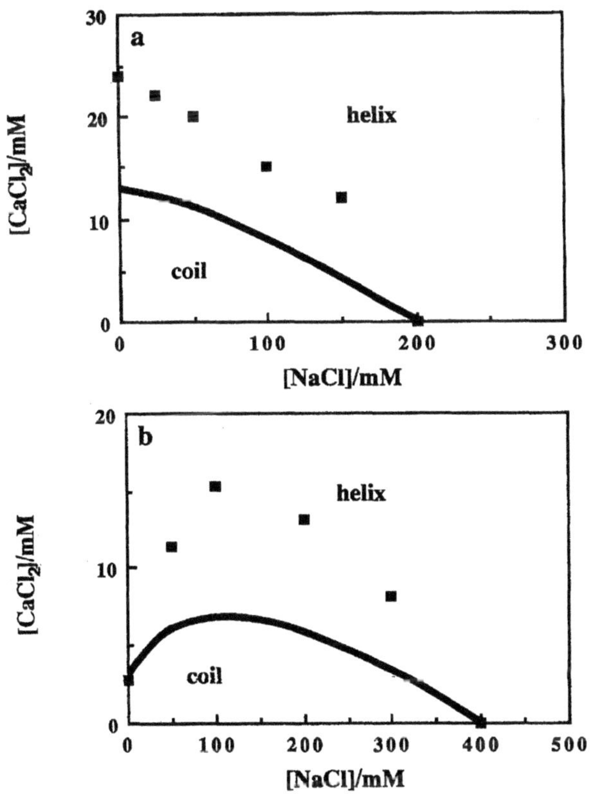

Fig. 4. Conformation stability diagrams for carrageenans in mixed NaCl/$CaCl_2$ solutions: a) 5 mM kappa-carrageenan at 18°C and b) 2.9 mM iota-carrageenan at 72.5°C (from [14]). Points show experimental data, solid lines show the PBCM predictions using as reference points the data obtained in the presence of only NaCl

pure monovalent salt systems as references. The relatively large helix stabilizing effect of divalent counterions is clearly predicted by the theory, without the need to invoke the existence of a different helical conformation or site-specific ion binding [2] when such ions are present.

A more remarkable feature of Fig. 4 is that both the theoretical and the experimental stability diagrams for iota-carrageenan contain maxima in the $CaCl_2$ concentrations at finite concentrations of NaCl, whereas both of the corresponding diagrams for the lesser charged kappa-carrageenan are monotonic. The results of Fig. 4b show that for iota-carrageenan in the presence of divalent ions, the addition of 1:1 electrolyte initially *destabilizes* the helix conformation. This is contrary to the effect of salt in a single salt solution (cf. Fig. 2). Similar "anomalous" salt effects in mixed counterion systems have, however, previously been observed for DNA [20, 33, 34].

The fact that the non-monotonic behavior of Fig. 4b is predicted also by the PBCM indicates that the phenomenon is of a generic electrostatic origin, and it may, in fact, be explained in rather simple terms. At high electrostatic potentials (high charge density, low salt concentration) in a mixed salt solution, the potential profile in the solution outside the polyion is largely determined by the divalent counterions; it is thus rather insensitive to additions of monovalent counterions. This insensitivity becomes more pronounced with increasing charge density which explains why, in this limit, the (more charged) helix conformation is less stabilized than the (less charged) coil conformation by the addition of monovalent cations. However, when sufficient amounts of monovalent counterions have been added, the electrostatic potentials will become lowered to the point where the normal salt dependence is regained. The non-monotonic behavior for iota-carrageenan is thus a general electrostatic high potential phenomenon; it will, however, not be captured by models which, unlike the PBCM, employ the linearized version of the Poisson-Boltzmann equation. Again it should be appreciated that, like the other electrostatic effects described in this section, the non-monotonic behavior of Fig. 4b is not, in principle, limited to coil-helix transitions.

V. Lyotropic effects

V.1. Physical interpretation

Even if certain manifestations of electrostatic salt effects on macromolecular association equilibria may be rather unexpected and/or subtle, as was demonstrated in the previous section, their physical origin is established at the outset by the very term "electrostatic effect" (cf. Section II.1). This is not at all the situation for the lyotropic salt effects which will be discussed in the present section. On the contrary, a wide range of phenomena or properties that are measurably sensitive to the identity of the ions present, and where the ions may be graded in order of effectiveness (approximately) according to the lyotropic series, are commonly regarded as manifestations of the lyotropic effect [9, 35]. In the present context, however, we shall under this heading restrict ourselves to the differing effects of various low-molecular cosolutes, ions in particular, on the association equilibria of (primarily uncharged) macromolecules. Our goals shall be to try to understand, in as general terms as possible, why such effects should exist at all and why they may be expected to differ, in general, for different cosolutes. We shall not try to explain the particular ordering of the ions in the lyotropic series; we will, however, discuss possible reasons why the *same* sequence of effectiveness of the ions may be expected to be found regardless of the chemical details of the macromolecular system. Already at the outset we also wish to point out that our physical interpretation of the lyotropic salt effects has much in common with ideas previously forwarded by Aveyard [36] in his discussion of the salting out of nonpolar nonelectrolytes.

A typical example of lyotropic salt effects on a macromolecular association process may be seen in Fig. 5, which shows the effects of different sodium salts on the temperature of onset of helix formation, T_0, in agarose solutions [16]. Both salting-in and salting-out effects are seen, and the magnitude of the effect clearly differs for the various anions. In the dicussion connected with Fig. 1 above, we stressed the fact that association processes in

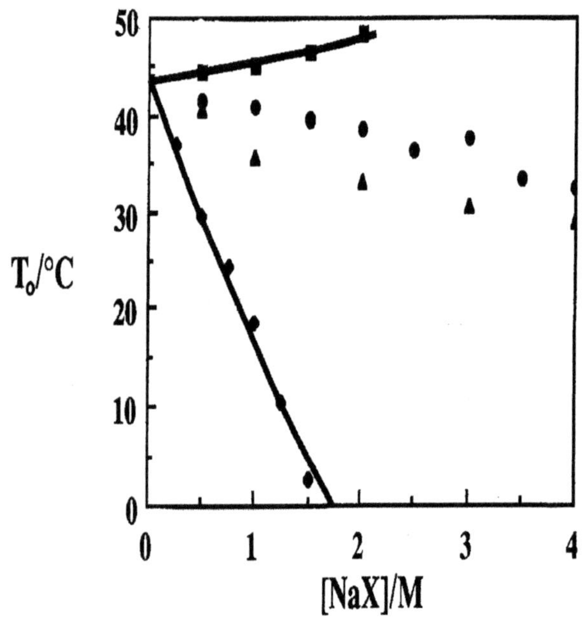

Fig. 5. Effects of various added sodium salts, NaX [X = Cl (■); Br (●); NO_3 (▲); SCN (◆);] on the temperature of onset of helix formation in 1% agarose solutions (from [16]). The solid lines show the predictions of Eq. (10) for K(NaCl) = 0.9 and K(NaSCN) = 2, with $\Delta H_{\text{oil}\rightarrow\text{helix}}$ = −5 kJ/mole and Δn = −10

general involve a change in the specific surface area which the macromolecule exposes to the solvent. Thus, there is a change in the interfacial energy of the system associated with the transition. We must then ask how added salts or other low-molecular cosolutes affect the interfacial energy of the solvent-macromolecule system. The answer is, of course, that the interfacial energy is affected when there is a (positive or negative) excess of the cosolute at the interface. From studies of macroscopic interfaces [37] we may then recall that a non-zero interface excess, i.e., a preferential adsorption, is the rule rather than the exception. Thus, the composition of a two-component solution at an interface is *generally* different from the composition in the bulk. This is true even for simple liquids [37], and we may be confident that preferential adsorption occurs, in general, also for an aqueous solution of a low-molecular solute in contact with a macromolecule.

For aqueous salt solutions in contact with macroscopic phases, evidence of preferential adsorption effects is given by the interfacial tension measurements of Aveyard *et al.*; their results pertaining to various salt solutions in contact with liquid alkanes [38] and decanol [39] are plotted in Fig. 6a and b, respectively. We note that interfacial excesses of salts may be either positive or negative, that the lyotropic sequence of the anions is observed, and that the interfacial excesses of all salts are shifted in the same direction when one polar liquid is replaced by another. All these features have their parallel in the effects of the various salts on the association equilibria of different macromolecules.

From the above considerations and general findings, we suggest that the stabilizing/destabilizing effects of a low-molecular cosolute on an associated state of a macromolecule in solution may be explained by a negative/positive excess of that cosolute at the macromolecule-solution interface. This is a general effect which should not be limited to aqueous solutions. Furthermore, from the results pertaining to the aqueous salt solutions in Fig. 6 it may be inferred that while the nature of the surface is of importance for the *overall tendency* of salts to adsorb, the *differences* between different salts in their tendency to adsorb at a given surface depends, at least to a large extent, on water-ion interactions. The latter conclusion, which offers an explanation for the quite common occurrence of the lyotropic sequence of the ions, is corroborated by the fact that this sequence is found also in the surface tensions of aqueous salt solutions against air [40].

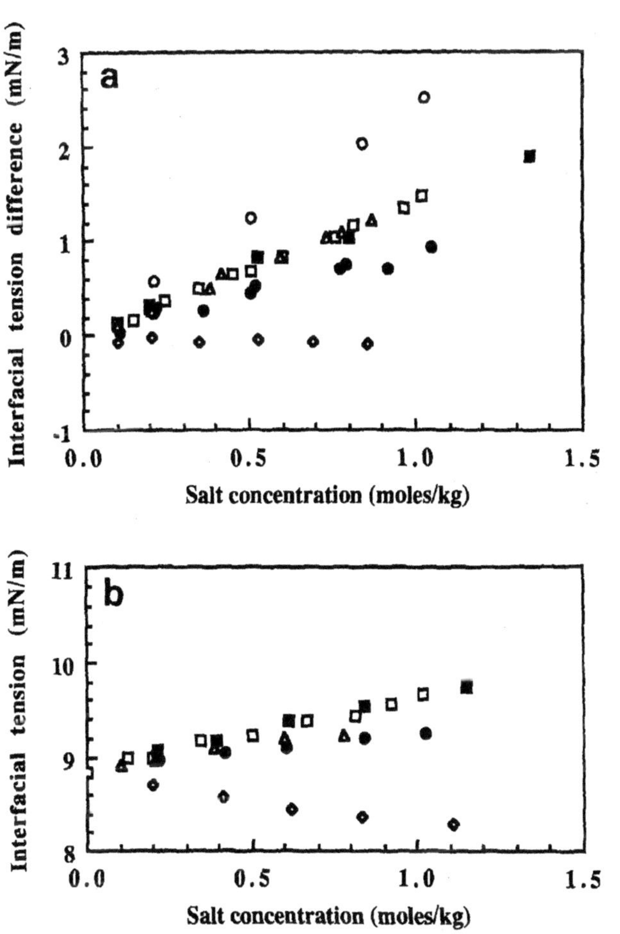

Fig. 6. Interfacial tensions at interfaces between aqueous salt solutions and a) dodecane/decane (data from [38] and b) decanol (data from [39]). Symbols refer to solutions of Na_2SO_4 (○), KCl (■), NaCl (□), LiCl (△), KBr (●) and KI (◇)

V.2. Thermodynamic model

To substantiate the above conclusions, it is obviously desirable with some quantitative estimate of the surface excesses required to yield, for instance, the salt effects observed for agarose in Fig. 5. In a simple thermodynamic model of the effect, $\Delta\mu_{\text{coil}\rightarrow\text{helix}}$ is written, as before, as a sum of two contributions

$$\Delta\mu_{\text{coil}\rightarrow\text{helix}} = \Delta\mu^0_{\text{coil}\rightarrow\text{helix}} + \Delta\mu_{\text{pa}} , \tag{7}$$

where $\Delta\mu^0_{\text{coil}\rightarrow\text{helix}}$ is the chemical potential difference in the pure solvent and $\Delta\mu_{\text{pa}}$ is the contribution due to the occurrence of preferential adsorption at the surface of the macromolecule. An expression for the latter contribution has been derived by Schellman [22] in his recent extension of the theory of effects of ligand binding on macromolecular association equilibria (cf. section II.3. above). Here Schellman explicitly considers the *exchange* process whereby a solvent molecule at a macromolecular adsorption site is replaced by a cosolute molecule according to the following scheme:

1(at site) + 3(in bulk) ↔ 3(at site) + 1(in bulk) .

Here, in accordance with conventional nomenclature [22], the main solvent component and the low-molecular cosolute are referred to as components 1 and 3, respectively. If the exchange reaction is characterized by an equilibrium constant K the theory yields the following expression for $\Delta\mu_{\text{pa}}$ [22]:

$$\Delta\mu_{\text{pa}} = -\Delta n RT \ln[1 + (K - 1)x_3] . \tag{8}$$

Here x_3 is the mole fraction of cosolute in the bulk (activity coefficients are neglected) and $-\Delta n$ represents the number of polymer adsorption sites (expressed per repeating unit) which are removed from solvent contacts as a result of the coil-to-helix transition. We now write

$$\Delta\mu^0_{\text{coil}\rightarrow\text{helix}} = \Delta H_{\text{coil}\rightarrow\text{helix}} - T\Delta S_{\text{coil}\rightarrow\text{helix}} , \tag{9}$$

and, as before (cf. section IV.1), assume that $\Delta H_{\text{coil}\rightarrow\text{helix}}$ and $\Delta S_{\text{coil}\rightarrow\text{helix}}$ are constants, independent of temperature or salt content. With the additional requirement [16] that the composite quantity $\Delta\mu_{\text{coil}\rightarrow\text{helix}}/RT$ should be a constant when evaluated at $T = T_0$, we arrive at the following expression for the salt dependence in T_0:

$$T_0 = \frac{\Delta H_{\text{coil}\rightarrow\text{helix}}/(\Delta n R)}{\Delta H_{\text{coil}\rightarrow\text{helix}}/(\Delta n R T_0^0) + \ln[1 + (K - 1)x_3]} . \tag{10}$$

It has been pointed out [4] that Eq. (10) (or, actually, the analogous equation which may be derived from Eq. (3) under the same simplifying assumptions of only one class of binding sites) reduces to an equation linear in the concentration of component 3 (i.e., an equation of the desired form of Eq. (2)) in cases when $(K - 1)x_3 \ll 1$ and when the relative changes in T_0 are small.

We have fitted [16] Eq. (10) to the NaCl and NaSCN data of Fig. 5; the fitted curves are also shown in Fig. 5. (Note that the fitted curves are almost linear.) With $\Delta H_{\text{coil}\rightarrow\text{helix}} = -5$ kJ/mole [16] and with $\Delta n = -10$ (estimating that roughly 10 hydration sites per repeating disaccharide unit are lost on creating the agarose double helix) these fits yielded the values K(NaCl) = 0.9 and K(NaSCN) = 2. These effective equilibrium constants include the (possibly opposing) effects of the cation and the anion [16]; however, the important conclusion of this analysis is that quite a small deviation from the bulk composition in the interfacial layer (remember $K = 1$ for a vanishing surface excess) suffices to explain the rather large effect of NaSCN on the agarose coil-to-helix transition temperature.

To complete our argument, we shall compare the increased interfacial concentration of NaSCN invoked in the case of agarose with the surface excesses of KI found at the aqueous soution-decanol interface by Aveyard et al. [39]. (To justify this comparison between different systems, we note that the effects of the iodide and thiocyanate ions typically are found to be rather similar [4]. Furthermore, it may be argued that the decanol-water interface should be reasonably similar to the agarose-water interface.) A linear fit to the interfacial tension data pertaining to KI in Fig. 6b yields a value of $\partial \gamma / \partial c = -0.5$ mNm^{-1} M^{-1}, and from the Gibbs adsorption equation

$$\Gamma = -(c/RT)/(\partial \gamma / \partial c) \tag{11}$$

we obtain an excess of 0.1 molecules of KI per nm^2 of interfacial area at [KI] = 1 M. Turning to the case of NaSCN and agarose, our thermodynamic analysis here yields an *excess* concentration of NaSCN in the interfacial layer which is of the order of 1 M at a bulk concentration of 1 M, since K(NaSCN) = 2. This corresponds to an excess molar fraction of NaSCN equal to 1/55.5 in the interfacial layer. If we estimate that a monolayer of water molecules contains 10 molecules per nm^2 at normal liquid density, we finally find that our thermodynamic analysis predicts an excess of NaSCN in the interfacial layer amounting to 10/55.5 $\approx$ 0.2 molecules per nm^2, a value which is of the same order of magnitude as the value we just deduced for KI in the macroscopic experiment by Aveyard et al.

We hasten to admit that the above analysis is quite crude: We have consistently used ionic concentrations instead of activities, the number Δn is not known with certainty, and so on. Consequently, the surface excesses quoted in the present section should be regarded as order-of-magnitude estimates only. However, we do maintain that our analysis demonstrates that surface excesses of salts, of a magnitude which is sufficient to explain the effects of salts on the association equilibria of uncharged macromolecules, have in fact been experimentally observed at macroscopic interfaces.

V.3. *Implications regarding molecular interactions*

Clearly, the above analysis of lyotropic salt effects says nothing about the nature of the molecular interactions which give rise to the adsorption/desorption of ions or uncharged cosolutes at the aqueous solution-macromolecule interface. This difficult problem, which involves consideration of the highly directional interactions of water molecules, has often been the main focus of discussions on lyotropic effects. In fact, we feel that detailed discussions of, in particular, solute effects on water structure have tended to unduly overshadow considerations of the type presented above, which do not rely on detailed analyses of molecular interactions but only on the general [37] occurrence of adsorption/desorption of cosolutes at interfaces.

Here we only wish to underline the fact that the net adsorption of one component at an interface is a result of *all* the component interactions in the system. For instance, it may be tempting to assume that the component which is adsorbed interacts more favorably with the interface than does the component which is desorbed, but this need not be the case. Lane has demonstrated in his analyses of regular solutions in contact with an interface [37, 41] that adsorption/desorption will occur even to indifferent surface as long as the *solution* is not ideal. And although aqueous solutions of ions are ill-suited for treatment in a regular solution model, it should nevertheless be clear that the preferntial adsorption/desorption of an ion at an interface could occur as a result of ion-water interactions.

We have, in fact, pursued this argument further in an analysis [42] of the effect of urea on the T_0 of agarose (Fig. 7). Urea is uncharged and we have, therefore, attempted to apply Lane's regular solution model to the system. The relative interactions between the different components of the system are described by a bulk interchange free energy w, which describeds the urea-water interactions, and an interface exchange free energy v, which describes the relative affinities of urea and water to the interface. With a value of w = 3.44 kJ/mole, obtained from fitting data on the osmotic coefficient of the binary urea/water system to the regular solutionmodel, we find [42] that with Δn equal to —10 or —1, reasonable fits to the experimental data of Fig. 7 require values of v amounting to 0.38 kJ/mole or —1.39 kJ/mole, respectively. Thus, the analysis yields an interface exchange free energy of a magnitude which is smaller than that of the bulk interchange free energy, and with a sign which depends on our estimate of the magnitude of Δn. This would suggest that the non-ideality of the urea-water mixture is the dominating contribution to the denaturing effect of urea agarose.

In light of the above analysis, we may also attempt a mechanistic distinction between lyotropic effects and effects of site-specific binding of the cosolute (section II.3 above) on macromolecular association equilibria. Formally, both effects may be treated in terms of Schellman's model [22], as was done previously in this section; the difference is then only reflected in the magnitude of the phenomenological equilibrium constant K, which is large in the case of site-binding, and of the order of unity when the effect of the cosolute is of the weak adsorption or lyotropic kind. The more detailed analysis of this subsection suggests that we identify situations where v dominates (leading to a pronounced sensitivity to macromolecule-cosolute interactions) as being of site-specific binding type. Conversely, situations where w dominates are insensitive to details of the macromolecule, but pronouneedly sensitive to the cosolute, which is typical of the lyotropic effect.

The above interpretation of the lyotropic effect is, of course, schematic and should not be taken too literally: In general, as we have seen, influences of *both* interchange free energies, v and w, must be considered in the weak adsorption situation. Furthermore, in rationalizing lyotropic sequences we should not exclude the very likely possibility that also variations in the interface exchange free energy may follow these sequences (e.g., for halide ions). Nevertheless, the interpretation of the lyotropic effect offered here brings into focus the solvent-cosolute interactions, which on intuitive grounds would seem to play a major role in view of the commonness of the lyotropic sequence, while still maintaining that the effect of lyotropic cosolutes on macromolecular association equilibria is best understood in terms of preferential adsorption phenomena.

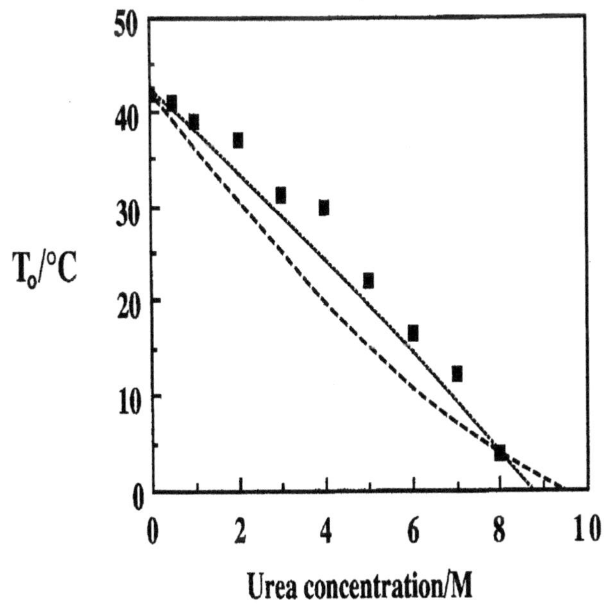

Fig. 7. Effects of added urea on the temperature of onset of helix formation in 1% agarose solutions (from [42]). Points show experimental data, lines show theoretical predictions from a regular solution model (see text) with w = 3.44 kJ/mole and with v = 0.33 kJ/mole, Δn = —10 (dashed line) or v = —1.39 kJ/mole, Δn = —1 (dotted line)

VI. Conclusions

On the basis of the results from our detailed experimental and theoretical studies of helix-forming linear biopolymers, we have here attempted to present an overview as well as a physical interpretation of the various effects of added salts on macromolecular association equilibria. The origin of two distinct types of general salt effects may be traced back to the fact that an association reaction generally results in i) a decrease in the exposed surface area of the macromolecule, and ii) a concomitant increase in the electrostatic potential of a charged macromolecule. The former of these features is proposed to be the origin of so-called lyotropic salt effects, since cosolutes, owing to the fact that they quite generally adsorb or desorb at interfaces, give rise to a change in the interfacial energy associated with the macromolecular reaction. Furthermore, the influence of the interactions between the cosolute and the solvent on the adsorption/desorption behavior is emphasized; important contributions from solvent/cosolute interactions may explain the general occurrence of the lyotropic series and, also, the relative insensitivity of the lyotropic effects to the details of the macromolecular surface.

Changes in the electrostatic potential profile resulting from the association of charged macromolecules are ultimately responsible for the occurrence of electrostatic salt effects on macromolecular association equilibria. Although such effects are well known in principle, some of their more non-trivial manifestations have here been demonstrated and rationalized within the Poisson-Boltzmann Cell Model, where both interpolyion and polyion-salt interactions are accounted for in a self-consistent way and where no linearization approximation is introduced. The effects discussed include an increase in the sharpness of the transition with added salt as a result of the screeing of electrostatic polyion-polyion interactions, and also the finding that in systems containing mixed mono- and divalent counterions the associated form may be either stabilized or destabilized on addition of monovalent salt, depending on the exact conditions.

Acknowledgement

This work has been supported by the Swedish Board of Technical Development (STU).

References

1. Clark AH, Ross-Murphy SB (1987) Adv Polym Sci 83:57
2. Morris ER, Norton IT (1982) Stud Phys Theor Chem 26:549
3. Sandford PA, Baird J (1983) In: Aspinall GO (ed) The Polysaccharides. Academic Press, New York, Vol. 2, p 411
4. von Hippel PH, Schleich T (1969) In: Timasheff SN, Fasman GD (eds) Structure and Stability of Biological Macromolecules. Marcel Dekker, New York, Chapter 6, p 417
5. Record MT, Anderson CF, Lohman TM (1978) Quart Rev Biophys 11:103
6. Manning GS (1978) Quart Rev Biophys 11:179
7. Anderson CF, Record MT (1982) Ann Rev Phys Chem 33:191
8. Painter T (1983) In: Aspinall GO (ed) The Polysaccharides. Academic Press, New York, Vol. 2, p 195
9. McBain JW (1950) "Colloid Science" Heath, Boston, Chapter 9, p 131
10. Zabik ME, Aldrich PI (1965) J Food Sci 30:795
11. Rochas C, Rinaudo M (1980) Biopolymers 19:1675
12. Grasdalen H, Smidsrød O (1981) Macromol 14:1842
13. Norton IT, Morris ER, Rees DA (1984) Carbohydr Res 134:89
14. Nilsson S, Piculell L, Jönsson B (1989) Macromol 22:2367
15. Nilsson S, Piculell L (1989) Macromol 22:3011
16. Piculell L, Nilsson S (1989) J Phys Chem 93:5596
17. Piculell L, Nilsson S (1989) J Phys Chem 93:5602
18. Piculell L, Håkansson C, Nilsson S (1987) Int J Biol Macromol 9:297
19. Piculell L, Nilsson S, Ström P (1989) Carbohydr Res 188:121
20. Dove WF, Davidson N (1962) J Mol Biol 5:467
21. Schellman JA (1955) Compt Rend Trav Lab Carlsberg, Ser Chim 29:230
22. Schellman JA (1987) Biopolymers 26:549
23. Bryce TA, Clark AH, Rees DA, Reid DS (1982) Eur J Biochem 122:63; Rees DA, Williamson FB, Frangou SA, Morris ER (1982) Eur J Biochem 122:71
24. Rees DA, Morris ER, Thom D, Madden JK (1982) In: Aspinall GO (ed) The Polysaccharides. Academic Press, New york, Vol. 1, p 195
25. Rees DA (1972) Chem Ind (London) 630
26. Rochas C, Rinaudo M (1984) Biopolymers 23:735
27. Belton PS, Morris VJ, Tanner SF (1986) Macromol 19:1618
28. Poland D, Scheraga HA (1970) "Theory of Helix-Coil Transitions in Biopolymers" Academic, New York
29. Zimm BH, Bragg JK (1959) J Chem Phys 31:526
30. Fuoss RM, Katchalsky A, Lifson S (1951) Proc Nat'l Acad Sci 37:579; Alfrey T, Berg PW, Morawetz H (1951) J Polym Sci 7:543; Gunnarson G, Jönsson B, Wennerström H (1980) J Phys Chem 84:3114
31. Peller L (1959) J Phys Chem 63:1199
32. Record MT (1967) Biopolymers 5:975
33. Record MT (1975) Biopolymers 14:2137
34. Krakauer H (1974) Biochemistry 13:2579
35. Collins KD, Washabaugh MW (1985) Quart Rev Biophys 18:323
36. Aveyard R (1982) Can J Chem 60:1317
37. Lane JE (1983) In: Parfitt GD, Rochester CH (eds) Adsorption from Solution at the Solid/Liquid Interface. Academic, New York, p 49
38. Aveyard R, Saleem SM (1976) J Chem Soc, Faraday Trans 1 72:1609
39. Aveyard R, Saleem SM, Heselden R (1977) J Chem Soc, Faraday Trans 1 73:84
40. Jarvis NL, Scheiman MA (1968) J Phys Chem 72:74
41. Lane JE (1968) Aust J Chem 21:827
42. Nilsson S, Piculell L, Malmsten M (1990) J Phys Chem 94:5149

Authors' address:

Dr. L. Piculell
Physical Chemistry 1
University of Lund
Chemical Center
Box 124
22100 Lund, Sweden

Charge fluctuations and microemulsion conductivity

B. Halle

Physical Chemistry 1, University of Lund, Sweden

Abstract: A theory for the conductivity of dilute ionic microemulsions of the water-in-oil droplet type is presented. In contrast to a previous theory proposed by Eicke et al., the present theory accounts for the observed sharp maximum in conductivity vs droplet size and predicts a strong dependence of the conductivity on the valence type of the electrolyte.

Key words: Microemulsion; conductivity; fluctuation; droplet; charge

Introduction

Within certain ranges of composition and temperature, water-in-oil microemulsions consist of discrete aqueous droplets dispersed in a continuous oil medium. If the system contains ionic species (ionic surfactants and/or simple inorganic ions), these tend to be confined to the aqueous droplet cores, where they are more favorably solvated than in the nonpolar surroundings. Moreover, due to the long range of the ion-solvent interaction, the ionic species tend to be distributed in such a way that each droplet contains an equal number of cations and anions (for a symmetric electrolyte). The droplets thus tend to be electroneutral and the (mean) electric field from the ions within the core does not penetrate into the low-permittivity oil region. (For spherical droplets, this follows from Gauss' law.) At any finite temperature, however, thermal fluctuations provide a driving force for spatial charge separation, leading to a distribution of net droplet charges. This distribution is generally established and maintained by a fusion-fission mechanism [1—3], which enables diffusive exchange of the constituents of two transiently coalesced droplets.

During the past decade of intensive research on microemulsions, the fact that microemulsion droplets may carry a net electric charge has rarely been appreciated, despite the potential implications for phase behavior and dynamics. Very recently, however, Eicke, Borkovec, and Das-Gupta [4] (referred to as EBD in the following) reported conductivity measurements for the microemulsion phase of the AOT-isooctane-water system [5] and interpreted their data in terms of droplet charge fluctuations. EBD made the intriguing observation that, as the droplet size is varied at a fixed low droplet volume fraction (well below the so-called percolation region), the conductivity exhibits a sharp maximum. While the theoretical treatment presented by EBD quantitatively explains the reduction of the conductivity for large droplets, it fails to account for the conductivity maximum. EBD attributed this deficiency in the theory to a breakdown of the dielectric continuum approximation used in their model. In the following, we present an alternative theoretical analysis which does account for the conductivity maximum, while retaining the dielectric continuum approximation. We adopt essentially the same model as EBD, but we use a more general statistical-mechanical approach for calculating the charge fluctuations.

Theory

Limiting conductivity

We consider a macroscopic microemulsion sample consisting of a very large number of aqueous electrolyte droplets dispersed in a continuous oil medium. For simplicity, we assume that the droplets are spherical in shape and monodisperse in size. If n denotes the total droplet number density and $n(z)$ the number density of droplets carrying a net charge ze, then

$$\sum_z n(z) = n , \tag{1}$$

where the summation index $z = \nu k$, with $k = 0, \pm 1, \pm 2, \ldots$, and ν is the charge number of the symmetric $\nu:\nu$ electrolyte in the droplets.

The zero-frequency conductivity κ is related to the droplet charge distribution $n(z)$ through

$$\kappa = \sum_z |z| e n(z) u(z) , \tag{2}$$

where $u(z)$ is the electric mobility of a droplet with net charge ze. In the infinite-dilution limit (droplet volume fraction $\varphi_d \to 0$), the electric mobility is independent of the charge distribution and is given by the Nernst-Einstein relation

$$u(z) = \frac{|z| e}{\zeta} , \tag{3}$$

where ζ is the translational friction coefficient of an isolated droplet.

For a spherical liquid droplet of diameter σ_d and viscosity η_d suspended in another (immiscible) liquid of viscosity η, the friction coefficient is [6, 7]

$$\zeta = 3\pi\sigma_d\eta \left[1 - \frac{\eta}{3(\eta + \eta_d)}\right] . \tag{4}$$

The first factor is the familiar Stokes' friction of a rigid ($\eta_d \gg \eta$) spherical particle, while the correction factor within brackets accounts for the reduction of the viscous dissipation resulting from the smaller velocity gradient in the external fluid when the tangential velocity component is continuous rather than zero at the interface. However, if the interface is covered by surfactant molecules which are insoluble in the two liquids, as in a microemulsion, then a surface flow cannot be sustained and the droplet core behaves essentially as a rigid sphere [6]. Further, although the outer shell of the droplet contains flexible hydrocarbon chains (surfactant tails), the hydrodynamic interaction within this dense shell should be sufficiently strong to justify a non-draining approximation, where the shell is regarded as rigid (cf the Rouss-Zimm theory of polymer dynamics [8]). In conclusion, it should be an excellent approximation to regard the entire droplet as rigid and to use the simple Stokes' friction, $\zeta = 3\pi\sigma_d\eta$.

Combination of Eqs. (2)—(4) now yields for the limiting specific conductivity

$$\Lambda \equiv \frac{\kappa}{\varphi_d} = \frac{2e^2}{\pi^2\eta\sigma_d^4} \langle z^2 \rangle , \tag{5}$$

where we have used the relation $\varphi_d = \pi\sigma_d^3 n/6$.

Self-energy

In order to calculate the mean square droplet valence $\langle z^2 \rangle$, we assume that the free energy per unit volume of microemulsion can be expressed as the following functional of $n(z)$

$$A[n(z)] = A_0 + A_S[n(z)] - TS[n(z)] . \tag{6}$$

A_0 contains all free energy contributions that are independent of the net droplet charge. A_S is the self-energy of the droplets, i.e., the electrostatic work of charging noninteracting droplets. Since our aim is to calculate the *limiting* ($\varphi_d \to 0$) conductivity, Eq. (6) contains no contributions from interactions among the droplets (cf Discussion). The last term in Eq. (6) represents the ideal entropy of mixing associated with the distribution $n(z)$.

The calculate the self-energy A_S, we model a microemulsion droplet as a spherical core of diameter σ_c (which may differ from the hydrodynamic droplet diameter σ_d) immersed in a dielectric continuum of relative permittivity ε_s. Confined to the core are N_+ cations and N_- anions. (In the case of surfactant ions, only the ionic headgroups are taken to be part of the core.) The self-energy is defined as the work of forming a droplet with $N_+ \neq N_-$, and, hence, a net charge $ze = (N_+ - N_-)\nu e$, from a neutral droplet with $N_+ = N_-$.

We decompose the self-energy into two contributions as

$$A_S = A_S^{\text{ext}}(z) + A_S^{\text{int}}(N_+, N_-) . \tag{7}$$

The first term represents the electrostatic work of charging a perfectly conducting droplet surrounded by an external dielectric medium. This depends on the net droplet charge, but not on the individual ion numbers:

$$A_S^{ext}(z) = \frac{(ze)^2}{4\pi\varepsilon_0\varepsilon_s\sigma_c} . \tag{8}$$

The second term in Eq. (7) accounts for changes in the configurational free energy of the internal charge distribution within the core as ions are transferred between the two droplets. It can be shown [9] that this internal contribution is negligible compared to the external one, provided that $|z| \ll N_+, N_-$. Discarding this term, we thus obtain for the self-energy per unit volume of microemulsion

$$\beta A_S[n(z)] = \frac{\lambda_B}{\sigma_c} \sum_z z^2 n(z) , \tag{9}$$

where $1/\beta = k_B T$ and λ_B is the Bjerrum length

$$\lambda_B = \frac{e^2}{4\pi\varepsilon_0\varepsilon_s k_B T} . \tag{10}$$

Entropy of mixing

We now derive an expression for the entropy of mixing per unit volume, $S[n(z)]$, associated with the droplet charge distribution $n(z)$. We consider a system containing N_0 cations (charge $+ve$) and N_0 anions (charge $-ve$) distributed over M droplets. As starting point, we take Boltzmann's formula

$$S = k_B \ln \Omega , \tag{11}$$

where Ω is the number of distinguishable configurations, i.e., the number of distinct partitions of the $2N_0$ ions over the M droplets (assumed to be of equal volume). Since the droplets as well as the ions (of either kind) are indistinguishable, we have

$$\Omega = (2N_0)! \left\{ \prod_{N_+=0}^{N_0} \prod_{N_-=0}^{N_0} [m(N_+,N_-)]! \times [(N_+)!\,(N_-)!]^{m(N_+,N_-)} \right\}^{-1} , \tag{12}$$

where $m(N_+,N_-)$ denotes the number of droplets containing N_+ cations and N_- anions. Inserting this into Eq. (11) and assuming that $m(N_+,N_-)$ and $N_\pm$ are large numbers (cf below) so that Stirling's approximation ($\ln x! = x\ln x - x$) may be invoked, we find

$$S = -k_B \sum_{N_+} \sum_{N_-} m(N_+,N_-) \ln m(N_+,N_-) - k_B \sum_{N_+} \sum_{N_-} m(N_+,N_-) \times [N_+ \ln N_+ + N_- \ln N_-] , \tag{13}$$

where constant terms (independent of the ion distribution) have been dropped. The first part of Eq. (13) is associated with the entropy of mixing droplets with different ion occupation numbers (N_+,N_-), while the second part refers to the entropy of mixing cations and anions within the droplets.

Next, we make a change of independent configurational variables from N_+ and N_- to $N = N_+ + N_-$ and $z = N_+ - N_-$. In terms of these variables, Eq. (13) takes the form (after dropping a constant term)

$$S = -k_B \sum_{z=-N_0}^{N_0} \sum_{N=0}^{2N_0} m(N,z) \ln m(N,z) - \frac{k_B}{2} \sum_{z=-N_0}^{N_0} \sum_{N=0}^{2N_0} m(N,z) [(N+z)\ln(N+z) + (N-z)\ln(N-z)] . \tag{14}$$

Now, on account of the large self-energy, only configurations with $|z| \ll N$ will be important. We thus expand the expression within brackets [ ] in Eq. (14) in powers of z/N, neglecting terms of order $(z/N)^2$. The second (intradroplet) part of Eq. (14) then becomes independent of $n(z)$ and so can be omitted.

Further simplification is possible if we make the reasonable assumption that fluctuations in the total number of ions per droplet (variation of N at fixed z) are independent of the net droplet charge. Then $m(N,z)$ may be factorized as

$$m(N,z) = m(z)p(N) , \tag{15}$$

where $m(z) = \Sigma_N m(N,z)$ and $p(N)$ is the probability of having N ions in an electroneutral droplet. Inserting this into the remaining first part of Eq. (14) and dividing by the system volume V, we obtain for the entropy density

$$S[n(z)] = -k_B \sum_z n(z)\ln n(z) - k_B n \sum_N p(N)\ln p(N) , \tag{16}$$

where we have dropped a constant term and introduced droplet number densities $n(z) = m(z)/V$. The first part of Eq. (16) is the desired entropic contribution from the droplet charge distribution $n(z)$, while the second part is associated with the total ion occupation number distribution $p(N)$, which, by assumption, is independent of $n(z)$, so that this part can be omitted.

Droplet charge distribution

Combination of Eqs. (6), (9), and (16) now yields for the free energy density

$$\beta A[n(z)] = \beta A_0 + \frac{\lambda_B}{\sigma_c} \sum_z z^2 n(z) + \sum_z n(z)\ln n(z) . \tag{17}$$

The equilibrium droplet charge distribution is obtained by minimizing $A[n(z)]$ with respect to variations in $n(z)$, subject to the constraint of constant overall droplet density as expressed by Eq. (1). Using the method of Lagrange, we thus arrive at the expected Boltzmann distribution

$$n(z) = \frac{n}{Q} \exp(-az^2) , \tag{18}$$

where we have introduced the partition function

$$Q(a) = \sum_z \exp(-az^2) , \tag{19}$$

and the dimensionless coupling parameter

$$a = \frac{\lambda_B}{\sigma_c} . \tag{20}$$

The mean square droplet valence, required to calculate the conductivity from Eq. (5), can now be obtained, using Eqs. (18) and (19), as

$$\langle z^2 \rangle = \frac{1}{n} \sum_z z^2 n(z) = -\frac{d\ln Q}{da} . \tag{21}$$

It may be noted that, in general, Q and $\langle z^2 \rangle$ depend on the charge number ν of the symmetric electrolyte contained in the droplets. This dependence is a consequence of restrictions in the z summation, as seen by rewriting Eq. (19) as

$$Q(a, \nu) = \sum_k \exp(-a\nu^2 k^2) , \tag{22}$$

where k runs overall all positive and negative integers.

It is instructive to examine the limiting behavior of Eq. (22). In the weak coupling regime, where $a \ll 1$ and the charge fluctuations are large, the exponential in Eq. (22) varies slowly with k and the sum may be approximated by an integral:

$$Q(a, \nu) = \int_{-\infty}^{\infty} dk \exp(-a\nu^2 k^2) = \frac{1}{\nu} \left(\frac{\pi}{a}\right)^{\frac{1}{2}} . \tag{23}$$

With Eq. (21) we then obtain

$$\langle z^2 \rangle = \frac{1}{2a} . \tag{24}$$

In this regime, $\langle z^2 \rangle$ is thus independent of the charge number ν. The divergence of $\langle z^2 \rangle$ in the limit $a \to 0$ is an artifact introduced by extending the integration limits in Eq. (23) to infinity. In practice, the internal part of the self-energy (cf Eq. (7)) set an upper limit (of the order N) for $|k|$. However, as long as $a \gg N^{-2}$, which should always be the case for water-in-oil microemulsions, k values approaching this limit do not contribute significantly to the integral, and Eq. (24) should be valid.

In the strong coupling regime, where $a \gg 1$ and the charge fluctuations are small, the sum in Eq. (22) may be truncated after $k = \pm 1$ and we obtain with Eq. (21)

$$\langle z^2 \rangle = 2\exp(-\nu^2 a) . \tag{25}$$

In this regime, $\langle z^2 \rangle$ thus depends strongly on the charge number ν. For example, with a droplet core radius of 3 nm and a typical Bjerrum length λ_B = 30 nm, $\langle z^2 \rangle$ and the conductivity (cf Eq. (5)) are reduced by more than six orders of magnitude in going from a 1:1 electrolyte to a 2:2 electrolyte.

Comparison with experiment

We shall now confront the theory with the experimental conductivity data reported by EBD [4] for the AOT-isooctane-water microemulsion phase. In this system, the droplet size can be varied by changing the water/AOT molar ratio x. If all AOT molecules reside at the interface of spherical droplets, it follows that the core diameter σ_c is a linear funtion of x. For the purpose of converting the experimental x data to σ_c values, to be used in Eq. (20) for calculating $\langle z^2 \rangle$, we shall adopt the empirical relation

$$\sigma_c/\mathrm{nm} = 1.2 + 0.32x \,, \tag{26}$$

which closely reproduces the experimental droplet dimensions determined by small-angle neutron and x-ray scattering and by fluorescence quenching [1, 2, 10]. The droplet diameter is then obtained as

$$\sigma_d/\mathrm{nm} = \sigma_c/\mathrm{nm} + 2\delta \,, \tag{27}$$

where δ is the effective thickness (cf above) of the shell of surfactant tails. For AOT, we take δ = 0.6 nm [11]. At the experimental temperature*), T = 303 K, isooctane has a relative permittivity [12] ε_s = 1.928, corresponding to a Bjerrum length λ_B = 28.60 nm, and a viscosity [13] η = 0.45 cP.

Our theoretical prediction for the specific conductivity is shown as the solid curve in Fig. 1. The agreement with the experimental data [4] is quantitative for large droplets and qualitative for small droplets. Most importantly, the observed sharp conductivity maximum is theoretically accounted for. The physical origin of this maximum is as follows. For small droplets ($\sigma_c \ll \lambda_B$), the self-energy is much larger than the thermal energy. Consequently, nearly all droplets are electroneutral and the conductivty is small. As the droplets grow in size, the self-energy becomes less prohibitive and the conductivity rises sharply (cf the exponential dependence in Eq. (25)). For still larger droplets, the increase in $\langle z^2 \rangle$ is not sufficiently strong to compete with the reduction of the number of charge carriers (at fixed droplet volume fraction) and the increasing hydrodynamic friction, which reduce the conductivity by factors σ_d^{-3} and σ_d^{-1}, respectively. As a result, the conductivity passes through a maximum and then decreases roughly as the inverse droplet volume (cf Eqs. (5), (20), and (24)).

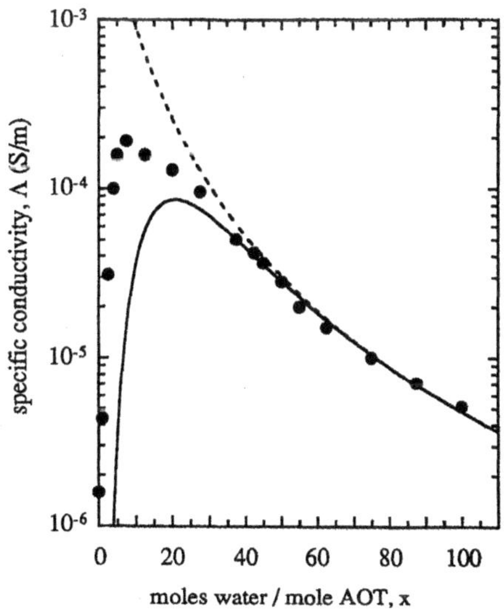

Fig. 1. Specific conductivity, $\Lambda = \kappa/\varphi_d$, of the AOT-isooctane-$H_2O$ microemulsion phase at 303 K versus the H_2O/AOT molar ratio, x. The points are experimental data ($\varphi_d \approx$ 0.05—0.15) obtained by Eicke et al. [4], the solid curve is our theoretical prediction, and the dashed curve corresponds to the weak coupling limit of the theory

Discussion

Since the physical model used here is essentially the same as that proposed by EBD [4], it may appear surprizing that EBD do not predict the conductivity maximum. However, the difference between the two treatments lies in the statistical-mechanical approximations invoked. While we consider the free energy of a macroscopic microemulsion sample, EBD regard a single droplet as a thermo-

*) The experimental Λ-data at the three largest x values, which were obtained at slightly lower temperatures, were corrected to T = 303 K using the experimental temperature dependence of ε_s and η [12, 13] and the proportionality $\Lambda \sim T\sigma_s/\eta$, which follows from Eqs. (5), (10), (20) and (24).

dynamic system. They then apply thermodynamic fluctuation theory to a grand canonical ensemble of such systems. Since the ion occupation numbers and, hence, the net droplet charge z, are regarded as thermodynamic variables in such an approach, one implicitly assumes that they can be varied continuously. This is a valid approximation under conditions where the droplets are highly charged, i.e., in the weak coupling regime. Indeed, in this regime, our result, Eq. (24), coincides with that of EBD, as illustrated by the dashed curve in Fig. 1. However, in the strong coupling regime, the EBD theory fails spectacularly: the 100-fold reduction of the conductivity for small droplets is not predicted, nor is the extreme sensitivity to the charge number of the electrolyte in this regime (cf above).

A further difference between our treatment and that of EBD is that the latter set $\sigma_c = \sigma_d$ in their model. Since the aqueous core and the hydrocarbon shell have widely different permittivities, this simplification severely overestimates the charge fluctuations and the conductivity for small droplets. Furthermore, EBD identifies σ_d with the hydrodynamic droplet diameter obtained from the (collective) diffusion coefficient determined by dynamic light scattering. Since charged droplets are involved in long-range interactions (cf below), such a procedure may overestimate σ_d. We have, therefore, chosen to take σ_c from experiments that are free from this complication, and then to obtain σ_d using known molecular dimensions.

Even though the conductivity maximum is predicted by our theory, there is a substantial quantitative disagreement for small droplets. Several factors are likely to contribute to this discrepancy [9]: i) We have taken no account of droplet interactions in our model. Both Coulomb interactions and induction forces (ion-induced dipole) are expected to be important under the experimental conditions of Fig. 1. These interactions stabilize charged droplets and, hence, increase the conductivity. ii) The conductivity increases if the droplets can adopt nonspherical shapes, since the effect of reduced self-energy is much larger than the effect of increased friction. In fact, recent spin relaxation studies [14] indicate that small droplets are distinctly nonspherical. The same result is expected if charged droplets aggregate (without fusing) with uncharged ones. iii) Droplet size polydispersity will also affect the charge fluctuations and may, for realistic size distributions [15], raise the conductivity for small droplets. iv) We have only presented a theory for the limiting conductivity. However, as the experimental data refer to droplet volume fractions of about 5—15%, one may be quite far from the infinite-dilution limit. Thus, we expect the conductivity to be reduced by the relaxation field and electrophoretic effects, particularly for the larger, more highly charged droplets.

In a forthcoming publication [9], we will show that inclusion of the refinements i)—iv) into the theory improves the agreement with experiment for small droplets and that the *quantitative* success of the present simple model for large droplets (cf Fig. 1) is partly fortuitous.

While there can be little doubt that the physical basis of microemulsion conductivity at low volume fraction is essentially captured in the charge fluctuation model proposed by EBD and elaborated here, the situation at higher volume fractions is more complex [16—19]. So far, only phenomenological treatments of this regime have been presented [16, 17]. It appears likely, however, that charge fluctuations will play an important role in a microscopic theory that accounts for the behavior of ionic microemulsions at large volume fractions.

Acknowledgements

Helpful discussions with Bengt Jönsson and Gunnar Karlström are gratefully acknowledged. This work was supported by the Swedish Natural Sciences Research Council.

Note added in proof

After the completion of this work, a closely similar analysis was published by Hall [20]. Using a slightly different approach, he arrives at an experession for the limiting conductivity which is identical to ours (Eqs. (5), (18)—(21)), except that, following EBD, he sets $\sigma_c = \sigma_d$. As a result, Hall's calculated conductivity *exceeds* the experimental one for small droplets. Hall's analysis is based on the assumption that the chemical potential of a droplet with net charge ze is given by (in our notation)

$$\mu(z) = \mu_0 + \mu_S(z) + k_B T \ln n(z) \,, \tag{28}$$

which is entirely equivalent to the free energy functional in Eq. (17). It should be noted, however, that although Eq. (28) has an intuitive appeal, it is not generally valid. This is easily seen by considering the limit $\mu_S = 0$, where Eq. (28) predicts that $n(z)$ is independent of z, which, considering the combinatorial factors, cannot be correct. As seen from our derivation of the entropy of mixing, Eq. (28) is valid only if the droplet valence is small compared to the total number of ions in the droplet, a condition that is fulfilled in the microemulsion systems under discussion.

In another recent contribution [21], Kallay and Chittofrati argue that the EBD theory can, in fact, account for the conductivity maximum, provided that σ_d is allowed to exceed σ_c. By using the σ_d values quoted by EBD and setting $\delta = 2.1$ nm in Eq. (27), these authors were able to reproduce the experimental Λ maximum. However, this δ value is unphysically large and leads to the absurd conclusion that Λ increases with droplet size only as long as the entire droplet core contains no more than a fraction of a water molecule! In any event, this modified EBD theory is open to a decisive experimental test, since it does not, like our theory, predict a dependence of Λ on the electrolyte charge number.

References

1. Fletcher PDI, Howe AM, Robinson BH (1987) J Chem Soc, Faraday Trans I 83:985
2. Lang J, Jada A, Malliaris A (1988) J Phys Chem 92:1946
3. Carlström G, Halle B (1988) Molec Phys 64:659
4. Eicke H-F, Borkovec M, Das-Gupta B (1989) J Phys Chem 93:314
5. AOT = Aerosol OT = sodium bis (2-ethylhexyl) sulfosuccinate, isooctane = 2,2,4-trimethyl pentane
6. Levich VG (1962) "Physicochemical Hydrodynamics", Prentice-Hall, Englewood Cliffs, ch 8
7. Batchelor GK (1967) "An Introduction to Fluid Dynamics", Cambridge University Press, Cambridge, ch 4.9
8. Doi M, Edwards SF (1986) "The Theory of Polymer Dynamics", Clarendon, Oxford, ch 4
9. Halle B, to be published
10. Luisi PL, Giomini M, Pileni MP, Robinson BH (1988) Biochim Biophys Acta 947:209
11. Kotlarchyk M, Huang JS, Chen S-H (1985) J Phys Chem 89:4382
12. Smyth CP, Stoops WN (1928) J Am Chem Soc 50:1883
13. Geist JM, Cannon MR (1946) Ind Eng Chem, Anal Ed 18:611
14. Carlström G, Halle B (1989) J Phys Chem 93:3287
15. Kotlarchyk M, Stephens RB, Huang JS (1988) J Phys Chem 92:1533
16. Laguës M, Sauterey C (1980) J Phys Chem 84:3503
17. Hilfiker R, Eicke H-F (1987) J Chem Soc, Faraday Trans I 83:1621
18. Howe AM, Mc Donald JA, Robinson BH (1987) J Chem Soc, Faraday Trans I 83:1007
19. Jada A, Lang J, Zana R (1989) J Phys Chem 93:10
20. Hall DG (1990) J Phys Chem 94:429
21. Kallay N, Chittofrati A (1990) J Phys Chem 94:4755

Author's address:

B. Halle
Physical Chemistry 1
Chemical Center
P.O. Box 124
22100 Lund, Sweden

Progress in Colloid & Polymer Science Progr Colloid Polym Sci 82:218—228 (1990)

Structure of inverse micelles and emulsion-gels with fluorinated nonionic surfactants. A small-angle neutron scattering study

J. C. Ravey and M. J. Stébé

Laboratoire de Physico-Chimie des Colloides, Lesoc, UA CNRS 406
University Nancy I, Vandoeuvre, France

Abstract: Viscoelastic, transparent, and amorphous materials can be obtained from certain mixtures of fluorinated hydrophobic nonionic surfactants, fluorocarbon, and large amounts of water. They may be termed as water-in (water-in-oil microemulsion) emulsions, and their structure has been investigated by small-angle neutron scattering (SANS), by making use of the contrast variation method. One important parameter is the hydration rate of the oxyethylene chains of the surfactant, as revealed by the high sensitivity of the neutron spectra towards the internal organization inside the micelles. Whether the size and/or the number concentration of these micelles are the important parameters governing the stability of these systems is a question naturally suggested by these recent findings.

Key words: Fluorinated surfactants; nonionic surfactants; gels; inverted micelles; neutron scattering; phase diagram

Introduction

By mixing oil, water, and nonionic surfactants, highly viscous and elastic systems can be obtained under certain experimental conditions (hydrophobicity of the surfactant, overall composition, temperature, etc.). Most of these so-called gels generally contain high amounts of surfactant and may be optically anisotropic; they are actually lyotropic phases (most generally lamellar) where the surfactant chains are in some frozen (crystalline or solid state) conformation. Such systems are true phases from the thermodynamic point of view.

However, when the surfactant content is very low (typically less than a few percent w/w), other viscoelastic systems can also be obtained; and, at least when they are not too opaque, they are found to be optically isotropic. They are not cubic lyotropic mesophases, since they exhibit no x-ray diffraction pattern characteristic of such a group of crystallinity, and since their surfactant content is fairly low. As a matter of fact, they constitute a peculiar type of emulsion where the volume fraction of the dispersed (internal) phase is so large that their aspect and consistency is that of gels. They are also termed as HIPRE, or high internal-phase-ratio emulsions [1—4]. Until now, the most considered "concentrated emulsions" have been oil-in-water emulsions, in view of their application in cosmetics, etc. [5, 6]. However, data concerning their stability and their actual structure remain rather sparse.

As far as water-in-oil concentrated emulsions are concerned, studies were practically nonexistent until recently. The existence of *transparent* W/O gels has been reported recently for a few hydrophobic *hydrogenated* ethoxylated alcohols [7, 8]. Although they are apparently stable over some range of (high) water content, they are transparent only for the highest water concentrations (about 98—99%).

However, viscoelastic and transparent materials have been obtained in our laboratory by certain mixtures of *fluorinated* nonionic surfactants and *fluorocarbons* which can accomodate water contents from about 50% to 98% [9]. As a general rule for any HIPRE, the larger the water content, the more viscous and more elastic these systems; their rheological behavior is that of a plastic flow with a yield value. But what is most striking is their perfect optical transparency in the whole range of composition. A macrostructure of the W/O type was readily suggested by electrical conductivity measurements, their conductivity being as low as 0.1 μS/cm for brine (0.5% NaCl) contents in the range of 60—80% [10]. The presence of concentrated emulsion globules has been confirmed by optical microscope observations, and also by electron microscope investigations, by using a freeze-etching technique [11]. One example of micrography is given in Fig. 1, where the diphasic state is clearly apparent. No other microstructure, globular nor anisometric (e.g., lamellar phase) can be distinguished.

As far as the stability of these gels is concerned, we have noticed that it may remain very good over long periods of time; for well chosen temperatures practically no demixing can be noticed for several months. Since that stability cannot be generally achieved by a simple and mild mixing, at least for very high water content (water must be added as the last component under extensive shaking [9]), these gels can hardly be termed as thermodynamically stable. However, their metastability being so remarkable, we could wonder whether it results from the presence of peculiar surfactant microstructures. Indeed, it is often argued that lamellar liquid crystals could serve as stabilizers for emulsion systems, e.g., by coating the dispersed droplets with a membrane made of this lamellar phase, or by the formation of a three-dimensional network of "plane" multilayers separating the (water) bubbles [12—14].

To tentatively answer that question, small-angle x-ray scattering and SANS experiments have been initiated. Both the x-ray and neutron spectra indeed suggest the presence of such microstructures. And, we have shown in a preliminary study that thick, multilayered domains constituting a three-dimensional network can be excluded [15]. Similarly, although they cannot totally be ruled out, we got strong evidence that thin, multilayered membranes around the water droplets cannot account for the peculiar shapes of the whole set of the experimental spectra.

The present paper must be considered as a mere introduction to these interesting systems. It will be shown how the microstructures present in these gels are certainly related to the phase behavior of the ternary mixtures (surfactant, oil, and water). Therefore, we present essential features concerning the phase diagrams of one typical system. Thereafter, we shall describe sets of neutron spectra; a few of them will be presented qualitatively, while others will be analyzed in a quantitative way.

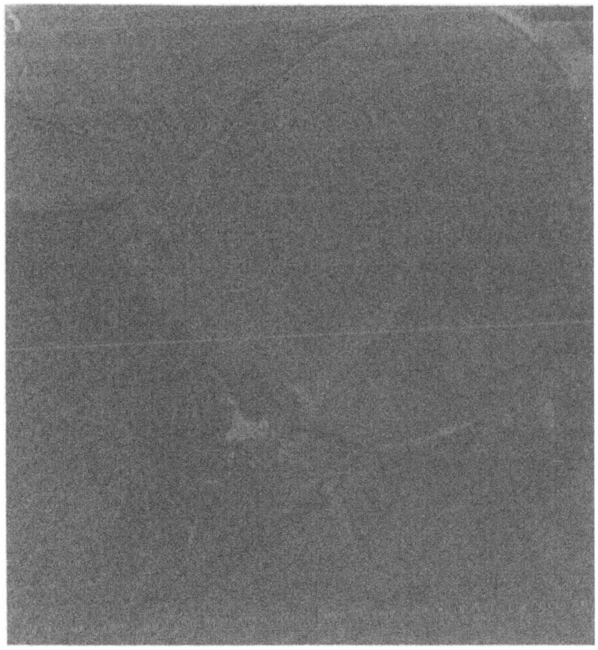

Fig. 1. Electron microscopic photography by the freeze-etching technique of a gel made of 70% water. The bar represents 1 μ

Experimental

Materials

Several series of fluorinated surfactants allow for the formation of the gels which have been synthesized according to an original method [16]. But the molecule of interest in the present paper is noted C_6F_{13}-C_2H_4-Σ-E_2, where E stands for OC_2H_4 and Σ represents SC_2H_4. The introduction of the Σ group makes easier the chemical synthesis of the highly pure amphiphile, which was purified by vacuum distillation.

The fluorinated oil is mainly perfluorodecalin, purchased from Ventron GmbH. A few data will also refer to other fluorocarbons: $C_4F_9CH{=}CHC_4F_9$ (noted C4) and $C_8F_{17}CH{=}CH_2$ (noted C8); they were generously supplied by Atochem (France) and were used as received. Note that all these oils contain the same number of carbon atoms; they differ essentially in their bulkiness, polarity, and free volume in the liquid state.

Neutron scattering

A part of the SANS measurements were performed at the Laue-Langevin Institute, Grenoble, France, using the

D17 camera. Others have also been carried out on the PACE instrument at the Laboratoire Léon Brillouin (CEA-CNRS, Saclay, France).

The experimental spectra were interpreted by using a best-fit procedure [17, 18], making use of the variation contrast method. For that purpose, theoretical spectra for model aggregates were calculated, taking into account the oil/water penetration into the surfactant palisade, the molecular characteristics (molar volumes, chain conformation, etc.), the interparticle effects (based on the Percus-Yevick solution for the equivalent hard, sticky spheres [19]), and all the orientations of the actual anisometric scatterers. Of course, the experimental data have to be evaluated on an absolute scale by using standard methods (use of the transmission of the samples and of water, and use of the scattering spectrum of water).

Results and discussion

Phase behavior of the ternary fluorinated systems

The gels are obtained only for systems whose HLB temperature is noticeably lower than the temperature for which they are intended to be formed. This can be achieved, for example, by using relatively hydrophobic surfactant, shorter fluorcarbon, etc. In that case, the micellar solubility of the amphiphile is quite low or nonexistent. More specifically, the binary aqueous phase diagram of the surfactant of interest in the present paper is shown in Fig. 2. There is no classical L_1 isotropic phase, and this dilute surfactant always appears "water insoluble", except for a very narrow salient region which is prominent from the strip that exists, whatever the temperature, for very high surfactant content; this domain is classically noted L_3 (or eventually keeps the L_2 denomination), and the corresponding structures are large, interconnected bilayers, i.e., made of two monolayers whose spontaneous mean curvature is toward water [20, 21]. The temperature of emergence of this phase is about 14 °C. In the whole region below L_2/L_3, the surfactant molecules aggregate into lamellar L_α phase, in equilibrium with monomers in water (W phase). Such a phase behavior is then typical of the strong tendency of such a surfactant to form large lamellar aggregates: bilayers in L_3, disordered stacks of multilayers in L_2, and organized multilayers in L_α.

Ternary phase diagrams of the surfactant-perfluorodecalin-water system are shown in Fig. 3 [22]. As expected, the diagrams are typical of systems well above their HLB temperature, limited to L_2 (and L_α) regions of restricted extent that shrink when temperature increases. For higher oil contents at 20 °C, water may be incorporated in the oil-surfactant mixture with a maximum of two water molecules per oxygen atom; and this limiting water/surfactant ratio markedly decreases with temperature. Incidentally, let us note that at 20 °C the L_2 phase is connected to the so-called L_3 phase of the binary water-surfactant system. The structures corresponding to this salient region are thought to be large, oil-swollen interconnected bilayers.

In the diagram are also represented very thin sector-like regions extending towards the water corner. They precisely correspond to the overall compositions for which gels of highest stability can be obtained. Therefore, as the most striking result concerning the formulation of these nonionic gels, we note that, for a given temperature, such systems can be formed only over a narrow range of oil/surfactant ratios. Thus, we can define an "optimum" mean oil/surfactant ratio ρ, which *decreases when temperature increases*. In other words, for a given composition defined by a certain oil/surfactant ratio, the gels can be built only for temperatures in the vicinity of an optimum temperature, noted T_{opt}.

When temperature departs from T_{opt} to a sufficient extent we can notice some "demixing" phenomenon, i.e., a layer of an oil phase at the bottom and/or traces of water on the top of the sample. The change of ρ with T is shown in Fig. 4, for the present system (R_fE_2-perfluorodecalin-H_2O): when temperature (T_{opt}) increases from 10° to 60 °C, ρ (expressed in w/w) decreases from 4.5 to 2. The values of T_{opt} depend on the chemical nature of both the surfactant and the oil: they are very different where perfluorodecalin, C4- or C8-oils are concerned, as it will be reported elsewhere [25].

But, quite surprisingly, this ratio also markedly depends on the *isotopic nature* of the water, as shown by the curves of Fig. 4, respectively, for pure H_2O and D_2O. As a first approximation, there is a nearly constant 20 °C-shift of T_{opt} when H_2O is replaced by D_2O, at fixed oil/surfactant ratio. Certainly, such an effect has to be ascribed to the difference in the hydrogen bonding strength of oxythylene groups with, respectively, H_2O and D_2O molecules in surfactant films or other (unknown) microstructures. Nevertheless, one may wonder whether such a slight difference in the H-bonding

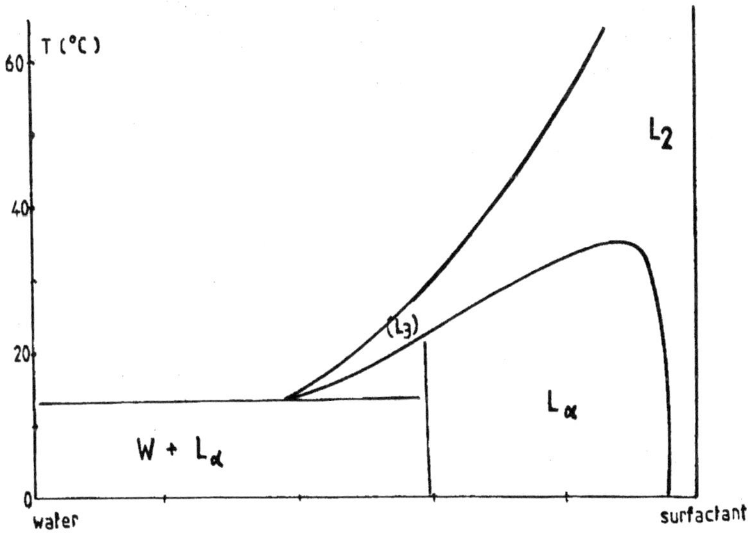

Fig. 2. Aqueous binary phase diagram of the fluorinated surfactant R_f-E_2. The isotropic phase L_2 may also be noted L_3 in its water-rich part, by reference to the bilayer structures present in this system

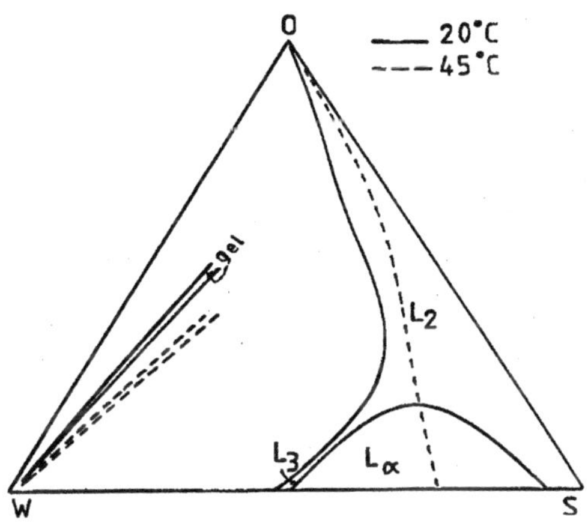

Fig. 3. Ternary phase diagram of the water-perfluorodecalin and the surfactant (*S*) whose aqueous behavior is described in Fig. 2. The sector-like domains of high stability transparent gels are also represented for 20 °C (——) and 45 °C (– – –)

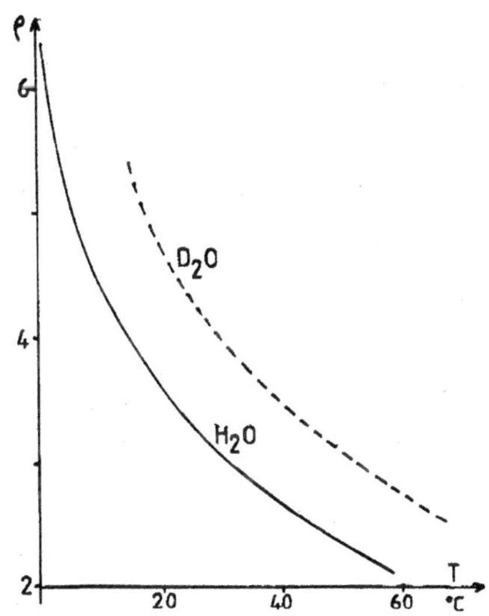

Fig. 4. Variation of the optimum oil/surfactant ratio with temperature giving rise to stable transparent gels (for the system described in Figs. 2—3) when water is pure H_2O (——) or pure D_2O (– – –)

strength effectively leads to such a noticeable difference of behavior.

Small-angle neutron spectra

The range of q-values was 0.007—0.3 Å^{-1} [$q = (4\pi/\lambda)\sin(\theta/2)$, λ = wavelength, θ = scattering angle] on the PACE instrument, and 0.01—0.25 Å^{-1} on the D17 spectrometer. To make use of the contrast variation method, gels were built from various D_2O/H_2O mixtures. As explained above, to each isotopic composition of the water (χ) and/or a given value of the oil/surfactant ratio, it will correspond a different temperature (T_{opt}) for the gels to exist. (The temperature shift is proportional to the

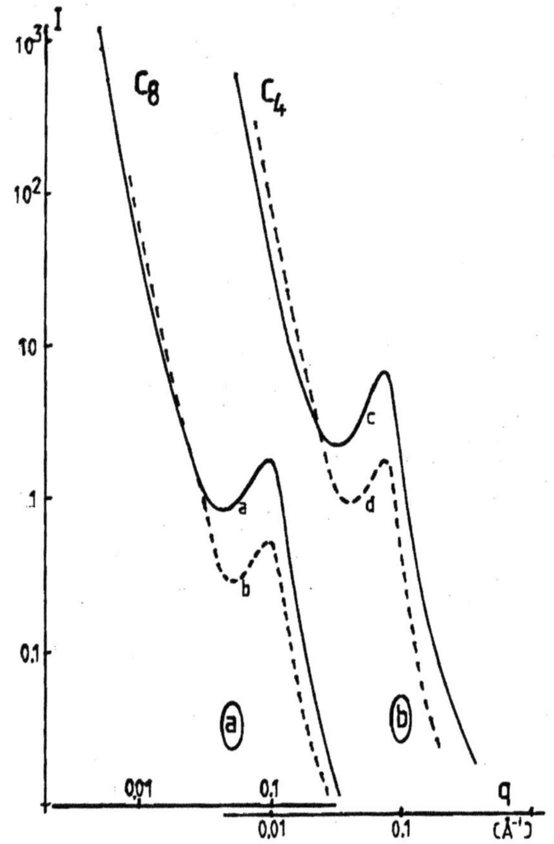

Fig. 5. Experimental neutron spectra for systems described in Table 1. a = C8-oil, 57% H_2O; b = C8-oil, 84% H_2O; c = C4-oil, 61% H_2O; d = C4-oil, 86% H_2O

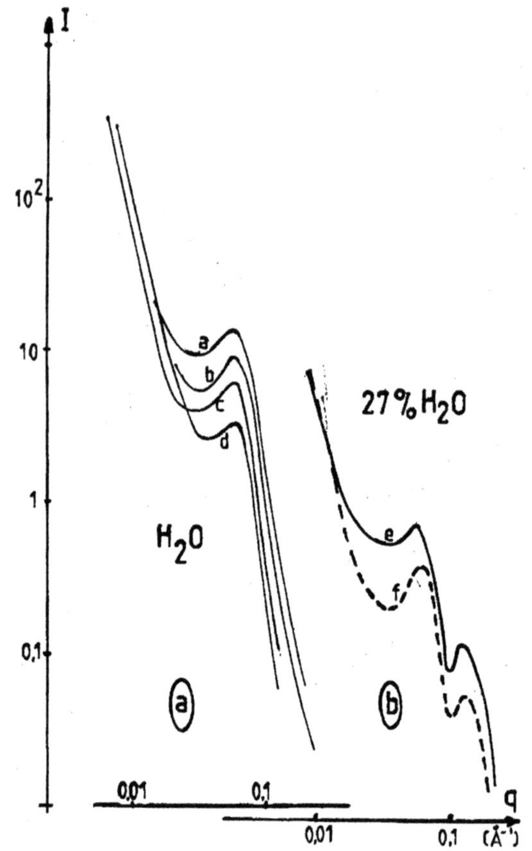

Fig. 6. Experimental neutron spectra with perfluorodecalin; a) corresponds to pure H_2O for the water component; a = 52%, b = 66%, c = 76%, d = 87% water, ρ = 3.8, T = 20 °C; b) corresponds to the isotopic composition 27% H_2O; e = 50%, f = 75% water, ρ = 4.5, T = 30 °C

D_2O content in the water component (i.e., to χ). Therefore, it can be anticipated that, according to that composition (χ) and for a given value of ρ (hence at a different T_{opt}), the microstructures present in the gels will be different from each other, unless there is an exact compensation effect due to the change in T_{opt}.

Most of the data presented here are relative to the R_fE_2-perfluorodecalin system, whose typical ρ ratio is between 2 and 4.5 w/w for temperatures between 10° and 60 °C. The results will concern samples with overall water volume fraction φ in the range 0.5—0.88. However, a few experimental spectra for samples made of the C4- and C8- (fluorinated) oils are also shown in Figs. 5a—5b, each for two values of φ. They can be compared to the spectra of the perfluorodecalin system, shown in Fig. 6a for four water concentrations φ. Data in Figs. 5 and 6a are relative to H_2O as water component. For those of Fig. 6b the isotopic composition χ is 27% H_2O w/w.

Let us emphasize that, for such an isotopic composition, both the water and the oil have an identical scattering length density; hence, in that case, the scattered intensity mainly results from "surfactant aggregates". Once again, let us recall that these data concern gels obtained at different T_{opt} temperatures. The corresponding values of ρ and T_{opt} for these systems are reported in Table 1.

As a general rule (i.e., except for χ close to 0.27), all the spectra exhibit a q^{-4} dependence for the smaller scattering angles ($7 \cdot 10^{-3} < q < 1.5\ 10^{-2}$ Å^{-1}) to which are superimposed series of broad minima and maxima whose amplitudes markedly depend on the χ value. Clearly, both the location (noted q_1) and the amplitude of the first maximum in spectra of Figs. 5—6a depend on the chemical nature of the oil: the larger the q_1, the lower the

Table 1. Corresponding values for ρ and T_{opt}

T_{opt} = 20°C, water = H_2O oil	% water (w/w)	$1 - \varphi$	φ_s	$R(1 - \varphi)$	$R(\varphi_s)$	$R(I)$
Perfluorodecalin	52 (*)	0.33	0.086	1	1	1
ρ = 3.8	65.5	0.23	0.067	0.70	0.71	0.70
	76	0.15	0.037	0.45	0.43	0.44
	87	0.078	0.018	0.24	0.21	0.23
C4-oil	61	0.29	0.12	3.2	3.3	3.6
ρ = 1.7	86 (*)	0.09	0.036	1	1	1
C8-oil	57	0.32	0.12	3.2	3.2	3.2
ρ = 2.1	84 (*)	0.10	0.037	1	1	1

Ratio of the scattered intensities at larger q, $R(I)$, compared to the ratios of $(1 - \varphi)$ and φ_s, φ and φ_s being respectively the volume fraction of the water and of the surfactant in gel-systems made of three different fluorinated oils. (*) designates the reference sample for the calculation of the ratios. ρ is the optimum oil/surfactant ratio at 20°C

amplitude of the peak; q_1 is twice as large for the C8-oil as for perfluorodecalin.

As a first set of observations, we can note that for a given system:

1) the location of the maxima (q_1, q_2) are independent of the amount of water (φ);
2) the intensity scattered at low q increases with this water content, i.e., it decreases with the surfactant concentration;
3) at higher q, it is almost proportional to $(1 - \varphi)$, and also proportional to the volume fraction of the surfactant (φ_s). This can be seen in Table 1 (values of $1 - \varphi$, φ_s, and ratio of the intensities scattered for any $q \geqslant q_1$).

Therefore, these observations suggest the spectra actually result from the superimposition of two components: the one at low q, related to the amount of water emulsified in the gel, and the one at larger q, which has to be correlated to surfactant microstructures. Now, since it can be shown that the large-q part of the spectra cannot be explained except by eventual, more or less thin, multilayered membranes around the water droplets, they must result from other surfactant aggregate structures.

As shown by the phase diagrams of Fig. 3, the binary oil-surfactant mixtures can incorporate a few percent water, probably by forming water-swollen micelles. Therefore, another possible explanation of the peculiar shape of the neutron spectra could be the presence of these inverted micelles (or microemulsions?) into the gels. Hence, in order to investigate the validity of such an hypothesis, we have also performed a structural determination of these micelles, that is, for systems containing only a few percent water, but characterized by the same oil/surfactant ratio as the gels at their optimal temperature T_{opt}.

Microstructures and macrostructures

A few experimental spectra are shown in Fig. 7 for the micellar samples at 13°C whose oil/surfactant ratio is 3.9. At this low temperature, relatively large water amounts can be incorporated (more than 10%). For the sample with 10% water, $\chi = 0.22$; the other data correspond to $\chi = 0$ (pure D_2O). For all these spectra the intensities may be safely extrapolated to $q = 0$, leading to a satisfactory determination of the size of the micelles, at least if interparticle effects are correctly taken into account. Quite interestingly, the larger the water (D_2O) content the lower the intensity at $q = 0$, although it is known that the water solubilization should induce a marked swelling and size increase of the micelles (i.e., an increase of the aggregation number N), without noticeable change of the disperse volume fraction [18]. This a result of the contrast variation: although the particles are heterogeneous (e.g., viewed as concentric multishell structures), by addition of D_2O their *mean* scattering length density

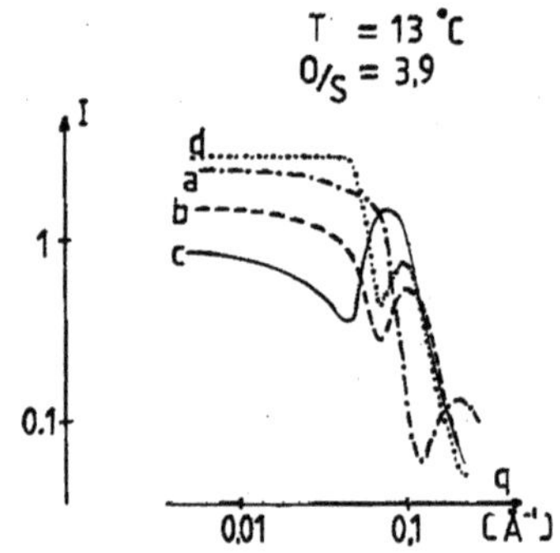

Fig. 7. Experimental neutron spectra for water swollen micelles, T = 13°C, ρ = 3.9; a = 3.6%, b = 5.9%, c = 8.8% D_2O; d = 10.4% water of isotopic composition χ = 0.22

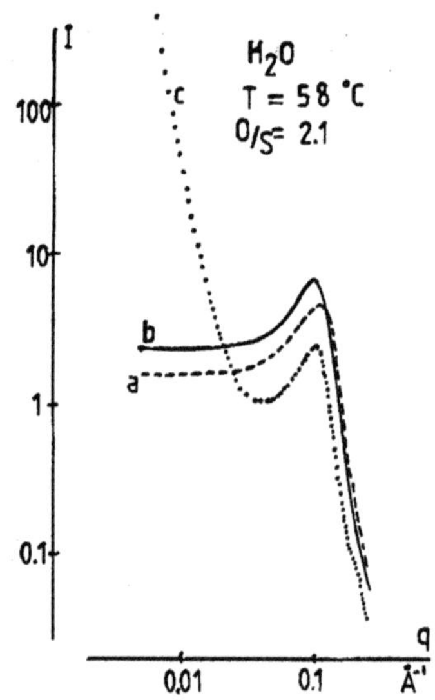

Fig. 8. Experimental neutron spectra for micelles (a = 3.5 and b = 4.5% H_2O) and a gel (c) containing 65% H_2O, with the same oil/surfactant ratio; ρ = 2.5, T = 58°C

changes in such a way that it tends to match that of the external medium (fluorocarbon). Correlatively, the spectra at larger q exhibit extrema which grow larger and larger and whose location shifts to smaller q. For example, for 3.6% D_2O the maximum is at $q \approx 0.2$ Å^{-1}; for the system with 9% water (D_2O) the intensity is maximum for $q \approx 0.08$ Å^{-1}, and the curve exhibits a well-pronounced minimum for $q \approx 0.04$ Å^{-1}. Obviously, such a behavior is not mainly due to interparticle effects in concentrated micellar phases, but to the peculiar value of the scattering length density of D_2O as compared to the one of the fluorinated oil.

However, one should not consider these interparticle interactions as negligible: they are simply (partially) concealed by the contrast effects. Indeed, if D_2O is replaced by pure H_2O, we get spectra like those of Fig. 8 (T = 58°C, oil/surfactant ratio = 2.1). In that case and for water contents of about 4%, we also observe a pronounced maximum for q = 0.09 Å^{-1} (to be compared to the corresponding value in Fig. 7 for D_2O), followed by a steep, but quite smooth decrease of the scattered intensity.

Quite interestingly, we note that for $q \approx 0.05$ Å^{-1} the spectra for these micelles lead to exactly the same spectra as for gels containing as much as 65% water, but with the exception of a constant multiplicative factor which is precisely equal to the ratio of the volume fractions of the surfactant (φ_s), respectively, in the gels and in the oil micellar phase. This result clearly suggests that the *continuous phase of the emulsion-gels is actually an inverted micellar phase (or microemulsion).*

Therefore, the analysis of the data related to the micelles must take into account the two types of parameters which have such a marked influence on the shape of the spectra, i.e., the interparticle effects which depend on the volume fraction of the dispersed micelles and the description, to a semi-molecular scale, of the aggregates: the mean size of the particles and of the (eventual) water core, the mean shape of a monodisperse micelle which could represent, at best, the experimental data (i.e., the mean curvature of the surfactant palisade, for example, a sphere-like or lamella-like aggregate, etc.), the most probable conformation of the surfactant chains and the penetration degree of water (oil) into that surfactant film. Of course, all the parameters are intercorrelated through geometrical constraints that are resolved in modeling the water-swollen micelles. As a matter of fact, the present spectra essentially depend on three independent "molecular" parameters: the mean aggregation number, the thickness of the hydrophilic part of the surfactant palisade and the degree of penetration of water into it.

As suggested by the relatively complex shape of the spectra in Figs. 7—8, which is highly dependent of the water incorporation into the micelles, the scattering intensities are very sensitive to the internal molecular organization inside the micelles. As a result, although a satisfactory fitting of data to spectra of model particles is not a trivial task, a consistent set of values for the various morphological parameters can be derived as a function of the isotopic nature of the water component, the water content, and the oil/surfactant ratio.

In brief, let us emphasize that the first maximum at small q (when it exists) results from interparticle effects; the other extrema (at larger q) depend on the internal morphology of the micelles. A detailed presentation and analysis of this structural investigation is out of the scope of the present paper. However, the main points may be stressed as follows.

— According to the maximum water content (i.e., to the demixing temperature) the largest aggregation number N of the swollen micelles is in the range 200—800.

— The maximum size of the water core is typically 30—50 Å. Its anisometry depends on the water content. Schematically, the lower the water/surfactant ratio the smaller and more anisometric the particle. Along the demixing lines (but only for the relatively large oil/surfactant ratios considered here), the swollen micelles may be considered as globular (i.e., nearly spherical).

— At a given temperature, the mean hydration rate (or room for water to penetrate the surfactant film) is about 1—2 water molecule per oxyethylene chain. As a striking result, we found a higher hydration rate when the water component is pure D_2O: certainly, this has to be related to the higher value of T_{opt} of the gels made of heavy water since, obviously, a light dehydration occurs when raising the temperature.

— The oxyethylene chain is in a rather extended conformation.

— The area per polar head (evaluated at the hydrophilic — hydrophobic interface within the surfactant film) is about 40 Å^2.

The part of the spectra in the low-q range can also be calculated, making the most likely assumption that it arises from the dispersion of the large water bubbles. The only important parameter to introduce is the specific area S_v. In fact, assuming the droplets are covered by a monolayer of surfactant, we also have to specify the morphology of the interfacial surfactant film. As a most obvious choice, we may suppose that the state of the surfactant molecule in this film and in the micelles must be similar: area per polar head of 40 Å^2, oxyethylene chain in its extended conformation inducing some water penetration along it, and hydrocarbon chain in a coiled conformation (no penetration of fluorinated oil into that hydrogenated part of the

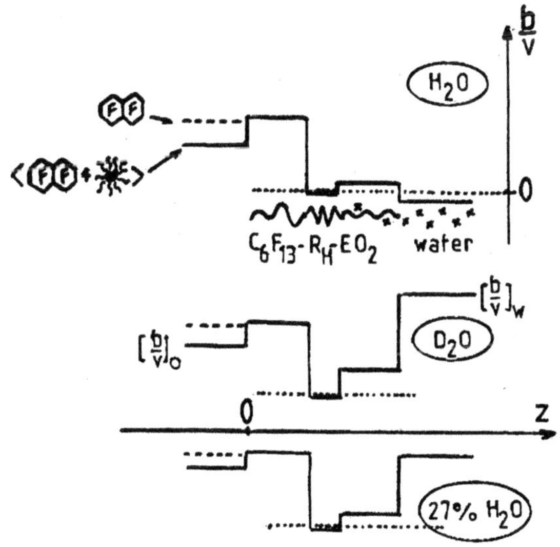

Fig. 9. Profiles of the scattering length density across the oil-water interface as a function of z, the distance from the oil phase surface, and for three isotopic compositions. $(b/v)_0$ is that of the pure oil (———) or the mean value for the oil micellar phase (——)

film). The conformation of the fluorinated part of the chain is practically irrelevant since both this chain segment and the perfluorodecalin have very similar neutron-scattering-length density (b/v).

The profiles of the scattering-length-density across the oil/water interface are shown in Fig. 9.

The calculations are performed by using the extended Porod's law, as derived by Auvray [23]:

$$q^4 I(q) = 2\pi S_v \{[(b/v)_0 - (b/v)_w]^2 + q^2 f_1 + \ldots\}$$

with:

$$f_1 = M_0^2 - 2\,[(b/v)_0 - (b/v)_w] M_1 \,,$$

$$M_n = \int z^n \cdot (b/v)_z \cdot dz \,.$$

$(b/v)_w$ is the scattering length density of the water and $(b/v)_0$ is that of the oil phase. Calculations for three isotopic compositions of the water have been performed; the resulting spectra for $S_v = 3\ \mu^{-1}$ are shown in Fig. 10. The radius R_e of the equivalent spherical droplet giving rise to the same specific area could be evaluated according to the relation:

$$S_v = 3\varphi_g / R_e \,,$$

φ_g being the volume fraction of the disperse globules.

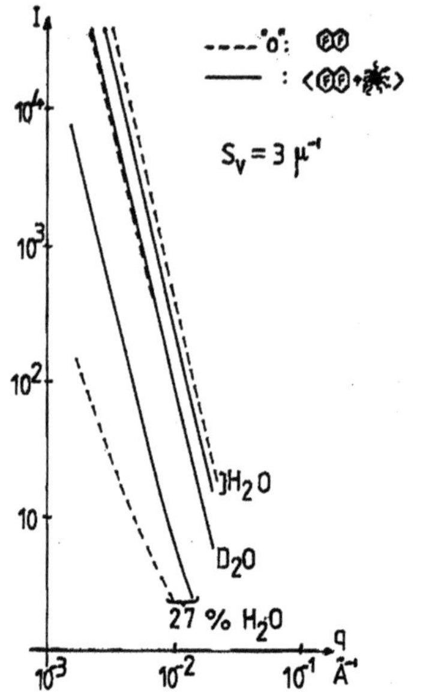

Fig. 10. Theoretical spectra at small q for dispersions of water globules according to the different values of $(b/v)_w$ and $(b/v)_0$ (—— for oil micellar phase and ––– for pure oil). The specific area is 3 μ^{-1}

A last question to be resolved concerns the evaluation of $(b/v)_0$. Two sets of calculations have been carried out: in the first one (dotted lines in Figs. 9—10) the external medium is assumed to be the pure oil; in the second one, a mean scattering density is used that takes into account the actual composition of the oil phase (i.e., the water swollen micellar phase):

$$(b/v)_0 = (\sum b_i)/(\sum v_i) ,$$

where i is the water, oil, and surfactant molecules present in this phase.

As a summary, a comparison of the data with the two sets of calculation shows that:

— the composite value of $(b/v)_0$ must be used to obtain coherent values of S_v as a function of the isotopic composition of the water;

— typical values of the apparent radius R_e are in the range 2—5 microns;

— S_v increases when φ increases at constant oil/surfactant ratio, as can be readily seen in the spectra of Figs. 5—6. (Indeed $I(q = 0.01)$ is proportional to S_v.) In other words, R_e decreases when more and more water is incorporated into the gel. This has been checked from optical microscope observations and can be understood as follows. Since the gels are stable for volume fractions φ_g larger than about 0.5, the systems are not far from close compact dispersions if spheres are rather monodisperse; the only way to introduce new water globules is to fill up the voids between the larger droplets with smaller ones:

the larger the (optimum) oil/surfactant ratio the smaller R_e, i.e., the larger the specific area S_v. This could explain why the gels cannot be formed for too large values of this ratio. A more complete presentation and discussion of the data related to the change of S_v with various parameters will be given elsewhere [25].

Conclusion

By superimposing the spectra of the two types of water globules (the large ones and the micelles), we can now fully understand the whole spectra of these emulsion-gels.

We have to determine the macrostructures and microstructures that simultaneously coexist in a given gel. Thanks to the studies separately carried out on the surfactant aggregates in the pure L_2 phase, a few morphological parameters may be considered as known quantities: the extended conformation of the oxyethylene chain and the interparticle attractive potential (typically —1.5 to —2 kT). The order of magnitude of the hydration rate of the oxyethylene groups is also known as a function of the temperature and the isotropic composition of the water; only small adjustments may be necessary for the best fitting.

Hence, the remaining parameters to be fitted are: the size of the water bubbles (about 5 microns), the aggregation number of the micelles, and the actual water content in the oil phase (5 to 10%). Indeed, there is no reason to believe, a priori, that the oil continous phase of the emulsion at a given temperature and a fixed oil/surfactant ratio must be indentical to the corresponding L_2 phase on its demixing boundary. Obviously, the procedure imposes that the sum of water and surfactant molecules in the water- and oil phases are equal to their overall content.

One example of curve fitting is shown in Fig. 11. It corresponds to samples with D_2O for which the oil/surfactant ratio is 4.1, and the overall water content is 60% w/w. Their optimum temperature

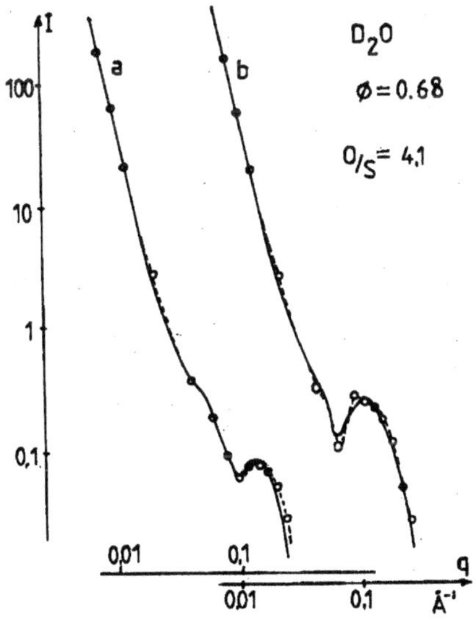

Fig. 11. Experimental spectra and theoretical (o—o—) best-fits for a gel made of D_2O; a = 25°C, b = 45°C (see text)

T_{opt} is 35°C. Although their optical transparency is no more perfect, such gels still exhibit a good stability at 25° and 45°C, temperatures for which the demixing is still negligible and does not prevent their structural determination to be performed.

Accordingly, the structural results for this gel may be summarized as follows. The size of the water bubbles R_e is about 2 microns, in accordance with the micrography (Fig. 1). For the micelles, the aggregation number N is 500, 350, 250, the mean diameter of the water core is 45, 35, 25 Å, and the water content of the external oil phase is 7.3, 6.8 and 5.3% when temperature is, respectively, 25°, 35°, and 45°C.

Qualitatively, these values are in agreement with the change in the location of the demixing line of the L_2 phase (see the phase diagrams in Fig. 3). However, a closer examination of these data (to be shown elsewhere [25]) would indicate that the external oil phase of the emulsion-gels actually contains 10-20% more water than the corresponding L_2 phases at their demixing boundary. Some data even suggest that the micellar *structures* themselves in the gels could be those present in the demixing L_2 samples, but at a *lower* temperature. In other words, the oil phase in the gels should be in some *surfusion state* as compared to the L_2 phase.

Much work remains to be done for a more explicit description of these interesting systems. Indeed, many correlations actually exist between the various phase and structural parameters, which will be investigated and/or published in the near future, including relations between the size of the water bubbles, the composition of the systems, the size and number density of the micelles, their hydration rate, the phase diagram delineations, and the isotopic composition of the water, etc.

At any rate, the driving parameter for the (meta) stability of these gels seems to be the optimum oil/surfactant ratio ρ, which is related to the temperature and to the structures and composition of the continuous oil phase.

Yet we have no explanation for the occurrence of this optimum ratio. However, in previous investigations on hydrogenated nonionics above the HLB temperature [18, 24], we had proposed a structural evolution from w/o globules to stacks of bilayers when increasing the surfactant concentration in the L_2 phase, as schematically represented in Fig. 12. And, for a certain (low) oil/surfactant ratio which depends on the temperature, one obtains the salient region (L_3) which extends toward the water corner or to the L_3 phase region of the binary water surfactant system (see Fig. 3). As already noted [20], the structures in such a phase are thought to be interconnected, oil-swollen surfactant bilayers made of monolayers whose spontaneous (low) curvature is towards water. Hence, we could suggest that our so-called optimum ratio would correspond to the onset of that tendency to form monolayers of low curvature towards water (water globules), induced by a tiny change in the surfactant packing in the interfacial film of the emulsion and micelles. As for the occurrence of the L_3 phase, this phenomenon (i.e., a slight decrease of the area per polar head or some contraction in the packing of the oxyethylene chains) could result from interparticle effects. But in the present case the liquid membrane in this "biliquid foam" would consist of a micellar phase, while in L_3 it consists of practically pure oil.

It is also interesting to note that the strong difference in the optical appearance of the gels in, respectively, hydrogenated and fluorinated systems probably results from some matching of the refractive index of the oil-surfactant mixture to that of the water when the oil is perfluorinated. Nevertheless, this matching phenomenon does not preclude the validity of our statements on the actual existence and stability of the gels since, from a *rheological* point of view (in default of the *optical* one), only one phase is observed, without noticeable demixing over a long period of time.

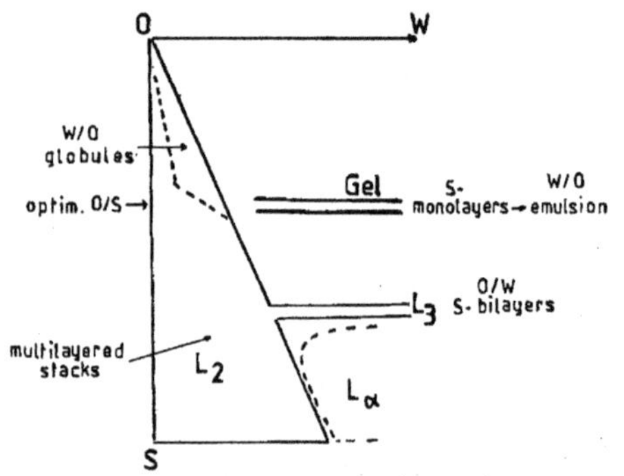

Fig. 12. Schematic localization of the various types of structures according to the oil/surfactant ratio for systems above the HLB temperature

Still, we must emphasize that the most perfectly transparent gels seem to be the most stable systems. Since the refractive indices are related to the Hamaker constant (A) of the materials (water and oil phases), their matching would lead to some equalization of their A constant; as a result, the attractive forces between water globules could be markedly reduced, preventing any coalescence phenomenon from taking place.

As a last suggestion, these transparent and viscoelastic systems could be termed water-in (water-in-oil microemulsion) emulsions.

References

1. Lissant KJ (1966) J Colloid Interface Sci 22:462
2. Lissant KJ (1970) J Soc Cosmetic Chemists 21:141
3. Princen HM (1983) J Colloid Interface Sci 91:160; ibid (1985) 105:150
4. Princen HM (1988) Langmuir 4:486
5. Attwood D, Florence AT in "Surfactant Systems", Chapman and Hall (1983) p 46
6. Sagitani H, Hattori T, Nabeta K, Nagai M (1983) Nippon Kagaku Kaishi: 1399
7. Solans C, Dominguez JG, Parra JL, Heuser J, Friberg SE (1988) Colloid Polym Sci 266:570
8. Kunieda H, Solans C, Shida N, Parra JL (1987) Colloids Surfaces 24:225
9. Delpuech JJ, Matos L, Moumni EM, Ravey JC, Selve C, Serratrice G, Stébé MJ, Cambon A, Thiollet G, French Patent 88-05229
10. Matos L, Thesis University Nancy I, France (1987)
11. The electron microgrpahies have been realized in the Biophysics Laboratory at Orléans (France) by M. Gervais
12. Barry BW (1975) Adv Colloid Interface Sci 5:37
13. Friberg SE, Mandell L, Fontell K (1969) Acta Chem Scand 23:1055
14. Groves MJ (1978) J Colloid Interface Sci 64:90
15. Ravey JC, Stébé MJ (1989) Physica B 156—157:394
16. Lampin JP, Cambon A, Szonyi F, Delpuech JJ, Serratrice G, Thiollet G, Lafosse L, French patent 8408376 (1985); European patent 844010451 (1985)
17. Ravey JC, Buzier M, Dupont G in "Microemulsions", Rosano H (ed) (1987) Marcel Dekker, New York, Surfactant Science Series 24:163
18. Ravey JC, Buzier M, Oberthur R (1987) Progr Colloid Polym Sci 73:113
19. Regnaut C, Ravey JC (1989) J Chem Phys 91:1211
20. Anderson D, Wennerström H, Olsson U (1989) J Phys Chem 93:4243
21. Porte G, Marignan J, Bassereau P, May R (1988) J Phys (Fr) 49:511
22. Cambon A, Delpuech JJ, Matos L, Serratrice G, Szonyi F (1986) Bull Soc Chim 965
23. Auvray L, Cotton JP, Ober R, Taupin C (1984) J Physique 45:913
24. Ravey JC, Buzier M, in "Surfactants in Solution", Mittal KL (ed) (1979) New-Dehli, 8:117
25. Ravey SC, Stébé MJ, to be published

Authors' address:

Dr. J. C. Ravey
Universite de Nancy I
Faculte des Sciences
Laboratoire de Physico-Chimi des Colloides
B.P. 239
54506 Vandoeuvre les Nancy Cedex,
France

Effect of alcohols on micelle formation in aqueous solutions of ethoxylated nonyl-phenols

Zs. Bedő[1]), E. Berecz[1]), I. Lakatos[2]) and J. Lakatos-Szabó[2])

[1]) Department of Physical Chemistry, University of Miskolc, Technical University of Heavy Industry
[2]) Research Laboratory for Mining Chemistry, Hungarian Academy of Sciences, Miskolc-Egyetemváros, Hungary

Abstract: The micelle formation of ethoxylated nonyl-phenols in the presence of short-chain alcohols was investigated within a temperature range of 293—333 K. On the basis of the actual CMC values and their temperature dependence the thermodynamic functions of micelle formation (ΔH^0_m, ΔS^0_m, ΔG^0_m) and the thermodynamic functions of transfer of surfactant from pure aqueous phase to water-alcohol mixtures (ΔH^0_{tr}, ΔS^0_{tr}, ΔG^0_{tr}) were also calculated. According to the experimental results the set of optimum conditions leading to the formation of a micellar structure with satisfactory stability can be determined. In the three-component systems the more stable the structure of micelles formed, the higher the temperature, the longer the alcohol chain length, the lower the alcohol content, and the shorter the polyoxyethylene chain of monomers. With respect to the practical EOR process, it may be stated that the addition of alcohols to the solutions of short-chain nonyl-phenols makes it possible to extend their application to hydrocarbon reservoirs having high formation temperature.

Key words: Non-ionic surfactants; ethoxylated nonyl-phenols; critical micelle concentration (CMC); cloud point; thermodynamics of micelle formation

Introduction

The application of non-ionic surfactants in enhanced oil recovery (EOR) has recently attracted considerable interest owing to their favorable characteristics under reservoir conditions [1—7]. Although the structure of micellar solutions may play a fundamental role in the shaping of interfacial phenomena in oil/water systems, the microscopic displacement efficiency of aqueous non ionics is, however, also affected by the addition of organic solubilizing agents. In order to widen the spectrum of utilization of surfactants, our previous investigations, which dealt with the properties of aqueous micellar solutions of ethoxylated nonyl-phenols [8, 9], were extended to the determination of the effect of alcohols in the process of micellization.

Scope of investigations

The micelle formation and, consequently, the practical features of non-ionic surfactants in an aqueous medium can be influenced by the addition of different organic compounds, such as short-chain alcohols. The study of the effects of these additives on the behavior of micellar solutions offers a good opportunity for the clarification of the various factors governing the micellization and stabilization of micellar structure in mixed solvents.

The experiments comprised the determination of the change in critical micelle concentration (CMC) and the cloud point of the ethoxylated nonyl-phenols characterized by various ethoxy group numbers (n_{EO} = 8—30), in the presence of short-chain alcohols with a different number of carbon atoms (n_C), such as methyl alcohol (MA), ethyl alcohol (EA), and iso-propyl alcohol (IPA), within a temperature range of 293—333 K. From the temperature dependence of CMC, the thermodynamic functions of micelle formation (ΔH^0_m, ΔS^0_m, ΔG^0_m) and the thermodynamic functions of the transfer of surfactant from pure aqueous phase to water-alcohol mixtures (ΔH^0_{tr}, ΔS^0_{tr}, ΔG^0_{tr}) were also calculated. Finally, conclusions were drawn for the optimum parameters leading to the formation of micellar structure with a high stability.

Experimental

The ethoxylated nonyl-phenyols with uniform polyoxyethylene chains ($NPEO_8$, $NPEO_{10}$, $NPEO_{12}$, $NPEO_{15}$, $NPEO_{20}$, $NPEO_{30}$) were the products of Dow Chemicals Co. (USA). Other chemicals used as additives were analytical grade reagents.

The critical micelle concentration (designated (c_M) in the formulae and figures) was determined by measuring the surface tension (γ) within a temperature range of 293—333 K. The surface tension of micellar solutions to be investigated was measured using a Cahn Electro Balance with surface tension accessories. The cloud point of the micellar solutions which contained various amounts of additives (c_A) under constant surfactant concentration (c_S) was obtained by an optical method.

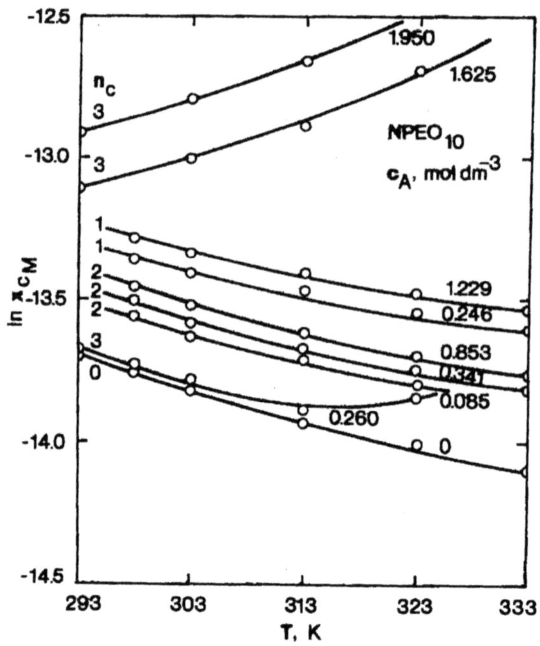

Fig. 1. Dependence of CMC on temperature and alcohol concentration for $NPEO_{10}$

Results and discussion

1. *Effect of alcohols and temperature on CMC*

The variation of the logarithm of c_M in mole fraction (x_{c_M}) vs temperature is shown for different alcohol concentrations and ethoxy group numbers in Figs.1 and 2.

As can be seen, the addition of all the investigated alcohols to aqueous solutions increases the CMC at constant temperature. That effect may be derived from changes in the solvent (water) properties. As a result of their favorable interaction with water, the structure of the solvents is modified, leading to the formation of a less polar medium and to the restriction of the micellization [10—12]. The observed CMC increasing effect of the investigated alcohols can be attributed to the enhanced solvent power of alcohol-water mixtures.

For the variation in the enthalpy of micelle formation (ΔH_m^0) represents no linear function of temperature, therefore the following approximation

$$\ln x_{c_M} = aT^{-2} + bT^{-1} \tag{1}$$

was fitted to the values of x_{c_M} obtained experimentally at various temperatures for most of the investigated systems.

But for the micellar solutions of $NPEO_{10}$ containing IPA as additive the increase in the CMC can be described by the following empirical relationship:

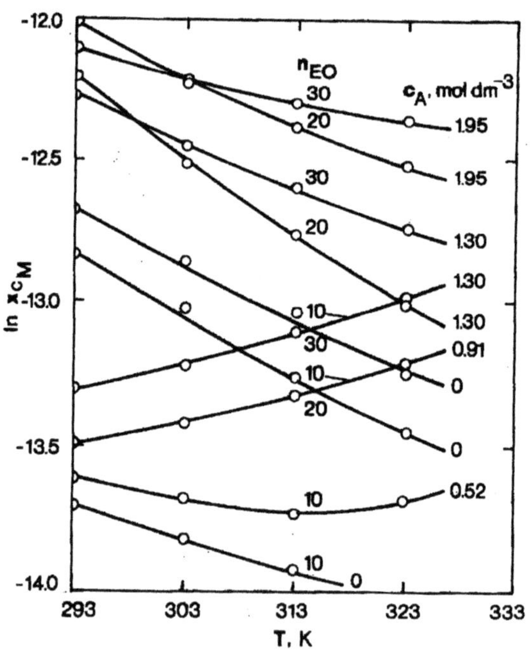

Fig. 2. Dependence of CMC on temperature and ethoxy group number using IPA as additive

$$\ln x_{cM} = aT - b\ln T \qquad (2)$$

above 0.70 mol dm^{-3} IPA content, while below this value it reaches a minimum at 313 K.

Above a given (0.70 mol dm^{-3}) IPA content, following the increase of temperature, the opposite trend of the $NPEO_{10}$-IPA system can be observed, and this is probably derived from the strenthening tendency of the solubilization process, concerned with the apolar group of monomer molecule and the mixed solvent.

Therefore the CMC-decreasing effect of temperature for the rest can be interpreted by the dehydration of the ethylene oxide chains, resulting in the solubility-decreasing influence of temperature.

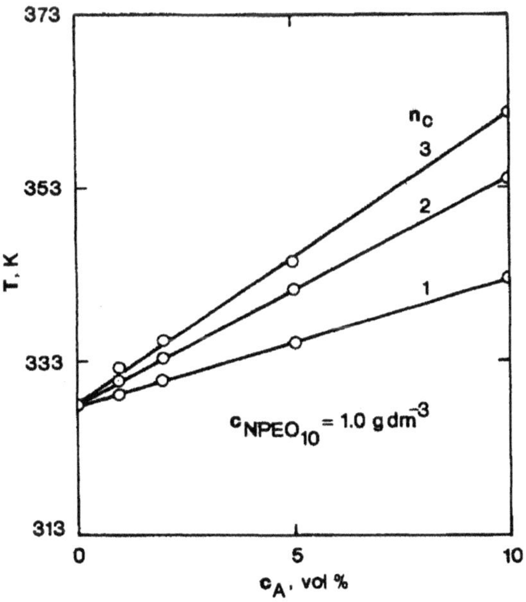

Fig. 4. Dependence of the cloud point on alcohol chain length

2. Effect of additives on cloud point

The cloud point of non-ionic surfactants in aqueous solutions is directly related to the hydrophilic character of the monomer, viz. the shorter the poly-oxyethylene chain in the surfactant molecule, the lower the cloud point. One of the most obvious macroscopic effects of alcohols on solubility of the non-ionic tensides is that the cloud point may be significantly increased by the addition of relatively small amounts of alcohol (<10 vol%).

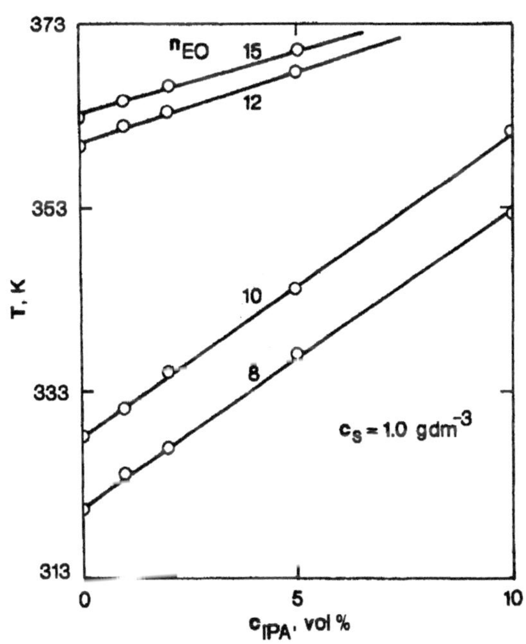

Fig. 3. Dependence of the cloud point on ethoxy group number using IPA as additive

The variation of the cloud points is dependent on the addition of alcohols for different ethoxy group numbers and alcohol chain lengths; this is shown in Figs. 3 and 4. As illustated, the greatest solubility enhancement was found for IPA; moreover, the lower the ethoxy group number, the greater the change in the cloud point.

3. Thermodynamic study of micelle formation

For non-ionic surfactants the variation in the standard free energy (ΔG_m^0) and enthalpy (ΔH_m^0) of micelle formation is given, under definite conditions [8], by the following relationships

$$\Delta G_m^0 = RT\ln x_{cM}\,, \qquad (3)$$

and

$$\Delta H_m^0 = -RT^2\left[\frac{d\ln x_{cM}}{dT}\right]p\,, \qquad (4)$$

where x_{cM} is the CMC in mole fraction.

Taking Eqs. (1), (2), and (4) into consideration, the standard enthalpy of micellization for the investigated systems can be calculated by the following formulas:

$$\Delta H_m^0 = 2aRT^{-1} + bR , \tag{5}$$

and

$$\Delta H_m^0 = bRT - a^2RT . \tag{6}$$

The dependence of enthalpy and entropy of micellization on temperature, alcohol chain length, alcohol concentration and ethoxy group number is shown in Figs. 5–8. As can be seen, at a given temperature, on the addition of a certain amount of methyl alcohol, ethyl alcohol, and iso-propyl alcohol to the aqueous solutions of surfactants, both the enthalpy (ΔH_m^0) and entropy (ΔS_m^0) of micelle formation decrease. The reduction of these thermodynamic functions is mainly due to the gradual cessation of the destruction of the ordered "iceberg" water structure around the hydrocarbon chains of alcohol during the micellization.

The fact that ΔH_m^0 and ΔS_m^0 become more positive with the decrease in temperature for all the investigated alcohols and surfactants indicates that,

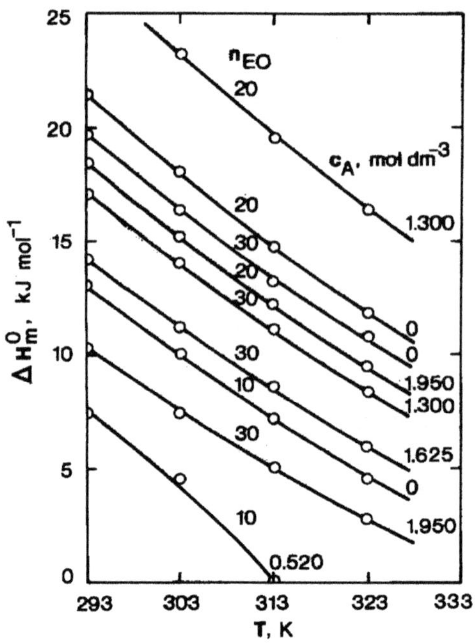

Fig. 6. Dependence of enthalpy of micelle formation on temperature and ethoxy group number using IPA as additive

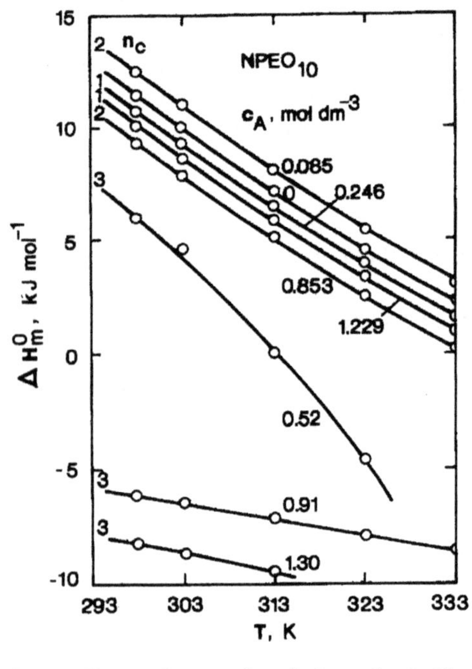

Fig. 5. Dependence of enthalpy of micelle formation on temperature and alcohol concentration for NPEO$_{10}$

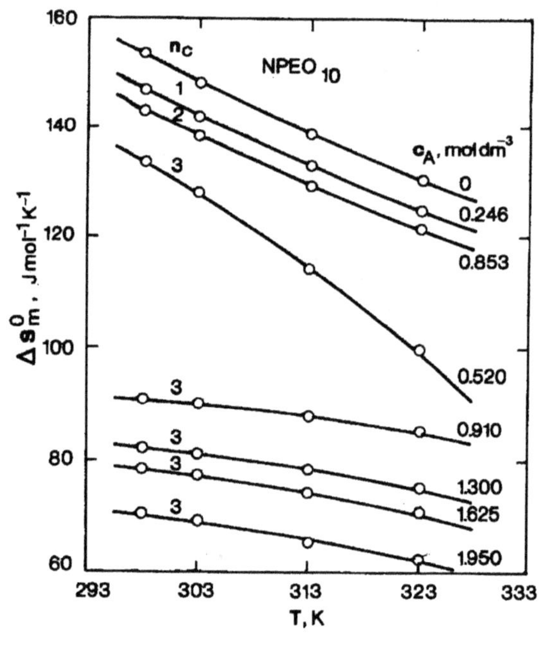

Fig. 7. Dependence of entropy of micelle formation on temperature and alcohol concentration for NPEO$_{10}$

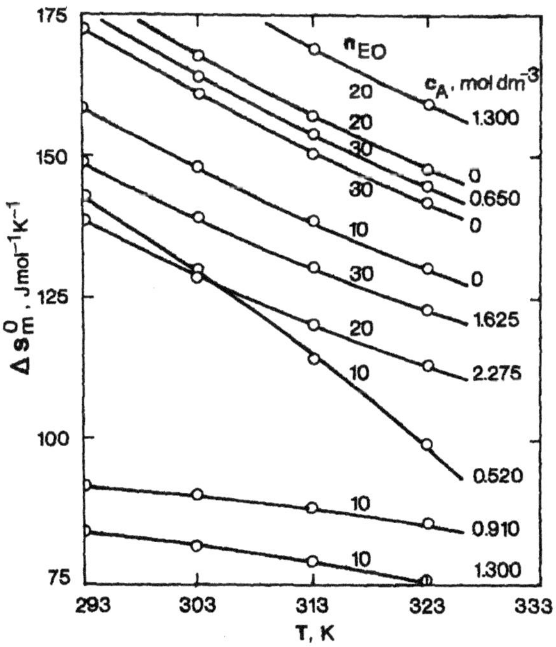

Fig. 8. Dependence of entropy of micelle formation on temperature and ethoxy group number using IPA as additive

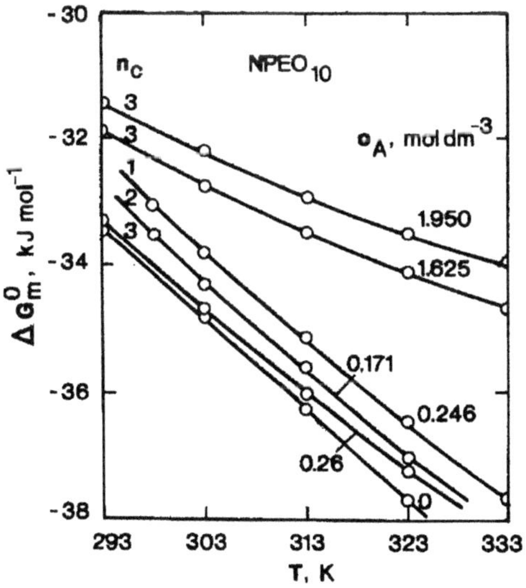

Fig. 9. Dependence of free energy of micelle formation on temperature and alcohol concentration for $NPEO_{10}$

at lower temperature, the added alcohols might barely affect the structure of water, consequently the penetration of alcohols into micelles exhibits a greater entropy dependence as was stated by Hayase et al. [13].

The thermodynamic study of micellization has shown that the micelle formation is governed by the dehydration of the poly-oxyethylene chains at low alcohol content in the solutions of $NPEO_{20}$ and $NPEO_{30}$, while the decreased hydrophobic interactions between the alcohol molecules and the micelles are predominant at increasing alcohol concentration in the systems of $NPEO_{10}$, $NPEO_{20}$, and $NPEO_{30}$. The conclusions drawn from the results of our study on thermodynamic behavior of the investigated systems have been confirmed by the publications dealing with the micellization of nonionic surfactants in the presence of short-chain alcohols [14].

The temperature-dependence of the standard free energy (ΔG^0_m) for different additive concentration, alcohol chain length, and ethoxy group number (Figs. 9 and 10) indicates that, in the micellar solutions investigated, the more stable the structure of micelles, the higher the temperature, the longer the alcohol chain length, the lower the alcohol content, and the shorter the poly-oxyethylene chain of the monomers.

The variation in the standard free energy of transfer of surfactant from pure aqueous phase to water-alcohol mixtures (ΔG^0_{tr}) is given by the following relationship:

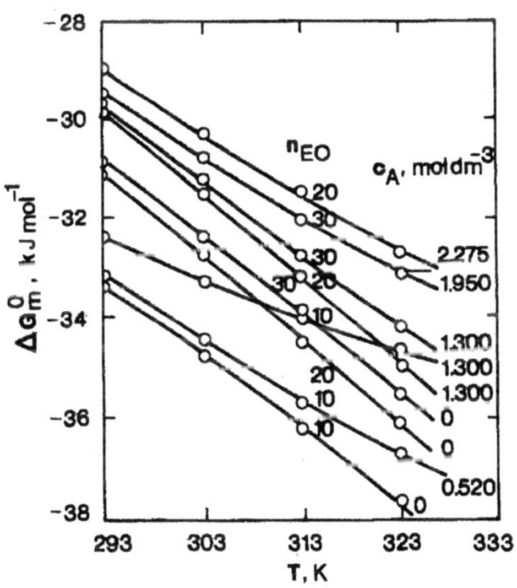

Fig. 10. Dependence of free energy of micelle formation on temperature and ethoxy group number using IPA as additive

$$\Delta G_{tr}^0 = RT \ln \frac{x_{c_{M,A}}}{x_{c_{M,W}}}, \tag{7}$$

where $x_{c_{M,A}}$ is the CMC in mole fraction in mixed solvent, and $x_{c_{M,W}}$ is the CMC in mole fraction in water.

The standard free energy of transfer is the most informative as regards the energy interchange of the structural modifications taking place during the micellization in mixed solvents.

The increase in the standard free energy of transfer ΔG_{tr}^0 with temperature (Fig. 11) is ascribed to the heat-induced structure-breaking effect of the additives near the hydrophobic co-sphere of the monomers [15, 16].

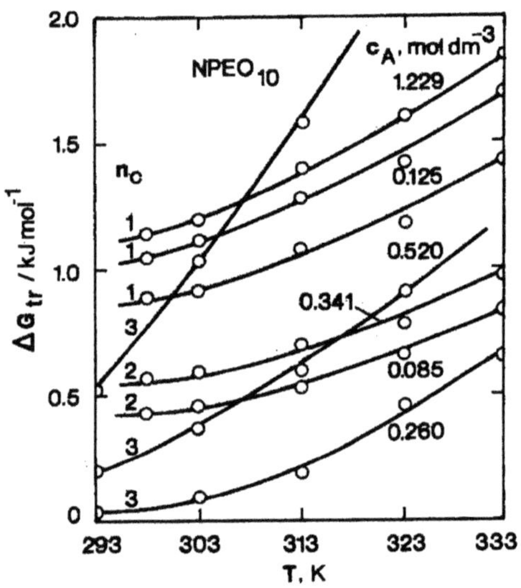

Fig. 11. Dependence of free energy of transfer on temperature and alcohol concentration for $NPEO_{10}$

The decrease in the standard free energy of transfer related to unity ethoxy group $\Delta G_{tr}^0/n_{EO}$ with poly-ethoxyethylene chain length of monomers (Fig. 12) suggests that, at a given temperature, the larger the standard free energy consumed by the formation of "swollen" micelles, the stronger the hydrophobic character of the surfactants. It may also be stated that the affinity of micelles towards alcohols is governed by the hydrophilic interactions between the surfactant's head-group and the alcohols's OH group, a suggestion that has also been made by De Lisi et al. [17].

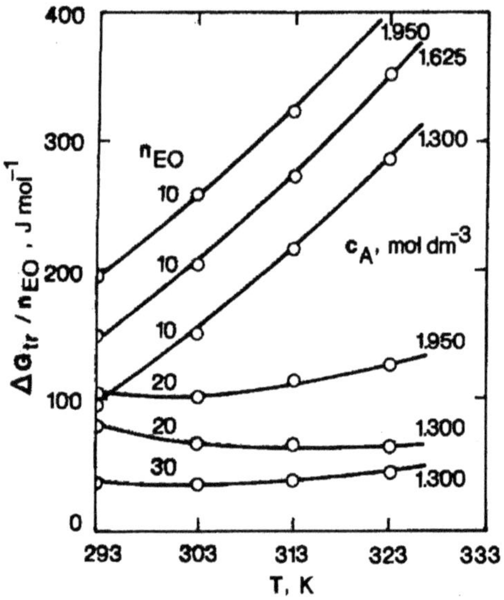

Fig. 12. Dependence of free energy of transfer for unity ethoxy group on temperature and ethoxy group number using IPA as additive

Conclusions

The investigation of micelle formation of ethoxylated nonyl-phenols in mixed solvents containing different short chain alcohols as a function of temperature (293—333 K) can be summarized as follows:

1) At constant temperature the investigated additives show a CMC-increasing effect which is attributed to the enhanced solvent power of these alcohol-water mixtures.

2) The logarithm of x_{c_m} dependent on the temperature generally decreases, while only for solutions of $NPEO_{10}$ in the presence of IPA it is an increasing function.

3) For constant ethoxy number (n_{EO} = 10) the greatest solubility enhancement was found for IPA; moreover, the greater the change in the cloud point caused by IPA, the lower the number of ethoxy groups in the monomer molecules.

4) In solutions containing the investigated additives, depending on the alcohol concentration and the ethoxy group number, the micelle formation is governed by the dehydration of poly-oxyethylene chains or by the decreased hydrophobic interactions.

5) In the three-component systems the following parameters favor the stability of the structure of

micelles: increase in temperature, increase in alcohol chain length, decrease in alcohol content, and decrease in poly-oxyethylene chain length.

6) At constant temperature the standard free energy consumed by the formation of "swollen" micelles is an increasing function of the strength of the surfactant's hydrophobic character.

As far as the practical EOR applications are concerned, it is extremely favorable that the addition of alcohols to solutions of nonyl-phenol ethoxylates with short poly-oxyethylene chains makes it possible to extend their application to hydrocarbon reservoirs having high formation temperature.

References

1. Trogus FJ, Sophany T, Schechter RS, Wade WH (1977) Soc Pet Eng J 17:337
2. Hayes Me, Bourrel M, El-Emary MM, Schechter RS, Wade WH (1979) Soc Pet Eng J 19:349
3. Ramirez WF, Shuler PJ, Freidman F (1980) Soc Pet Eng J 20:430
4. Ziegler WM, Handy LL (1981) Soc Pet Eng J 21:210
5. Capelle A, Littmann W (1982) Oil-gas European Magazine 1.53
6. Capelle A, Littmann W (1982) SPE Preprint 10601
7. Lakatos I, Milley Gy, Tóth J, Kristóf M, Capelle A, Wagner O, Lakatos-Szabó J, Felián B (1985) Acta Geodaet Geophys et Montanist 20:425
8. Bedő Zs, Berecz E, Lakatos I (1986) Colloid Polymer Sci 264:267
9. Bedő Zs, Berecz E, Lakatos I (1987) Colloid Polymer Sci 265:715
10. Nishikodo N, Moro Y, Uehara H, Matuura R (1974) Bull chem Soc Jpn 47:2634
11. Birdi KS, Backlund S, Sorensen K, Trag T, Dalsager S (1978) J Colloid Interface Sci 66:118
12. Rao IV, Ruckenstein E (1986) J Colloid Interface Sci 113:375
13. Hayase K, Hayano S, Tsubota H (1984) J Colloid Interface Sci 101:336
14. Ray A, Némethy G (1971) J Phys Chem 75:809
15. Perron G, Desrosiers N, Desnoyers JE (1976) Can J Chem 54:2163
16. Perron G, De Lisi R, Davidson I, Genereux S, Desnoyers JE (1981) J Colloid Interface Sci 79:432
17. De Lisi R, Milioto S (1985) J Solution Chem 17: 245

Authors' address:

Dr. Zsuzsa Bedő
University of Miskolc,
Technical University of Heavy Industry,
Department of Physical Chemistry
3515 Miskolc-Egyetemváros, Hungary

On the variation of microstructure within surfactant solutions

S. T. Hyde

Applied Mathematics Dept., Research School of Physical Sciences,
Australian National University, Canberra, Australia

Abstract: Experimental and theoretical studies of the microstructure of surfactant solutions within both microemulsion and cubic "single phase" regions of the phase diagram are producing clear evidence of first- and second-order phase transitions within these regions. We can understand this behaviour in terms of the competition between local and global constraints on the surfactant interfacial geometry. The local curvatures of the surfactant film are set by the preferred configurations of the surfactant chains. The global constraint on the film geometry — the surface-to-volume ratio — is set by the composition of the solution and the preferred head-group area. These constraints allow us to predict a schematic pseudo-binary phase diagram. The sequence of microstructures is shown to be high topology monolayers (sponge-like) transforming to lower topology, less intraconnected monolayers, through cylinders and spheres. Upon further dilution, the interface consists of a bilayer, whose topology increases upon addition of the internal medium. This model is consistent with standard phase diagrams, with the exception of the hexagonal phase. We suggest that this phase actually consists of a periodic minimal surface of hexagonal symmetry. A series of "mesh" structures are also expected to exist: these may well correspond to the tetragonal and rhombohedral crystalline phases occasionally observed.

Key words: Surfactant solutions; microemulsions; cubic phases; curvature

Introduction

It is a long established view that surfactants can aggregate in solution to form classical Euclidean geometric structures; *viz.* spheres (micellar phases), cylinders (hexagonal phases) and planes (lamellar phases). More recently, it has been suggested that surfactant interfaces may adopt more exotic geometries. In particular, the microstructure of *bicontinuous* surfactant phases — where both the oil and water domains are intra-connected — is presently under scrutiny. Such bicontinuous structures have been inferred from small-angle scattering studies, freeze-fracture microscopy and nuclear magnetic resonance (NMR) investigations. They have been found to exist within cubic phases of surfactant solutions [1—5, 7, 11, 15] as well as certain microemulsions [6, 8—10].

Analyses of the cubic phases reported so far indicate that such phases consist of bilayers of surfactant, lying on periodic minimal surfaces of cubic symmetry [4, 5, 11]. Studies of ternary microemulsions formed by the cationic quarternary ammonium surfactant didodecyldimethyl ammonium bromide (DDAB) in the presence of water and short-chained olefins, have led to the suggestion that in these systems the surfactant forms a monolayer. The geometry of the monolayer resembles a set of spheres arranged randomly in space, connected by cylinders [8—10]. It is apparent that the topology of the monolayer varies continuously throughout the single-phase microemulsion region of the ternary phase diagram [6, 8, 9]. Recent small-angle x-ray scattering and calorimetric studies of the cubic phase of DDAB, water and cyclohexane mixtures have revealed that here too the microstructure changes within a so-called single-phase region of the ternary phase diagram [12, 13], confirming the suggestion of Fontell and Jansson [14]. This polymorphism within a cubic phase has also been observed within a binary lipid-water solution [4, 15].

These observations call into question the standard procedure for determing phase diagrams of surfactant solutions, and challenge our accepted notions of a single phase. In what follows we attempt to analyse the reasons for this polymorphism. We suggest that the microstructure adopted within a surfactant system is that which resolves the often conflicting requirements set by local and global constraints. Local constraints determine the microstructure of the surfactant interface at a molecular scale, and are due to the preferred conformation of surfactant molecules. Global constraints determine the overall interfacial geometry and topology (roughly speaking, the embedding of the surface in our space); these constraints are set by the composition of the solution.

Analysis of geometric constraints

The local constraints on the interfacial geometry and topology refer to the preferred dimensions of the surfactant molecule (head-group area, tail length, etc.). The determination of local constraints can be carried out ab initio from statistical mechanics [16], however, we resort instead to the semiempirical "effective surfactant parameter", defined by the hydrophobic chain volume, plus any oil absorbed between the surfactant chains, v divided by the product of the head-group area a and the chain length l [17].

For the purpose of the present analysis, we adopt the simplifying assumption that these dimensions are set by the surfactant molecule alone, and do not vary significantly with surfactant concentration. We shall consider the situation where the surfactant is preferentially curved towards the polar regions of the mixture-equivalent to constraining the effective surfactant parameter to be greater than one. This constraint corresponds to Type-II structures, as defined by Luzzati et al., for which the observed generic phase behaviour upon water dilution is reverse micelles: reverse cylinders: lamellae [18].

Under these assumptions, we can derive estimates of the microstructural variation throughout the phase diagram. The geometric constraints can be mathematically expressed in terms of the curvature at the surfactant interface.

It has been established elsewhere [9, 11] that the effective surfactant parameter is related to the mean curvature H and Gaussian curvature K of the interface by:

$$\left(\frac{v}{al}\right)_{\text{eff}} = 1 + Hl + \frac{Kl^2}{3}\,. \tag{1}$$

It is crucial to appreciate that — with the exception of planes, cylinders, and spheres — these curvatures vary over an arbitrary surface. In order to account for this variation, we replace the (pointwise defined) mean and Gaussian curvatures by their surface-averaged values:

$$\langle H\rangle = \frac{\iint_{\text{surface}} H\,da}{\text{surface area}} \quad \text{and} \quad \langle K\rangle = \frac{\iint_{\text{surface}} K\,da}{\text{surface area}}\,. \tag{2}$$

If we take these surface-averaged values of the interfacial curvatures, the righthand side of Eq. (1) yields the average value of the effective surfactant parameter throughout the interface. Alternatively, if the value of the effective surfactant parameter is known, the (surface-averaged) curvature of any interfacial structure formed by the surfactant mixture must obey Eq. (1). This quantifies the local packing constraint of the interface.

The global constraint — defined by the surface to volume ratio of the mixture — is less readily quantifed. In general, it is impossible to derive a general relation between the relative volumes on either side of an interface and the interfacial curvatures. (For example, it is easy to see that partitioning of space into two equal volume fractions can be achieved by parallel planes, spheres, cylinders or minimal surfaces.) Later, we derive a link between global constraints and the surface geometry for bicontinuous structures.

We can exhaust all interfacial geometries by classifying interfaces according to their topology. We quantify the topology of an interface by the Euler-Poincaré characteristic χ, which is related to the genus of the (oriented) surface g by

$$\chi = 2 - 2 \cdot g\,. \tag{3}$$

The Euler-Poincaré characteristic is related to the integral curvature by the Gauss-Bonnet theorem:

$$\iint_{\text{surface}} K da = 2\pi\chi \ . \qquad (4)$$

This topological indexing of interfacial structures means that all structures which are related by stretching or bending (without gluing or tearing) are contained within the same topological class. (Hence the description of topology as "rubber-sheet geometry".) For example, spheres and ellipsoids are topologically equivalent. The resulting topological inventory of surface structure is given in Table 1. We now seek a description of the variation of the interfacial geometry with composition in terms of the topological description of surface structures.

Table 1. A catalogue of surface structure according to surface topology

Topology: genus	Euler characteristic	Surfaces	Surface properties
0	2	spheres ellipsoids	closed surfaces
1	0	cylinders Delaunay surfaces	open along one axis
2	−2	"mesh" surfaces	open along a 2-d net of tunnels
3, higher	−4, lower	"strut" surfaces	bicontinuous 3-d tunnel network

An ideal binary phase diagram at constant temperature

For simplicity, we define the composition of a surfactant mixture in terms of the "internal volume fraction" (Φ_{int}). Since we are dealing with surfactants whose effective surfactant parameter is larger than one, this volume fraction refers to the volume fraction of water (plus polar head-groups). Thus, here water dilution refers to increasing internal volume fraction. The variation in interfacial geometry as a function of the internal volume fraction corresponds to a pseudobinary phase diagram of a surfactant-water mixture.

The situation for low internal volume fractions has been calculated extensively elsewhere [8, 9] as a model for the microstructure of ternary cationic microemulsions. The results can be summarised as follows:

No uniform interface can be formed until the internal volume fraction is large enough to contain the surfactant head-groups without undue crowding (essentially water of hydration). Once this volume fraction is attained, the surfactant interface forms a surface of high (positive) mean curvature — tightly curved towards the polar volume. From Eq. (1), we see that this implies a large negative Gaussian curvature, in order to achieve the requisite effective surfactant parameter. This in turn implies that the Euler-Poincaré characteristic is a large negative number (high genus); so the interfacial topology is large (from Eq. (4)), giving a highly interconnected surface. In short, at low polar volume fractions, the ideal microstructure is bicontinuous, very bicontinuous.

As the internal volume fraction is increased, the mean curvature decreases in magnitude, curving less thightly towards the polar regions. In order to compensate for this change, whilst maintaining the effective surfactant parameter at its preferred value, the Gaussian curvature must become less negative. As a consequence, the genus *decreases* smoothly to zero. During this dilution, the interface transforms from a highly interconnected sponge morphology (genus two and up), through "baubled spaghetti" (genus one), to closed polar globules and finally spheres (both genus zero). The composition at which spheres (reverse micelles) form is clearly a function of the effective surfactant parameter. To lowest order it is easy to show from geometrical considerations that this internal volume fraction is given by

$$\Phi_{int} \approx \frac{1}{\frac{3v}{al}\left[\frac{v}{al} - 1\right] + 1} \ , \qquad (5)$$

where v refers to the molecular volume of the paraffin moiety of the surfactant solution.

For typical double-chained surfactants (which have a surfactant parameter greater than one), reverse micelles are expected to form at polar volume fractions between about 20% and 50%.

Now, as more water is added to the mixture, further increasing the internal volume fraction, this decrease of interfacial topology cannot proceed further: the nature of Euclidean three-space prohibits lower topologies.

The resolution of this apparent problem is simple: the surfactant interface at higher internal volume fractions becomes a *bilayer*. We describe the bilayer structure by the geometry (symmetry) and topology of the interface at the centre of the bilayer. Since the bilayer must exhibit symmetry with respect to both sides of this central surface (assuming a "monodisperse" surfactant distribution within the mixture), we expect the mean curvature of this surface to vanish; even though the mean curvature at the surfactant head-groups in each monolayer is nonzero. Thus, minimal surfaces — which are defined to be surfaces of vanishing mean curvature — are the natural structures for symmetric bilayers.

The local and global demands on a bilayer structure can be quantitatively expressed as follows. We assume that the bilayer structure is crystalline. This assumption is made at this stage in order to simplify the analysis. We conjecture the validity of this assumption later on. For a translationally periodic surface, a useful index of the surface is the normalized surface to volume ratio

$$\sigma = \frac{S}{V^{2/3}} = \frac{S}{c^{2/3}a^2}, \tag{6}$$

where S is the surface area per unit cell, v is the unit cell volume and a is the lattice parameter of the unit cell. (c is a constant which is related to the symmetry of the lattice; e.g. if the surface is cubic, c is equal to one.) From parallel surface theory, it can be shown that the lattice parameter is related to the surfactant head-group area per unit cell (A) and the bilayer topology per unit cell, χ by [11]

$$a^2 = \frac{A}{2\sigma}\left(1 - \frac{4\pi\chi l^2}{A}\right). \tag{7}$$

Combining this equation with Eq. (4), we get

$$\frac{\sigma^3}{\chi} = \frac{\pi(1 - 2\langle K\rangle l^2)}{\langle K\rangle}\left(\frac{S}{V}\right)^2. \tag{8}$$

The interfacial area at the centre of the bilayer (which is equal to that of the periodic minimal surface), S is related to the head group area by the following formula [11]:

$$S = \frac{A}{2}(1 + \langle K\rangle l^2)^{-1} \approx \frac{A}{2}(1 - \langle K\rangle l^2). \tag{9}$$

Now the surface-to-volume ratio S/V is related to the internal volume fraction of the surfactant mixture:

$$(1 - \Phi_{\text{int}}) = \frac{n_{\text{paraffin}} v_{\text{paraffin}}}{n_{\text{paraffin}} v_{\text{paraffin}} + n_{\text{polar}} v_{\text{polar}}} = \left(\frac{v_{\text{paraffin}}}{a}\right)\frac{A}{V}, \tag{10}$$

(where n and v denote the number of polar and paraffin molecules per unit cell and the paraffin/polar volumes per molecule respectively).

Combining these equations gives

$$\sqrt{\frac{\sigma^3}{-\chi}} \approx \left\{\sqrt{\frac{\pi(1 - 2\langle K\rangle l^2)}{-\langle K\rangle l^2}} \cdot \frac{al(1 - \langle K\rangle l^2)}{2v}\right\} \times (1 - \Phi_{\text{int}}). \tag{11}$$

Note that the surface averaged value of the Gaussian curvature is approximately related to the effective surfactant parameter by [11]

$$\langle K\rangle \approx \frac{3}{2l^2}\left(1 - \frac{v}{al}\right). \tag{12}$$

Equation (11) relating the bilayer geometry (σ) and topology (χ) to the composition of the surfactant mixture is very powerful. The complicated bracketed expression on the righthand side is constant (assuming the effective surfactant parameter does not vary) and the critical index of the bilayer interfacial structure is the simple ratio on the lefthand side [19]. The value of this index for various periodic minimal surfaces is given in Table 2 in order of decreasing magnitude (expected to occur at successively higher polar volume fractions). The microstructure of many bicontinuous cubic phases consists of genus three (and, possibly higher) minimal surfaces. But the analysis presented here suggests that other microstructures can also form in the vicinity of these cubic phases.

The behaviour of this index is clear from the limited number of examples known to date. The topology per unit cell of the minimal surface generally increases as the magnitude of the index decreases. This means that the topology of an ideal bilayer structure must *increase* with increasing water content. Unfortunately, we do not know the actual

Table 2. Indices for some periodic minimal surfaces. We expect the bilayer interface to form these structures upon polar/paraffin dilution for surfactant interfaces curved preferentially towards the polar/paraffin moiety

Periodic minimal surface	Symmetry	Genus per unit cell	$\{\sigma^3/-\chi\}^{1/2}$
H-surface	hexagonal	3	?
CLP-surface	tetragonal	3	1.9430
Gyroid	cubic	3	1.9213
D-surface	cubic	3	1.8796
I-WP surface	cubic	4	1.8612
P-surface	cubic	3	1.7956
Neovius surface	cubic	9	1.6643
F-RD surface	cubic	6	1.6487

values of this index for mesh surfaces (Table 1). However, it is clear that they must be larger than those for genus three surfaces (struct surfaces). Thus, bilayers forming mesh structures (parallel layers of mesh), should form at intermediate composition to those which yield reverse micelles and those which yield standard cubic phases. Further, the indices for tetragonal and hexagonal surfaces (the so-called CLP-surface and H-surface of Schwarz [20]) are expected to lie between those for mesh surfaces and the cubic genus three structures.

This theoretical phase behaviour is drawn schematically in Fig. 1.

We now summarise the deliberations. Assuming only adherence to the dual demands placed upon an interface by the local surfactant geometry and

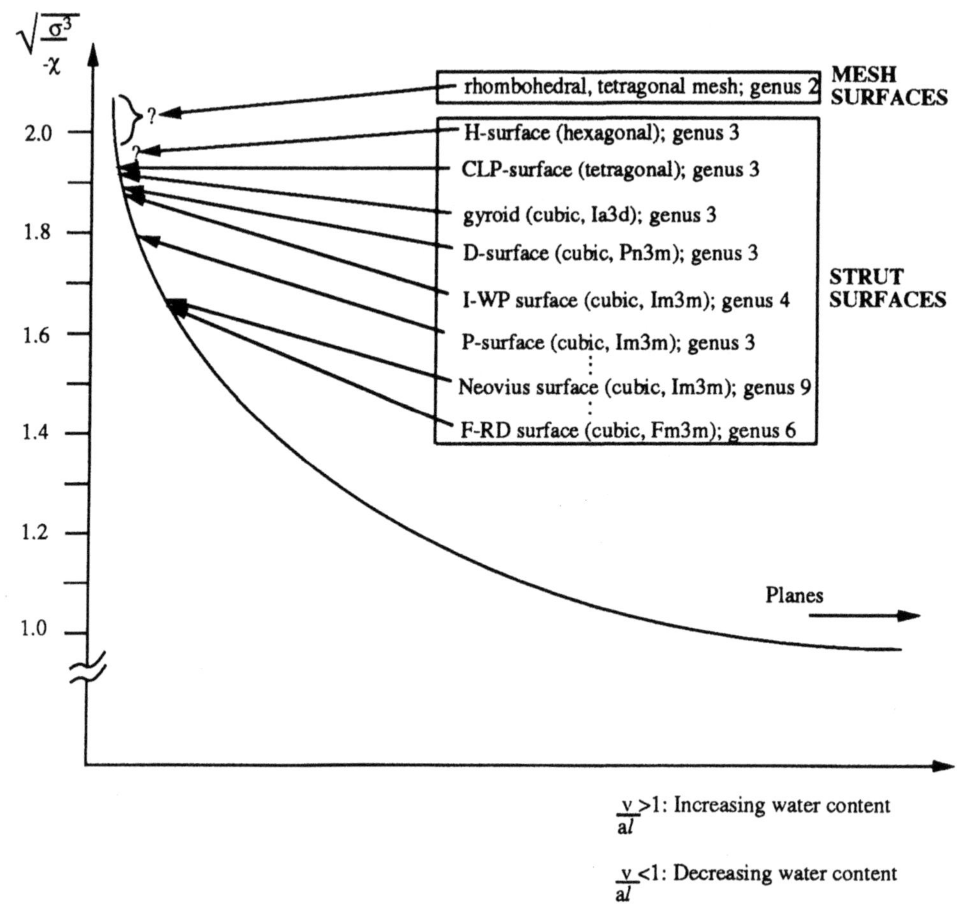

Fig. 1. Schematic illustration of the variation of surface characteristics defined by the normalised surface-to-volume ratio and the Euler-Poincaré characteristic per unit cell (σ and χ resp.) as a function of the water volume fraction. Values of this index for known minimal surfaces of strut geometries are displaced. The lower topology mesh structures are expected to lie above those of strut geometries. (Mesh structures are characterised by a two-dimensional tunnel system, while strut geometries define a three-dimensional network). Surfactants are expected to form bilayers lying on these surfaces. (At lower volume fractions a sequence of monolayer structures occur; see text)

the composition (and a preferred curvature towards the polar moiety of a surfactant mixture), the succession of microstructures for a surfactant mixture upon water dilution is:

- high topology monolayers ($H \gg 0$, $K \ll 0$): interconnected polar tubules;
- low topology monolayers (H, $K > \approx 0$): polar cylinders, globules, spheres;
- low topology bilayers ($H > 0$ (at the headgroups), $K < 0$): mesh surfaces;
- higher topology bilayers ($H \gg 0$, $K \ll 0$): strut surfaces: H-surface
  T-surface
  gyroid
  D-surface
  I-WP surface
  P-surface
  ⋮
  Neovius surface
  ⋮
  F-RD surface
  ⋮

(Other, as yet undiscovered periodic minimal surfaces may well intervene after the P-surface.)

Note that this order is reversed for surfactant mixtures where the surfactant molecule is preferentially curved towards the paraffin regions.

Comparison with acutal data

The actual range of structures adopted by a particular surfactant solution cannot be predicted a priori, although the appearance of various microstructures must follow the series outlined above. In fact, since the curvatures vary over most surfaces, the maximum topologies achievable for monolayers (i.e. minimum polar fraction) and bilayers (maximum polar fraction) depends critically upon the bending moduli for the surfactant films. This is because the variation of curvatures over a surface appears to increase as the topology of the surface increases. Furthermore, the width of the outer volumes subtended by an interface exhibits a larger variation for high topology interfaces. (This latter problem can be mitigated by adding some hydrophobic liquid — if the surfactant parameter is greater than one — thereby filling in the voids created by the surfactant chains, hence the high topologies achieved by ternary microemulsion monolayers at low water content.)

These properties of high topology surfaces tend to diminish the ubiquity of these more exotic microstructures. In fact, for all surfactant mixtures, we expect that a sufficiently high polar volume fractions the bending energy cost will result in the preferred formation of planes — at the expense of an average value of the surfactant parameter equal to one rather than the preferred value — over high topology bilayers of low curvatures. Thus, lamellar phases fit into this scenario if we include the bending energy of the surfactant film.

We have yet to account for the existence of crystalline phases. This is a difficult problem, since little data is available on disordered surface structure. However, our experience with bicontinuous periodic surfaces suggests that "symmetric" structures minimise the variation of curvatures over the surface. Certainly, any perturbation away from a symmetric structure must increase the variation of curvatures (since this results in non-uniform stretching and compressing). If this thesis is accepted, it follows that the crystallinity of structures is due to the saving of bending energy. Given the high, bending energy cost associated with bilayers (compared to monolayers), bilayer interfaces should exhibit a more marked tendency to form symmetric interfaces than monolayers.

Our analysis to accounts for some of the phase behaviour of actual surfactant mixtures. Nevertheless, the high topology monolayers predicted to exist at low internal volume fractions are often not seen. This is not surprising, for the reasons outlined in the previous paragraph. However, they are known to occur in microemulsions [8—10].

The expected succession of cubic phases has, in fact, been observed to a limited extent. The transition from the gyroid structure to the D-surface has been observed in glycerol monooleate lipids upon water dilution [4]. Recent studies of the cubic region of the DDAB/cyclohexane/water system have revealed a transition from the D-surface to the P-surface [12, 13].

The analysis predicts the occurrence of some exotic structures. Liquid crystalline phases of tetragonal symmetry (of genus three)-predicted here to occur at a slightly lower internal volume fraction to cubbic phases — have, in fact, been reported in the literature [7, 21]. Luzzati et al. have also reported the existence of a mesh surface of rhombohedral symmetry at high temperature [22] — their so-called R phase.

A striking omission from our schematic phase diagram is the existence of cylinders between spheres and cubic phases, viz. hexagonal phases. Is it possible that the microstructure of hexagonal phases consists of the H-surface, which appears at a composition intermediate to spheres and cubic phases, rather than rods? The tunnel networks are very similar, although the tunnel morphology of the H-surface is bicontinuous.

Acknowledgements

These analyses would have been impossible without the substantial corpus of structural data collected by Krister Fontell, Leo Mandell and Per Ekwall. This meeting is a particularly apt occasion to thank these people, who, along with the group of Luzzati in France, pioneered phase studies of surfactant mixtures. I am also indebted to my colleagues (both past and present) at the Australian National University: Ian Barnes, Philippe Barois, Ewa Radlinska, Barry Ninham and Thomas Zemb.

References

1. Luzzati V, Mariani P, Gulik-Krzywicki T (1987) In: Meunier J, Langevin D, Boccara N (eds) Physics of Amphiphilic Layers. Springer Verlag, Berlin, p 131
2. Fontell K, Larsson K, Krogh N (1980) Chem Phys Lipids 27:321
3. Lindblom G, Larsson K, Johansson L, Fontell K, Forsén S (1979) J Am Chem Soc 101:5465
4. Hyde S, Andersson S, Ericsson B, Larsson K (1984) Z Kristallogr 168:213
5. Charvolin J, Sadoc J (1987) J Phys (Les Ulis, France) 48:1559
6. Evans D, Mitchell D, Ninham B (1986) J Phys Chem 90:2817
7. Kékicheff P, Cabane B (1987) J Phys (Les Ulis, France) 48:1571
8. Barnes I, Hyde S, Ninham B, Dérian P, Drifford M, Zemb T (1988) J Phys Chem 92:2286
9. Hyde S, Ninham B, Zemb T (1989) J Phys Chem 93:1465
10. Jahn W, Strey R (1988) J Phys Chem 92:2301
11. Hyde S (1989) J Phys Chem 93:1458
12. Barois P, Hyde S, Ninham B, Dowling T (1990) Langmuir 6:1136
13. Radlinska E, Hyde S, Ninham B (1989) Langmuir 5:1427
14. Fontell K, Jansson M (1988) Progr Colloid Polym Sci 76:169
15. Larsson K (1983) Nature 304:664
16. Ben-Shaul A, Szleifer L, Gelbart W, In: Meunier J, Langevin D, Boccara N (eds) Physics of Amphiphilic Layers. Springer Verlag, Berlin, p 2
17. Mitchell J, Ninham B (1981) J Chem Soc Faraday Trans 2 77:601
18. Luzzati V, Tardieu A, Gulik-Krzywicki T, Rivaz E, Reiss-Husson F (1968) Nature 220:485
19. A related expression is given in Ref. [11] (Eq. (36))
20. Schwarz H (1890) In: Mathematische Abhandlungen, Springer
21. Kékicheff P, Tiddy G (1989) J Phys Chem 93:2520
22. Luzzati V, Gulik, Krzywicki T, Tardieu A (1968) Nature 218:1031

Authors' address:

S. T. Hyde
Department of Applied Mathematics,
Research School of Physical Sciences,
Australian National University,
Box 4, Canberra, 2601, Australia

Appendix

For a quantitative analysis of the stability of various cubic phases, see J. Physique (Colloque "Geometry and Interfaces"), to appear.

Progress in Colloid & Polymer Science Progr Colloid Polym Sci 82:243—248 (1990)

Aggregation behavior of sodium dodecylsulfate in water-1,2-ethanediol mixtures

A. M. Blokhus[1]), H. Høiland[1]), M. I. Gjerde[1]), S. Backlund[2]), M. Ruths[2]), and G. Douhéret[3])

[1]) Department of Chemistry, University of Bergen, Bergen, Norway
[2]) Department of Physical Chemistry, Åbo Akademi, Åbo, Finland
[3]) Laboratoire de Thermodynamique et Cinétique Chimique, URA. au CRNS No. 434, Université Blaise Pascal, Aubière, France

Abstract: The phase diagram of the system water-1,2-ethanediol-sodium-dodecylsulfate has been determined at 298.15 K. Conductivity, density, and compressibility measurements have been used to determined the critical micelle concentration (CMC) in the solution phase. Below a 1,2-ethanediol mole fraction of about 0.05 the CMC ist not significantly affected by ethanediol. In the range 0.05—0.3 the CMC increases moderately. Above 0.3 it increases steeply, and it becomes difficult to detect the break point that is associated with the CMC. Partial molar volumes and isentropic partial molar compressibilities of sodium dodecylsulfate have been calculated both below and above the CMC. Differences in these quantities between the micellar and the monomer states tend towards zero at high 1,2-ethanediol mole fractions. The self-diffusion coefficient of the micelles increase significantly, relative to the ethanediol self-diffusion coefficient. At high ethanediol mole fractions the self-diffusion coefficient of the aggregates and the monomers only differ by a factor of approximately 2. The results suggest that micelles are formed, but only up to an ethanediol mole fraction of about 0.3. In this range the aggregation number decreases continuously. Above an ethanediol mole fraction of 0.3 aggregation may still take place, but proper micelles are not formed. Instead the aggregates form as a result of a progressive association process involving only a few monomers.

Key words: Surfactant aggregation; critical micelle concentration; phase diagram; molar volume; molar compressibility; self-diffusion coefficient

Introduction

The behavior of surfactant molecules in non-aqueous solvents has attracted interest for two reasons. Firstly, there is the possibility of using non-aqueous micellar solutions or non-aqueous micro-emulsions as media for specific chemical reactions [1]. Secondly, it may be used to study aggregation processes and the mechanisms of micelle formation, particularly with respect to the so-called hydrophobic effect [2].

Presently, we shall focus on ionic surfactants and their aggregation processes in non-aqueous and mixed aqueous-non-aqueous solutions. Such studies have normally been carried out by looking for a critical micelle concentration (CMC). Break points in curves of electrical conductivity, surface tension, and other easily measurable quantities *vs* surfactant concentration have been taken as evidence for a micellization process [3—8]. However, Binana-Limbele et al. [9] have recently shown that such break points are ambiguous with respect to micelle formation. Break points may be observed even if the association process involves a progressive association of a few monomers and not a co-operative micellization process. It is, therefore, important to distinguish between micelle formation as a co-operative aggregation phenomenon where the size distribution is narrow and displays a minimum, and other association phenomena normally involving a few monomers [10]. In the latter case break points may be observed, but they are usually not distinct, and various experimental techniques yield CMC values that differ appreciably. Typically, claims have been made of micelles forming in a variety of pure polar solvents including dimethylsulfoxide and 1,2-ethanediol [3, 6], while others report no micelle formation of the same surfactants in the same solvents [4, 11].

On this basis it seemed interestingly to study the aggregation process in a mixed aqueous-non-aqueous solvent over the entire composition range, and for this purpose 1,2-ethanediol (ED) was chosen as the solvent and sodium dodecylsulfate (NaDDS) as the surfactant. 1,2-ethanediol is a polar, hydrogen-bonding solvent. It has a cohesive energy density of 12×10^{-7} J cm^{-3} mol$^{1/3}$ which is slightly above the value quoted as required for micelle formation [8], and the water-1,2-ethanediol system is one of the best characterized systems in the literature [12].

Materials and methods

Sodium dodecylsulfate (NADS) was supplied by BDH, ,,specially pure" quality. It was dried in vacuo at 315 K overnight. 1,2-ethanediol (ED) was supplied by Aldrich (99.5% pure) or by Fluka, *purest* quality. It was used as received. Distilled water was filtered through a Millipore Q filter before use; it's conductivity was about 1 μS cm^{-1}.

The phase diagram of the water-1,2-ethanediol-sodium dodecylsulfate system was constructed by visual inspection of mixed samples after equilibration in a thermostatted bath kept at 298.15 K. The possible occurrence of liquid crystalline phase was detected by looking at the samples through two crossed polaroids. The electrical conductivity was measured by a Wayne-Kerr bridge automatically recording the resistances. Density measurements were carried out with a Paar density meter (DMA 601), and the isentropic compressibilites were calculated by measuring the speed of sound by the ,,sing-around" method as previously described [13]. The temperature was controlled to better than ± 0.005 K in these experiments.

Self-diffusion coefficients were measured on a Bruker 100 MHz NMR instrument by the method of Stilbs and Moseley [14]. The temperature was controlled within ±0.2 K.

Results

The water-1,2-ethanediol-sodium dodecylsulfate system forms an extensive solution phase *L* at 298.15 K, as shown in Fig. 1. A hexagonal liquid crystalline phase *E* is also found.

No further study of the *E*-phase has been carried out, but the *L*-phase has been studied with respect to micelle formation. At constant H_2O-ED mass fractions, conductivities, densities, and compressibilities were measured as a function of the surfactant molality. When plotted vs surfactant molality, all three sets of data exhibited distinct breaks in the slopes when the ED mole fraction was less than 0.3. This is illustrated in Fig. 2 where the isentropic compressibilities have been plotted vs the NaDDS molality. The conductivity and density data are not shown, but they exhibit the same features. The CMC values, taken as the break points, agree very well. At $x_{ED} > 0.3$ break points are still observed for conductivities, densities, and compressibilities. However, as seen from Fig. 2 they are not as distinct as before, and there are significant differences between the apparent CMC values as determined by the various methods. For instance, at $x_{ED} = 0.53$ density yields an apparent CMC of 7.5×10^{-2} and compressibility 6.5×10^{-2} mol kg^{-1}. Figure 3 shows the CMC values, defined by the break points of conductivity, density, and compressibility as a function of the ED mole fraction. It appears to be divided into three regions. For $x_{ED} \leqslant 0.05$ the CMC does not change significantly. In the range 0.05 to 0.3 there is a gradeful increase of the CMC, and above $x_{ED} = 0.3$ it increases dramatically. Test were made on solutions with

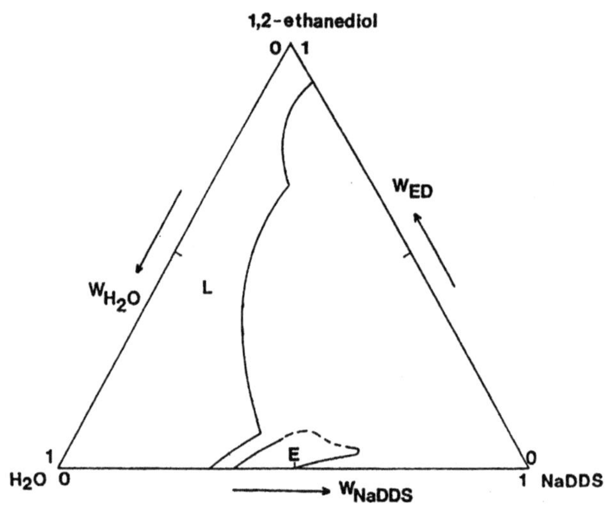

Fig. 1. Phase diagram of the system water-1,2-ethane-diol-sodium-dodecdylsulfate at 298.15 K. *L* denotes isotropic solution phase and *E* hexagonal liquid crystalline phase

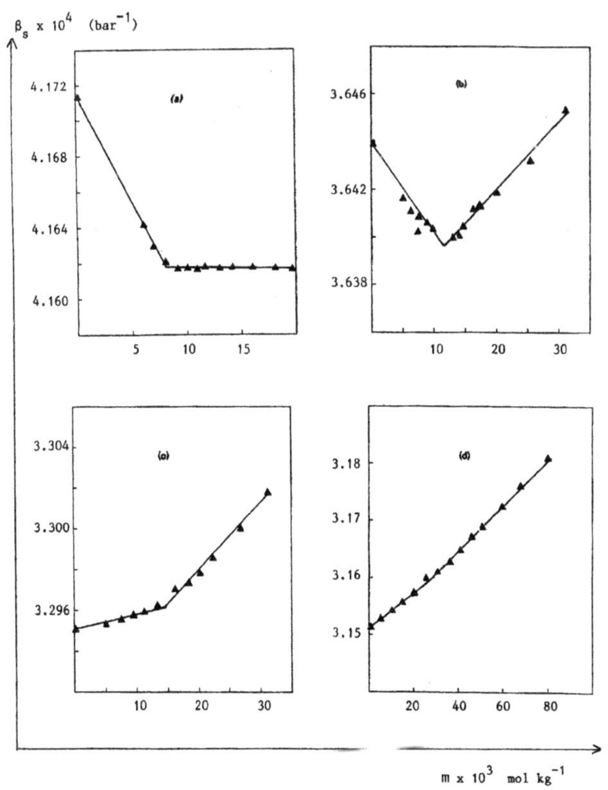

a) x_{ED} = 0.031; b) x_{ED} = 0.111; c) x_{ED} = 0.226;
d) x_{ED} = 0.403

Fig. 2. The isentropic coefficient of compressibility βs as a function of added surfactant, at different 1,2-ethanediol mole fractions

x_{ED} = 0.72 and in pure 1,2-ethanediol, but not breaks in the density or compressibility curves could be seen.

The density and isentropic compressibilities are linear functions of the NaDDS molality at all ethanediol mole fractions, both below and above the apparent CMC. Thus, the partial molar volumes can then be calculated as [15]

$$V_{m,2} = M_2/\rho - \{(w_1 + w_2)/\rho^2\}(\partial\rho/\partial n_2)_{n1} . \qquad (1)$$

Here the subscript 2 refers to sodium dodecylsulfate and the subscript 1 to the mixed solvent (water and 1,2-ethandiol). M_2 is the molar mass of NaDDS, w is the mass, and n is the amount of substance. The partial molar volumes of sodium dodecylsulfate below and above the CMC have been extrapolated to infinite dilution, and the results are shown in

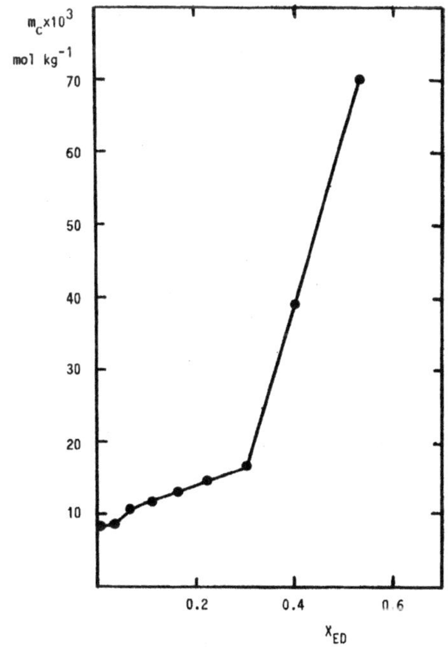

Fig. 3. The surfactant molality m_c at the CMC as a function of the 1,2-ethanediol mole fraction

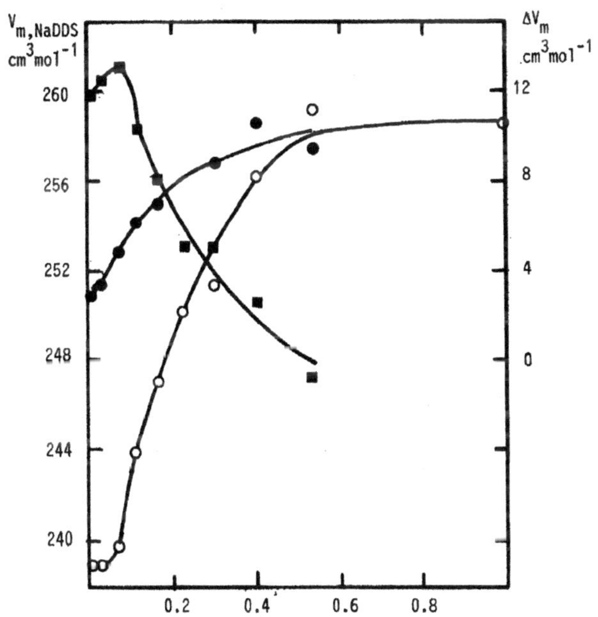

Fig. 4. The partial molar volume of NaDDS in the micellar state (□) and in the monomer state (△) at infinite dilution as a function of the ethanediol mole fraction. ΔV_m is the change in the partial molar volume of NaDDS due to micelle formation (●)

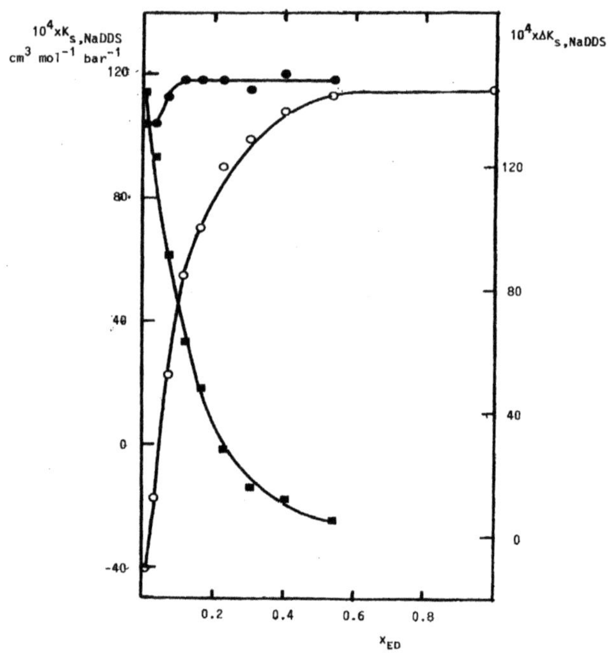

Fig. 5. The isentropic partial molar compressibility of NaDDS in the micellar state (△) and in the monomer state (□) at infinite dilution as a function of the ethanediol mole fraction. $\Delta K_{s,\mathrm{NaDDS}}$ is the change in the isentropic partial molar compressibility due to micelle formation (●)

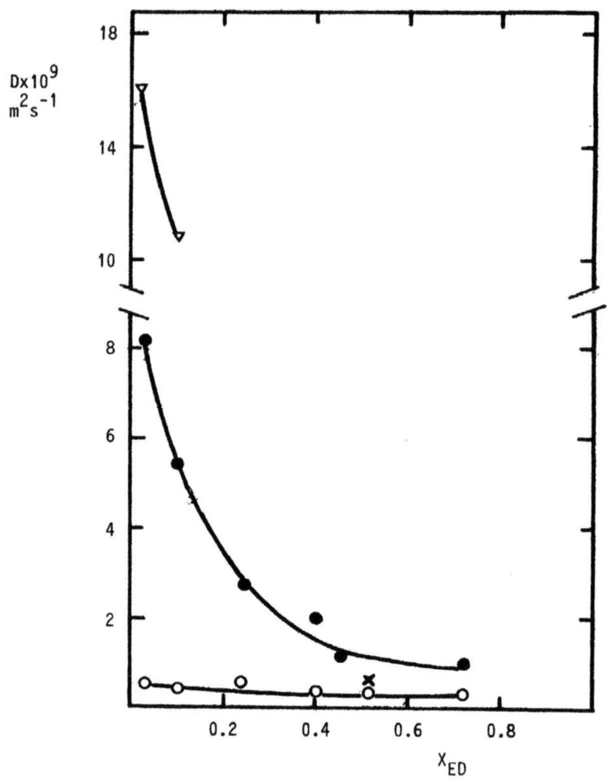

Fig. 6. NMR self-diffusion coefficients of water (■), ethanediol (○), NaDDS micelles (▲), and NaDDS monomers (◇) as functions of the ethanediol mole fraction

Fig. 4. ΔV_m, the partial molar volume change of micelle formation has also been plotted in Fig. 4. It goes through a small maximum at $x_{\mathrm{ED}} \approx 0.07$ and then decreases continuously with the ED content until it becomes zero at $x_{\mathrm{ED}} \approx 0.5$. Backlund et al. [11] also observed a small maximum for ΔV_m of tetradecyltrimethylammonium bromide at the same ethanediol mole fraction. The maximum reflects that the partial molar volume of NaDDS in the micellar state increases more rapidly than in the aqueous state at the lowest ethanediol mole fractions. Partial molar compressibilities can be calculated by differentiating Eq. (1) with respect to pressure. The results is [15]

$$K_{s,2} = \{(w_1 + w_2)/\rho\}\partial\beta_s/\partial n_2) + \beta_s V_{m,2} \, . \quad (2)$$

Here the symbols and the subscripts are the same as in Eq. (1). $K_{s,2}$ is the isentropic partial molar compressibility of NaDDS and β_S is the isentropic compressibility of the solution. $K_{s,2}$ values of NaDDS have been determined both below and above the CMC and extrapolated to infinite dilution. The infinite dilution data have been plotted vs the 1,2-ethanediol mole fraction in Fig. 5. They show the same general features as the partial molar volumes. ΔK_m, the change of the isentropic partial molar compressibilities of micelle formation, has also been plotted in Fig. 5. It decreases continuously, but does not reach zero at the highest 1,2-ethanediol mole fraction measured.

Self-diffusion coefficients of NaDDS, ED, and H_2O have been measured for $m_{\mathrm{NaDDS}} = 0.2$ mol kg^{-1}. This high surfactant molality was chosen to ensure that we were above the CMC at all ED mole fractions. The measured self-diffusion coefficients of NaDDS are average values of the monomer and micellar contributions. At low and moderate ED contents the fraction of monomers is small, being at $x_{\mathrm{ED}} = 0.3$ only 0.08, and the measured values can be regarded as the micellar diffusion coefficient without making significant errors. At higher ED contents the fraction of monomers is small, being at x_{ED} = 0.3 only 0.08, and the measured values can be regarded as the micellar diffusion coefficient without making significant errors. At higher ED contents the monomers may contribute significantly to the measured value. Therefore, the self diffusion coefficient of the monomers was measured at x_{ED} = 0.5. It is, thus, possible to correct for the monomer contribution. This has been done, and the results have been plotted in Fig. 6. The water and ED self-diffusion coefficients decrease rapidly with increasing ED content, due to the increasing viscosity of the solutions. The NaDDS self-diffusion coefficient remains surprisingly constant. At high ED contents the difference between the monomer and the aggregate self-diffusion coefficients is small, 6.5×10^{-10} and 3.1×10^{-10} m^2 s^{-1}, respectively.

Discussion

The water-1,2-ethanediol system has recently been reviewed by Huo et al. [12]. From a detailed thermodynamic study over the entire composition range they conclude that the system can be divided into three distinct regions. The region $0 < x_{ED} < 0.1$ can be characterized by progressive loss of the characteristic co-operative fluctuations of liquid water. At the upper limit of this region minimal $(H_2O)_n$ co-operative units, apparently pentamers, remain.

The second composition region is found for $0.1 < x_{ED} < 0.3$. It is characterized by the loss of three-dimensional H_2O-H_2O hydrogen-bonding connectivity and the onset of significant ED-ED connectivity. The third region at $x_{ED} > 0.3$ exhibits pseudoideality between ED and H_2O, without the characteristic hydrogen-bonded network and cooperative fluctuation behavior found in water.

Huo *et al.* [12] have shown that bulk properties and first derivative properties (with respect to Gibb's energy) do not reveal the three regions. Second derivative properties generally do, with one notable exception, the partial molar compressibility.

It is, thus, not surprising that the volume and compressibility data, as shown in Figs. 4 and 5, do not reveal these changes in the solvent properties. The CMC values, however, provide evidence for the two transition points. The first region, where the CMC remains practically constant, extends to x_{ED} = 0.05, only half the value quoted by Huo *et al.*, [12]. The second region, where the CMC increases moderately, extends to x_{ED} = 0.3 in excellent agreement with the quoted value. The discrepancy at low x_{ED} could be an effect of the added NaDDS. The three-component system will necessarily be different from the two-component system, but with the small amount of NaDDS added, this will only be important at low x_{ED} values.

The next question concerns micelle formation. The significant break points in density and compressibility curves combined with large values of ΔV_m and ΔK_m suggest that proper micelles form in this mixed solvent, at least up to $x_{ED} \approx 0.3$. It is tempting to conclude that above this mole fraction co-operative association, i.e., micelle formation, no longer takes place. Instead, aggregates form as a result of a non-cooperative self-association process involving only a few monomers. Unfortunately, neither the volumes nor the compressibilities show any direct evidence for this; they are both smooth functions of x_{ED} in this range.

The self-diffusion coefficient of the NaDDS aggregates changes but little with the ED content, but the ED coefficient does change. The large drop in ED diffusibility is mainly due to increasing viscosity of the solution. Thus, relative to ED the micellar self-diffusion increases signficantly, suggesting that the micellar aggregation number decreases correspondingly. This agrees well with the data of Almgren et al. [16] and Hashimoto and Thomas [17]; they find that the aggregation number of NaDDS in ED-H_2O mixtures decreases from 70 in pure water to 55 at x_{ED} = 0.08 (w_{ED} = 0.2) and 29 x_{ED} = 0.23 (w_{ED} = 0.5). At $x_{ED} > 0.3$ there is only a factor 2 between the monomer and aggregate self-diffusion coefficients, which is further evidence for small aggregates at these ED contents.

All methods suggest that there is a continuous decrease in aggregation number with addition of ethanediol. However, no method except, possibly, the plot of CMC values vs the ED content, distinctly shows when the cooperative micellization process is replaced by a progressive association process. It may, thus, be interesting to look at criteria for micelle formation. It has been suggested that a high cohesive energy density and hydrogen bonding ability of the solvent is a prerequisite for micelles to form [9, 18, 19]. The cohesive energy density can be represented by the parameter $\gamma / V_m^{1/3}$, where γ is the surface tension [8, 20]. Binana-Limble et al. [9] have reviewed micellization data in pure organic solvents, and conclude that the minimum $\gamma / V_m^{1/3}$

value (cohesive energy density) required for ionic surfactants, is about 17×10^{-7} J cm^{-3} $mol^{1/3}$. The $\gamma/V_m^{1/3}$ parameter is not known for mixtures of water and ethanediol. However, one may take the data of Huot et al. [12] for the cohesive energy density calculated from the enthalpy of vaporization and assume the same functional relationship for the parameter $\gamma/V_m^{1/3}$. In so doing the $\gamma/V_m^{1/3}$ parameter drops from the value in water, 27×10^{-7} J cm^{-3} $mol^{1/3}$, to 17×10^{-7} J cm^{-3} $mol^{1/3}$ at $x_{ED} \approx 0.3$.

In conclusion, it seems that sodium dodecylsulfate forms micelles in water-1,2-ethanediol mixtures up to an ethanediol mole fraction of approximately 0.3. Above this mole fraction progressive association rather than co-operative micellization takes place. However, thermodynamic data like partial molar volumes and isentropic partial molar compressibilities show a continuous change with 1,2-ethanediol mole fraction, and a transition point where the aggregation changes from micelles to progressive association is not observed. Self-dissufion coefficients combined with data for aggregation numbers suggest that the micellar size decreases continuously with 1,2-ethanediol addition until, presumably, co-operative association is replaced by progressive association processes.

Acknowledgements

This research has been supported by VISTA, a research cooperation between the Norwegian Academy of Science and Letters and Den Norske Stats Oljeselskap (Statoil). S.B. thanks Nordiska Forskarkurser, and G.D. thanks *CNRS* for financial support.

References

1. Lattes A, Rico I (1989) Colloids and Surfaces 35:221
2. Evans DF (1988) J Amer Chem Soc 91:6511
3. Ray A (1969) J Amer Chem Soc 91:6511
4. Gopal R, Singh JR (1973) J Phys Chem 77:554
5. Rico I, Lattes J (1986) J Phys Chem 90:5870
6. Singh HN, Salem SM, Singh RP, Birdi KS (1980) J Phys Chem 84:21
7. Ionescu L, Fung D (1981) Bull Chem Soc Jpn 54:2503
8. Ramadan M, Evans DF, Lumry R, Philson S (1985) J Phys Chem 89:3405
9. Binana-Limbele W, Zana R (1989) Colloid Polym Sci 267:440
10. Wennerström H, Lindman B (1979) Phys Rep 52:1
11. Backlund S, Bergenståhl B, Molander O, Wärnheim TJ, Colloid Interface Sci In Press
12. Huot J-Y, Battisel E, Lumry R, Villeneuve G, Lavalle JF, Anusiem A, Jolicoeur C (1988) J Solution Chem 17:601
13. Høiland H, Vikingstad E (1976) J Chem Soc Faraday Trans I 72:1441
14. Stilbs P, Moseley ME (1980) Chem Scr 15:176, 215
15. Vikholm I, Douhéret G, Backlund S, Høiland H (1987) J Colloid Interface Sci 116:582
16. Almgren M, Swarup S, Löfroth JE (1985) J Phys Chem 89:4621
17. Hashimoto S, Thomas JK (1983) J Amer Chem Soc 105:5230
18. Gopal R, Singh JR (1973) J Ind Chem Soc 49:4
19. Beesly A, Evans DF, Laughlin R (1988) J Phys Chem 92:791
20. Ritchie CD (1969) In: Coetzee JF, Ritchie CD (eds) Solute-Solvent Interactions, Marcel Dekker, New York

Authors' address:
H. Høiland
Department of Chemistry
University of Bergen
Allegt. 41
5007 Bergen, Norway

Solution properties of selected functionalized surfactants*)

B. Burczyk and K. A. Wilk

Institute of Organic and Polymer Technology, Technical University of Wrocław, Wrocław, Poland

Abstract: The surface-active functionalized surfactants, i.e., hexadecyl(2-hydroxyethyl)dimethylammonium, hexadecyl(2-hydroxypropyl)dimethylammonium, hexadecyl(2-hydroxybutyl)dimethylammonium, and 2-hydroxyhexadecyltrimethylammonium bromides have been synthesized and characterized. Surface properties of these substances have been investigated in aqueous solution and in solutions of sodium bromide of 0.1 M total ionic strength at 25 °C. The properties measured were maximum surface excess concentration Γ_{max}, minimum area per molecule at the aqueous solution — air interface A_{min}, efficiency of surface tension reduction $p\,C_{20}$, effectiveness of surface tension reduction π_{CMC}, standard free energy of adsorption ΔG^0_{ad}, and critical micelle concentration CMC, the latter being determined by means of surface tension, conductivity, and ion-selective electrode measurements. The obtained results indicate that the surface properties of studied compounds depend on the nature and position of the hydroxy group.

Key words: Hexadecyl(2-hydroxyalkyl)dimethylammonium bromides; 2-Hydroxyhexadecyltrimethylammonium bromide; cationic surface-active substances; adsorption at the aqueous solution — air interface; micelle formation

Introduction

Strong interest in the use of functional micelles has focused attention on the synthesis and properties of functionalized surfactants [1—7]. The functionalized quaternary ammonium surfactants are the most common catalysts in functional micellar catalysis [8, 9]. The cited literature provides only information comprising the critical micelle concentration (CMC) values of the functionalized surfactants. The surface activity characteristics of these derivatives have not been previously published.

In the present contribution, we report surface and micellar properties of a number of hydroxy-functionalized surfactants having the structures

$$[R\overset{+}{N}(CH_3)_2R^1]Br^- ,$$

where

$$R = C_{16}H_{33}\,,\quad R^1 = CH_2CH_2OH\,; \tag{1a}$$

$$R = C_{16}H_{33}\,,\quad R^1 = CH_2CH(CH_3)OH\,; \tag{1b}$$

$$R = C_{16}H_{33}\,,\quad R^1 = CH_2CH(C_2H_5)OH\,; \tag{1c}$$

and

$$R = C_{14}H_{29}CH(OH)CH_2\,,\quad R^1 = CH_3\,. \tag{2}$$

The functional micelles of compound (1a) have increasingly been investigated in catalysis of many organic reactions [1, 10]. Surfactants (1b) and (2) are known to catalyse the basic hydrolysis of carbamates [11]. The surfactant (1c) has not yet been described.

The main purpose of the present work was to study the effect of both the nature of the hydroxy group in the functionalized surfactant (i.e., primary vs secondary) and the position at which the hydroxy group is bonded to the surfactant molecule on the surface, and micellization behavior of the selected substances.

---

*) Part XXIII of the series: Chemical Structure and Surface Activity.
Part XXII. A. Sokolowski, B. Burczyk, J. Beger (1990) Colloid Surface 44:89—100

The surfactants (1a—c) were prepared and purified by the method described in [12, 13] and the compound (2) was prepared according to the procedures given in [14] and [15].

The following physicochemical properties of the surfactants' aqueous solutions were determined: surface tension by the drop weight method, bromide ion activity and electric conductivity, both according to known procedures.

The surface properties of the investigated functionalized ammonium surfactants, calculated according to the Rosen approach [16, 17] are summarized in Table 1. The set of pC_{20} values, defined by Rosen [18] as a measure of the surface-tension-reduction efficiency of a surfactant, shows that the class of compounds being investigated has very similar efficiency, compound (1c) being the most surface active. This is supported by the values of π_{CMC}, which differ only slightly among themselves.

The values of A_{min} indicate that the position of hydroxy group in surfactants' molecules significantly influences their surface packing. There appears a marked difference between compounds (1a) and (1b), carrying a primary and secondary hydroxy group, respectively. The effect of the alkyl chain length of the substituent R^1 on the A_{min} value is of minor importance. However, the hydroxy group attached to the hexadecyl chain does not seem to influence the surfactant (2) molecule packing at the aqueous solution — air interface. The ΔG^0_{ad} values of the studied compounds (1a—c) and (2) are almost identical. However, ΔG^0_{ad} for compound (2) is less negative than others which means that surfactant (2) is more weakly adsorbed at the interface than are compounds (1a—c). This result seems to be reasonable as the hydrophobic character of the hexadecyl chain is somewhat weakened by the presence of the hydroxy group at the C-2 atom.

The values of CMC determined by surface-tension measurements, electric conductivity, and ion-selective potentiometry are tabulated in Table 2. Typical curves for these methods are shown in Fig. 1. The CMC values are comparable with those reported in [11]; they show comparatively slight dependence on the nature of the ionic head group, as we found previously for sulphonium surfactants [19]. The CMC value for compound (2) is higher in relation to the surfactants (1a—c), which is in accord with the conclusion previously stated for the ΔG^0_{ad}

Table 2. Values of the critical micelle concentration (CMC) of $[R\overset{+}{N}(CH_3)_2R^1]Br^-$ in H_2O at 25 °C

Compd	CMC × 10^4, M		
	Surface tension isotherms	Electric conductivity	Ion selective potentiometry
(1a)	8.7	9.3	8.1
(1b)	8.3	8.4	7.6
(1c)	7.6	8.0	7.2
(2)	12.1	11.5	12.5

Table 1. Surface properties of $[R\overset{+}{N}(CH_3)_2R^1]Br^-$ in H_2O and in aqueous 0.1 M NaBr at 25 °C

Compd	Γ_{max} (mol cm^{-2} × 10^{10})	A_{min} (nm^2 × 100)	pC_{20}	π_{CMC} (mN m^{-1})	$-\Delta G^0_{ad}$ (kJ mol^{-1})
			H_2O-air		
(1a)	3.14	52.9	3.52	35.2	35.1
(1b)	2.06	80.8	3.72	35.4	35.4
(1c)	1.92	86.6	3.80	36.3	35.8
(2)	3.34	49.7	3.38	34.2	33.3
			0.1 M NaBr(aq)-air		
(1a)	3.28	50.6	4.63	41.9	42.3
(1b)	2.20	75.5	5.34	42.5	44.9
(1c)	1.96	84.7	5.71	42.8	45.6
(2)	3.32	50.0	4.44	40.9	41.7

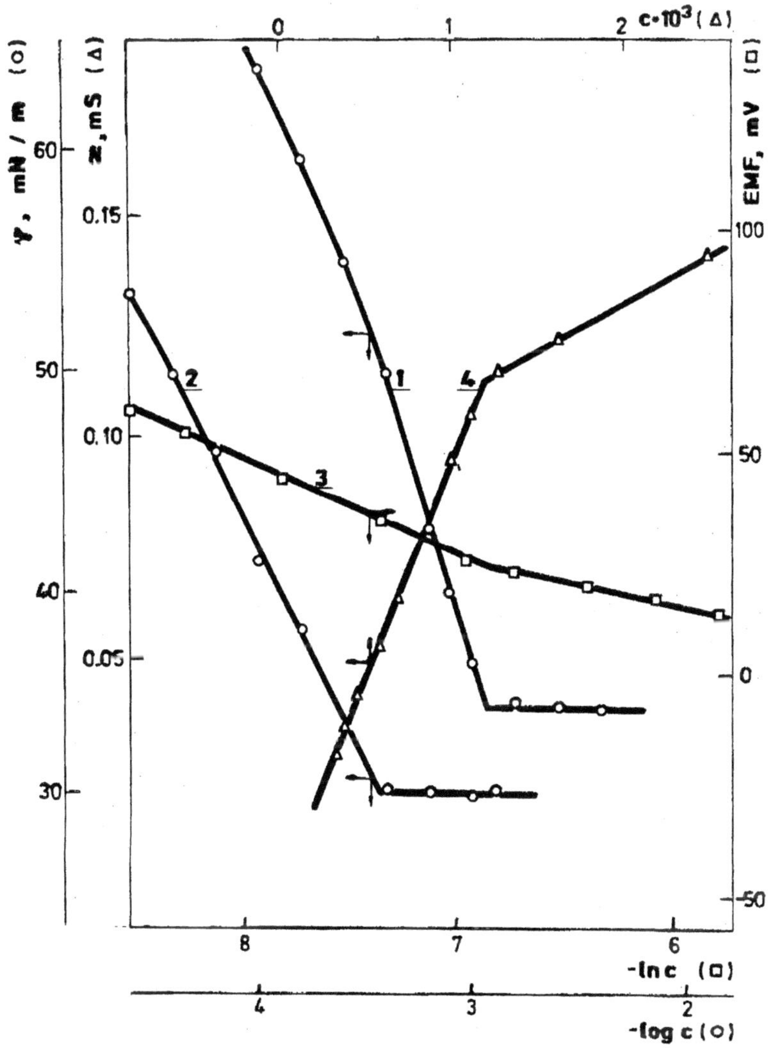

Fig. 1. Determination of CMC for $[C_{14}H_{29}CH(OH)CH_2\overset{+}{N}(CH_3)_3Br^-$; surface-tension measurements: 1) aqueous solution; 2) 0.1 M NaBr(aq); 3) ion-selective potentiometry; 4) electric conductivity

findings, because the hexadecyl chain hydrophobic character is diminished by the presence of the hydroxy group. According to the cited literature [20] the length of the hydrophobic part of an ionic surfactant primarily determines its CMC and an increase of the number of carbon atoms in alkyltrimethylammonium bromides causes a sizable decrease in the CMC. The change of one methyl group to 2-hydroxyethyl in the surfactant molecule involves litle effect on the CMC value (i.e., 9×10^{-4} M for hexadecyltrimethylammonium bromide and 8.7×10^{-4} M for 2-hydroxyethyl derivative (1a)). This corresponds to recent observations by Blackmore and Tiddy [21] related to hexadecyl (2-hydroxyethyl)ammonium bromide-water systems.

Conclusion

Our results show that the nature and position of the hydroxy group in quaternary bromides have little affect on their surface and micellar properties, contrary to the known effects of alkyl chain prolongation.

Acknowledgement

This work was supported by the Polish Academy of Sciences, Research Grant No. RPBP 01.10.

References

1. Tonellato U (1979) In: Mittal K (ed) Solution Chemistry of Surfactants. Plenum, New York, Vol 2, pp 541–558
2. Menger FM, Whitesell LG (1985) J Am Chem Soc 107:707–708
3. Moss RA, Kim KT, Swarup S (1986) J Am Chem Soc 108:788–793
4. Bunton CA, Ionescu LG (1973) J Am Chem Soc 95:2912–2917
5. Tonellato U (1976) J Chem Soc Perkin II:771–776
6. Moss RA, Lukas TJ, Nohas RC (1977) Tetrahedron Lett 3851–3854
7. Moss RA, Bizzigotti GO, Lukas TJ, Sanders WJ (1978) Tetrahedron Lett 3661–3664
8. Biresaw G, Bunton CA, Quan C, Yang Z-Y (1984) J Am Chem Soc 106:7178–7184
9. Biresaw G, Bunton CA, Savelli G (1985) J Org Chem 50:5374–5376
10. Bunton CA, Savelli G (1986) Adv Phys Org Chem 22:213–309
11. Broxton TJ, Chung RP-T (1986) J Org Chem 51:3112–3115
12. Bunton CA, Romsted LS, Sepulveda (1980) J Phys Chem 84:2611–2618
13. Tonellato U (1977) J Chem Soc Perkin II:821–826
14. Langman AW, Dalton RD (1980) In: Coates RM (ed) Organic Syntheses. Wiley, New York, Vol 59, p 16
15. Scott AB, Tartar HV (1943) J Am Chem Soc 65:692–698
16. Rosen MJ, Dahanayake M, Cohen AW (1982) Colloid Surface 5:159–172
17. Rosen MJ, Aronson S (1981) Colloid Surface 3:201–208
18. Rosen MF (1974) J Am Oil Chem Soc 51:461–465
19. Chlebicki J, Wilk KA (1990) Colloid Surface 44:71–79
20. Zana R (1980) J Colloid Interface Sci 78:330–337
21. Blackmore ES, Tiddy GJT (1988) J Chem Soc Faraday Trans II 84:1115–1127

Authors' address:

Prof. B. Burczyk
Instytut Technologii Organicznej
i Tworzyw Sztucznych Politechniki Worclawskiej
Wybrzeze St. Wyspiańskiego 27
50-370 Wrocław, Poland

Progress in Colloid & Polymer Science Progr Colloid Polym Sci 82:253—262 (1990)

Nonionic surfactant phase behavior. The effect of CH_3 capping of the terminal OH. Accurate measurements of cloud curves

J. P. Conroy, C. Hall, C. A. Leng, K. Rendall, G. J. T. Tiddy, J. Walsh, and G. Lindblom[1])

[1]) Dept of Physical Chemistry, University of Umeå, Sweden

Abstract: Phase diagrams have been determined by optical microscopy for several polyoxyethylene surfactants where the terminal OH group is replaced by OMe (surfactants denoted C_nEO_mOMe). The derivatives studied were $C_{12}EO_4OMe$, $C_{12}EO_6OMe$, and $C_{12}EO_8OMe$. In addition, a phase diagram is given for the conventional nonionic $C_{12}EO_2$. Changing OH to OMe has a drastic effect on phase behavior at higher temperatures (>50 °C), reducing mesophase upper temperature limits and the cloud temperature (T_c) by 20—30 °C. Phase transitions that occur below 20 °C and the mesophase sequence with increaseing surfactant concentrations are almost unaltered by Me substitution. Accurate cloud-curve measurements are reported for C_8EO_4, $C_{12}EO_4$, $C_{12}EO_8$, $C_{16}EO_6$, $C_{16}EO_8$, and C_{22}=EO_6 (hexaethylene glycol-*cis*-13-docosenyl ether, erucyl EO_6). From these curves, accurate values of the critical surfactant concentration and temperature for the lower consolute boundary are derived. All the data support the hypothesis that two very different processes exist for nonionic surfactant clouding, one due to critical de-wetting of EO chains (similar to EO polymer clouding) and the second arising from a rod/disc micelle shape change. Finally, surfactant self-diffusion measurements are reported for the two reversed bicontinuous cubic (V_2) phases of $C_{12}EO_2$. These are similar for both phases and to not allow any structural difference to be determined.

Key words: Nonionic surfactant; polyoxyethylene surfactant; cloud temperature; cloud-curve measurement; critical de-wetting of EO chains

1. Introduction

In recent years there has been a considerable interest in the properties of polyoxyethylene surfactants ($C_nH_{2n+1}(OCH_2CH_2)_mOH$; C_nEO_m), because of their partial miscibility in water and the wide variety of liquid crystal phases formed. The existence of the closed-loop partial miscibility, and particularly, the lower consolute temperature (the cloud point) have been thoroughly investigated, particularly, to discover the link between this behaviour and micelle size [1—18]. Most authors ignore the existence of the terminal OH of the head group. Any change in head group behavior with EO number is attributed solely to the ethylene oxide moiety. However, for low m-values the terminal OH does make a significant contribution to head group polarity. The major part of this report describes the phase behaviour of three polyoxyethylene surfactants where the terminal OH is replaced by OMe ($C_{12}EO_mOMe$, m = 4, 6, 8). Prior to this a general discussion is given of nonionic clouding mechanisms, micelle structure, and liquid crystal structures. This is used to predict the changes expected on substitution of OMe for OH. Finally, we include accurate data on the shape of the cloud curve for a series of EO surfactants, and a previously unreported phase diagram and self-diffusion data for the $C_{12}EO_2$/water system. Taken together, the results contribute to a much clearer picture of surfactant solution clouding.

2. Mechanisms of nonionic cloud points

There is almost universal agreement that the closed-loop phase behaviour of polyoxyethylene surfactants [18] is closely related to the clouding of polyoxyethylene polymers. Flory-Huggins theory was first employed in an early attempt to calculate cloud curves [19]. The theory was also used by Kjellander [20], who required the presence of rod micelles which become longer on approaching the critical temperature (T_c) to give the experimentally observed critical volume fractions (φ_c). Unfortunately, for several systems there is unequivocal evidence that the micelles remain small up to T_c [1]. Even for those that do form rods, what happens close to T_c is not clear. In some systems the rod length *decreases* towards T_c [21, 22]. Hence we need to review the factors responsible for micelle size and shape, and then to relate this to the behaviour of nonionic surfactants close to the cloud point.

Micelle shape is determined by the relative sizes of head groups and alkyl chains [23]. Large head groups are required for spherical micelles, while with smaller head groups the micelles are rod- or disc-shaped. It is easy to show [23] that for alkyl chains in water-free micelle interior having a smooth interface with the aqueous region, the area per chain at the micelle surface (a) must be larger than ca. 70 Å^2 for spherical micelles, ca. 47 Å^2 for rod micelles, and ca. 23 Å^2 for bilayer micelles (discs) [10, 23, 24]. Surface curvature and entropy always act to ensure that the smallest micellar aggregation number possible will occur.

One can obtain an indication of the micelle shape from the mesophases that occur at concentrations just above the micellar solution boundary [10]. Thus, one generally expects [10, 24]:

spherical micelles ⟶ I_1-cubic phase[25] ;

rod micelles ⟶ H_1-hexagonal phase ;

disc micelles ⟶ L_α-lamellar phase .

The simple theory predicts that, once formed, rod micelles will grow very long *via* intermediate sizes [23, 26]. The same theory predicts that disc micelles do not do this, but on growth, immediately form very large bilayers with no intermediate-size micelles. For linear mono-alkyl surfactants increasing CH_2 group number has no effect on micelle shape, because chain volume and micelle radius both depend linearly on chain length [24]. Hence, the mesophase formed should be independent of the chain length.

Experimentally, it is found that solutions containing long rod micelles (>10^3 Å) form viscous and viscoelastic solutions. Moreover, the solutions are bounded by an H_1 phase at high concentrations. Our measurements on surfactants that form H_1 at the L_1 boundary show that the short-chain derivatives do not give the viscous/viscoelastic dilute solutions indicative of long rod micelles, as predicted by the above theory. Thus, the micelles are smaller than expected. This may well be a consequence of micelle surface roughness. If the micelle surface is no longer smooth, then the maximum micelle radius (assumed above [23, 24] to be the all-trans chain length, l_t) is increased by the thickness of the "rough" layer (Δl). This reduces the *a* values for a micelle-shape transition by a factor $l_t/(l_t + \Delta l)$.

Clearly, the factor is smaller for long chains; hence, if it occurs, increasing the CH_2 number will eventually cause a transition from small micelles to long rods [27]. We have observed that *all* surfactant types (ionic, zwitterionic, nonionic) where L_1 is bounded by H_1 do form viscous/visco-elastic solutions (indicative of long rod micelles) if the alkyl chain length is increased enough (usually C_{18} derivatives show this behaviour to some extent) [28]. In many cases the increasing insolubility of the surfactant with increasing CH_2 number limits the observation of this trend. Thus, if long rod micelles play a role in clouding, then T_c should be strongly dependent on chain CH_2 number for surfactants where H_1 is formed at the L_1 boundary, particularly for the range where viscoelastic micellar systems begin to occur. In the second part of this report we present cloud point data for such surfactants.

To assess the influence on clouding of OCH_3 or OH as the EO terminal groups we first need a molecular explanation for poly-EO clouding, and then need to judge the contribution of OH hydration. Many authors [20, 29, 30] use the concept of "water structure" to account for the decrease of EO solubility with increasing temperature. This concept is high in descriptive capability but low on predictive power, at least for molecular properties. Recently, an alternative description has been given by Karlstrom [31, 32]. He proposes that only a few particular EO chain conformations are stabilized in aqueous solutions, these being ones with *gauche* C—O orientations about C—C bonds and *trans* conformations about C—O bonds. These conformations have a large segment dipole moment which leads to water solubility. Increasing temperature reduces the fraction of these conformations, reducing water solubility. For a long polymer, altering OH to OMe will have little effect on the chain conformations, and hence, a miniscule effect on clouding. For a short EO chain the effect may be much larger, since the OH group contributes by increasing head group polarity as well as by modifying the EO conformation. Micelle formation will reduce the probability of gauche conformers close to the micelle surface, because of inter-EO steric effects (say for the initial 1—2 EO groups), otherwise the chain conformation should be relatively independent of EO number. A hydrated OH group is not expected to have a much different steric effect to OCH_3, so the effect of this change on EO conformations is expected to be small.

It is difficult to estimate the contribution of the terminal OH group to EO surfactant solubility without complex calculations, for example, similar to those of Karlstrom [31, 32]. Such a study is beyond the scope of this work and, indeed, the capability of most of the authors. Very crudely, we can estimate the effect of OH polarity by comparing the phase behavior of poly-OH surfactants with poly-EO derivatives. It is well known that at least two OH substituents are required for a head group to have enough polarity for mesophase formation by mono-alkyl derivatives [33]. The mesophase formed for short chain surfactants (C_n, $n < 12$) is L_a, while reversed phases (V_2, H_2) also occur with longer chain derivatives above ca. 70°—80°C. (For examples, see monoglyceride phase diagrams [34, 35].) Hence, a single OH contribution is equivalent to ca. 3 EO units at 25°C, since $C_{12}EO_3$ and $C_{12}EO_4$ both give L_a dispersions at this temperature.

We have previously argued that there are two distinctly different clouding phenomena with poly-EO-surfactants [10], albeit that both mainly arise from the dehydration of EO groups on increasing temperature. One, where mesophases do not occur above T_c (e.g., $C_{12}EO_8$, Fig. 1a), is due to critical dewetting of EO groups and has small micelles (close to spherical) at T_c. The second, where L_a + W dispersions are observed above T_c (e.g., $C_{12}EO_4$, Fig. 1c) occurs because disc micelles are formed from rods in the temperature range below T_c [10]. In these systems an L_3 phase that contains disc micelles [36] occurs at temperatures very much above T_c (40—50° for $C_{12}EO_4$). Both the mesophase melting and the formation of L_3 are attributed to the critical dewetting of EO groups. Because both L_1 at T_c and L_3 where thought to contain disc micelles, some explanation is required as to why the micelles phase separate at T_c and then briefly redissolve in water to form L_3, followed by a second phase separation. An additional contribution to inter-micelle attractions at the much lower T_c from van der Waals forces between disc micelles was suggested previously to explain why T_c is well below the critical dewetting region.

In the second case, the key argument of *disc* rather than rod micelles at T_c was based on the observation that T_c never occurs below the H_1 melting temperature. Recent theoretical work has shown that disc micelles have attractive forces an order of magnitude larger than rod micelles of the same volume [37], hence, this additional attractive mechanism (van der Waals attraction) may not be required. None of the supporters of rod micelles at T_c have an explanation as to why the rods do not form an H_1 phase at higher concentrations. Indeed, most of them ignore the fact that L_a and L_3, an aqueous micellar solution with 90% or more water, occur well above T_c. This is hardly a satisfactory state. (Note that $C_{12}EO_6$ does form an L_a + W dispersion above T_c in the presence of additives [38], hence, it represents a borderline case between the two mechanisms.) Very recently, we have reported that long rod micelles (>2000 Å) occur in $C_{16}EO_6$ [21, 22]. These decrease in length on increasing temperature close to T_c. We have suggested that the rods develop an elliptical cross-section on approaching T_c. They can be regarded as flexible "rulers" with the disc character arising from the ruler faces. The aggregation number of these micelles could continue to increase through T_c, although the length decreases. This shape has not been considered for other systems (e.g., $C_{12}EO_4$, etc.), but appears to be consistent with the available evidence.

Changing OH to OMe is expected to have a dramatic effect on the first clouding mechanism (critical dewetting) since it reduces the equivalent EO length by 3 EO units. Consider $C_{12}EO_8$ as an example; the W + L_1 co-existence curve is rather flat with $C_{12}EO_8$, hence inter-micellar attractions are increasing rapidly with temperature [10]. Note that T_c for $C_{12}EO_{12}$ is only 20° above the value for $C_{12}EO_8$, while the value for $C_{12}EO_6$ is 30° below [10]. The OH/OMe change might be expected to reduce T_c by 20°—30°C at most. The effect of critical de-wetting on the mesophases at T_c will be to lower their melting temperatures also by the

20 °—30 °C, as discussed above. However, because mesophase type is determined by a, and this mainly arises from EO-EO steric repulsions, mesophase structures will be hardly changed.

For surfactants clouding via the second mechanism (rod/disc-ruler micelle shape transition) the important parameter is the value of a. It is expected to be almost unaltered by substitution of the OMe groups for OH, because this end of the head group is farthest from the micelle surface (where the strongest head group repulsions occur). Hence, we predict a very much smaller decrease in T_c for $C_{12}EO_4$, much less than the effect on $C_{12}EO_8$. This is despite the fact that the $C_{12}EO_4$ cloud-curve changes much less strongly with temperature than that for $C_{12}EO_8$, indicating that the intermicellar attractions increase more slowly with temperature. The region where critical de-wetting does occur (L_3 formation, L_a melting) will be markedly altered, possibly by 20 °—30 °C or more. Again, meosphase structures will be unchanged.

However, if we assume that clouding arises from critical de-wetting of *rod* micelles then we judge that the OH contribution to EO_4 solubility (43%) is significantly larger than with EO_8 (~27%). Hence, substitution of OMe will reduce T_c for $C_{12}EO_4$ by more than for $C_{12}EO_8$, particularly, because the EO_4 attractions vary more slowly with temperature than those of EO_8.

3. Experimental

a) Surfactants

The surfactants C_8EO_4, $C_{12}EO_4$, $C_{12}EO_8$, $C_{16}EO_6$, and $C_{16}EO_8$ were those used in previous studies (stored below 0 °C) [10] or were used as received from Nikkol Chemicals (Japan). The methyl-capped nonionics $C_{12}EO_mOMe$ (m = 4, 6, 8) were prepared by reacting $C_{12}EO_m$ (m = 4, 6, 8, ex. Nikkol) with sodium hydride, followed by addition of methyl iodide in dry dimethoxyethane.

Hexaoxyethylene glycol erucyl ether (C_{22} = EO_6) was prepared by reacting erucyl mesylate (Sigma, USA) with monosidum hexaoxyethylene glycol (Lancaster Synthesis, U.K.). All the synthesised surfactants give single spots on TLC and were estimated to be >98% pure from the 360 MHz high resolution proton NMR spectra.

b) Cloud point and phase diagram determinations

Most cloud points were determined by noting the onset of turbidity in slowly heated, stirred samples. The samples were placed in a stirred, enclosed water bath contained within an outer jacket through which additional heated water flowed. This allowed temperature homogeneity of ~0.01 °C to be obtained, giving an accuracy of ±0.02 °C or better. Each cloud temperature is the average of two or more determination. The latter measurements on C_{22}=EO_6 were made with visual observation of a sample contained in a heated water bath. In this case, temperatures were measured to ±0.1 °C.

Phase diagrams were determined using a combination of the optical microscope "penetration" technique and observations on bulk samples. Twenty to 40 samples were used to construct the phase boundaries to an accuracy of 2% by wt. and ±1 °C.

c) Self-diffusion measurements

Nuclear magnetic resonance (NMR) self-diffusion measurements of $C_{12}EO_2$ were made using a Bruker Bkr 322S spin-echo spectrometer, as described previously [39]. The reference used was glycerol.

4. Results and discussion

a) Phase diagrams of $C_{12}EO_m$ OMe/water and $C_{12}EO_2$/water systems

Phase diagram of $C_{12}EO_8$, EO_6, and EO_4 from previous work [10] are shown in Fig. 1 for comparison with those determined here for the OMe derivatives (Fig. 2). Comparing the EO_8 derivatives, we see that the most extensive mesophase in both cases is H_1, with I_1 also having a reasonably large existence range. Note the two different structures discernable within the I_1 region for $C_{12}EO_8$ OMe. In fact, two regions are also present in the original $C_{12}EO_8$. They are distinguished by the occurrence of a refractive index discontinuity within the I_1 region during a microscope penetration scan. Similar observations have also been reported by Finkelmann et al. [40] for $C_{14}EO_8$. The cloud temperature of $C_{12}EO_8$ OMe (52 °C) falls midway between values reported by Yeates et al. [41] for $C_{10}EO_8OMe$ (58 °C) and $C_{16}EO_8OMe$ (45°, 47 °C), showing excellent agreement with these data. Examination of Table 1 shows that the major difference between the two EO_8 derivative is in the reduction of T_c and the H_1 upper temperature limit ("melting temperature"). The I_1 melting temperature is unaltered by the change of OMe for OH. The V_1 and L_a phases of $C_{12}EO_8$ are both bounded by the liquid surfactant region ("L_2"), a

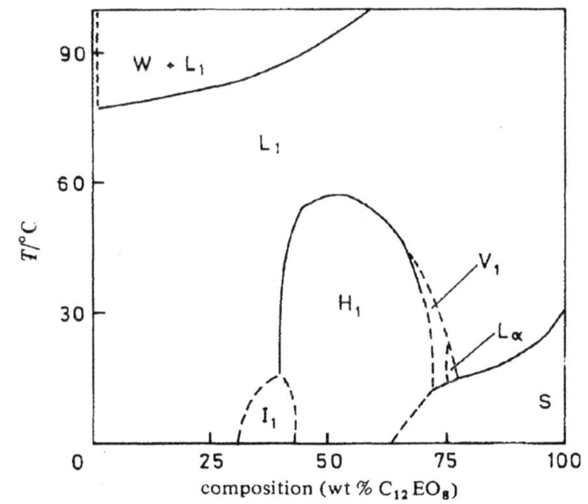

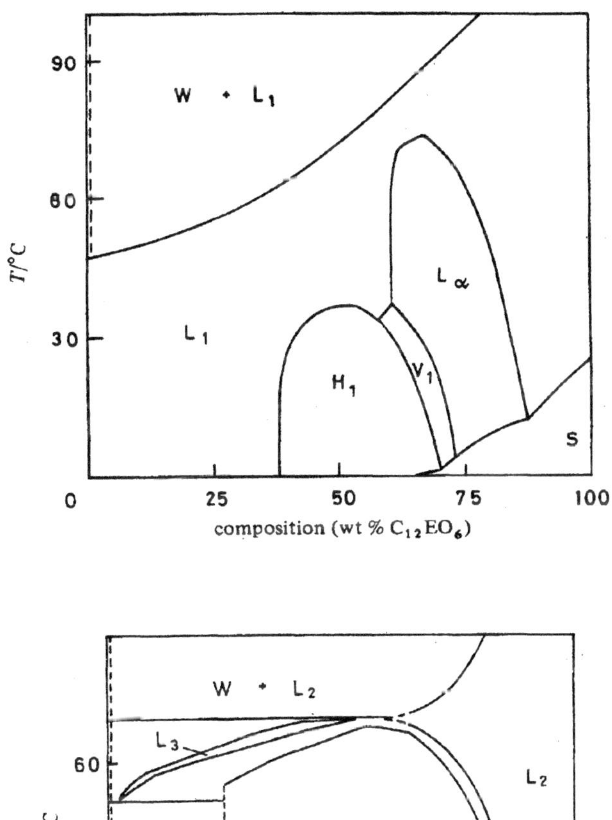

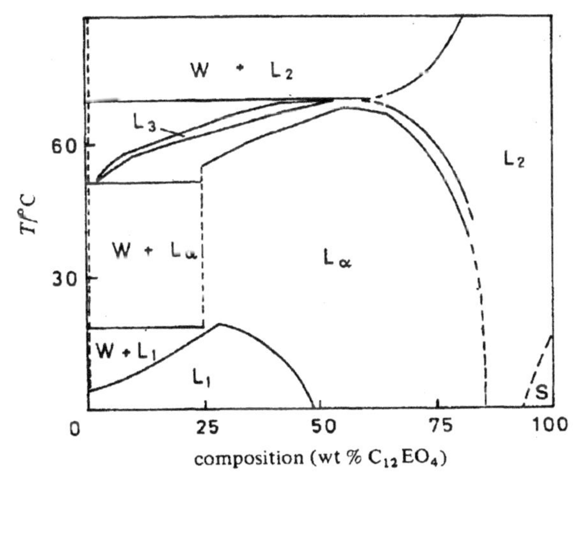

transition previously assigned to critical de-wetting [10]. While it is not possible a priori to predict the effect of the substitution on their stability, a shift of the L_2-region to lower surfactant concentration is expected. Hence, the absence of the V_1 and L_α phases for $C_{12}EO_8OMe$ is consistent with the previous ideas.

For the EO_6 and EO_4 surfactant pairs, we see a continuation of the trends in that transition temperatures in the range 50°–80°C for the OH surfactants are reduced by 20°C or more in the OMe derivatives, while those below 20°C for OH compounds are much less altered for OMe. We were unable to observe V_1 or H_1 phases for $C_{12}EO_4OMe$, possibly because of rapid ice formation. Note that the phase diagram for $C_{12}EO_4OMe$ is for mixtures with ^2H_2O, while normal water was used to obtain the data listed in Table 1. Substitution of ^2H_2O for ^1H_2O generally lowers mesophase temperature limits and cloud temperatures by 1–2°C [1, 28, 42].

Perhaps the most remarkable feature of the data in Table 1 is the small change in T_c for the EO_4 derivatives (3°C) compared to that of the EO_8 surfactants (25°C). This, together with the mesophase melting temperatures, represents strong evidence in favor of there being two distinct clouding processes as described in Section 2.

For completion, we include details of the phase behavior of $C_{12}EO_2$ (Fig. 3, Table 1) to compare with $C_{12}EO_4OMe$. This surfactant forms L_α and two V_2 phases, similar to the behavior of $C_{16}EO_3$ [42]. Using the microscope, it is easy to distinguish between the two V_2 phases by allowing the sample to dry at the edge or by adding water. The $V_2(1)$ region shows a refractive index discontinuity on drying while the $V_2(2)$ phase gives one in excess water. (See photograph in [43] for illustration.) It is now known that there are three distinct V_2 phases [44]. In an effort to distinguish between the two phases of this system, surfactant self-diffusion (D) measurements were made using the NMR pulsed gradient technique. The values listed in Table 2 show no difference between the two phases nor is

◆ Fig. 1. Phase diagrams of surfactant/water systems (from [10]); a) $C_{12}EO_8$/water; b) $C_{12}EO_6$/water; c) $C_{12}EO_4$/water. Symbols: I_1 = small micelle cubic phase; H_1 = hexagonal phase; V_1 = bicontinuous cubic phase; L_α = lamellar phase; L_1 = micellar solution; L_2 = concentrated surfactant liquid; L_3 = isotropic micellar solution; W = dilute surfactant solution

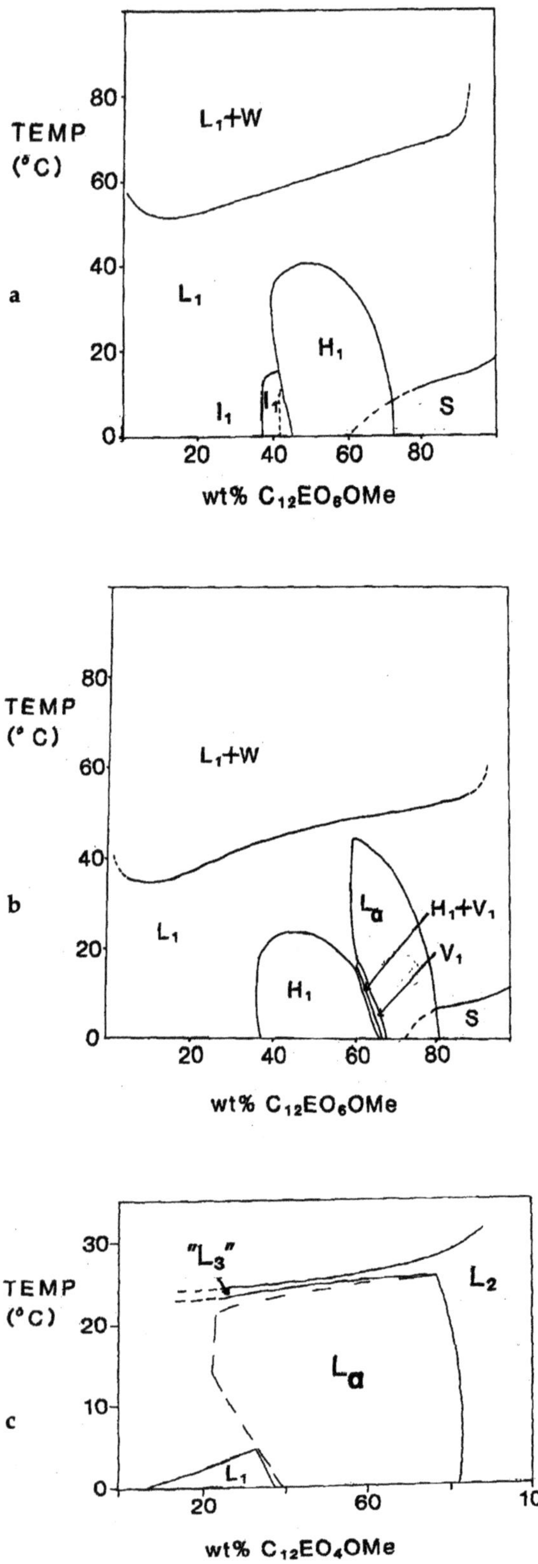

Fig. 2. Phase diagrams of methyl-capped surfactant/water systems; a) $C_{12}EO_8OMe$/water; b) $C_{12}EO_6OMe$/water; c) $C_{12}EO_4OMe$/water (^2H_2O). Symbols as in Fig. 1

Table 1. Cloud Temperatures (T_c) and Mesophase, upper temperature limits for $C_{12}EO_m$ and $C_{12}EO_mOMe$/water systems. (Temperatures in °C (±1°) for surfactnats in normal water)

Surfactant	T_c	I_1	H_1	V_1	L_α	V_2	L_3[a])
$C_{12}EO_4$ OH	3.6	—	−2[b]	?0	68	—	51.5—70
OMe	0.2	—	—	—	27	—	24—27[c])
$C_{12}EO_6$ OH	48	—	37	38	73	—	—
OMe	35	—	24	18	43	—	—
$C_{12}EO_8$ OH	77	16	59	43	22	—	—
OMe	52	15	41	—	—	—	—
$C_{12}EO_2$ OH	—	—	—	—	27.5	24—33[d] 27—33	15—34[c]

Notation: [a]) Indicates complete temperature range of L_3 region; [b]) Metastable — equilibrium structure is Ice + L_α; [c]) L_3 is continuous with L_2; [d]) Indicates complete temperature range of both V_2 phases.

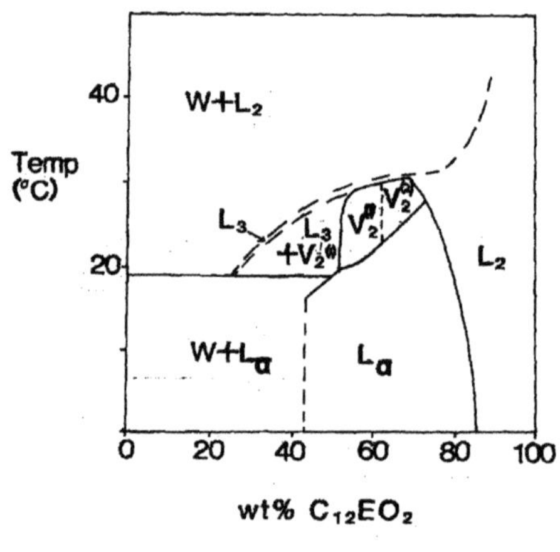

Fig. 3. Phase diagram of $C_{12}EO_2$/water (^2H_2O) system. Symbols as in Fig. 1; V_2 = reversed bicontinuous cubic phase

there any dependence on concentration. The magnitude is similar to that found for other systems [39]. One does expect D to decrease with increasing surfactant composition, but possible changes in EO conformation and water-binding make any detailed analysis very complicated. From the similarity of the mesophase upper temperature limits for $C_{12}EO_2$ and $C_{12}EO_4OMe$, one can judge that the terminal OH of the EO chain is equivalent to ca. 2 EO groups. The formation of V_2 rather than L_α

Table 2. Self-diffusion coefficients (D) for the cubic (V_2) phases of diethylene glycol dodecyl ether ($C_{12}EO_2$) at 25.5 °C

Composition (wt % $C_{12}EO_2$) (in ^2H_2O)	Phase	D (10^{-11} m^2 s^{-1})
52.3	$V_2(1)$	4.8 ± 0.2
55.0	$V_2(1)$	4.7
57.4	$V_2(1)$	5.1
59.8	$V_2(1)$	5.0
62.5	$V_2(2)$	4.9

reflects the smaller size of hydrated EO_2 compared to EO_4 OMe. As pointed out in our previous work [10] the fact that L_2 has boundaries with V_2, L_a, V_1, and H_1 rather than always with the same phase does strongly indicate a catastrophic change in EO interactions at the L_2/mesophase boundary.

b) Lower consolute boundaries

The detailed lower consolute boundaries for the various nonionic surfactants are shown in Fig. 4, while Table 3 lists the critical concentrations and temperatures. Visual observation of samples showed that 0.5% $C_{16}EO_6$ and $C_{22}{=}EO_6$ (hexa-ethylene glycol-cis-13-docosenyl ether, erucyl EO_6) solutions were highly viscous and viscoelastic (indicative of long rod micelles). Note that $C_{22}{=}EO_6$ was chosen for study because unsaturation reduces the efficiency of packing in the crystal, hence lowering the Krafft boundary without causing large changes in micelle structure. Only a few points were obtained for $C_{12}EO_4$OMe solutions because of limited availability of material, and because the cloud-curve paralleled that of $C_{12}EO_4$.

The onset of solution turbidity was independent of heating rate for most of the surfactants studied. However, for $C_{22}{=}EO_6$ solutions it was necessary to employ heating rates of ca. 10 °C/h or less to obtain reproducible results. Otherwise, temperatures of ~1° or more higher were obtained. we attribute this effect to possible slow equilibration of (large) micelle size on increasing temperature, although no detailed studies of the phenomenon were made.

The only unusual feature of the data is the change in slope of the surfactant-rich solution boundary for C_{22} = EO_6. Otherwise, surfactants clouding via the critical de-wetting mechanism have φ_c in the range 4–7%, well above that for clouding via the rod/disc-ruler transition (~1%). For one surfactnat where long rod micelles are present ($C_{16}EO_6$) we have previously suggested [22] that "ruler-shaped" micelles occur above T_c. To check that the cloud temperatures recorded really did represent a liquid-liquid phase separation process, solutions of $C_{12}EO_4$ were held at a temperature just above T_c (~0.05°) to allow separation of the two liquid phases. Two layers were formed after about 24 h. These gave a single phase on lowering the temperature by ~0.1° and mixing the sample.

The unusual shape of the $C_{22}{=}EO_6$ boundary referred to above is further proof of a change in micelle shape. The steep initial portion indicates slowly increasing inter-micellar attractions, while the flatter portion indicates more rapidly increasing attraction. This clearly points to a change in aggregate shape. At higher concentrations an H_1 phase occurs which melts at 12 °C. This is the first observation of an H_1 phase above T_c. However, at ca. 8° a low viscosity, birefringent phase occurs between H_1 and L_1. Low angle x-ray diffraction measurements indicate that it is a rod micelle phase with ordered elliptical micelles (i.e., rulers) and it is closely related to the L_a phase (rather than being similar to H_1) [45]. It is separated from H_1 at the low concentration boundary by L_1 (i.e., on increasing temperature the sequence is $H_1 \rightarrow L_1 \rightarrow$ ordered rulers $\rightarrow L_a$). Again, we have indirect evidence for ruler (disc) micelles at T_c.

5. Conclusion and general comments

The data described above give firm evidence to support the view that clouding of surfactants where no L_a + W dispersion occurs above T_c (Type 1) differs significantly from that of surfactants where this is observed (Type 2). Our hypothesis is that the EO-EO attractive interactions vary with temperature for spherical, circular/rod, and disc/ruler micelles (of the same *constant* size) as shown in Fig. 5. There is a slow, gradual increase in attraction at low temperature, followed by a very sharp increase at high temperatures. The latter correspondens to the "catastrophic dehydration" of EO groups leading to the disappearance of surfactant aggregates. Hence, spherical, rod, and ruler micellar attractions lie on a common curve at high

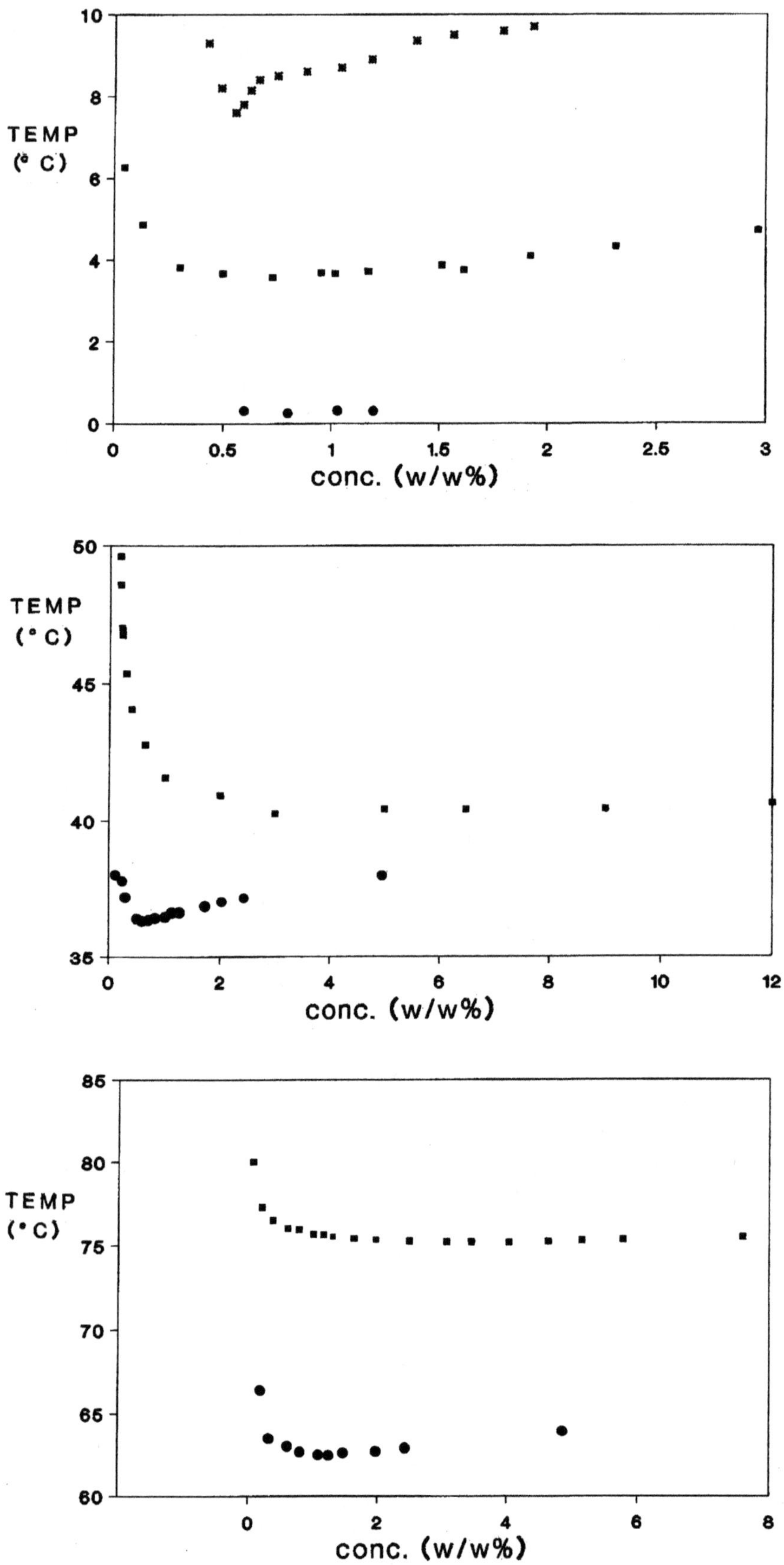

Fig. 4. Lower consolute boundaries for: a) $C_{12}EO_4$ (■), $C_{12}EO_4OMe$ (●) and $C_{22} = EO_6$ (*); b) C_8EO_4 (■) and $C_{16}EO_6$ (●); c) $C_{12}EO_8$ (■) and $C_{16}EO_8$ (●)

Table 3. Critical concentration, (φ_c) and critical temperatures (T_c) for nonionic surfactants

Surfactant	φ_c (wt %)	T_c (°C)
C_8EO_4	7 ± 1.5	40.4°
$C_{12}EO_4$	0.8 ± 0.1	3.6
$C_{12}EO_4$ OMe	0.8	0.3
$C_{12}EO_8$	4.0 ± 0.5	75.2
$C_{16}EO_6$	0.64 ± 0.05	36.4
$C_{16}EO_8$	1.2	62.5
$C_{22}{=}EO_6$	0.55 ± 0.03	7.6

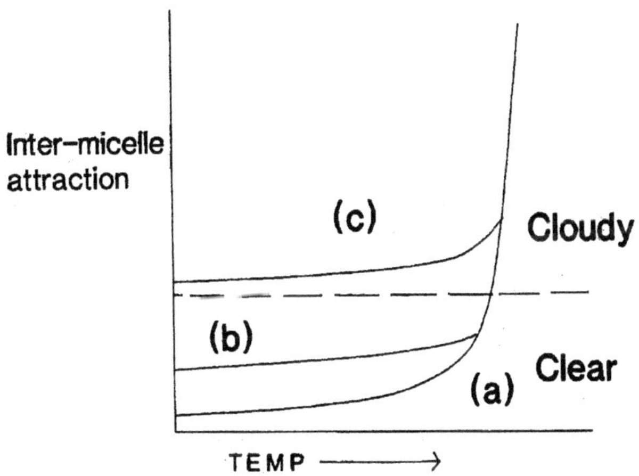

Fig. 5. Schematic illustration of the variation of intermicellar attraction vs temperature for a) spherical micelles, b) circular/rod micelles, c) disc-ruler micelles. The dotted line represents the strength of attraction required for phase separation

temperature, but vary in the sequence sphere < rod < ruler at low temperatures. Assuming the necessity of a roughly equal attractive force for phase separation of each type of micelle then T_c (clouding) clearly occurs at a much lower temperature for ruler micelles than for rods and spheres. Thus, the type (1) clouding arises from the "catastrophic dehydration" of EO groups. It usually involves small micelles, and micellar growth has no influence on T_c. In the second case (Type 2) the gradual dehydration of EO groups leads to micellar growth and, eventually, a micelle-shape change from circular rods to disc-ruler micelles. During this process clouding occurs. In practice the details of the process will be affected by the particular surfactant chemical structure, with the micelle aggregation number being generally larger, the longer the paraffin chain length. Micelle aggregation numbers could continuously increase through the transition, although some micelle-shape polydispersity must occur. This is because circular rod micelles have a diameter equal to about two all-trans alkyl chain lengths, while the ruler micelles have a diameter for the flat face of about one all-trans chain (a values of ~47 Å^2 and 44 Å^2, respectively). What exactly the detailed shapes are remains an unsolved problem.

In principle there is no reason why circular-rod micelles or spherical micelles cannot have T_c in the temperature interval well below the "catastrophic dehydration" region. In practice the attractive forces with currently available polyoxyethylene surfactants are just not large enough. It is possible that such behavior might occur with surfactants having long alkyl chains and multiple, short, EO head groups. A possible candidate would be a surfactant of general structure $C_n(EO_m)_p$, with $p \simeq 3$, $m \simeq 3$, 4, $n \simeq 16$, also containing CH_3 branching or unsaturation to reduce the Krafft boundary.

Given that the EO-EO attraction follows the form given in Fig. 5, we require an additional repulsion force critically dependent on temperature in the EO "catastrophic dehydration" region to make L_3 form from L_α or V_2. During this dehydration process it is likely that there are large fluctuations in hydration of EO groups, hence, the alkyl chain/water interface will become ill-defined. In this event the interface will be very elastic, giving rise to a weak repulsion from Helfrich-type undulation forces. Such interactions have already been invoked to account for L_3 formation in microemulsion systems [46]. These interactions could be highly dependent on temperature, having a maximum contribution just prior to the loss of definite surfactant aggregate structures.

References

1. Degiorgio V (1985) In: Degiorgio V, Corti M (eds) Physics of Amphiphiles: Micelles, Vesicles and Microemulsions. North Holland, Amsterdam, p 303 and refs therein
2. Nilsson PG, Wennerstrom H, Lindman B (1985) Chem Scripta 25:67
3. Degiorgio V, Piazza R (1987) Progr Colloid Polymer Sci 73:76
4. Nash ME, Jennings BR, Tiddy GJT (1987) J Colloid Interface Sci 120:542

5. Lum Wan JA, Warr GG, White LR, Grieser F (1987) Colloid and Polymer Sci 265:528
6. Herrington TM, Sahi SS (1988) J Colloid Interface Sci 121:107
7. Leng CA (1985) J Chem Soc Faraday Trans 2, 81:145
8. Evans H, Tildsey DJ, Leng CA (1987) J Chem Soc Faraday Trans 2, 83:1525
9. Zulauf M, Weckstrom K, Hayter JB, Degiorgio V, Corti M (1985) J Phys Chem 89:3411
10. Mitchell DJ, Tiddy GJT, Waring L, Bostock T, McDonald MP (1983) J Chem Soc Faraday Trans 1, 79:975
11. Nilsson PG, Wennerstrom H, Lindman B (1983) J Phys Chem 87:1377
12. Brown W, Johnson R, Stilbs P, Lindman B (1983) J Phys Chem 87:4548
13. Corti M, Minero C, Degiorgio V (1984) J Phys Chem 88:309
14. Binana-Limbelé W, Zana R (1988) J Colloid Interface Sci 121:81
15. Degiorgio V, Piazza R, Corti M, Minero C (1985) J Phys Chem 82:1025
16. Weckstrom K (1985) Chem Phys Lett 119:503
17. Weckstrom K, Zulauf M (1985) J Chem Soc Faraday Trans 1, 81:2947
18. Lang JC, Morgan RD (1980) J Chem Phys 73:5849
19. Goldfarb J, Sepulveda L (1969) J Colloid Interface Sci 31:454
20. Kjellander R (1982) J Chem Soc Faraday Trans 2, 78:2025
21. Cummins PG, Hayter J, Penfold J, Staples EJ (1987) Chem Phys Lett 138:436
22. Baxandall LG, Lal M, Rendall K, Tiddy GJT (1986) Proc Int Symp "Surfactants in Solutions" New Delhi, to be published ... eventually!
23. Israelachvili JN, Mitchell DJ, Ninham BW (1976) J Chem Soc Faraday Trans 2, 72:1525
24. Tiddy GJT (1985) in Eicke H-F (ed) "Modern Trends of Colloid Science in Chemistry and Biology", Birkhauser, Basel, p 148
25. I_1 cubic phase can contain short rods rather than spheres, but even here the *a* value is expected to be >60 Å^2
26. Wennerstrom H (1979) J Colloid Interface Sci 68:590
27. Other mechanisms that could also play a role are entropy effects associated with micelles having small aggregation numbers, and surface curvature forces which depend on micelle radius
28. Tiddy GJT et al., unpublished results
29. Ninham BW, Evans DF (1986) Faraday Disc Chem Soc 81:1
30. Claesson PM, Kjellander R, Stenius P, Christenson HK (1986) J Chem Soc Faraday Trans 1, 82:2735
31. Andersson M, Karlstrom G (1985) J Phys Chem 87:4957
32. Karlstrom G (1985) J Phys Chem 87:4762
33. Lawrence ASC (1969) Mol Cryst Liquid Cryst 7:1
34. Lutton ES (1965) J Amer Oil Chem Soc 42:1068
35. Larsson K (1967) Z Phys Chem (Frankfurt) 56:173
36. Nilsson PG, Lindman B (1984) J Phys Chem 88:4764
37. Evans H: URPSL, personal communication
38. Carvell M, Leng CA, Leng F, Tiddy GJT (1987) Chem Phys Lett 137:188
39. Lindblom G, Larsson K, Johansson L, Fontell K, Forsen S (1979) J Amer Chem Soc 101:5465
40. Jahns E, Finkelmann H (1987) Colloid Polymer Sci 265:304
41. Yeates SG, Craven JR, Mobbs RH, Booth C (1986) J Chem Soc Faraday Trans 1, 82:1865
42. Adam CD, Durrant JA, Lowry MR, Tiddy GJT (1984) J Chem Soc Faraday Trans 1, 80:789
43. Tiddy GJT, Rendall K, Galsworthy P (1982) Mol Cryst Liquid Cryst 72:147
44. Andersson S, Hyde ST, Larsson K, Lidin S (1988) Chem Rev 88:221
45. Funari S, Holmes M, Tiddy GJT, unpublished results
46. Porte G, Appel J, Bassereau P, Marignon J, paper presented at 7th S.I.S. Conference, Ottawa 1988

Authors' address:

G. J. T. Tiddy
Unilever Research Port Sunlight Laboratory
Port Sunlight
Merseyside, United Kingdom L63 35W

Physical organogels: solvent-induced structural variations in rod-like surfactant aggregates

P. Terech

Institut Laue-Langevin, 156X, Grenoble Cedex, France

Abstract: Small-angle scattering (neutrons and x-rays) and optical polarizing microscopy are used to study the influence of the nature of the solvent upon the aggregation reaction of a steroid surfactant in hydrocarbons in the related gel phase. Firstly, on a macroscopic scale, it is shown that the stability of the gels is better in cyclohexane or trans-decalin than in other hydrocarbons. secondly, on a microscopic scale, the diameter of the rod-like aggregates increases up to 50% from cyclohexane (Ø = 105 Å) to methyl-cyclohexane (Ø = 157 Å) while the aggregates become much more monodisperse. Absolute intensity measurements, radii of gyration determination, model fittings of the full scattering curves and Hankel transformation yield a coherent picture of the structural modifications of the gel network. A conformational rod-like solute/solvent matching is correlated with the observations.

Key words: Gel; scattering; neutron; x-ray; solvent

I. Introduction

Colloid science in non-aqueous media is involved in numerous fundamental and applied fields. For instance, petroleum and derivative industries commonly deal with apolar media from oil recovery problems to the delicate methods of surfactant adjunction in oil, lubricating greases [1, 2] or fuels. Pharmaceutic industries try to entrap biocompatible molecules and drugs in organic solvents [3]. Numerous industrial applications (agricultural, food, medicinal, paint and asphalt industries [4]) from ternary or quaternary systems require a definite knowledge of the influence of the nature of the solvent on the structure and the stability of the colloidal systems. This question constitutes the fundamental topic of the paper.

In organic media, inverse structures are obtained where the polar or ionic head (in case of ionic surfactant) builds up the polar core of the aggregate while the apolar hydrocarbon tails form the outer shell of the structure. This point emphasizes the basic differences between surfactant aggregation in water and organic solvents. In aqueous solutions, the hydrophobic interactions explain the transfer of the alkyl part of the surfactant molecules from the bulk water to a hydrocarbon environment. By contrast, in apolar solutions the dipole-dipole interactions between polar heads are the driving force of the aggregation process. When compared with the aqueous systems, the situation in organic solvents is not so definite. Among the various factors influencing the standard free energy of the aggregate formation [5], it has been felt that the influence of solvent could not be restricted to the solvent polarizibility or solubility parameter for instance, but should include some specific solute/solvent sterical and molecular packing terms. Demonstratively, we can mention the example of isotactic polystyrene physical gels in decalin [6]. In this case, the solvent dependence of the polymer chain trajectory and of the relaxation modulus in compression is clearly established for cis and trans-decalin solvents.

In this paper, we are concerned by physical gels from a non-ionic surfactant in hydrocarbons (organogels). We study the influence of the solvent both on the macroscopic behaviors of these colloidal systems and the structural modifications of the aggregates. This surfactant is a steroid derivative (SNO) in which the D-ring has been modified [7] to incorporate a nitroxide spin label NO (Fig. 1). In various apolar hydrocarbons, supersaturated solutions of SNO give rise to gel phases. In cyclohexane they have been extensively studied by us [8—10]. The gel phase is constituted by long (up to several microns) rod-like aggregates, the principal diameter of which is $\varphi \sim 100$ Å as demonstrated by SANS experiments [8] and confirmed by freeze-fracture and etching electron microscopy (EM) experiments [9]. These aggregates are fibers constituting an interconnected infinite network which ensure the viscoelastic behavior of the gel phases. EM has revealed two main fibrillar "reticulation" modes in the network. The bodily contact of fibers on variable axial fiber distances constitute a first way of fiber connection. The chiral structure of individual fibers, originating from the molecular asymmetry of the SNO derivative, provides a second fibrillar entanglement mechanism by a helical exchange of some of the constitutive filaments.

In the present study, we consider the stability of SNO organogels and the microscopic structure as a function of the molecular shape of various hydrocarbon solvents (see Fig. 1). These solvents are cyclohexane, methyl-cyclohexane, ethyl-cyclohexane, heptane, octane, and decalins which all have nearly the same dielectric constant (ca. 2.0). X-ray and neutron small-angle scattering and optical polarizing microscopy will be used.

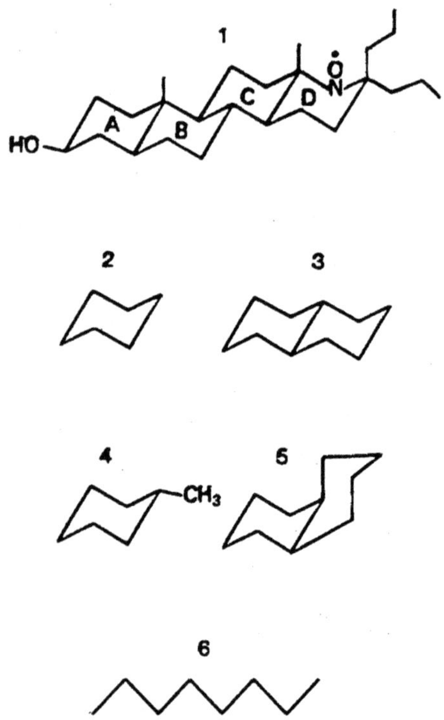

Fig. 1. Chemical formulae of the steroid nitroxide SNO (1) and some of the solvents used: cyclohexane (2), trans-decalin (3), methyl-cyclohexane (4), cis-decalin (5) (only one of the two conformations is represented), and n-octane (6)

II. Experimental

Details concerning the synthesis and preparation of gel samples are described in [7, 10]. Solvents were used as received from Aldrich; the water content is less than 0.01%. The same quality of the solvents was used for this study. In the following, cyclohexane and methyl-cyclohexane solvents will be referred as H12, H14 or D12, D14 according to if they are normal or deuterated, respectively. Most spectra were recorded at room temperature for a $10^{-2}/5 \times 10^{-2}$ gcm^{-3} concentration range in the gel state for each solvent.

An Olympus BH-2 microscope equipped with crossed polarisers is used for optical characterization of the gel phases.

SANS experiments were carried out with the D11 spectrometer at the high flux reactor of the Institute Laue Langevin (ILL Grenoble-France) with only D12 and D14 gels because of deuteration constraints of the solvents. Samples were held in 1-mm path QS quartz cells and two-dimensional detectors were used for neutron data collection. Analysis was achieved as usual [8].

SAXS experiments used the synchrotron x-ray monochromatic radiation (Ge crystal, wavelength λ = 1.62 Å of LURE-Orsay (France)). Samples in protonated solvents were tightly kept in aluminium and copper cells equipped with capton windows and teflon spacers. A linear position sensitive detector was used. Corrections for background scattering and transmission have been performed. We have studied six samples: cyclohexane, ethyl-cyclohexane, heptane, octane, trans-decalin, and the mixture cis/trans decalin.

The experimental momentum transfer range studied was 0.002—0.14 Å^{-1} and 0.006—0.5 Å^{-1} for SANS and SAXS experiments, respectively.

III. Optical observations

The stability of cyclohexane gels is such that even over several years the colloidal phase is stable and no phase separation or crystallization can be seen. The medium is optically isotropic in the concentration range studied as shown by the micrograph of Fig. 2a. These gels have been optically studied in detail in reference [10].

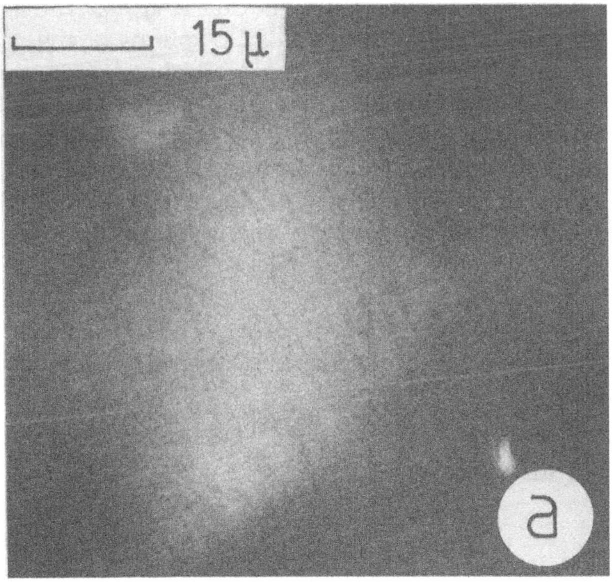

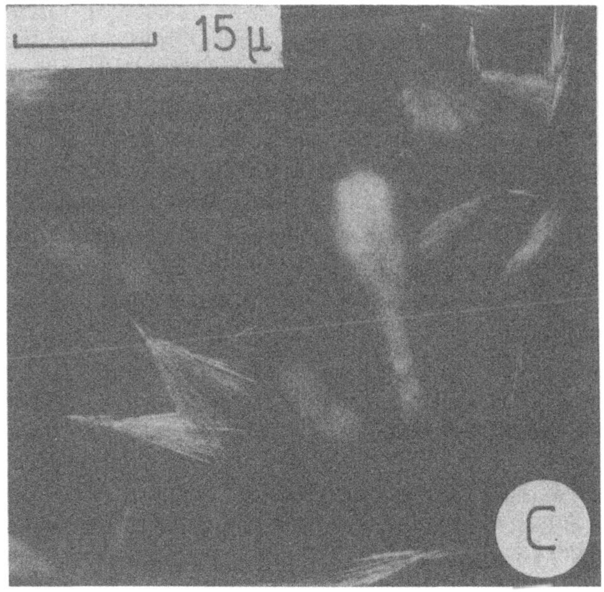

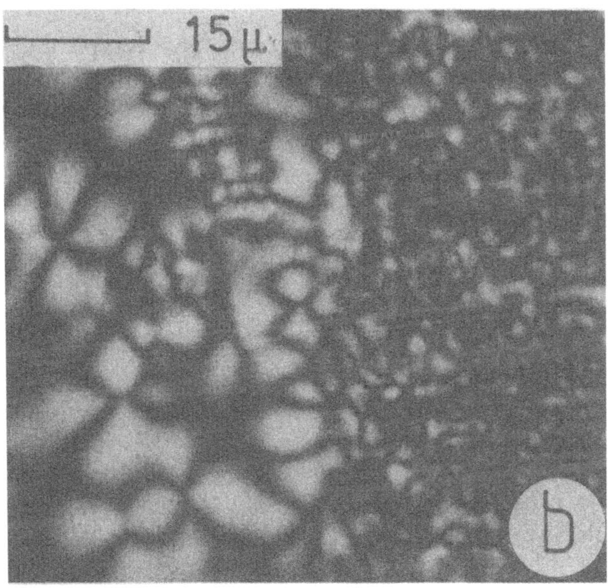

Fig. 2. Optical polarizing microscopy of SNO gel samples. (a) cyclohexane; (b) methyl-cyclohexane; (c) ethyl-cyclohexane; $C_0 = 4 \times 10^{-2}$ gcm^{-3}

By contrast, the stability of the gel phases in the other hydrocarbons is generally poorer and strongly dependent upon the nature of the hydrocarbon and the steroid concentration. The gel phases are metastable and evolve with a very variable time scale (from hours to weeks). For example, if we change the solvent from cyclohexane to methyl-cyclohexane, at a given concentration, large anisotropic domains (from microns to millimeters) appear. Birefringent textures such as Schlieren textures or extinction crosses confined in droplets are observed (see Fig. 2b). Despite this new situation, the gel state is retained. If we change the solvent from cyclohexane to ethyl-cyclohexane, the gel phase gives a growing dispersion of germination sites. The crystallites grow as fine needles (see Fig. 2c) while a phase separation process finally destroys the gel phase. The heterogeneous final state is then constituted by a solid SNO deposit in a nearly pure solvent phase.

Samples with decalins give another striking example of solvent differentiation by the system. Cis-decalin gels ($C \sim 0.035$ gcm^{-3}) are very strongly birefringent, opaque. The situation is completely different with trans-decalin where the gel samples are transparent and isotropic for concentrations up to ca. 0.12 gcm^{-3} at least.

At this stage, two classes of solvents can be distinguished: the first, cyclohexane and trans-decalin which give rather stable isotropic gel states, and the second, methyl-cyclohexane, ethyl-cyclohexane, cis-decalin, heptane, and octane, which give rather unstable and birefringent gel states. We note that trans-decalin is the most stable isomer in which the two chair forms are joined together with the ring-junction hydrogens being on opposite sides of the molecule (Fig. 1). Trans-decalin is a relatively rigid system and presents the same trans chair-chair conformation as that of the A, B, C, D steroid ring-junctions (see Fig. 1) of the SNO solute. On the contrary, we remark that in cis-decalin solvent which is a relatively flexible system where a dynamical motion of ring inversion occurs in the chair conformation, gels are only kinetically stable and evolve toward a biphasic system. A qualitative correlation can be drawn between the shape of the SNO solute itself and that of solvents: a conformational aggregate/solvent matching should be implied.

IV. Structural study

We first focus our attention on the neutron results. Notations used are those of a previous SANS study [8] which has emphasized the unidimensional character of the objects constituting the gel network. The conclusion of this work was that the best simulation of the neutron-scattering curves is obtained with a symmetrical double-helix model. However, the much simpler cylindrical rod model gives a satisfactory agreement. We shall adopt this latter model for sake of simplicity because we are not interested here in details of the cross-section of the steroid fibers.

a) Theoretical background

For an isotropic dispersion of non-interacting homogeneous cylindrical objects of radius r_0 of infinite length L, the neutron scattered intensity is given by:

$$\left(\frac{d\Sigma}{d\Omega}\right) = \left(\frac{d\Sigma}{d\Omega}\right)_0 \left[2\,\frac{J_1(Qr_0)}{Qr_0}\right]^2 \tag{1}$$

$$\left(\frac{d\Sigma}{d\Omega}\right)_0 = \frac{\pi c}{Q}\,\Delta b^2 M_L\,, \tag{2}$$

where $(d\Sigma/d\Omega)$ is the so-called absolute differential scattering cross-section, Q the momentum transfer, J_1 the first order Bessel function of the first kind, c the rod concentration ($g \cdot cm^{-3}$), M_L the mass per unit length of the rod, and Δb (cm/g) is the specific contrast given by:

$$\Delta b = b_2 - \rho_s v_2\,, \tag{3}$$

where b_2 is the specific scattering length of the aggregate (cm/g), v_2 is the specific volume of the aggregates (cm^3/g) and ρ_s the scattering length per unit volume of solvent (cm/cm^3).

If we take into account a Gaussian distribution of the rod radius centered around r_0 and characterized by the polydispersity ε where $\Delta R_{1/2}$ is the full width at half maximum of the distribution:

$$\sigma = \Delta R_{1/2}/2r_0\,, \tag{4}$$

the expression of the scattered intensity becomes:

$$\left(\frac{d\Sigma}{d\Omega}\right) = \left(\frac{d\Sigma}{d\Omega}\right)_0 \cdot \frac{1}{\varepsilon} \cdot \sqrt{\frac{\ln 2}{\pi}} \int \left[2\,\frac{J_1(Qr_0x)}{Qr_0x}\right]^2 \times \exp\left(-\frac{\ln 2}{\varepsilon^2}(x-1)^2\right) dx\,, \tag{5}$$

where $x = r/r_0$.

Equations (1) and (5) account for the typical Q^{-1} dependence of the scattered intensity at low angles and show that the appropriate graphic representation for long filaments is a plot of the product of the scattered intensity by the momentum transfer Q vs Q.

b) Results

Figures 3 and 4 show typical scattering curves for cyclohexane and another of the most representative solvents for SANS and SAXS techniques, respectively. The two techniques offer a very useful complement for accurate, absolute intensity measurements and for a thorough comparative investigation of the protonated hydrocarbons, respectively. For cyclohexane gels, the scattering curve reveals a poorly defined shoulder in the 0.08 $Å^{-1}$ momentum transfer region. In the case of SANS, these D12 gels are described in detail in [8]. For the other hydrocarbons gels, the deuterated methyl-cyclohexane gelling solutions D14 have been studied by SANS, while we have investigated the normal ethyl-cyclohexane, heptane, octane, trans-decalin, and the 50/50 mixture cis/trans decalin organogels with SAXS. For the normal hydrocarbons mentioned above, the scattering curve profile is different from that of cyclohexane and always reveals two well-defined oscillations (Fig. 4). The first maximum is located at $Q \sim 0.06$ $Å^{-1}$ while a second maximum

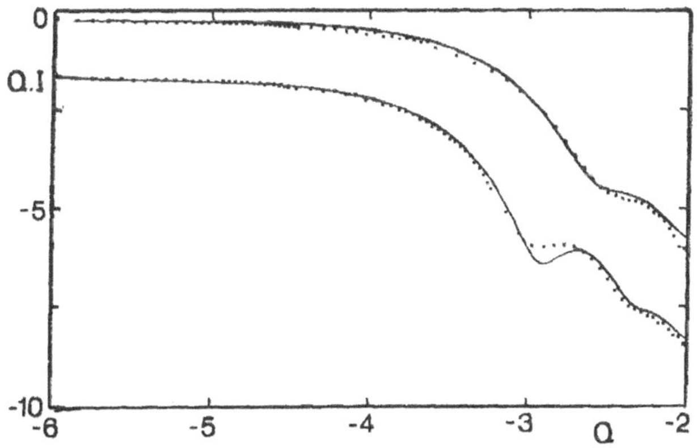

Fig. 3. Double logarithmic (Ln) plots $Q \cdot I(Q)$ vs Q for SANS experiments. The ordinates are in arbitrary units and the scattering curves have been vertically translated for sake of clarity. Each curve is obtained by three different experiments at L = 20 m, 6 m, 1.6 m sample-detector distances at λ = 10.0 Å. Points are experimental values and full lines are the corresponding best adjustments according the full homogeneous cylindrical rod model. Upper curve: D12, $C_0 = 2.3 \times 10^{-2}$ gcm^{-3}, fit $R_0 = 54$ Å, $\varepsilon = 0.2$; lower curve: D14, $C_0 = 2.1 \times 10^{-2}$ gcm^{-3}, fit $R_0 = 75$ Å, $\varepsilon = 0.12$

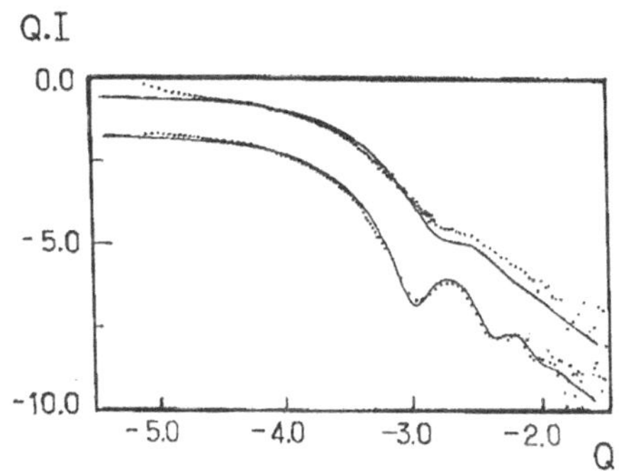

Fig. 4. Double logarithmic (Ln) plots $Q \cdot I(Q)$ vs Q for x-ray experiments. The ordinates are in arbitrary units and the scattering curves have b een vertically shifted for sake of clarity. Each curve results from experiments at L = 2.17 m and 0.506 m sample-detector distances at λ = 1.62 Å. Points are experimental values and full lines are the corresponding best adjustments according the full homogeneous cylindrical rod model. From top to bottom: (1) cyclohexane, $C_0 = 0.975 \times 10^{-2}$ gcm^{-3}, fit $R_0 = 66$ Å, $\varepsilon = 0.22$; (2) ethyl-cyclohexane, $C_0 = 1.17 \times 10^{-2}$ gcm^{-3}, fit $R_0 = 79$ Å, $\varepsilon = 0.12$

can be seen at $Q \sim 0.108$ Å^{-1}. Nearly superposable curves are obtained by SAXS when whichever of the above-mentioned solvent was used. With the deuterated methyl-cyclohexane D14 samples, the SANS scattering curve is comparable to that obtained by SAXS (see Fig. 3) and also displays two well-resolved oscillations.

The following structural features can be extracted:

1) Low Q region

i) A typical flat plateau in the range $Q\langle l \rangle \geqslant 1$ and $QR_c \ll 1$ is clearly observed down to very low Q-values, confirming the rod-like character of the aggregates. R_c is the radius of gyration of the normal section of the rod and l is the persistence length, i.e., the average length below which the rod is assumed to be rigid.

ii) For the central part of the scattering, the corresponding "Guinier exponential decay" of intensity is governed by (6)

$$\left(Q \frac{d\Sigma}{d\Omega}\right)_{Q\to 0} = \left(Q \frac{d\Sigma}{d\Omega}\right)_0 \exp(-R_c^2 Q^2/2) \,. \quad (6)$$

For deuterated solvents, the corresponding "Guinier plots" Log($Q \cdot d\Sigma/d\Omega$) vs Q^2 for $Q <$ 0.03 Å^{-1} provide the radii of gyration mentioned in Table 1. A strikingly large variation of the gyration radius of about 50% from 37.5 Å (cyclohexane D12) to 55.4 Å (methyl-cyclohexane D14) is revealed. If we assume a homogeneous circular cross-section for the constitutive fibers, the related geometrical diameters (Ø) increase from 105 Å (D12 gels) to 157 Å (D14 gels) according to [7]

$$r = R_c\sqrt{2} \,. \quad (7)$$

Table 1. Results from "Guinier plots" and absolute intensity determinations. D exponent refers to deuterated solvents of SANS experiments, C_0 is the total SNO concentration (gcm^{-3}), R_c is the radius of gyration of the rod cross-section (Å), R_0 is the related geometrical radius (Å), $(Q \cdot d\Sigma/d\Omega)_0$ the extrapolated intensity (Å^{-1} cm^{-1}), and $n_L p$ is the number of SNO molecules per unit length of rod (molecules Å^{-1})

	$C_0 \times 10^2$	R_c	R_0	$(Q \cdot d\Sigma/d\Omega)_0$	$n_L p$
Cyclohexane (D12)	2.3	37.5	53	0.700	4.5
Methyl-cyclohexane (D14)	2.1	55.4	78.5	1.431	10.6
	7.1	54.8	77.5	4.522	9.8

For cyclohexane gels, a spread of the experimental gyration radii values (see [8]) is noticed according to the concentration and history of the gel samples. The mean value retained is ca. 40 Å. (This point will be discussed in section V.)

iii) Absolute intensities $(Q \cdot d\Sigma/d\Omega)_0$ are deduced from extrapolations to $Q = 0$ of the "Guinier plots" and are analyzed with Eqs. (2) and (6).

The specific contrast of steroid rods in cyclohexane b_{D12} ($-5.7358 \cdot 10^{10}$ cm/g) is nearly identical to that in methyl-cyclohexane b_{D14} ($-5.6612 \cdot 10^{10}$ cm/g) as calculated with $b_2 = 0.349 \cdot 10^{10}$ cm/g, $b_{D12} = b_{D14} = 7.52236 \cdot 10^{10}$ cm/g, $v_2 = 0.909$ cm^3/g, $\rho_{D12} = 0.890$ g/cm^3, and $\rho_{D14} = 0.879$ g/cm^3 [8]. Consequently, the observed variations of the central intensities are due on the one hand, to differences in the fraction p (relation (8)) of the initial steroid concentration C_0 taking part in the rods (phase diagram origin), and on the other hand to differences in the number of steroid molecules per unit length of the rod n_L (molecular packing origin)

$$p = c/c_0 \tag{8}$$

$$M_L = n_L M/N \tag{9}$$

where N is the Avogadro number and M is the molecular mass of SNO.

As shown in Table 1, the main feature is that the quantity pn_L is increased by at least a factor 2 from cyclohexane to methyl-cyclohexane gels. With $p_{D12} \sim 0.75$ (determined from the related phase diagram [8]) $\sim p_{D14}$, we can retain $n_L(D12) \sim 6$ molecules Å^{-1} and $n_L(D14) \sim 13.6$ molecules Å^{-1}. This increase of molecular weight per unit length from D12 to D14 is compatible with the observed growth of geometrical radius since the following equality is fulfilled:

$$(\varnothing_{D14}/\varnothing_{D12})^2 = (157/105)^2 = 2.24$$

$$\# \ (n_L(D12)/(n_L(D14)) = 13.6/6 = 2.27 \ .$$

The diameter growth occurs at constant molecular packing from D12 to D14 gels.

2) Large Q region

For higher Q-values ($Q > 0.005$ Å^{-1}) the more or less damped intensity oscillations are representative for the cross-section geometry and its related polydispersity ε. As can be seen from Figs. 3 and 4, these extrema are much less pronounced in cyclohexane than for other solvents. These differences are significant for both SANS and SAXS techniques.

Although the initial assumption of the rod-like character of the SNO aggregates (expression (5)) is sufficient to extract directly the structural information from the Q-space of the scattering curves, we find it in practice more convenient and explicit to treat the SANS data in real r-space while the SAXS data will be treated in Q-space. This choice is justified by the extended low-Q range of SANS data and their related quality resulting from corrected and calibrated bidimensional arrays of intensity values.

i) SANS data.

Extraction of the structural information concerning the shape and internal structure of the diffracting aggregates is done by computing the SANS distance distribution function $p(r)$. Previous contrast variation experiments [8] have indicated that the fibers are homogeneous with respect to the neutron-scattering densities. In these conditions, $p(r)$ has a clear significance in real space and represents the distribution of distances r in the scattering fibers. The peculiar rod-like geometry of the steroid aggregates which is that of a cross-section of maximum dimension ($\varnothing \sim 100$ Å) considerably smaller than the mean length of the fibers (microns) argues for the use of the Hankel [11] transformation (see expression (10)) rather than the usual Fourier transformation for three-dimensional objects:

$$P_c(r) = \frac{1}{2\pi} \int_0^\infty \left(Q \frac{d\Sigma}{d\Omega}\right) J_0(Qr) Qr dQ \ , \tag{10}$$

where $P_c(r)$ is the distance distribution function within the cross-section of the rods.

Because the Q-range of SANS is large enough, especially in the low Q region, the artefacts due to Q truncation are avoided. The quality of corrected calibrated SANS data prevent the use of any data extrapolation. Comparisons between experimental data and theoretical adjustments are performed on the same experimental Q-values set. Figure 5 clearly confirms the structural modification of the rod-like aggregates from cyclohexane to methyl-cyclohexane gels. A good agreement is obtained with a homogeneous cylinder model. Table 2 compares the results of the best theoretical adjustments with the

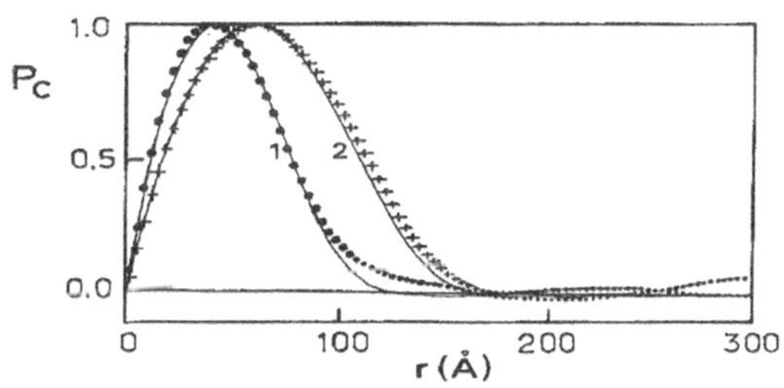

Fig. 5. Distance distribution function of the cross-section (Hankel transformation) for SNO gels (SANS experiments). Points are experimental values and full lines are the corresponding best fits of Eq. (5). (1): D12 gels; (2): D14 gels (see Table 2 and text)

Table 2. Comparison between structural parameters deduced from the best Hankel fits for D12 and D14 SANS experiments with data from "Guinier plots". R_0, R_c^H, and R_c^G are, respectively, the geometrical rod radius (Å), the theoretical gyration radius (Å), and the experimental radius of gyration. *H* and *G* exponents refer to Hankel and "Guinier plots", respectively; ε is the polydispersity of radii (see text)

	R_0^H	ε	R_c^H	R_c^G
D12	54	0.2	42	$37 \ll 44$
D14	72	0.12	54	55.5

Hankel experimental data and with Guinier analysis for SANS experiments. A very good agreement is found for methyl-cyclohexane gels while for cyclohexane samples, a noticeable spread of gyration radii and polydispersity is again observed [8]. Fibers in D14 samples appear to be much more monodisperse ($\varepsilon = 0.12$) than in D12 gels ($\varepsilon = 0.2$). At this stage, we observe simultaneously an increase of the geometrical radius from D12 to D14 gels and a decrease of the related polydispersity ε. From these Hankel fits, we retain the geometrical radius r_0 and polydispersity σ values to calculate the corresponding scattering curves (*Q*-space) for isolated full homogeneous cylinders. Comparison with the experimental scattering curves is shown in Fig. 3 for SANS data.

ii) SAXS data.

Using Eq. (5), a direct adjustment on the scattering curves is made (Fig. 4). The *r* values are slightly different from that of SANS because of a slight departure from the Q^{-1} behavior in the low *Q* region of the protonated samples. The structural modifications (concerning *r* and ε) are qualitatively confirmed. We can add that the rod-like geometrical model is acceptable as a first approximation, despite minor discrepancies in the large *Q*-range due to the real helical substructure [8] of the fibers.

V. Discussion

The importance of the nature of the solvent on the characteristic SNO gel features has been demonstrated both on a macroscopic and microscopic scales.

The poor resolution of the small-angle scattering technique (especially neutrons) is not adapted to detail the internal cross-section structure. For this purpose, oriented gel samples are required. The reader is referred to [12] for a preliminary example of such a study.

On the basis of the SAXS spectra of the different solvents studied, we note the very specific role of cyclohexane in the elaboration of the gel network structure.

An increase of the rod diameter and a simultaneous decrease of polydispersity of the aggregate are found when cyclohexane is replaced by methyl-cyclohexane in gel samples.

The ability of methyl-cyclohexane to give anisotropic gels is pointed out by the obtention of birefringent textures of Fig. 2b and of anisotropic neutron scattering (detected as elliptic signals instead of circular isotropic signals). The excluded volume effects between rods define an isotropic/nematic-like transition for a concentration which is as low as the persistence length of the rods is large. We assume that this tendency to rod orientation when comparing methyl-cyclohexane or cis-decalin gels and cyclohexane or trans-decalin gels is consistent with an increase of the rod diameter and persistence lengths. We note that collagen [13] also exhibits a liquid-crystalline behavior which can be seen by the spontaneous occurrence of birefringence. The isotropic to mesophase transition is induced when the rigidity of the fibers is increased by a pH increase while oriented collagen gels can be obtained in magnetic fields.

An additional consequence of the rigidity increase is the modification of the probability of entwining for fibers in the "reticulation zones" of the gel network. Lower rigidity is associated with higher polydispersity degree of the gyration radii:

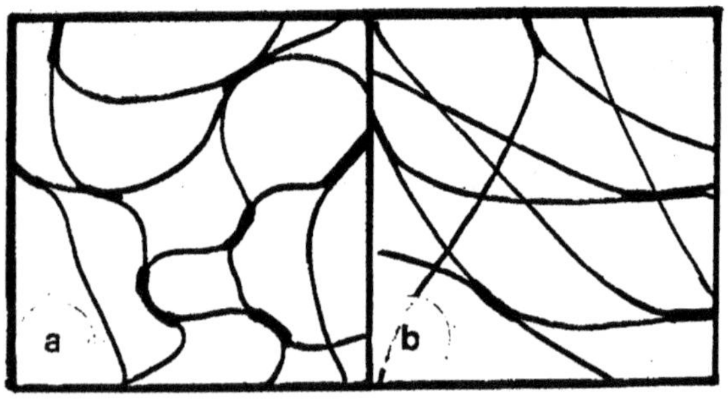

Fig. 6. Schematic representation of the SNO gel network and "reticulation zones" in cyclohexane (a) and ethyl-cyclohexane (b). In cyclohexane the rigidity of the fibers is such that the network is rather entangled and, consequently, the polydispersity values are found to be higher. The opposite situation is found in ethyl-cyclohexane gels for which the increased diameter is not represented in the scheme. In this case, the increased rigidity of the fibers is a limitation of the entanglement process and the polydispersity values are decreased

this is the case of cyclohexane gels. Figure 6 is a modelization of the topography of the gel network in a solvent conformationally identical to the steroid solute (cyclohexane, trans-decalin) and in a sterically different solvent (methyl-cyclohexane, cis-decalin, octane and others).

An analogy is found with isotactic polystyrene (iPS) gelling solutions. In this polymeric system, crystal growth, structure of the constitutive coils, ageing and the related mechanical properties are dependent upon the thermal kinetic story of the sample. SANS studies of the chain trajectory in the gel phases demonstrate a decrease of rigidity of the polymer chains from cis to trans decalin. As long as we consider the network topographies, the crystal growth process, the quenching procedure for preparing the gel phases, the influence of the nature of the solvent, both on macroscopic properties (mechanical and optical behaviors) and microscopic properties (persistence lengths and geometrical features), the similarities between SNO and iPS gels in organic solvents indicate a more general relationship between surfactant-built chains and polymers in solution. A conformational or sterical matching between solvent and rod-like aggregates in physical gels could be implied.

Apart from the fundamental point of view, the applied consequences are numerous, as mentioned in the introduction. We recall the case of lubricating greases manufactured from oils of various hydrocarbon compositions. It is easily understandable that the fibrillar structural modifications induced by the nature of oil can have drastic effects on the tribologic properties of the greases for which these systems are used.

Acknowlegements

The author is very grateful to Dr. F. Volino for his continuous interest and helpful discussions and Dr. R. Ramasseul for providing the steroid derivative.

References

1. Polishuk AT (1977) J Am Soc Lub Eng 33:133
2. Uzu Y (1975) J Japan Oil Chem Soc 24:261
3. Scartazzini R, Luisi PL (1988) J Phys Chem 92:829
4. Lissant KJ (1974) In: Emulsions and emulsion technology, M Dekker, New York
5. Eicke HF (1980) In: Topics in Current Chemistry: Micelles, vol 87, Springer-Verlag, New York, pp 91
6. Guenet JM (1987) Macromolecules 20:2874; ibid, Guenet JM, Lotz B, Wittmann JC (1985) 18:420
7. Martin-Borret O, Ramasseul R, Rassat A (1979) Bull Soc Chim Fr II:401
8. Terech P, Volino F, Ramasseul R (1985) J Phys France 46:895
9. Wade RH, Terech P, Hewat EA, Ramasseul R, Volino F (1986) J Colloid Interface Sci 114:442
10. Terech P (1989) Mol Cryst Liq Cryst 166:29
11. Glatter O, Kratky O (1982) In: Small angle X-ray scattering, Academic Press, London
12. Terech P (1989) Prog Colloid Polym Sci 79:81
13. Murthy NS (1984) Biopolymers 23:1261

Author's address:

P. Terech
Institut Laue-Langevin
156X
38042 Grenoble Cedex, France

Phase diagrams for cationic surfactants in polar solvent systems

T. Wärnheim[1]), A. Jönsson[1]) and M. Sjöberg[1,2])

[1]) Institute for Surface Chemistry, Stockholm, Sweden
[2]) Department of Physical Chemistry, Royal Institute of Technology, Stockholm, Sweden

Abstract: Phase diagrams are presented for cationic surfactants, and alkyltrimethyl ammonium bromide in polar solvent systems, such as ethylene glycol, formamide, glycerol, and their mixtures with water. Provided that the solvent is sufficiently lipophobic, and that the hydrocarbon moiety of the surfactant is sufficiently large, liquid crystalline phases form according to the normal association pattern for ionic surfactants. In, for example, ethylene glycol, liquid crystals are only formed with surfactants with a longer alkyl chain length than dodecyl. — The formation of didodecyldimethylammonium bromide microemulsions containing polar solvents — mainly formamide, water and their mixtures — and hydrocarbon has been studied. The extension of the solution phase region may decrease, although the mutual miscibility of polar solvent and the hydrocarbon increases. NMR self-diffusion measurements reveal that this is accompanied by a corresponding loss of structure, i.e., a smaller distinction between polar and apolar regions.

Key words: Phase equilibria; polar solvents; alkyltrimethylammonium bromide; didodecyldimethylammonium bromide; non-aqueous microemulsions

Introduction

Much attention has recently been directed to the study of aggregation, phase equilibria, and formation of lyotropic liquid crystals in systems where the water has been replaced with another highly polar solvent [1, 2]. Liquid crystalline phases with lecithin as amphiphile, and water replaced by polar solvents such as diols or glycerol were detected some 10 years ago [3, 4], and more recently, liquid crystals have been found for single chain surfactants [5—7]. However, the properties of the solvent needed to promote surfactant aggregation has not been completely clarified [7—9] and further experimental and theoretical work along these lines is anticipated.

For some years now, we have systematically studied the phase behavior and the aggregation pattern in surfactant systems where the water has been completely or partially replaced by other polar solvents such as short-chain alcohols, diols, glycerol, formamide, etc. [7, 10—13]. From these investigations a rather complete phenomenological picture has emerged: the effects on the phase equilibria, when using strongly polar solvents such as formamide or glycerol instead of water, are gradual and not always dramatic. Provided that the solvent is sufficiently polar — or more appropriately, immiscible with hydrocarbon — and the size of the hydrocarbon moiety of the surfactant is large enough, formation of liquid crystals occur.

In this paper some examples are given to illustrate the role of the lipophobicity of the solvent: it is possible to gradually change the phase behavior from normal, salt-solvent equilibria to typical, surfactant-water phase diagrams. This is, in the first part of the paper, reviewed for a series of polar solvents (and in one case, in mixtures with water) and alkyltrimethylammonium bromide surfactants [7, 13]. In the second part, the formation of non-aqueous microemulsions with added hydrocarbon is investigated for the particular case of a double-chain cationic surfactant didodecyldimethylammonium bromide.

Experimental

The phase diagrams were determined as described previously [7, 10, 11], mainly by visual observation of samples in a thermostatted bath, optical microscropy, differential thermal analysis, low-angle x-ray diffraction, and ^{2}H NMR measurements. Self-diffusion measurements were carried out with the FT-NMR PGSE method on a JEOl FX-100 spectrometer at the ambient temperature of 25 °C. Surfactants were of highest purity commercially available, and were only dried in vacuum prior to use, with the exception of didodecyldimethylammonium bromide, which was recrystallized in ethylacetate. Solvents were used as received or dried with molecular sieves, depending on water contents as checked by Karl Fischer titration. Water was twice distilled.

Results and discussion

Phase diagrams for the surfactant hexadecyl trimethylammonium bromide and some different solvents are shown in Fig. 1 [7]. The surfactant associates in the aqeuous system [14] into micelles at low concentrations; a solutions phase L is stable until the liquid crystalline phases-hexagonal (E), cubic (I), and lamellar (D)-form with increasing surfactant concentration (Fig. 1a). (It is possible, and even probable, in this system that phases other than the cubic forms intermediate to the hexagonal and the lamellar phase in narrow concentration regimes [15]).

Changing solvent to glycerol increases the melting point for solid, solvated surfactant and decreases the existence region for the liquid crystalline phase. The decrease is more marked with formamide as solvent [6]; and with ethylene glycol the hexagonal phase has almost disappeared. Finally, with N-methylformamide, no hexagonal phase occurs (Fig. 1b—e).

In a very simple rotationalization of these results it can be stated that the solvophobic effect is the most important factor for determining the extent of surfactant aggregation. In Table 1 the interfacial tension between hydrocarbon and different polar solvents is shown; the magnitude of the interfacial tension — and thus, presumably, of the solvophobic interaction — follows the same trend as the extension region of, for example, the hexagonal liquid crystalline phase. Although equally important for modelling association of ionic surfactants [16], the dielectric properties of the solvent, reflected in the dielectric constant ε_r, show no such simple correlation. This is unexpected, since the formation of the hexagonal phase depends on repulsive aggregate interactions. However, the electrostatic inter- and intraaggregate interactions lead to a more complex pattern than was predictable from the numerical value of ε_r alone.

The difference in the driving force for aggregation (the solvophobic interaction) for a variation in surfactant alkyl chain length is illustrated in Fig. 2 where ethylene glycol is used as solvent. For dodecyltrimethylammonium bromide, no hexagonal phase is formed (Fig. 2a), while increasing the alkyl chain length to tetradecyl leads to its formation, but only in a narrow temperature and concentration region (Fig. 2b). With hexadecyl this region grows and, in addition, a cubic phase is formed at higher concentration of surfactant (Fig. 2c). Comparisons between water and other polar solvents, prior to declaring whether or not aggregation takes place in a solvent, must consider this simple point of surfactant chain length.

Also, the stability region for solid, solvated surfactant has a much stronger variation with temperature and concentration than in aqueous systems, i.e., the straight Krafft-point line of Fig. 1a is not seen in the corresponding non-aqueous systems. This gradual melting of the solid surfactant should correspond to a gradual change in chemical potential of the surfactant in solution (rather than reflecting changes in the crystalline state). The simple phase separation model for micellization is thus not valid, either reflecting that the aggregation varies with concentration, i.e., a gradual change in aggregation number, or that interaggregate interactions become more important at lower concentrations than in water.

The Krafft-point is normally lowered in aqueous systems by addition of alcohols, forming mixed aggregates and making the crystalline state relatively less stable. This effect on the melting point of solid, solvated surfactant can also be seen in the ternary system ethylene glycol — hexadecyltrimethylammonium bromide — decanol (Fig. 3). Although no liquid crystals form below 50 °C in the binary system (Fig. 1d), there is an extensive liquid crystalline region at 25 °C with added decanol. The main features of the system in Fig. 3 are also found, for example, with sodium dodecylsulphate — glycerol — decanol [5].

Our final example for the alkyltrimethylammonium bromide surfactants is the gradual change of the phase diagram by gradually varying the composition of the solvent from pure water to pure

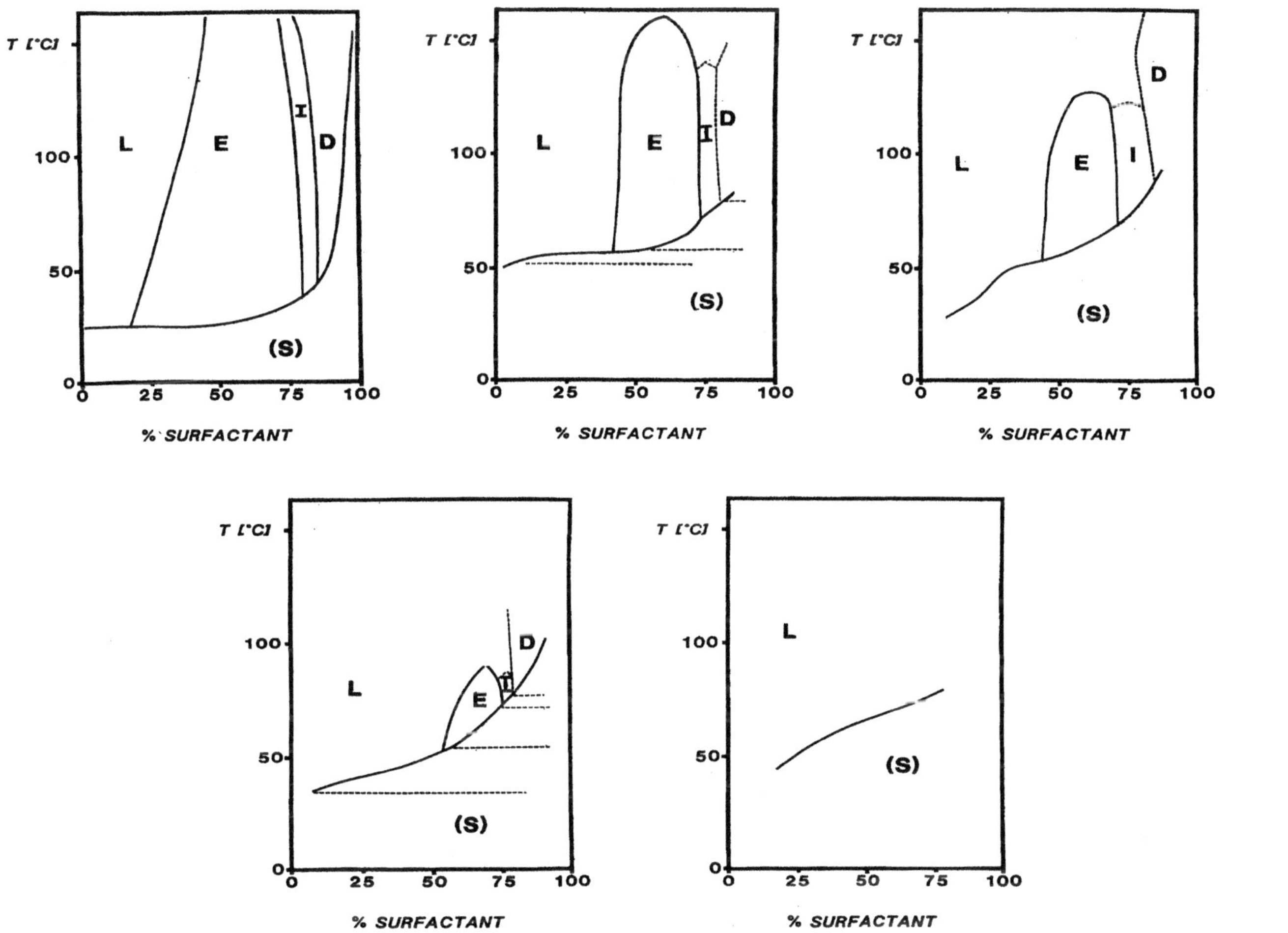

Fig. 1. Phase diagrams at varying temperature for hexadecyltrimethylammonium bromide and different polar solvents [7]. a) water [14]; b) glycerol; c) formamide [6]; d) ethylene glycol; e) N-methyl formamide. L denotes solution phase, E, I, and D denote hexagonal, cubic , and lamellar liquid crystals, respectively, and (s) denotes solid surfactant. The existence regions for the anisotropic liquid crystalline phases also comprise two phase regions: isotropic + anisotropic phases

Table 1. Properties for various polar solvents [7, 13, 25]

Solvent	Dielectric constant[a])	Interfacial tension[b]) [mN/m]
Water	78	50
Water/ethylene glycol 0.8/0.2	74	42
Water/ethylene glycol 0.6/0.4	67	32
Water/ethylene glycol 0.4/0.6	58	27
Water/ethylene glycol 0.2/0.8	48	24
Ethylene glycol	38	17
Glycerol	43	30
Formamide	109	27
N-methylformamide	182	13

[a]) At 20 °C; [b]) Against hexadecane (water) or dodecane at 20 °C

ethylene glycol (Fig. 4). When the weight fraction ethylene glycol exceeds 0.6 the decrease in the regions of the hexagonal and cubic phases becomes more marked. This is again (cf. Figs. 1 and 2) accompanied by a marked increase in temperature stability of the solid, solvated surfactant, and a more pronounced slope of that phase boundary.

Thus, single chain cationic surfactants follow the same association pattern in polar solvents other than water. In the absence of any detailed model, the tendency to form aggregates when comparing different solvents can be discussed in terms of the solvophobic interaction expressed as the interfacial tension between solvent and hydrocarbon. This quantity, at least if considered at the interface of a surfactant aggregate, is conveniently amended for

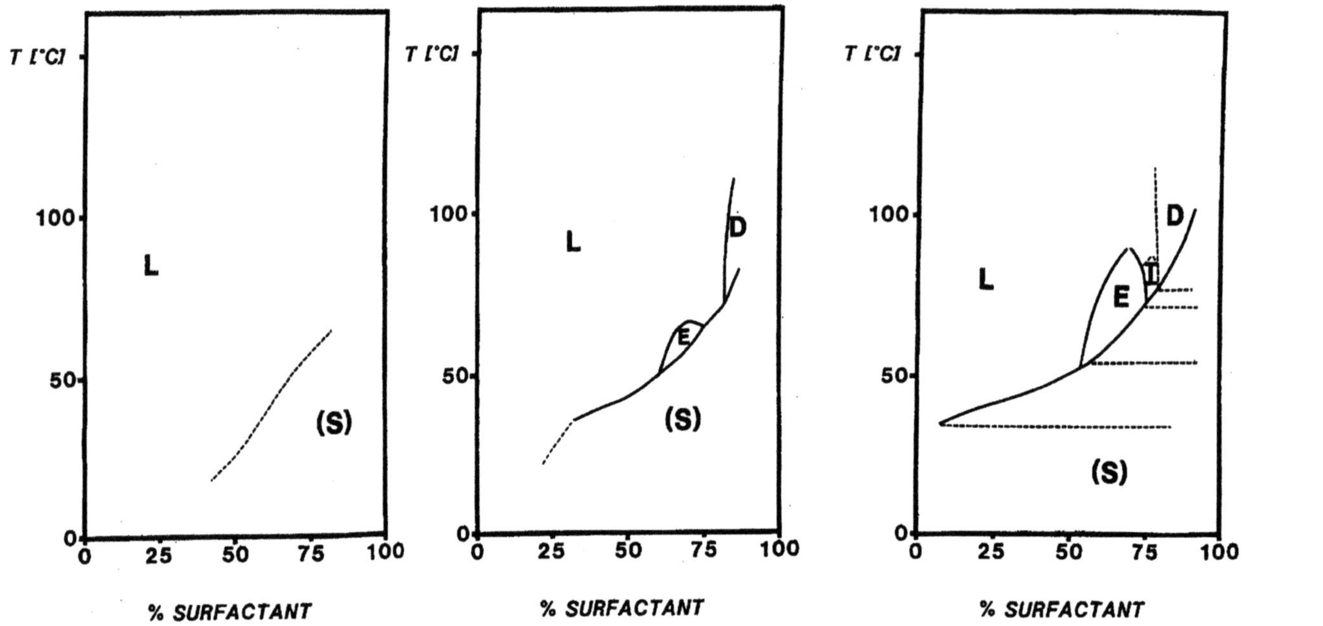

Fig. 2. Phase diagrams at varying temperature for alkyltrimethylammonium bromide of varying alkyl chain length and ethylene glycol [7]. a) dodecyltrimethylammonium bromide; b) tetradecyltrimethylammonium bromide; c) hexadecyltrimethylammonium bromide. Notations as in Fig. 1

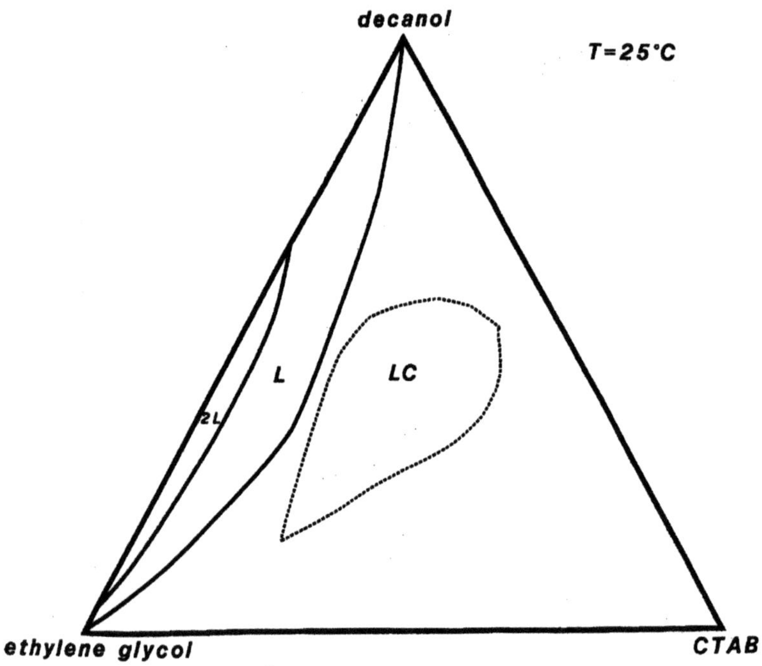

Fig. 3. Phase diagram for the system ethylene glycol-hexadecyltrimethylammonium bromide at 25°C. Notations as in Fig. 1

theoretical modelling [16], although the literature on aggregation in non-aqueous solvents discusses the matter in terms of parameters more directly reflecting the cohesive energy of the solvent.

Let us now consider the possibility of forming structured solution phases containing hydrocarbon microemulsions, using a polar solvent other than water. Double-chain cationic surfactants have, for

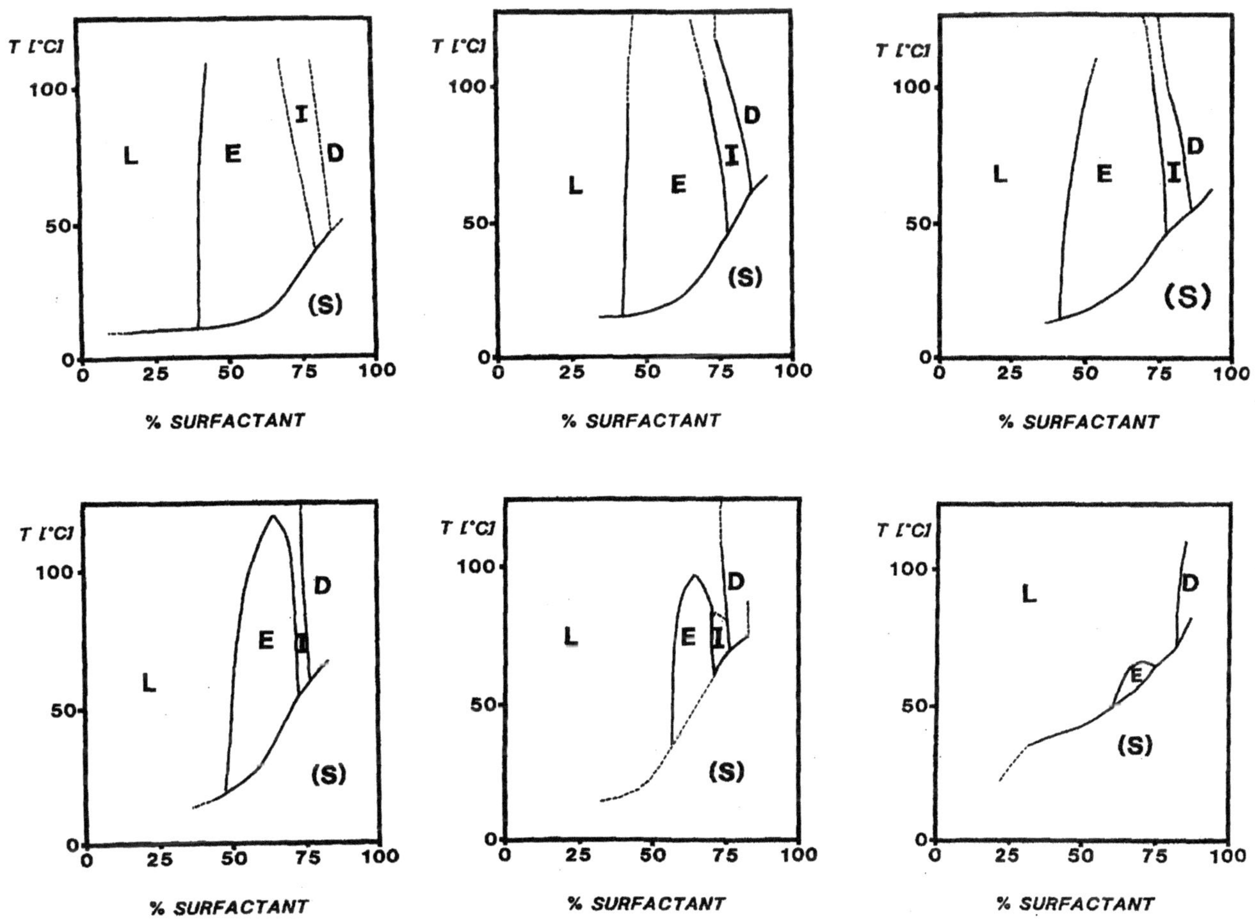

Fig. 4. Phase diagrams at varying temperature glycol-hexadecyltrimethylammonium bromide and water, ethylene glycol, and their mixtures as solvent [13]. a) water; b) water/ethylene glycol 80/20 (by weight); c) water/ethylene glycol 60/40; d) water/ethylene glycol 40/60; e) water/ethylene glycol 20/80; f) ethylene glycol. Notations as in Fgi. 1

packing reasons, a tendency to form lamellar structures; an interesting feature is their capacity to form microemulsions without addition of cosurfactant [18]. Also, the hydrophobic part of the molecule should definitely be large enough to stabilize surfactant aggregates in different polar solvent (considering Figs. 1 and 2), without raising the Krafft-point too much [18]. In Fig. 5, the effect when replacing water with formamide is shown in a number of isothermal phase diagrams of the quaternary system water/formamide/didodecyldimethylammonium bromide/dodecane. The surfactant is a swelling amphiphile and forms two different lamellar phases [19], a quite rare phenomenon [20]. It is noteworthy that these two lamellar phases are stable with didodecyldimethylammonium bromide up to equal mass ratio water/formamide in the solvent mixture.

The swelling of the solvent rich lamellar phase decreases somewhat at increasing formamide ratio. The maximum swelling is affected by the efficiency of screening the electrostatic repulsion between the surfactant lamellae. The decrease in extension could be due to the gradual increase in dielectric constant, but could also be due to the increase in monomer solubility with decreasing solvophobicity of the solvent; this leads to a higher ionic strength in the interlamellar solution which works in the same direction. At high ratios of formamide only a solution phase is stable up to high surfactant concentrations at 25 °C. The area per polar group, determined from low-angle x-ray diffraction (Table 2) reveals that there is a gradual increase in area with increasing weight fraction of formamide in the solvent: the area increases from around 65 Å^2 to 79 Å^2 in pure water and 1:2 formamide:water, respectively, at 25 wt% surfactant. This should be interpreted as a gradual dissolution (and possible penetration) of the lamellar phase.

Table 2. Areas per polar group S in the D-phase of the system water/formamide/didodecyldimethylammonium bromide at constant concentration surfactant, 25 wt%, and varying amounts of formamide and water at 20 °C

Formamide [wt%]	Water [wt%]	S [Å^2]
0	75	65
5	70	68
10	65	70
15	60	73
20	55	75
25	50	79

Similar differences occurs in the quarternary systems with added hydrocarbon. Didodecyldimethylammonium bromide is a balanced surfactant; microemulsions form readily upon addition of any hydrocarbon of appropriate size to penetrate the surfactant alkyl chains [21]. Less than 10 wt% surfactant is needed to enter the one-phase microemulsion region at equal ratio water/dodecane [19, 22]. At equal amounts of water/formamide as polar solvent the amount of surfactant needed is slightly higher: 15 wt%. At higher amounts of formamide, surfactant efficiency detoriates, and in the pure formamide system only a limited dissolution of the dodecane in the formamide/surfactant solution is possible.

Replacing dodecane with toluene in the formamide/didodecyldimethylammonium bromide system gives a considerably more extensive microemulsion region; this increase can, of course, be due to higher mutual miscibility from the increase in polarity of the hydrocarbon (Fig. 6a). However, this contrasts to the corresponding phase diagrams with ethylene glycol/didodecyldimethylammonium bromide/toluene or dodecane. While the phase diagram with dodecane (Fig. 6b) is very similar to the one with formamide as polar solvent, the phase diagram for ethylene glycol/didodecyldimethylammonium bromide/toluene (Fig. 6c) shows a more predominant miscibility gap than with formamide.

It has been shown [19, 22] that the microemulsion in the aqueous system is a highly structured solution; at low water content the system is bicontinuous, while there is a rather distinct transition to an oil-continuous system at higher water contents. It has been suggested that this corresponds to the transition from a tubular, bicontinuous structure to swollen inverse spheres; such inverse structures are stable at lower areas per polar group. Replacing water with formamide could, of course, lead to definite changes in structured and phase behavior, both due to a general disordering effect and changes in the preferred packing behavior.

We have performed some NMR self-diffusion measurements to compare systems with differences in the extension of the solution phase. Unfortunately, since the extensions of the solution phases vary when changing components, no direct comparisons can be made. However, in Fig. 7a, measurements in the system formamide/didodecyldimethylammonium bromide/toluene at a constant concentration of surfactant and varying mass ratio α = (toluene/toluene + formamide) are shown as D/D_0, relative to the self-diffusion for the pure components. At low concentrations of surfactant (13 wt%) a solution phase is stable over a wide composition region. Indeed, the self-diffusion is similar to typical aqueous microemulsions of the bicontinuous type: the formamide diffusion decreases gradually and the toluene diffusion increases with increasing α, while the surfactant diffusion remains constant and rather low throughout ($\sim 0.060 \cdot 10^{-9}$ m^2 s^{-1}). Thus, the solution phase is microstructured (this is also true for the system at higher surfactant contents).

For the corresponding ethylene glycol system (at 50 wt% surfactant) the situation is somewhat different (Fig. 7b). There are no different trends in the diffusion of ethylene glycol and toluene; the ratio D/D_0 is constant up to high values of α (mass ratio α = toluene/(toluene + ethylene glycol)). At α = 0.8, the self-diffusion coefficients increase, toluene more so than ethylene glycol. The surfactant diffusion is lower than with formamide, $\sim 0.020 \cdot 10^{-9}$ m^2 s^{-1}, but increases at α = 0.8 with a factor of two. The main features in Fig. 7b suggest a considerably lower degree of structure; the slower diffusion in absolute numbers can be correlated to the increase to micro- or macroviscosity. Thus, for these ternary microemulsions stabilized by didodecyl-

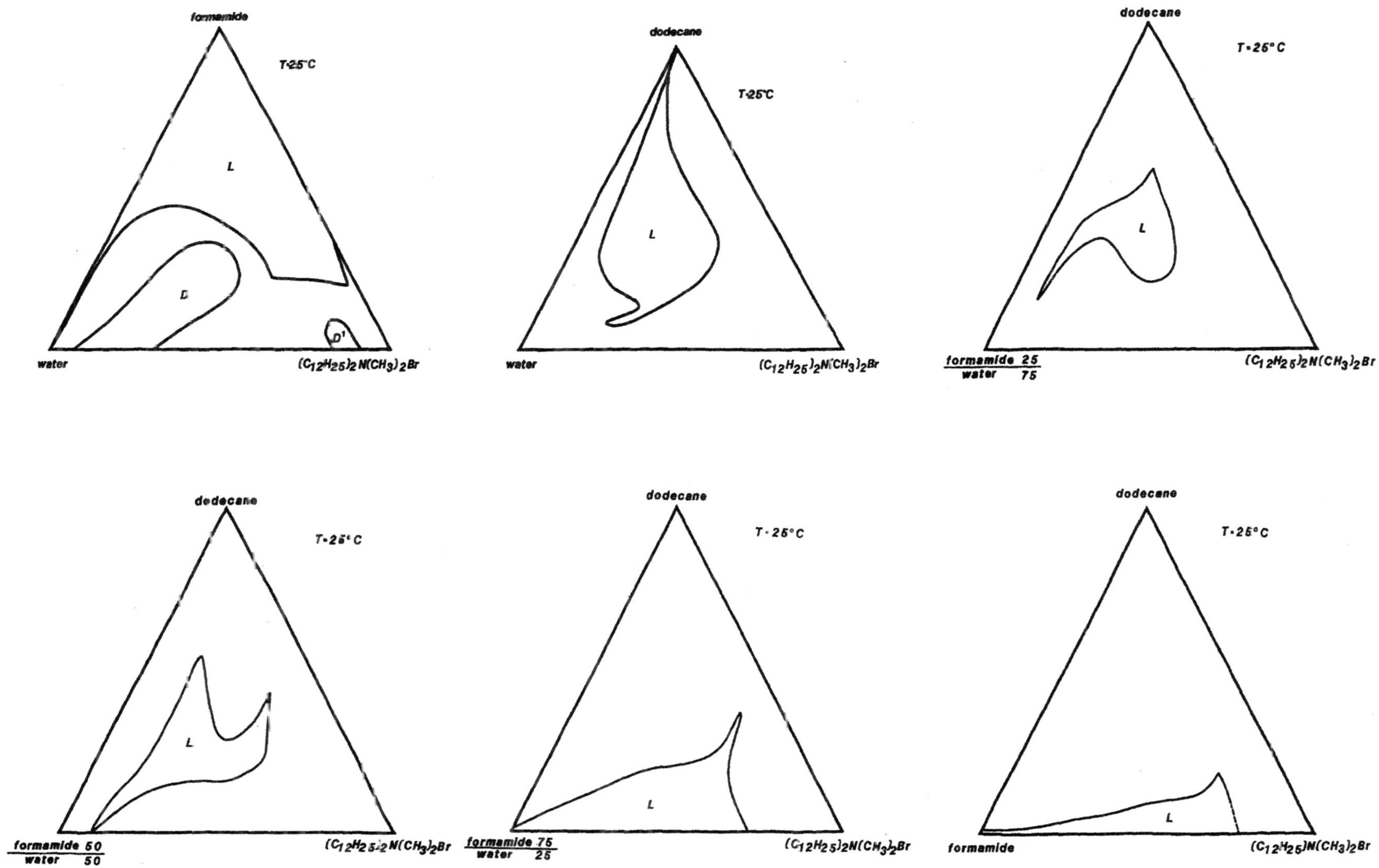

Fig. 5. Partial isothermal phase diagrams (25°C) for water/formamide/didodecyldimethylammonium bromide ($(C_{12}H_{25})_2N(CH_3)_2Br$)/dodecane. a) water/formamide/$(C_{12}H_{25})_2N(CH_3)_2Br$; b) water/$(C_{12}H_{25})_2N(CH_3)_2Br$/dodecane [19]; c) water:formamide 75:25/$(C_{12}H_{25})_2N(CH_3)_2Br$/dodecane; d) water:formamide 50:50/$(C_{12}H_{25})_2N(CH_3)_2Br$/dodecane; e) water:formamide 25:75/$(C_{12}H_{25})_2N(CH_3)Br$/dodecane; f) formamide/$(C_{12}H_{25})_2N(CH_3)_2Br$/dodecane. Notation as in Fig. 1, except that D and D′ denote two different lamellar phases

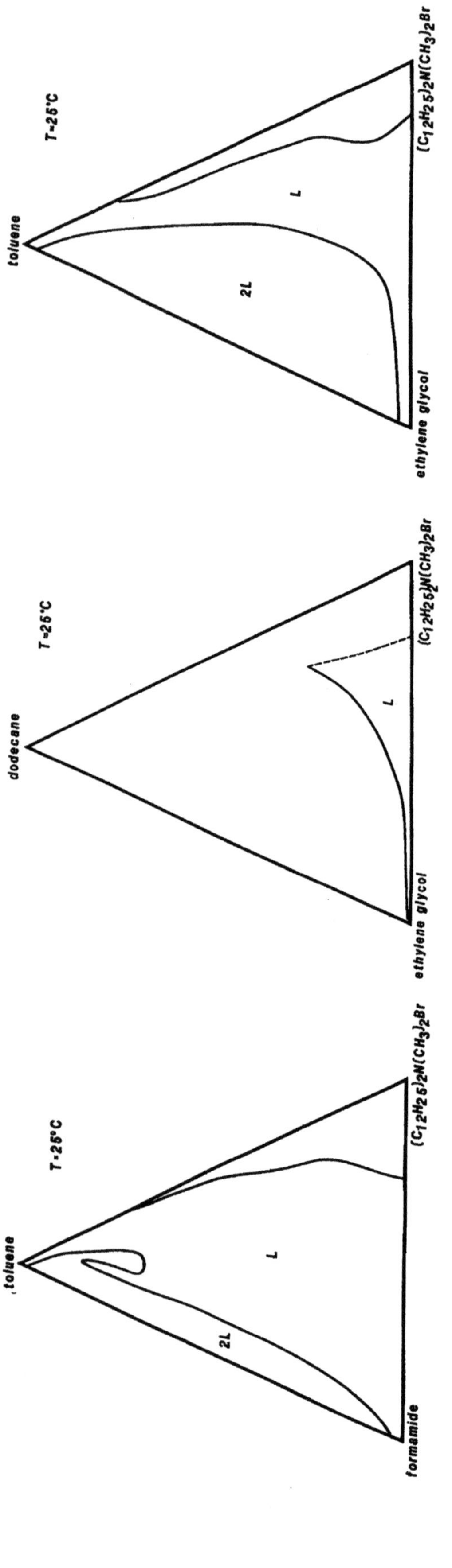

Fig. 6. Partial isothermal phase diagrams (25°C) for polar solvent/didodecyldimethyl ammonium bromide ($(C_{12}H_{25})_2N(CH_3)_2Br$)/hydrocarbon. a) formamide/$(C_{12}H_{25})_2N(CH_3)_2Br$/toluene; b) ethylene glycol/$(C_{12}H_{25})_2N(CH_3)_2Br$/dodecane; and c) ethylene glycol/$(C_{12}H_{25})_2N(CH_3)_2Br$/toluene.

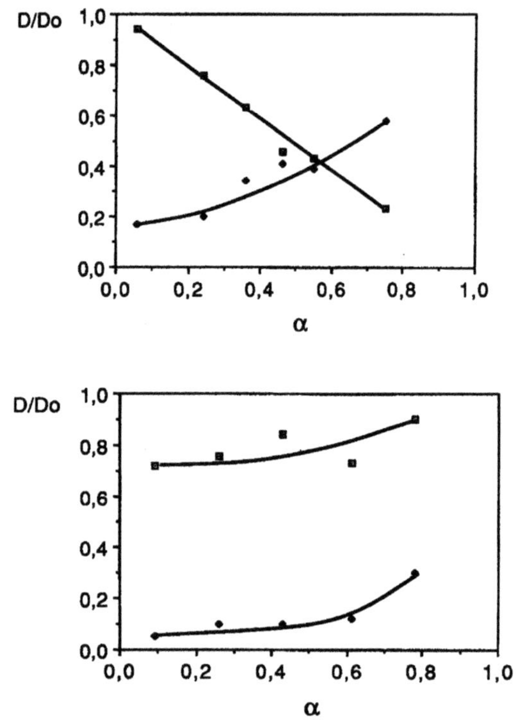

Fig. 7. Self-diffusion measurements in the systems
a) (■) formamide,
(◆) toluene
at constant concentration of surfactant = 13 wt%. D_0 for formamide and toluene is 0.47 and 2.40 · 10^{-9} m^2 s^{-1}, respectively.
b) (■) ethylene glycol
(◆) toluene
at constant concentration of surfactant = 50 wt%. D_0 for ethylene glycol is 0.12 · 10^{-9} m^2 s^{-1}

dimethylammonium bromide, structure and segregation into solvophilic and lipophilic domains seems to be a prerequisite for forming microemulsions at low surfactant contents, as intuitively anticipated.

A lower order (i.e., absence of extensive surfactant aggregation) of the nonaqueous microemulsions has also been noted for quaternary ionic microemulsions of the cosurfactant type [23]. However, microemulsions lead to a dramatic decrease in the existence region of the microemulsion. In systems where the mutual solubility of the components is sufficiently high, such as quaternary microemulsions with short or medium chain alcohol cosurfactant, the situation is closer to a molecular solution, even in the aqueous phase [24]. A further loss of structure, i.e., a decreasing difference between polar and apolar domains, in this case, does not imply a destabilization of the solution phase, while this does seem to occur with didodecyldimethylammonium bromide, as noted. A similar decrease in the extension of the microemulsion region also occurs for balanced nonionic surfactant systems with formamide instead of water [12].

Acknowledgements

This research was financially supported by the Research Council of the Technical Board of Development (STUF). Collaboration with Sune Backlund, Åbo Akademi, and Björn Bergenståhl is gratfully acknowledged.

References

1. Friberg SE, Liang YC (1988) Microemulsions. CRC Press, OH 1988, Ch 3, p 79
2. Evans DF (1988) Langmuir 4:3
3. Moucharafieh N, Friberg SE (1971) Mol Cryst Liq Cryst 49:231
4. Bergenståhl B, Stenius P (1987) J Phys Chem 91:5944
5. Friberg SE, Liang P, Liang YC, Greene B, van Gilder R (1986) Coll Surf 19:249
6. 6a) Belmajdoub A, Marchal JP, Canet D, Rico I, Lattes A (1987) Nouv J Chim 11:415. b) Auvray X, Anthore R, Petipas C, Rico I, Lattes A (1988) CR Acad Sci Paris II, 306:695
7. Wärnheim T, Jönsson A (1988) J Coll Interf Sci 125:627
8. Beesley AH, Evans DF, Laughlin RG (1988) J Phys Chem 92:791
9. Lattes A, Rico I (1989) Coll Surfaces 35:221
10. Bergenståhl B, Jönsson A, Sjöblom J, Stenius P, Wärnheim T (1987) Progr Colloid Polymer Sci 74:108
11. Wärnheim T, Bokström J, Williams Y (1988) Coll Polymer Sci 226:562
12. Wärnheim T, Sjöberg M (1989) J Coll Interf Sci 131:402
13. Backlund S, Bergenståhl B, Molander O, Wärnheim T (1989) J Coll Interf Sci 131:393
14. Wolff T, Bünau GV (1984) Ber Bunsenges Phys Chem 88:1098
15. Blackmore ES, Tiddy GJT (1988) JCS Faraday Trans 284:1115
16. Jönsson B, Wennerström H (1987) J Phys Chem 91:338
17. Almgren M, Swarup S, Löfroth JE (1985) J Phys Chem 89:4621
18. Kunieda H, Shinoda K (1978) J Phys Chem 87:538
19. Fontell K, Ceglie A, Lindman B, Ninham B (1986) Acta Chem Scand A 40:247
20. Khan A, Jönsson B, Wennerström H (1985) J Phys Chem 89:5180
21. Angel LR, Evans DF, Ninham BW (1983) J Phys Chem 87:538
22. Blum FD, Pickup S, Ninham BW, Chen SJ, Evans DF (1985) J Phys Chem 89:711
23. Das KP, Ceglie A, Lindman B, Friberg S (1987) 116:390
24. Stilbs P, Lindman B, Moseley ME (1981) J Coll Interf Sci 83:564
25. Handbook of Chemistry and Physics (1986) CRC Press, Ohio

Authors' address:

T. Wärnheim
Institute for Surface Chemistry
P.O. Box 5607
114 86 Stockholm, Sweden

Phase behavior of some nonionic polymers in nonaqueous solvents

A. A. Samii[1]), B. Lindman[1]) and G. Karlström[2])

[1]) Physical Chemistry 1 and [2]) Theoretical Chemistry Chemical Center, University of Lund, Sweden

Abstract: The phase behavior of a set of block copolymers of polyethyleneoxide-polypropyleneoxide type has been studied in several different solvents. It is found that the polymers are soluble in most solvents, both at low and high temperatures; for formamide and water a clouding behavior is observed, i.e., the solubility decreases with temperature. In this work we also report on the phase behavior of ethylhydroxyethylcellulose (another nonionic polymer) in formamide when an ionic surfactant and salt are added. The presented data indicate that it is the same clouding mechanism that is operating in the polymer — formamide systems as in the polymer — water systems.

Key words: Non aqueous solvents; nonionic polymers; solubility; clouding temperature; surfactants

Introduction

The phase behavior of aqueous solutions of nonionic polymers and surfactants containing the CH_2—O—CH_2 segment is a problem that has both theoretically and experimentally received much attention in recent years, perhaps mainly due to the phase separation occurring in such systems at elevated temperatures [1—6]. The obvious explanation to this decrease in solubility at higher temperatures is a rapid increase in the effective attraction between different solute molecules with temperature. The origin of this effect is, however, not fully understood today, and three different mechanisms behind this behavior are suggested in the literature. In the oldest of these models from Kjellander and Florin [7], it is suggested that water forms an ordered structure around the ethyleneoxide (EO) chain at low temperatures and that this highly ordered structure, which is entropically unfavorable, is destroyed at higher temperatures.

The second model [8] is an adaptation of early work by Hirschfelder et al. [9], to systems containing EO groups. In this model the decrease in solubility is explained by hydrogen-bonding between the EO groups and water; the hydrogen bonds are destroyed with increasing the temperature.

In the last model the decrease in solubility is explained as being due to the change in conformational structure of the EO chains. At lower temperatures the segments of EO chains are mainly in a polar conformation that interacts favorably with water, but at higher temperatures entropy makes the less polar structures more probable. These less polar conformers interact unfavorably with water [10].

The phase behavior of this type of polymer or surfactant in other types of solvents is less well studied. However, Saeki et al. [2] have investigated the phase behavior of polyethyleneoxide (PEO) in tertiary butylacetate and they found a similar behavior as in water, and Wärnheim et al. [6] recently reported about clouding for a nonionic surfactant of EO type in formamide. Baily and Koleske [1] refer to several other solvents which are known to dissiolve PEO in a more normal way. In the first part of this work we will report on the phase behavior of two different polyethylene-oxide-polypropylene-oxide block copolymers in 19 different solvents.

Addition of an ionic surfactant to an aqueous solution of a nonionic polymer or surfactant of this type does, in general, strongly raise the cloud point [11, 12]. However, it was recently shown that in the presence of even low concentrations of simple electrolyte the first additions of ionic surfactant dramatically lower the cloud point of certain nonionic polymers [12]. For understanding the underlying mechanism it is essential to examine if this effect is restricted to aqueous systems or if it is also displayed for other solvents. Thus, the combined effect of ionic surfactant and sodium chloride on the cloud point of a nonionic cellulose ether, ethylhydroxyethylcellulose (EHEC), was investigated for solutions in formamide.

Materials

The two different poly(ethyleneoxide-propyleneoxide) samples supplied by Polysciences, Inc. Warrington, PA 18976, USA, had a molecular weight of 2917 and 3438. Both were block copolymers with one block of polyethylene oxide and one block of polypropylene oxide. The former polymer had a molar ratio of EO/PO of 0.8, and the latter had a ratio of 0.33. Both polymers were used as received. The EHEC polymer (Bermocoll E 351 x) was obtained from Berol Kemi AB, Stenungsund, Sweden. On the average, each anhydroglucose unit in the cellulose molecule was substituted with 1.4 ethyl groups and 0.9 ethylene oxide groups. A 1% aqueous solution has a cloud point of 40 °C. The EHEC polymer

Table 1. C.P. and solubility of P.E.O.-P.O. 2917 (ratio PEO/PO = 0.8:1) and P.O.E.-P.O. 3438 (ratio PEO/PO = 0.33:1) in different solvents

Solvent	Dielectric const.		Cloud point		Solubility	
			P.E.O.-P.O. 2917	P.E.O.-P.O. 3438	P.E.O.-P.O. 2917	P.E.O.-P.O. 3438
1-Butanol	17.8	(20 °C)	no C.P. up to 110 °C	no C.P. up to 110 °C	Soluble at R.T.	Soluble at R.T.*
Cyclohexanone	18.3	(20 °C)	no C.P. up to 150 °C	no C.P. up to 150 °C	Soluble at R.T.	Soluble at R.T.
1-Propanol	20.1	(25 °C)	no C.P. up to 90 °C	no C.P. up to 90 °C	Soluble at R.T.	Soluble at R.T.
Acetone	20.7	(25 °C)	no C.P. up to 60 °C	no C.P. up to 60 °C	Soluble at R.T.	Soluble at R.T.
Benzoyl chloride	23.0	(20 °C)	no C.P. up to 140 °C	no C.P. up to 140 °C	Soluble at R.T.	Soluble at R.T.
Ethanol	24.3	(25 °C)	no C.P. up to 70 °C	no C.P. up to 70 °C	Soluble at R.T.	Soluble at R.T.
2,4-Pentane-dione	25.7	(20 °C)	no C.P. up to 100 °C	no C.P. up to 100 °C	Soluble at R.T.	Soluble at R.T.
1,2-Propane diol	32.0	(20 °C)	no C.P. up to 180 °C	no C.P. up to 180 °C	Soluble at R.T.	Soluble at 70 °C
Methanol	32.6	(25 °C)	no C.P. up to 70 °C	no C.P. up to 70 °C	Soluble at R.T.	Soluble at R.T.
1,3-Propane diol	35.0	(20 °C)	no C.P. up to 200 °C	no C.P. up to 200 °C	Soluble at 100 °C	Not soluble even at high temperature
Di-Methyl-formamide	37.0		no C.P. up to 150 °C	no C.P. up to 150 °C	Soluble at R.T.	Soluble at R.T.
Glycol	37.7	(25 °C)	no C.P. up to 200 °C	no C.P. up to 200 °C	Not soluble even at high temperature	Not soluble even at high temperature
Glycerol	42.5	(25 °C)	no C.P. up to 135 °C	no C.P. up to 135 °C	Soluble at low conc. (1% weight)	Soluble at low conc. (1% weight)
D.M.S.O.	49.0		no C.P. up to 150 °C	no C.P. up to 150 °C	Soluble at low conc. (1% weight)	Soluble at 100 °C
Water	78.54	(25 °C)	see Fig. 2a	see Fig. 1a	Soluble at low conc. (1% weight)	Not soluble at R.T.
Formamide	109.5	(20 °C)	see Fig. 2b	see Fig. 1b	Soluble at R.T.	Soluble at R.T.
N-Methyl-acetamide	175.7		no C.P. up to 160 °C	no C.P. up to 160 °C	Soluble at R.T.	Soluble at R.T.
N-Methyl-formamide	182.4		no C.P. up to 180 °C	no C.P. up to 180 °C	Soluble at R.T.	Soluble at R.T.
t-Butylacetate	5.0	(20 °C)	no C.P. up to 96 °C	no C.P. up to 96 °C	Soluble at R.T.	Soluble at R.T.

* R.T. = room temperature

was dialyzed against a membrane-filtered water (millipore) to remove salt (NaCl) present in the supplied polymer. All other chemicals were of analytical grade and were used as supplied.

Methods

All samples were prepared by weighing the components. Polymer and NaCl concentrations are expressed in percent by weight and surfactant concentrations are expressed in moles per kilogram of solvent (molal). The phase separation temperature and solubility were determined for different concentrations from 0.1% to 10.0%. The appearance of the phase separation is different in character compared to the aqueous systems. In formamide systems a rather indistinct haziness occurred instead of a distinct cloudiness appearing in aqueous solutions. In water and formamide solutions, after heating the samples above the clouding temperature, the cloud point temperature was taken as that temperature when the last visible sign of clouds disappeared on cooling. We estimate that the accuracy of the cloud point measurements is 0.5°C.

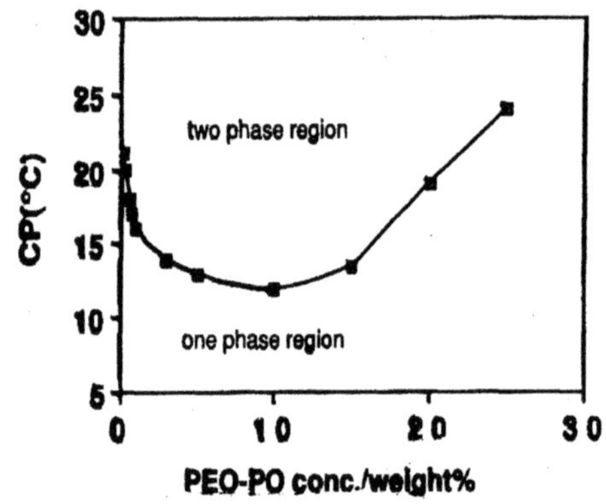

Fig. 1a. Partial phase diagram for the system PEO-PPO (MW = 3438) in water

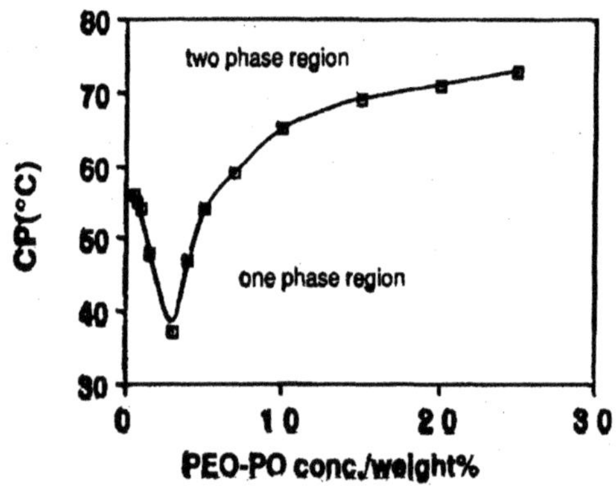

Fig. 1b. Partial phase diagram for the system PEO-PPO (MW = 3438) in formamide

Results and discussion

In Table 1 we present the results from the cloud point measurements made on the two different PEO-PPO block copolymers. As is easily seen from this table, clouding occurs only in water and formamide. Similar results have been obtained by Wärnheim for nonionic surfactants of the EO-type [6]. In Figs. 1a and b and 2a and b, we present the phase diagrams for the two studied polymers in water and formamide. Both these solvents are hydrogen-bonding and polar, but there were other hydrogen-bonding polar solvents investigated, shown in Table 1, both with higher and lower dielectric permittivity which does not induce clouding at higher temperatures. One may thus ask why clouding occurs in these two systems? According to theory, the dielectric permittivity (for a hard-sphere dipolar solvent) depends only on the (dipole moment)^2 (molecular radius)^3. The interaction energy between a solvent molecule and a solute molecule is, in the same approximation, dependent on: dipole moment of solvent* dipole moment of solute/intermolecular distance**^{3}. (See Table 2 for some relevant numbers). As can be seen from Table 2, μ/v has a larger value for water and formamide than for the other solvents. In practice this means that the solvation energy obtained for a solute molecule in two different hard-sphere solvents with the same dielectric permittivity, but with different radius and dipole moment is larger in the solvent with the smaller molecules. From this point of view it looks as if it is those solvent molecules that are small and polar that will induce clouding when the temperature is increased.

In Fig. 3 we show the phase diagram obtained for the substituted cellulose ether EHEC in formamide; a very similar phase diagram is obtained in water, but the clouding in water occurs for this particular EHEC polymer at 40°C. In Figs. 4 and 5 we show the phase behavior when a cationic surfactant cetyltrimethylammonium bromide (CTAB) is added to 1.5% EHEC solution for different salt concentrations. In Fig. 4 we present a similar part of the phase diagram for the situation when an anionic surfactant sodium dodecyl sulphate (SDS) is added to a 1.5% solution with different salt concentrations. It should be noted that salt alone in these amounts hardly effects the cloud point temperature of the formamide-EHEC system, and that, when

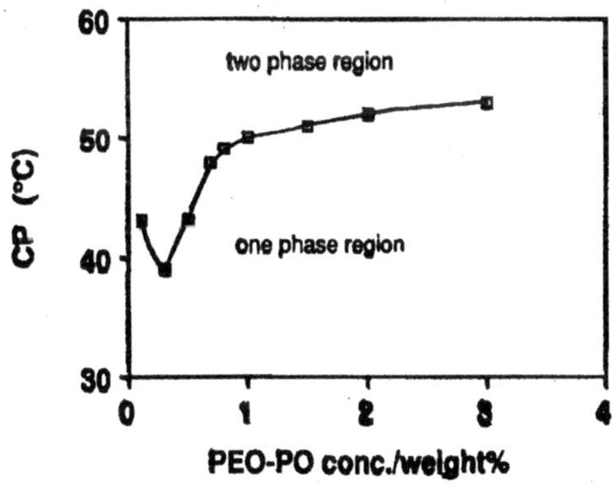

Fig. 2a. Partial phase diagram for the system PEO-PPO (MW = 2917) in water

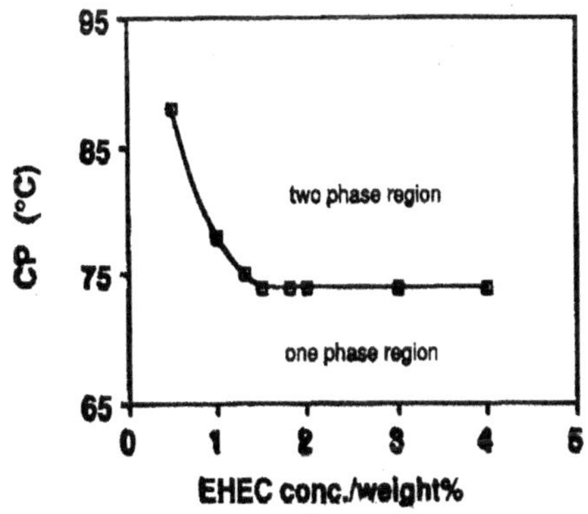

Fig. 3. Partial phase diagram for the EHEC-formamide system

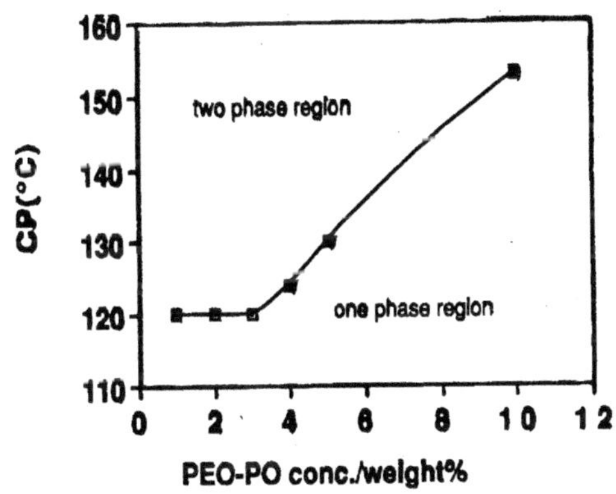

Fig. 2b. Partial phase diagram for the system PEO-PPO (MW = 2917) in formamide

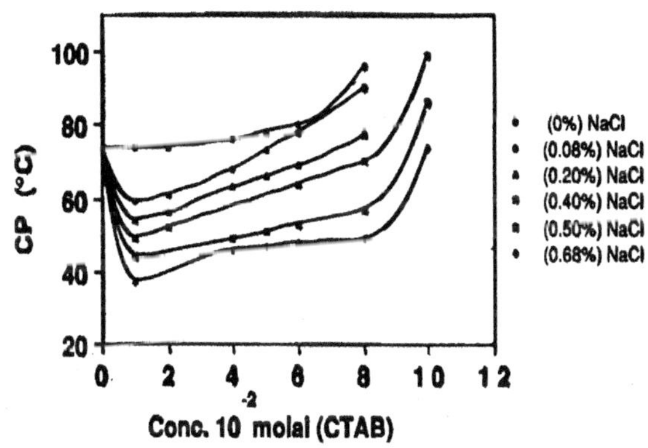

Fig. 4. Partial phase diagram for a 1.5% EHEC solution in formamide as a function of CTAB concentration. Different curves correspond to different NaCl concentrations. CTAB has a CMC of 0.09 m at 60 °C in formamide

Table 2. Some relevant properties for the solvents used

Molecule	Molecular dipole-moment μ (D)	Molecular volumes V (Å)	$\frac{\mu}{V}$	$\frac{\mu^2}{V}$	Dielectric permittivity ε	Hydrogen bonding
Water	1.85	29.89	0.0619	0.1145	78.5	Yes
Formamide	3.73	65.98	0.056	0.2109	109	Yes
N-Methyl formamide	3.83	97.01	0.039	0.1512	182.4	Yes
NN Dimethyl formamide	3.82	127.91	0.0292	0.1141	37	No
DMSO	3.96	117.78	0.034	0.1331	49	No
N-methylacetamide	3.73	126.81	0.294	0.1097	175.7	Yes

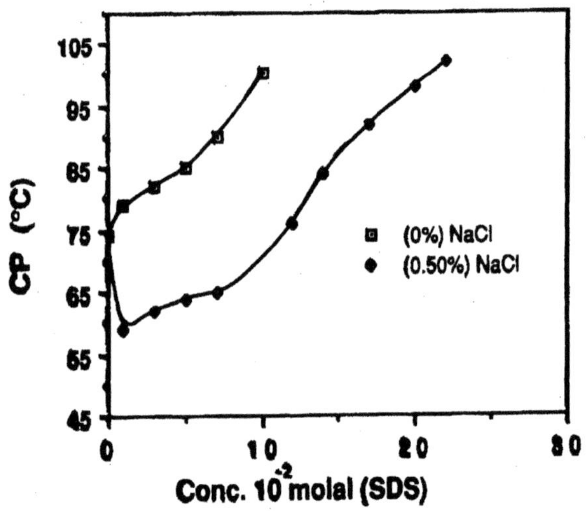

Fig. 5. Partial phase diagram for a 1.5% EHEC solution in formamide as a function of SDS concentration. Different curves correspond to a salt-free and two different NaCl concentrations. (SDS has a CMC of 0.22 m at 60 °C in formamide)

only the ionic surfactant is present, this leads to an increase in cloud point temperature. Thus, we may see that, just as for the EHEC-water system, there is a large synergistic effect of salt and ionic surfactant for the phase behavior in the EHEC-formamide system. An important difference should be noted, i.e., the effect of the cationic surfactant is larger in formamide than the effect of anionic surfactant; in water the trend is the opposite. A possible explanation is that water solvates the EHEC polymer with the water protons pointing, on the average, toward the polymer and, consequently, creating a weak, positive surface-charge density around the polymer. This positive surface-charge density favors binding of anionic surfactants. When formamide is used as solvent the situation may be the opposite.

The experiments presented above indicate remarkable similarities in the phase behavior of nonionic polymers containing EO-groups in water and formamide. This suggests that the mechanisms behind the clouding phenomena in these two solvents are the same. Since both water and formamide are polar hydrogen-bonding solvents the observations are consistent with both the hydrogen-bond and the EO-conformational model, but not with the water-structure model, since that would mean that a similar structure would exist in formamide, and this seems highly unlikely.

References

1. Bailey Jr FE, Koleske JV (1976) (Poly(ethylene oxide) Academic Press, New York
2. Saeki S, Kuwahara N, Nakata M, Kaneko M (1976) Polymer 17:685
3. Nakagawa T, Shinoda K (1963) In Colloidal Surfactants, Academic Press, New York, Chapter 2
4. Nakagawa T (1967) In Nonionic Surfactants, Shcik MJ ed., Marcel Dekker, New York, Surfactant Sci Ser, Vol. 1, Chapter 17
5. Lindman B, Karlström G (1987) Z Phys Chem 155:199
6. Wärnheim T, Bokström J, Williams Y (1988) Colloid Polym Sci 266:562
7. Kjellander R, Florin E (1981) J Chem Soc Faraday Trans 1 77:2053
8. Goldstein RE (1984) J Phys Chem 80:5340
9. Hirschfelder J, Stevenson D, Eyring H (1937) J Chem Phys 5:896
10. Karlström G (1985) J Chem Phys 89:4962
11. Valaulikar BS, Manohar C (1985) J Colloid Interface Sci 108:403
12. Carlsson A, Karlström G, Lindman B (1986) Langmuir 2:536

Authors' address:

A. A. Samii
Physical Chemistry 1
Chemical Center
University of Lund
P.O. Box 1 24
22100 Lund, Sweden

Progress in Colloid & Polymer Science Progr Colloid Polym Sci 82:285—289 (1990)

Phase behavior and structure in the soybean phosphatidylcholine-ethanol-water system

I. Söderberg

Department of Food Technology, University of Lund, Lund, Sweden

Abstract: The ternary phase diagram of soybean phosphatidylcholine (SPC)-ethanol-water has been determined at room temperature (21 °C). The structure of the different phases was investigated by x-ray diffraction, D_2O-NMR, and polarization microscopy. The phase diagram consists of an extensive liquid solution and a lamellar (L_α) phase. The liquid transforms gradually from a L1 type of solution near the water-ethanol line to an oil continuous solution of L2 type near the SPC-ethanol line at high SPC concentrations. At 40% (w/w) ethanol the L_α phase is not able to form a liposome dispersion. The liposomes separate after they have been formed, giving a coherent L_α phase.

Key words: Phase diagram; SPC; ethanol; water; x-ray diffraction; D_2O-NMR

Abbreviations

PC:	phosphatidylcholine
SPC:	soybean phosphatidylcholine
DPPC:	dipalmitoyl phosphatidylcholine
PE:	phosphatidylpentanolamin ethanolamine
L_α:	lamellar phase
H_{II}:	reversed hexagonal phase

Introduction

Ethanol exhibits a wide spectrum of physiological effects [1], and many effects are related to the interaction with membrane lipids. One example is the general anesthetic properties. General anesthetics are reported to induce phase transition at pharmacological concentrations in lipid bilayers [2, 3].

The anesthetic potential is directly related to lipid solubility, according to the Meyer-Overton rule [4, 5]. Ethanol has a rather low membrane solubility, expressed as membrane/buffer partition coefficient, compared with other anesthetic substances, which is consistent with its weak anesthetic effect.

Even very low ethanol concentrations have an effect on neurons, which was reported recently [7]. Ion current activated by N-methyl-D-aspartate (NMDA) is inhibited at 5—50 mM ethanol; NMDA is one receptor agonist to glutamate, a major nerve transmittor in CNS of mammals.

Higher concentration of ethanol exhibits various effects on membrane functions, for example desentization of the acetylcholine receptor [6].

A number of physico-chemical studies have been performed on ethanol and membrane lipids. Several methods have shown that ethanol and other alcohols affect the transition temperature for gel to liquid crystalline state [8—11]. Long chain alcohols increase the transition temperature [12]. When ethanol is added to saturated PC, the transition temperature is decreased at low ethanol concentrations and increased at higher concentrations [13—15].

Ethanol is also reported to stabilize synthetic PE membranes in the lamellar phase [16]. In these cases ethanol concentration exceeds what is pharmacologically relevant.

The effect of ethanol on membrane fluidity has been studied extensively [17—19]. Ethanol increases membrane fluidity, even at low concentrations [20]. A selective perturbation hypothesis has been suggested to explain physiological effects of ethanol. (In the hypothesis ethanol should preferentially perturb the liquid crystalline domains of the membrane [21].)

Here the phase behavior, in equilibria of PC, water, and ethanol is reported in order to improve our understanding of the physico-chemical interactions between the components. It is also obvious that knowledge of this system is useful for many technical applications, particularly for foods, because ethanol is the only organic solvent which can be accepted toxicologically.

Material and methods

Materials

Pure phosphatidylcholine from soybean, SPC (Epicuron 200) was obtained from Lucas Meyer (Hamburg, FRG). Thin-layer chromatography showed a major band of SPC and trace amounts of lyso-SPC. The SPC was used without further treatment. Due to a high degree of unsaturation (see Table 1), the chain melting temperature is below 0°C in the investigated region.

Table 1. The fatty acid pattern of SPC (Epicuron 200), % (w/w) [22]

C_8	C_{16}	$C_{16:1}$	C_{18}	$C_{18:1}$	$C_{18:2}$	$C_{18:3}$
0.8	12.2	0.4	2.7	16.7	67.2	6.0

The water was double distilled. Pure deuterated water (D_2O), >99.7% isotopic purity, was obtained from NÖrsk Hydro (Rjukan, Norway). Absolute ethanol was used.

Samples were prepared by weighing appropriate amounts of SPC, water, and ethanol into ampoules which were flame sealed, centrifugated for 30 min, and then stored at 60°C for 2 h, and thereafter at room temperature. Centrifugation was repeated during storage until no change in macroscopic behavior could be seen.

All samples were repeatedly examined between crossed polarizers for birefringence and texture.

When the samples had reached equilibrium, the structure of the different phases was investigated by low-angle x-ray diffraction, D_2O-NMR, and polarization microscopy.

Methods

For the x-ray measurements a DPT camera [22], a Guinier camera after Luzzati [24] with a quarts monochromator, and a Kiessig camera with point collimation [26] were used.

In the NMR experiments a custom built spectrophotometer [25] was used. The magnetic field strength was 6 Tesla with a frequency of 39.14 MHz. Deuterium (^2H, I = 1) has a quadrupolar nuclei and in anisotropic phases quadrupolar splittings are observed in the NMR spectra. With several phases coexisting, isotropic as well as anisotropic, the signals from each phase are superimposed, provided the deuterium exchange between the phases is low.

In the ideal case the L_a phase gives twice as large splitting as the H_{II} phase. D_2O-NMR is a well established method and is thoroughly described elsewhere [27, 28].

For polarization microscopy an Olympus Vanox microscope was used.

Results and discussion

The ternary phase diagram of SPC, water, and ethanol is shown in Fig. 1. The phase diagram shows an extensive isotropic liquid phase and a lamellar phase, less extended than in the binary system SPC and water [29].

In the polarizing microscope the texture of the L_a phase is characterized by a course mosaic texture and by oily streaks [30]. The fluidity and transparency of the L_a phase increase with increasing ethanol concentration.

The liquid phase continuously transforms to an isotropic "oily" and highly viscous solution of L2 type near the SPC-EtOH line and at high SPC concentration (>70% w/w). Near the ethanol-water line there is a thin zone of almost pure water-ethanol, L1 type of solution, with very low lipid monomer concentration. Between these two zones, the major part of the liquid phase exists.

Almost the whole two-phase region consists of lamellar phase in equilibrium with the water-ethanol continuous liquid, but in a small part of this region a lamellar phase is in equilibrium with an L2 phase. The coexistence of a lamellar phase in a large region in equilibrium with an L2 phase is unusual in lipid systems.

With an ethanol concentration, <48% (w/w), the lamellar phase is in equilibrium with an almost pure ethanol-water mixture, L1 type of phase. With an ethanol concentration >48% (w/w), L_a is in equilibrium with a liquid solution, in which a large amount of SPC is solubilized.

The repeat distance in the lamellar phase, at a constant ethanol-water ratio, is shown in Fig. 2. (The repeat distance is plotted as a function of volume fraction water, assuming the partial specific volumes of water, ethanol and SPC equal to 1). The repeat distance shows the increase in interbilayer

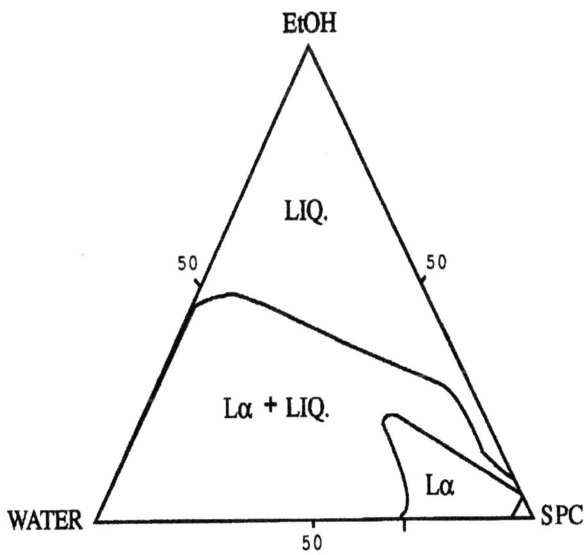

Fig. 1. The phase diagram for the ternary system of soybean phosphatidylcholine (SPC), ethanol, and water at 21 °C. The "SPC" corner ⩾95% (w/w) SPC is not investigated

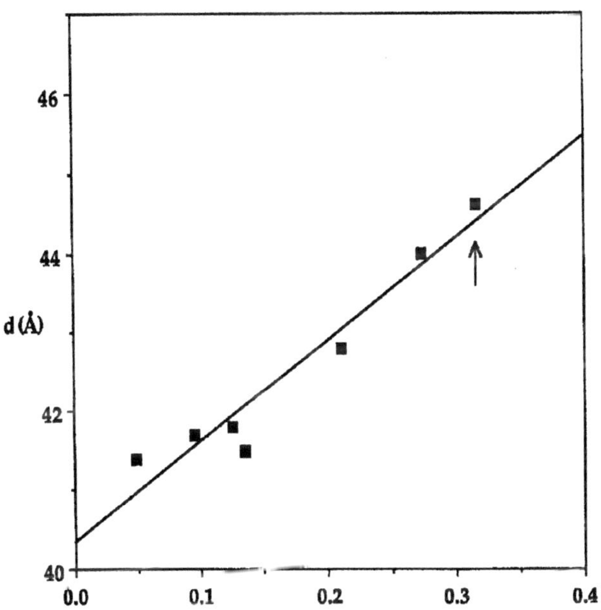

Fig. 2. The repeat distance d(Å) vs volume fraction water. The ethanol water ratio is 1:1. The phase boundary towards water is marked with an arrow

distance due to swelling. The maximum swelling is reached at 44 Å, which is less than that obtained with pure water as solvent [29].

Already at low ethanol concentration, the x-ray diffraction lines except first order are lost. This is probably due to increased static disorder at the head group plane. To investigate if other liquid crystalline phases exist, D_2O-NMR was used.

The D_2O-NMR spectra gave one quadrupole splitting of the same magnitude through the liquid crystal region, from low ethanol concentrations, where L_a phase was confirmed from x-ray diffraction to maximum ethanol concentration at the phase boader. Only one phase, the lamellar, is thus present.

At constant water concentrations in the L_a phase, the repeat distance decreases when the amount of ethanol increases as can be seen in Fig. 3. The cross-sectional area of the polar head group as a function of ethanol concentration, at constant water concentration (as in Fig. 3), increases from 66 $Å^2$ to 102 $Å^2$ at the phase boarder towards excess of water/ethanol. (In the calculation it was assumed that all ethanol is in the aqueous phase. The aqueous phase includes the polar headgroup plane). In the SPC/water system, the maximal molecule area is 70 $Å^2$ [29].

The increased head group area is balanced by a decreased thickness of the hydrocarbon part, showing an increased mobility of the chains occupying a corresponding larger cross sectonal chain area.

According to the membrane/buffer partition coefficient, ethanol is soluble in the hydrocarbon part of the bilayer to a limited extent.The major part of ethanol will be in the water phase. The overall partition coefficient of deuterated ethanol in DPPC liposomes, in the liquid crystalline state, has been determined by ^{2}H-NMR [31]. The result is interpreted as a two site binding of ethanol to the model membrane. The first coefficient, representing binding to the hydrocarbon region, had a low constant value when ethanol concentration increased. The second site was assumed to be surface binding of ethanol, which is a cooperative process. Other methods have reported similar results [13, 32].

Since the head group region is somewhat disordered, and the surface area of the head groups is increased given from x-ray measurement, one possible explanation is that ethanol might interact with the polar head group (ethanol can act as donator of hydrogens for hydrogen bonding between the polar groups). With increased concentration of ethanol, this effect lead to gradual disordering of the L_a phase.

A general conclusion from this work is that the ratio between the molecular packing parameters

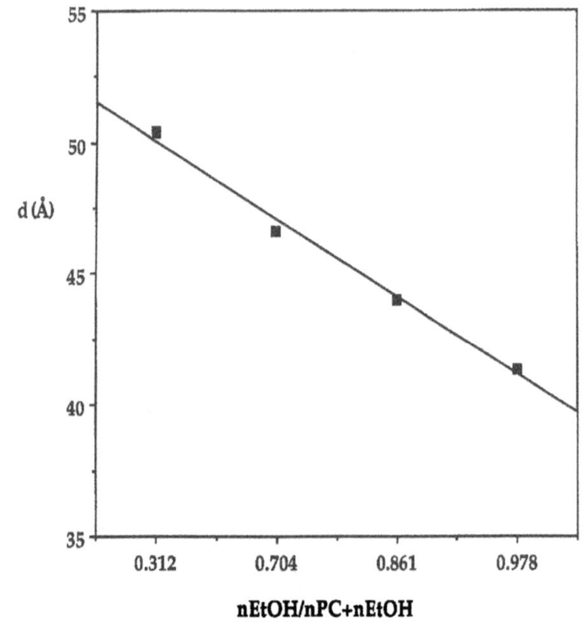

Fig. 3. The repeat distance d(Å) at constant water concentration, 22% (w/w) mol ratio of ethanol to ethanol + lipid

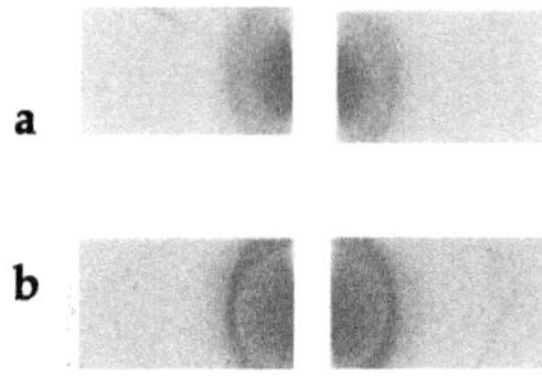

Fig. 4. a) A sample with the composition 76% SPC, 17% EtOH, and 7% H_2O, showing an L_α phase; b) A sample with the composition 69% SPC, 26% EtOH, and 5% H_2O, showing an L2 phase. (A DPT-type camera [22] was used.)

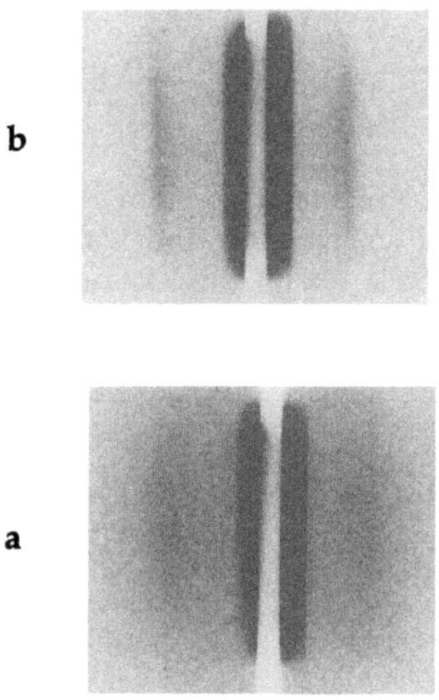

Fig. 5. a) A sample with the composition 60% SPC, 33% EtOH, 7% H_2O, showing an ethanol-water continuous solution; b) A sample with the composition 92% SPC and 8% EtOH showing an L2 type of solution. (A Luzzati camera [23] was used.)

[33] is not changed when water is exchanged with ethanol. The molecule form is preserved and, thus, the L_α phase.

The L2 phase is closely related to the lamellar phase, according to its x-ray diffraction characteristics. The slight increase in line broadening is considered to be due to a reduction of the size of the lamellae (see Fig. 4). The reduction in repeat distance is 4 Å going from the L_α phase to L2 along the SPC-ethanol line. Only a small amount of water, 5% (w/w), can be incorporated. This indicates a lamellar type of structure, where the water will form a thin film between the polar interfaces. The x-ray diffraction pattern from L2 and from solutions is compared in Fig. 5. There is a drastic increase in line broadening and a decrease in intensity when L2 gradually transforms to a water-ethanol continuous solution. The changes in x-ray pattern correspond to a considerable decrease in number of aggregates and a size reduction as well. This is known from other systems where an L2 phase transforms to L1 type of solution. From vapor pressure studies of PC-ethanol mixtures, PC was found to form aggregates in ethanol [34].

Introducing a solvent like ethanol affects the dispersion behavior of the lamellar phase. This is illustrated in Fig. 6 by a series of liposome dispersions (10% (w/w) SPC) with increasing ethanol concentration. In any liposome dispersion, size growth is a spontaneous process, but it is usually extremely slow. Such dispersions may be stable for years. With gradually increasing ethanol to water ratio, the liposomes ultimately form visible aggregates and form a coherent phase, which clearly can be seen at 40% ethanol (Fig. 6). The density gradient of the solution does not explain this drastic effect. Thus, increasing ethanol concentration causes a large decrease in area to volume ratio for the L_α phase in the two-phase region.

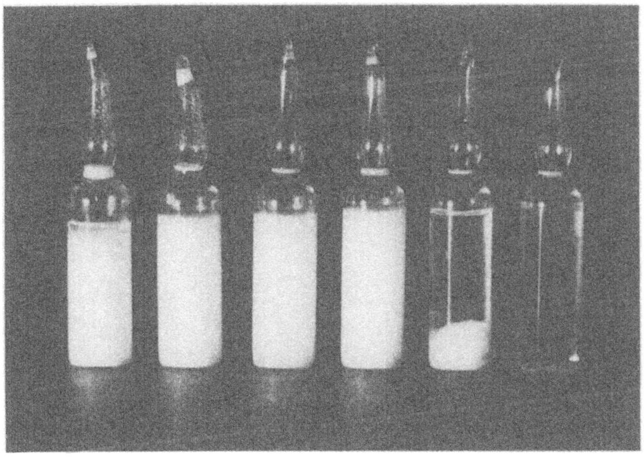

Fig. 6. Liposome dispersions at constant lipid concentration, 10% (w/w). Ethanol concentrations from left 0, 10, 20, 30, 40 and 50% (w/w). Temperature ~21°C

Acknowledgements

Grants supporting this work were obtained from Karlshamns AB Research Foundation. Dr Ali Khan is thanked for helping with the D_2O-NMR measurements. Professor Kåre Larsson is thanked for valuable discussions and comments on the manuscript. Tomas Land is thanked for help with sample preparation.

References

1. Alkohol som sjukdomsorsak (1988) In: Nordén Å (ed) Almqvist & Wiksell
2. Larsson K (1986) Acta Chem Scand A40:313–314
3. Larsson K (1988) Langmuir 4:215–217
4. Meyer HH (1906) Harvey Lect Lippinott, Philadelphia, pp 11–17
5. Overton E (1901) Studien über die Narkose. Fisher, Jena
6. Miller KW, Braswell LL, Dodson BA, Forman SA (1986) In: Roth SH, Miller KW (eds) Molecular and Cellular Mechanisms of Anesthetics. Plenum Medical Book Company, New York, NY, pp 125–137
7. Lovinger DM, White G, Weight RR (1989) Science 143:1721–1724
8. Jain MK, Wu NM (1977) J Membr Biol 34:157–201
9. Krishnan KS, Brandts JF (1979) Mol Pharmacol 16:181–188
10. Hill MW (1974) Biochim Biophys Acta 356–117–124
11. Rowe ES (1980) Clin Exp Res 4:227
12. Richards CD, Martin K, Gregory S, Keightley CA, Hesketh TR, Smith GA, Warren GB, Metcalfe JC (1978) Nature (London) 276:775–779
13. Rowe E (1983) Biochemistry 22:3299–3305
14. Rowe E (1985) Biochim Biophys Acta 813:321–330
15. Simon SA, McIntosh TJ (1984) Biochim Biophys Acta 773:169–172
16. Cullis PR, Hornby AP, Hope MJ (1980) In: Fink BR (ed) Molecular Mechanisms of Anesthesia. Raven Press, New York, pp 397–403
17. Chin JH, Goldstein DB (1977) Science 196:684–685
18. Taraschi TF, Rubin E (1985) Laboratory Investigation 52:120–131
19. Harris RA, Hitzemann RJ (1981) In: Galanter M (ed) Currents in Alcoholism. Grune & Stratton, New York, vol 8, pp 379–404
20. Harris RA, Schroeder F (1981) Mol Pharmacol 20:128–137
21. Klausner RD, Kleinfeld AM, Hoover RL, Karzowsky MJ (1980) J Biol Chem 255:1286–1295
22. Rydhag L (1979) Fette, Seifen, Anstrichmittel 81:168–173
23. Stenhagen E (1951) Acta Chem Scand 5:805–814
24. Luzzati V, Mustacchi H, Skoulios A, Husson F (1960) Acta Crystallogr 13:660–667
25. Drakenberg T, Forsén, Lilja H (1983) J Magnetic Resonance 53:412–422
26. Kiessig H (1942) Kolloid-Z 98:213–221
27. Khan A, Fontell K, Lindblom G, Lindman B (1982) J Phys Chem 86:4266–4271
28. Wennerström H, Lindblom G, Lindman B (1974) Chem Scr 6:97–103
29. Bergenståhl B, Fontell K (1983) Progr Colloid & Polymer Sci 68:48–52
30. Rosevear FB (1954) J Am Oil Chem Soc 31:628–639
31. Kreishman GP, Brittain CG, Hitzemann RJ (1985) Biochem Biophys Res Comm 130:302–305
32. Seeman P (1972) Pharmacol Rev 24:583–654
33. Israelchvili JN, Marcelja S, Horn RG (1980) Q Rev Biophys 13:121–260
34. Kaneko T, Shinoda K (1988) Yukagaku (Japan) 37:709–710

Author's address:

I. Söderberg
Department of Food Technology
University of Lund
Box 124
22100 Lund, Sweden

Partial phase equilibria and partition of alcohol cosurfactants in ionic microemulsions

S. Backlund[1]), G. Boström[1], K. G. Nordli[2]) and T. Wärnheim[3])

[1]) Department of Physical Chemistry, Åbo Akademi, Åbo, Finland
[2]) Department of Chemistry, University of Bergen, Norway
[3]) Institute for Surface Chemistry, Stockholm, Sweden

Abstract: The partial phase behaviour of quintinary microemulsions containing water (0.5 m NaCl), sodium dodecylsulfate, octane and alcohol, where the alcohol is a butanol or pentanol isomer, has been determined. In addition, the partition of the cosurfactant between the phases has been investigated, as well as in detail for the 1-butanol system. The efficiency of the cosurfactant for forming microemulsions is found to exhibit no simple relationship to its solubility properties. In addition, while branched alcohol cosurfactants promote the formation of solution phases at high surfactant content, they destabilize the microemulsion at low surfactant contents.

Key words: Microemulsion; partial phase equilibria; cosurfactant partition; sodium dodecylsulfate; short chain alcohols

Introduction

Microemulsions were hailed decades, ago as potentially useful in a large number of technical processes, e.g., for cleaning, in tertiary oil recovery, as processing aids, or as reaction media [1—3]. In particular, large scale operations such as tertiary oil recovery are strongly dependent on two factors: minimizing the amount of surfactant needed to create the microemulsions, since the surfactant normally is the most expensive component in the system, and the availability of information about how the microemulsion will react upon changes in external parameters, such as temperature, pressure, contact with excess oil and aqueous phases, and behaviour upon changes in concentration in the components due to, for example, adsorption on solid surfaces. Thus, sufficient knowledge about phase behaviour, including the composition of the different phases, is definitely a prerequisite for any application of microemulsions.

We have chosen to study microemulsion systems stabilized by sodium dodecylsulfate (NaDDS) and an alcohol cosurfactant. The reason for our choice is that these microemulsions are often used as model systems, and knowledge about the phase behaviour over a wide composition range is quite extensive [4—6] and has been complemented by structural investigations for some specific cosurfactants, especially butanol [7—10].

We have chosen to concentrate on a few factors of fundamental and practical importance that are not too well investigated. We have studied the effect on the phase diagram of varying the molecular structure of the alcohol cosurfactant in these composition regions. The cosurfactants studied are 1-pentanol, 3-methyl-1-butanol, 1-butanol, 2-methyl-1-propanol, 2-pentanol, and 3-pentanol. It is known that a branching of the alcohol cosurfactant at high surfactant content normally serves to decrease the order of surfactant packing, which influences the phase behaviour: e.g., extension regions of ordered lamellar phases are decreased in favour of solution phases [11—13]. The surfactant content of the systems is deliberately kept low (mass fraction w of NaDDS, $w_{\mathrm{NaDDS}} = 0.01$), in contrast to much of the previous work. In addition, we have attempted to accurately determine the partition of the alcohol cosurfactant in the 1-butanol system, since the partition of the cosurfactant is a second factor of prime importance for determining its efficiency [1]. The aim of the work has been to, as far as possible, discuss the question of cosurfactant efficiency in terms of either molecular structure, or more simply, from its solubility properties.

Experimental

Chemicals

Sodium dodecylsulfate (NaDDS) (BDH, specially pure) and sodium chloride (Merck, p.a.) were dried in an evacuated dessicator before use at 330 K and 390 K, respectively. 1-butanol (Merck, p.a.), 2-methyl-1-propanol (Merck, p.a.), octane (Fluka, puriss), 1-pentanol (Fluka, p.a.), 3-methyl-1-butanol (Aldrich, p.a.), 2-pentanol (Merck, zur synthese) and 3-pentanol (Fluke, purum) were used as supplied. The water was twice distilled immediately before use. All samples were prepared by weight.

Phase behaviour

The phase behaviour was determined by preparing samples of fixed composition and allowing them to equilibrate in a thermostatted bath at 298.2 K with subsequent visual determination of the number of phases present. In most cases, the equilibration was very rapid and equilibrium was attained within a few minutes. Around 250 samples were prepared for each partial phase diagram with a specific cosurfactant. It should be noted that the Krafft-point for NaDDS is around 298 K in 0.5 m NaCl. However, the Krafft-point is rapidly lowered upon addition of alcohol and no precipitation of surfactant occurred in any system, even with long-term storage.

Solubilities

The solubilities of the alcohols in water and 0.5 m NaCl were determined from density measurements as previously described [14]. Densities of the solution were measured by an Anton Paar DMA 602 density meter (298.15 K).

Gas chromatography

The partition of the alcohol cosurfactant was determined on separated phases by gas chromatography on a Perkin Elmer with Apiezon column or on a Hewlett Packard 5720 A with Porapak Q column.

Results and discussion

The partial phase diagrams for the system water (0.5 m NaCl)-octane-alcohol-sodium dodecylsulfate, where the alcohols are 1-butanol, 1-pentanol and 2-methyl-1-propanol are shown in Fig. 1. The mass fraction of surfactant is constant $w_{NaDDS} = 0.01$ (the molality at the c.m.c. of NaDDS in 0.5 m NaCl is 0.52 mmol kg^{-1}, i.e., far below the surfactant content used in this work [15]). The relative mass fraction $a = w_{octane}/(w_{octane} + w_{0.5\ m\ NaCl})$ is varied between 0 and 1. Without cosurfactant, the system rapidly separates into two phases, NaDDS in 0.5 m NaCl, and almost pure octane. Adding cosurfactant to the system results first in the extension of the solution phase in the water-rich part of the phase diagram; upon further addition of alcohol, the system enters the three-phase region, i.e., an aqueous phase, a microemulsion phase, and an oil-phase. On further addition the system again separates into two phases. An frequently chosen representation in these systems is the change in phase behaviour with increasing salinity, but an equivalent pattern is, as seen, obtained for increasing mass fraction of cosurfactant as well.

Comparing the performance of the different investigated cosurfactants, 1-pentanol is by far the most efficient: $a = 0.5$ and $w_{NaDDS} = 0.01$, the required $w_{alcohol}$ for entering the three-phase region is 0.05, 0.10, and 0.17 for 1-pentanol, 1-butanol and 3-methyl-1-propanol, respectively. However, the 1-pentanol system gives three phases in a rather narrow region; i.e., it is considerably more sensitive to changes in the composition.

Since the investigated systems contain five components, it becomes nearly impossible to map-out the effects when varying the content of each component. However, when comparing cosurfactant efficiency it is important to check that the effects are not coincidental, due to the choice of a specific salinity or surfactant concentration. In Fig. 2, the effect of varying the initial molality of NaDDS (as calculated on 0.5 m NaCl) is shown for the three different cosurfactants, together with two more pentanol isomers, 3-methyl-1-butanol and 2-pentanol. The required mass fraction of alcohol to enter the three-phase region varies with m_{NaDDS}, but the trend $w_{1\text{-pentanol}} < w_{3\text{-methyl-1-propanol}} <$ 1-butanol $< w_{2\text{-methyl-1-propanol}} < w_{2\text{-pentanol}}$ remains in the investigated interval $0.025 < m_{NaDDS} < 0.15$ mol kg^{-1}. The system with 3-pentanol has no three-phase region whatsoever at these surfactant contents and

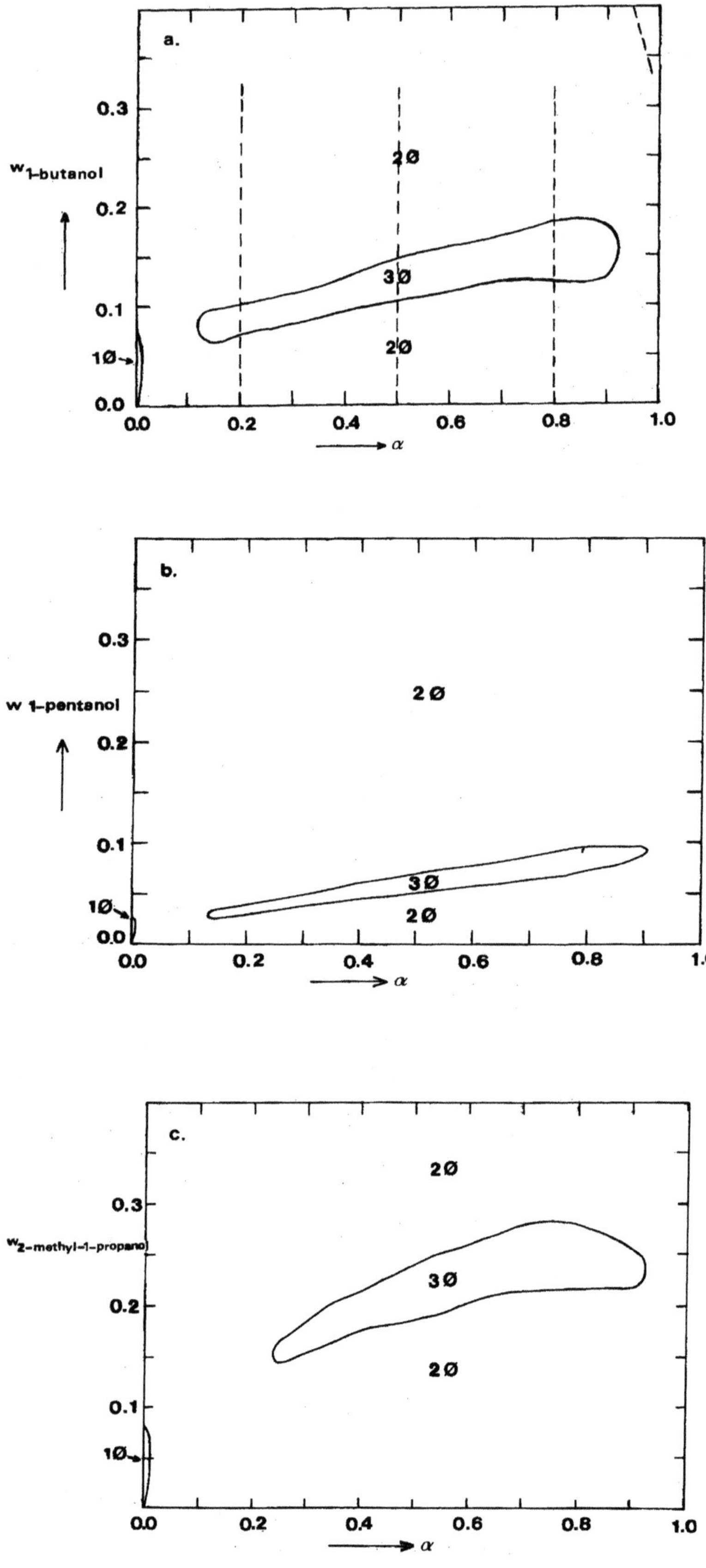

Fig. 1. Partial phase diagram for 0.5 m NaCl-octane-alcohol-sodium dodecylsulfate (NaDDS) at mass fraction $w_{NaDDS} = 0.01$ and varying $w_{alcohol}$ at 298.2 K. The mass ratio $\alpha = w_{octane}/(w_{octane} + w_{0.5\ m\ NaCl})$ is varied from 0 to 1. The alcohols used are: a) 1-butanol; b) 1-pentanol; c) 2-methyl-1-propanol. $n\Phi$ denotes number of phases

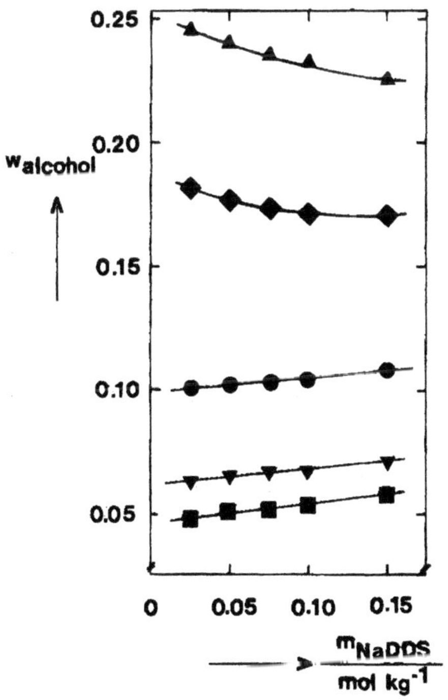

Fig. 2. The mass fraction of alcohol at 298.2 K required for separation into three phases as a function of initial molality of sodium dodecylsulfate (NaDDS). The mass ratio $a = w_{octane}/(w_{octane} + w_{0.5\ m\ NaCl}) = 0.5$. ■ = 1-pentanol; ▼ = 3-methyl-1-butanol; ● = 1-butanol; ◆ = 2-methyl-1-propanol; ▲ = 2-pentanol

$w_{3\text{-pentanol}} < 0.5$. It should be noted from Fig. 2 that the differences between the mass fractions of the cosurfactants are larger at low m_{NaDDS}, an effect that we will comment on later.

In order to consider a possible correlation between cosurfactant efficiency and solubility properties, the appropriate comparison would be to determine the partition of the cosurfactant between the aqueous and oil phases in the presence and absence of surfactant. Such a study was performed in detail for the 1-butanol system; in addition, a more limited study of the solubility properties and the partition for the rest of the alcohol cosurfactants was made. There is no direct correlation between their solubility in water or 0.5 m NaCl and efficiency as cosurfactant, i.e., 2-pentanol and 3-pentanol deviate from the trend when comparing Fig. 2 and Table 1. Also, the partition of the different alcohols between phases was determined at compositions just before the three-phase region appears. Although the systems are thus compared at different compositions, the trend in the partition coefficients could serve for the comparison. Defining

Table 1. Solubility of the alcohol cosurfactants used at 298.2 K in water and 0.5 m NaCl, respectively

Alcohol	Solubility	
	in water [mol/kg]	in 0.5 m NaCl [mol/kg]
1-pentanol	0.26	0.20
3-methyl-1-butanol	0.31	0.25
2-pentanol	0.52	0.39
3-pentanol	0.62	0.43
1-butanol	1.07	0.84
2-methyl-1-propanol	1.19	0.92

$K = w_{alcohol\ (oil\text{-}phase)}/w_{alcohol\ (aqueous\ phase)}$, K varies as 1-butanol < 1-pentanol < 2-methyl-1-propanol < 3-methyl-1-butanol < 2-pentanol, which shows no relation to Fig. 1. Thus, differences in the trend, both with respect to aqueous solubility, and the partition of the alcohol are found.

For the 1-butanol system, a detailed study of the partition was performed for two- and three-phase systems corresponding to the compositions shown in Fig. 1c, and the results are collected in Figs. 3a—c for $a = 0.20$ (Fig. 3a), 0.50 (Fig. 3b), and 0.80 (Fig. 3c), respectively. Typically, for $a = 0.50$, the butanol is at low mass fractions, preferentially dissolved in the aqueous phase; increasing $w_{1\text{-butanol}}$ will lead to approaching and exceeding of the solubility limit in the aqueous phase, 0.84 mol kg^{-1}, until, finally, the level drops off. Entering the three-phase region, it should be noted that $m_{1\text{-butanol}}$ in the three phases in most cases is not, even as an approximation, constant, since we are dealing with a quintinary system. At even higher contents of 1-butanol, in the two-phase region, it all dissolves in the oil-phase, and the molality in the aqueous phase decreases far below that of the value in 0.5 m NaCl.

Considering the rather curious partition behaviour at low $w_{1\text{-butanol}}$ values, we have investigated that region in more detail and compared the data for systems with and without surfactant (Fig. 4). The addition of small amounts of surfactant do not, at low surfactant content ($w_{NaDDS} = 0.01$), cause any larger redistribution of alcohol. Adding surfactant does, of course, increase the solubility in the aqueous phase, but the solubilization limit is not far above the limiting value in 0.5 m NaCl.

There are two points to be considered from this investigation. First, it suggests that phase equi-

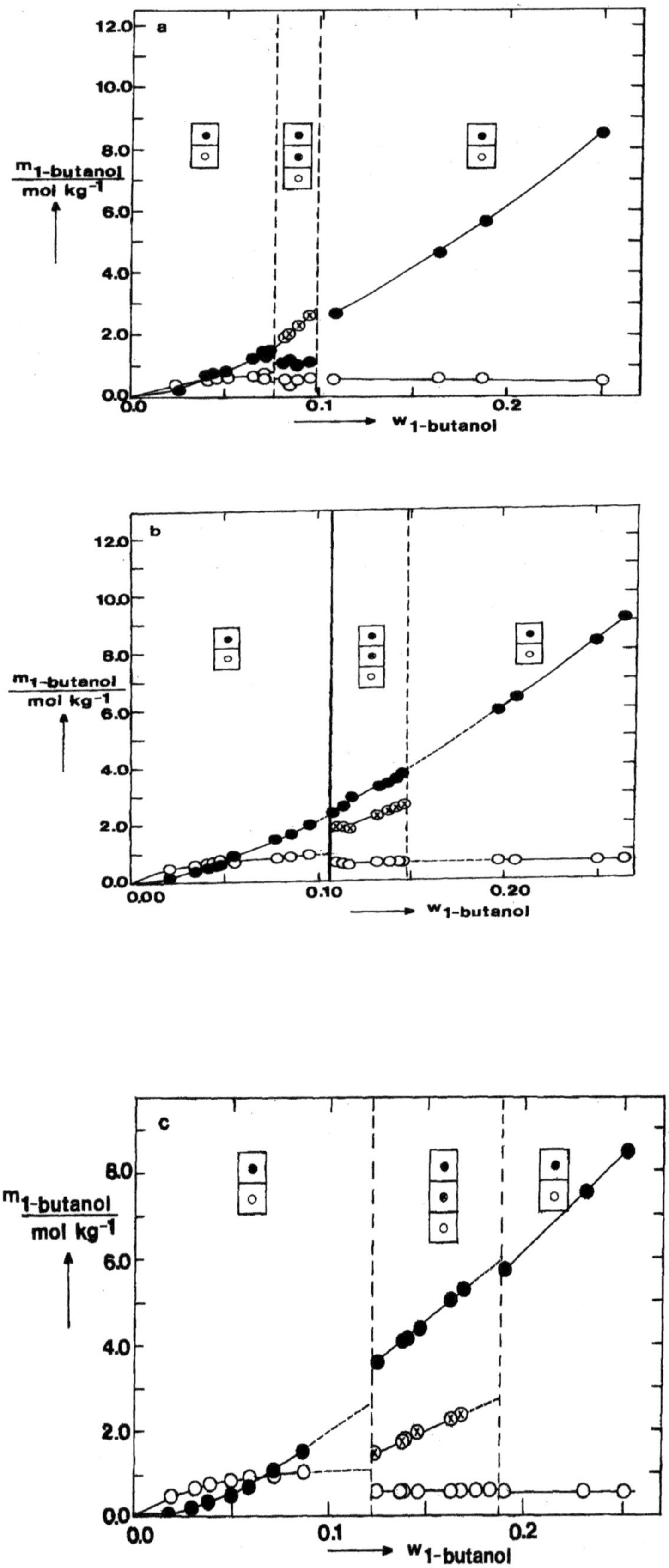

Fig. 3. The molality of 1-butanol at 298.2 K in the different phases for varying mass fractions $w_{1\text{-butanol}}$. Three different mass ratios $a = w_{\text{octane}}/(w_{\text{octane}} + w_{0.5\text{ m NaCl}})$ are used and the mass fraction of sodium dodecylsulfate is constant, $w_{\text{NaDDS}} = 0.01$. a) $a = 0.2$; b) $a = 0.5$; and c) $a = 0.8$

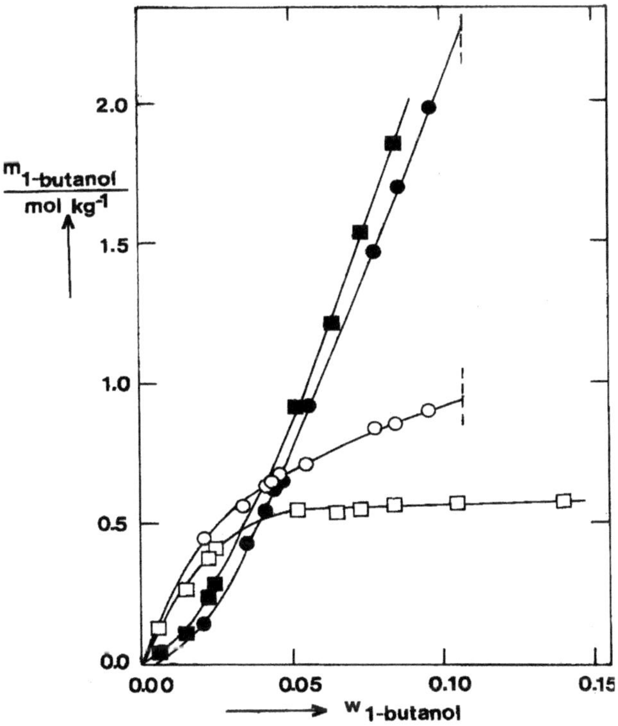

Fig. 4. The molality of 1-butanol at 298.2 K in the aqueous and oil phases for varying mass fractions $w_{1\text{-butanol}}$. The mass ratio $\alpha = w_{\text{octane}}/(w_{\text{octane}} + w_{0.5\ m\ \text{NaCl}}) = 0.5$. The symbols denote aqueous ○ and oil ● phases at mass fraction of sodium dodecylsulfate $w_{\text{NaDDS}} = 0.01$ and aqueous □ and oil ■ phases at $w_{\text{NaDDS}} = 0$, respectively

libria is far from being completely determined by the solubility properties of the components. It is clear that the branched alcohol cosurfactants can favour, as well as disfavour the microemulsion formation. At high alcohol and surfactant content where the lamellar liquid crystalline phase competes with the solution phase, a disordering of the lamellar phase instead gives a solution phase. However, at lower surfactant content, the trend seems reversed. here the microemulsion competes with the formation of water-rich or oil-rich solution phases. For microemulsions stable at very low surfactant content, NMR self-diffusion measurements clearly show that these systems have a bicontinuous structure consisting of intertwined oil and water lamellae separated by a surfactant film with low or zero mean curvature [7, 8], and such structures need to maintain some order within the interfacial film. As mentioned earlier the differences between the cosurfactants are most marked at low surfactant contents (Fig. 2) and decrease with increasing m_{NaDDS}, which is consistent with our interpretation.

As a second point, we would emphasize that the actual macroscopic composition of the phases in these systems is not easily predicted, neither from solubility measurements nor from determinations of compositions at a few chosen points. While a complete evaluation of the composition in a multiphase, multi-component microemulsion is indeed a cumbersome task, our efforts could be useful for the general discussion on the requirements for optimization of microemulsion components.

Acknowledgements

This research has been supported by VISTA, a research cooperation between the *Norwegian Academy of Science and Letters* and *Den Norske Stats Oljeselskap A/S (Statoil)*. S.B. thanks *Nordiska Forskarkurser* and G.B. thanks the *Research Institute at Åbo Akademi* for financial support.

References

1. Robb ID (1981) Microemulsions, Plenum Press, New York
2. Gillberg G (1984) In: Lissant KJ (ed) Emulsions and Emulsion Technology Part III. Marcel Dekker, New York
3. Stenius P (1984) In: Cazabat AM, Veyssie M (eds) Colloides et Interfaces Les Ulis
4. Bellocq A-M, Biais J, Clin B, Gelot A, Lalanne P, Lemanceau B (1980) J Colloid Interface Sci 74:311
5. Cazabat AM, Langevin D, Meunier J, Pouchelon A (1982) Adv Colloid Interface Sci 16:175
6. Bellocq A-M, Biais J, Bothorel P, Clin B, Forche G, Lalanne P, Lemaire B, Lemanceau B, Roux D (1984) Adv Colloid Interface Sci 20:167
7. Guering P, Lindman B (1985) Langmuir 1:464
8. Olsson U, Ström P, Söderman O, Wennerström H (1989) J Phys Chem 93:4572
9. Bellocq A-M, Biais J, Clin B, Lalanne P, Lemanceau B (1979) J Colloid Interface Sci 70:524
10. van Nieuwkoop J, Snoei G (1985) J Colloid Interface Sci 103:417
11. Wärnheim T, Sjöblom E, Henriksson U, Stilbs P (1984) J Phys Chem 88:5420
12. Wärnheim T, Henriksson U (1987) J Colloid Interface Sci 115:583
13. Peyrelasse J, Boned C, Heil J, Clausse M (1982) J Phys C 15:7099
14. Høiland H, Ljosland E, Backlund S (1984) J Colloid Interface Sci 101:467
15. Emerson MF, Holtzer A (1965) J Phys Chem 69:3718

Authors' address:

S. Backlund
Department of Physical Chemistry
Porthansgatan 3—5
20500 Åbo, Finland

Progress in Colloid & Polymer Science Progr Colloid Polym Sci 82:296—310 (1990)

Effects of pressure, temperature and salinity on the multiphase behavior of the surfactant/methane and n-decane/NaCl brine system

T. Austad*), H. Hodne*) and G. Staurland

*) Rogaland University Center, Department of Petroleum, 4001 Stavanger, Norway
Rogaland Research Institute, Department of Petroleum, 4001 Stavanger, Norway

Abstract: The multiphase behavior of a live model oil system, consisting of a one-component anionic surfactant (n-decane-ortho-dimethyl-benzene sulfonate), 44.5 mole% methane and 55.5 mole% n-decane, and NaCl-brine, has been studied at different temperatures, pressures, and salinities. The solubilization parameters of oil and water in the middle phase have been determined, and the values were adjusted for thermal expansion and compressibility of the excess phases. — At salinities toward the WII(—) state relative to the optimum, complicated phase behavior was observed at certain temperatures and pressures, resulting in the formation of two middle phases. In general the system will move towards the WII(—) state as the pressure is increased. The effect of temperature in the 50°—100°C range is complicated, but at higher temperatures the system will probably move towards a WII(—) state as the temperature is further increased. — At salinities towards the WII(+) state relative to optimum, an increase in the temperature will move the system through optimum and towards the WII(—) state. Changes in the pressure appear to have rather small effects on the phase behavior at the various temperatures. Only one middle phase was observed at different pressures and temperatures studied. — For comparison, the dead oil system containing only n-decane was studied, and the phase behavior was in line with previously published data. Both increases in temperature and pressure move the system towards the WII(—) state.

Key words: Pressure; temperature; salinity; surfactant; multiphase

Introduction

It is well documented that the oil recovery from a reservoir is related to the capillary number, and that the capillary number is mainly determined by the interfacial tension between the various phases present in the porous media [1]. The interfacial tension can be drastically lowered by means of surfactants that are able to solubilize large amounts of oil and water. *The condition for optimal oil recovery has been described as the state where equal amounts of oil and brine are solubilized in the microemulsion phase* [2].

Different reviews have been published that describe the sensitivity of microemulsion phase behavior to both compositional and intensive parameters such as surfactant concentration, alcohol concentration, brine-oil ratio, surfactant struture and molecular weight, ionic strength and type of cations, pressure, and temperature [3—6]. It is important that these variables do not change during the chemical flooding, or that the selected surfactant system will maintain phase properties close to optimum, even though the intensive variables change a little.

The effects of pressure and temperature on micelles and microemulsions in water have been reviewed by Offen [7]. The literature is rather limited in this area, and, indeed, very few papers have been published dealing with pressure and temperature effects on multiphase microemulsions [3, 8—12]. In most cases, mixtures of surfactants and alcohols have been applied, and only a few of these papers deal with live oil systems [9, 10]. Usually the live oil systems are studied by compressing the tank oil or model oil with methane to a certain gas oil ratio GOR, and then the changes in the phase behavior are recorded. These phase changes can, however, be related to changes both in composition and pressure, and it is difficult to discriminate between them. *Systematic studies on the phase behavior using one-component surfactant without any alcohol as co-surfactant and live oil are lacking in the literature.*

In order to simplify a time-consuming laboratory work in selecting the best surfactant system for a given oil reservoir, methods have been described for preparing model oils and model brines to simulate the reservoir fluids [4, 9]. Surfactant systems, have been successfully constructed by mixing different commercial surfactants and using relative large amounts of co-surfactants. *More sophisticated, one-component surfactants may be more sensitive to small compositional changes and changes in temperature and pressure.*

It has been reported that the Chun Huh relation [2] predicts acceptable values for the interfacial tension between the excess phases and the middle phase, applying solubilization parameters, even though the system is not at the optimum.

The scope of this investigation is to study the behavior of multiphase microemulsions by measuring the solubilization parameters above the bubble point at different *T*, *P*, and slinities for a model oil system, made by mixing 55.5 mole% n-decane and 44.5 mole% of methane, resulting in an average molecular weight equivalent to hexane. The surfactant system is a one-component surfactant without any alcohol. A rather complex phase behavior was observed and, for comparison, the dead oil n-decane has also been studied at different *P* and *T*.

Experimental

Surfactant

The Exxon product RL-3011 (n-C_{12}-o-xylene sulfonate) was purified by multiple extraction with hexane. The product was dried under vacuum at 100 °C, and a solid material resulted. This was kept in closed bottle due to its hydroscopic nature. Two-phase titration resulted in an average molecular weight of 376.5 g/mole. The calculated value is 383.4 g/mole. Proton NMR studies suggest that the surfactant is not isomerically pure. This is further confirmed by the HPLC chromatogram which clearly showed two peaks.

Oil

Two model oils have been applied: n-decane and a live oil containing 44.5 mole% methane and 55.5 mole% n-decane, resulting in an average molecular weight equivalent to hexane.

Brine

Three different brines were applied: 3.80, 4.30, and 4.76 wt% NaCl dissolved in distilled water.

PVT-Equipment

A JERFI PVT window cell supplied from D. B. Robinson & Associates, Ltd. was used. The phase volumes were determined by means of a cathetometer. The cell was placed in a temperature-controlled air bath.

Measurements

Prior to the microemulsion phase studies the excess phases were calibrated with respect to compressibility and thermal expansion in the pressure range 150—600 bar and the temperature range 20°—120°C. Separate experiments were conducted with each of the excess phases in the cell, and with both of the excess phases in the cell at the same time. In both cases the values of the compressibility and the thermal expansion agreed within the experimental error.

To avoid precipitation of the surfactant, the cell was loaded in the following manner: first, n-decane and the calculated volume of methane were injected into the cell resulting in a volume of about 35 ml at specified *P* and *T*. Then half of the water volume containing the surfactant was injected, and finally, the second half of the water volume containing the salt. The total volume was about 70 ml. The injection pressure was kept at 250 bar and the temperature was 23 °C. At lab conditions the water oil ratio WOR was equal to 1. In all experiments the surfactant concentration was 2.0 wt% based on the aqueous phase.

Three different salinity scans were conducted: 3.80, 4.30, and 4.76 wt% NaCl solution. After each pressure or temperature step the mixture was carefully mixed and the phase volumes were determined when no changes in the volumes took place.

Solubilization parameters were determined by supposing the surfactant to have the same compressibility and thermal expansion as the n-decane, and a density at laboratory conditions equal to 1.0.

Results and discussion

Compressibilities and thermal expansion of the excess phases

Expansion and compressibility data for the brine at different temperatures and pressures was determined for 3.00 and 4.30 wt% NaCl solutions. In the latter case the brine was in equilibrium with the live oil (C_1 + n-C_{10}). Only small differences between the two sets of data were observed. Furthermore, the data were consistent with the results of Long and Chierici [13].

The data for the live oil (5.55 mole% n-C_{10} and 44.5 mole% C_1) have been determined, both in the pure state, and in equilibrium with an equal amount of brine. Only small differences in the expansion and compressibility factors were observed. In line with Sage and Berry [14], the thermal expansion factor has a maximum in the range 70—100 °C at the different pressures. The compressibility factors increase nearly linearly with the temperature in the studied pressure intervals.

The volumetric calibrations for T and P were made without any surfactant added to the system, supposing that the excess phases of the multiphase microemulsion system were very low in surfactant concentration.

The surfactant/n-decane/brine system

The solubilization parameters of oil and brine in the middle phase are determined as a function of pressure at different temperatures (Fig. 1). The system studied contained 2.0 wt% surfactant and 3.8 wt% NaCl solution; both values are related to the amount of water. The water/oil ratio, WOR, was equal to 1 at lab conditions.

The solubilization parameters are defined as the volume of oil or water in the middle phase divided by the volume of surfactant. The curves with a positive slope in Fig. 1 represent the solubilization parameters of the water in the middle phase. The horizontal curves or the curves with a slightly negative slope give the solubilization of the oil. At 60 °C an optimum in the phase behavior is found at about 80 bar. In the temperature range 60—100 °C, an increase in the pressure results in an increase in solubilization of water, leaving the solubilization of the oil (n-decane) nearly constant. Thus, the system changes towards the WII(—) state upon increasing the pressure. These results are in line with what Skauge and Fotland observed for systems containing model oil and alcohol [3].

At 110 °C no middle phase was observed at any pressure, i.e., the system was in the WII(—) state. Thus, an increase in the temperature results in a lower pressure at which the system turns from a

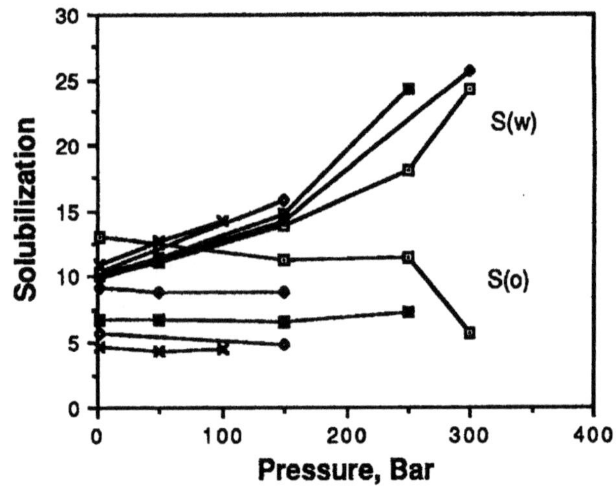

Fig. 1. Solubilization of oil and brine as the function of pressure at different temperatures. Surfactant: RL-3011, 2.0 wt%; Brine: 3.8 wt% NaCl; Oil: n-decane; WOR: 1.0 at lab. conditions. —□— 60°, —◆— 70°, —■— 80°, —◇— 90°, —×— 100°C

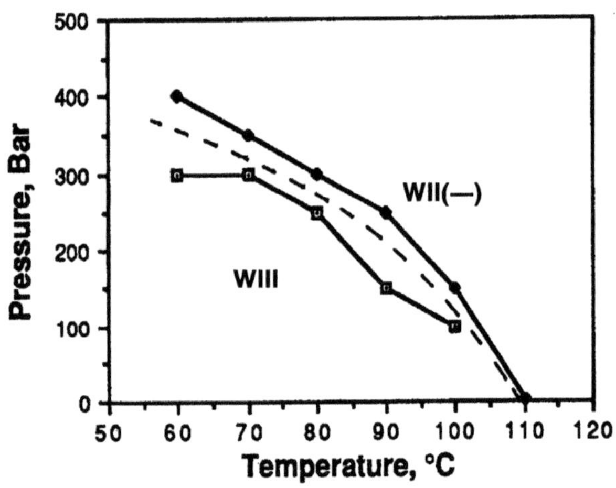

Fig. 2. The pressure — temperature relation for phase transition between WIII and WII(—), stipped curve. Surfactant: RL-3011, 2.0 wt%, brine: 3.8 wt% NaCl; Oil: n-decane; WOR: 1.0 at lab. conditions. —□—: P-T values in the WIII state, —◆—: P-T values in the WII(—) state

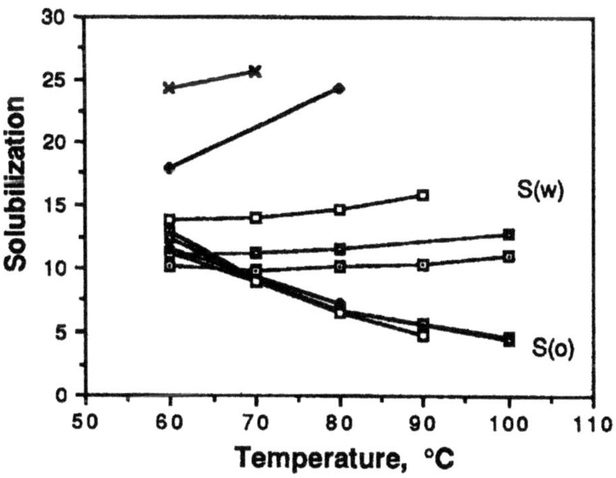

Fig. 3. Solubilization of oil and brine as the function of temperature at different pressures. Surfactant: RL-3011, 2.0 wt%; brine: 3.8 wt% NaCl; Oil: n-decane; WOR: 1.0 at lab. conditions. —□ 1 bar, —■ 50 bar, —□ 150 bar, —● 250 bar, — × 300 bar

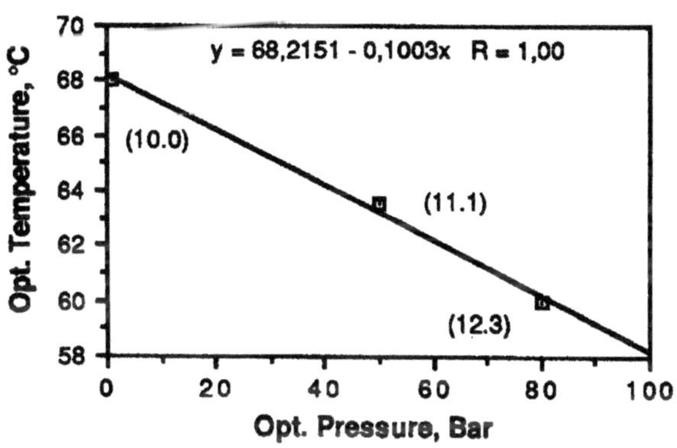

Fig. 4. Relation between T(opt) and P(opt) for the microemulsion system. Surfactant: RL-3011, 2.0 wt%; Oil: n-decane; Brine: 3.8 wt% NaCl; WOR: 1.0 at lab conditions. The solubilization parameters are listed

WIII state to a WII(—) state. Figure 2 shows the relation between pressure and temperature and the transition between the two mentioned phase states.

The solubilization parameters as the function of temperature at different pressures are shown in Fig. 3. The curves with a negative slope represent the solubilization of oil. In the temperature range studied, 60—100 °C, the solubilization of water into the middle phase is nearly constant in the pressure range 1—150 bar, while at higher pressures, 250—300 bar, it increases with the temperature. Thus, at the present salinity the system moves towards WII(—) as the temperature is increased. This is also in line with the results of Skauge and Fotland [3] for anionic surfactant systems containing alcohol. *Both the pressure and the temperature have the same effect on the phase behavior of this multiphase system.*

Two of the solubilization curves, 1 and 50 bar, cross each other and the optimal temperatures are 68° and 63.5 °C, respectively. The relation between the intensive variables T and P, satisfying the optimum condition for the system with a salinity of 3.8 wt% NaCl, is shown in Fig. 4. It appears that T(opt) shows an inverse linear relationship to P(opt). Furthermore, the solubilization at optimum increases with a decrease in T(opt).

Phase properties of the live oil

The model live oil used in these experiments consists of 55.5 mole% n-decane and 44.5 mole% methane. The two-phase region was calculated directly using Robinsons "Equi Phase" program without any adjustments. Two experimentally determined bubble points were found to agree very well with the simulated values. The scope of the present paper is to look into the multiphase behavior of a microemulsion at different P and T in the unsaturated area of the phase diagram. The multiphase microemulsions were made by using the RL-3011, a one-component surfactant combined with the following salinities:

a) 3.80 wt% NaCl, giving a multiphase behavior mostly toward the WII(—) state, relative to optimum;
b) 4.30 wt% NaCl, giving a multiphase behavior close to optimum;
c) 4.76 wt% NaCl, giving a multiphase behavior toward the WII(+) state, relative to optimum.

The low salinity multiphase behavior

This live oil microemulsion system consisting of 2.0 wt% surfactant and 3.80 wt% NaCl solution both related to the aqueous phase, and 55.5 mole% n-C_{10} and 44.5 mole% C_1 did, at some pressures and temperatures, show complicated phase behavior, resulting in two middle phases. Relatively large

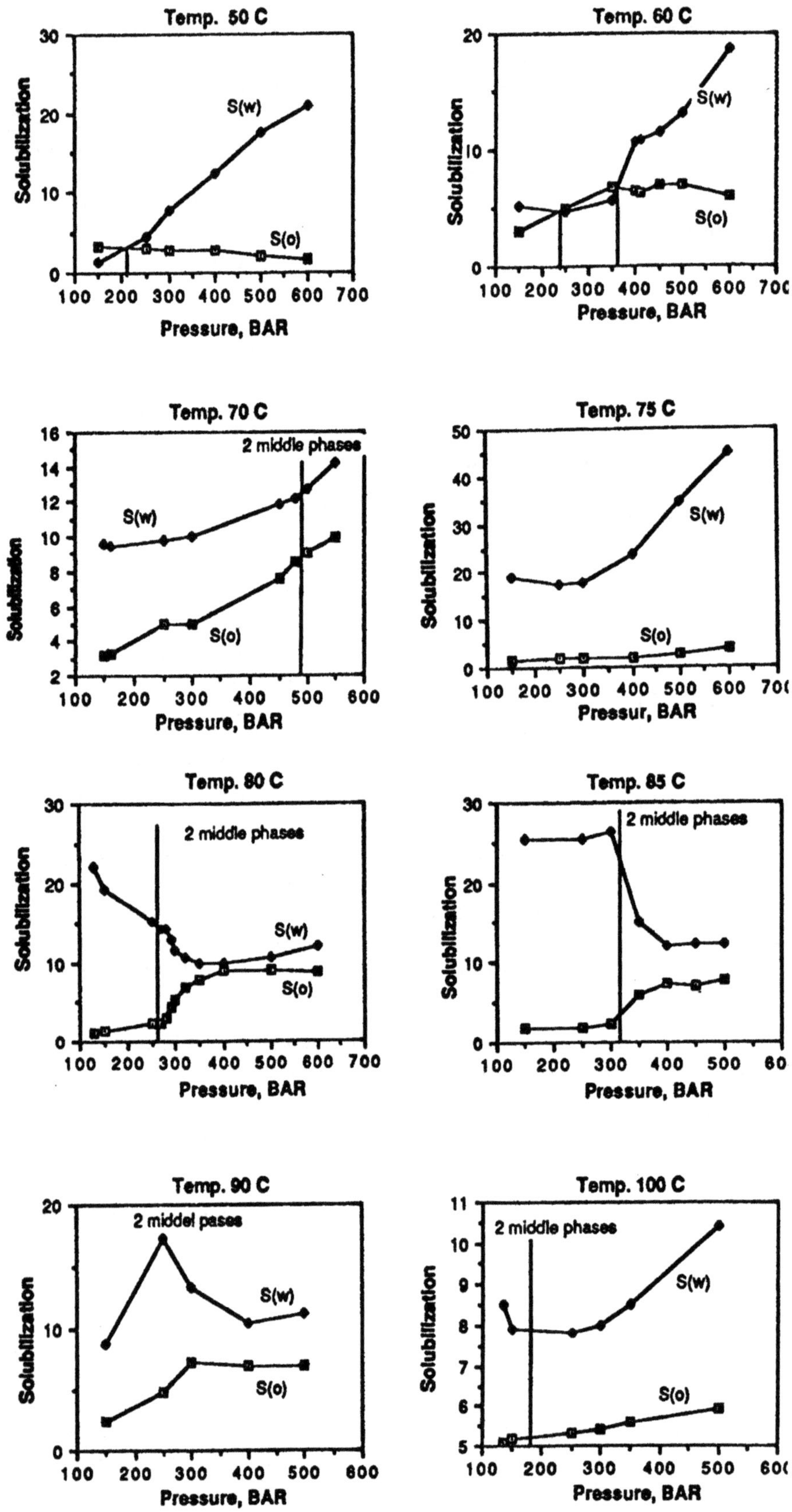

Fig. 5. The solubilization of oil and water in the middle phase(s) as the function of pressure at different temperatures surfactant; RL-3011, 2.0 wt%; Oil: 55.5 mole% n-C10 and 44.5 mole % C1; Brine: 3.8 wt% NaCl; WOR: 1.0 at lab. conditions

variations in the solubilization parameters with small changes in P and T were observed. In these cases the solubilization parameter is regarded as an average value for both of the middle phases. The solubilization parameters of oil and brine in the middle phase(s) as the function of pressure at different temperatures are shown in Fig. 5.

At 50 °C an optimum was found at 210 bar with $S(w) = S(o) = 3.15$. $S(w)$ increases drastically, and $S(o)$ is low and decreases slightly as the pressure is increased. The phase behavior is similar to the dead oil system (Fig. 1).

At 60 °C two optimums were observed at 240 and 360 bar with solubilization parameters equal to 4.7 and 6.7, respectively. At higher pressures the phase behavior appears to move towards the WII(—) state.

At higher temperatures, 70° and 75 °C, no optimums in phase behavior were observed. In fact, at 75 °C $S(o)$ is extremely low compared to $S(w)$ during the whole pressure range, and the middle phase will mainly contain water. At 70 °C two middle phases were observed at pressures above 480 bar. The ratio between the upper middle phase volume, $U(m)$, and the lower middle phase volume $L(m)$ varied between 5 and 4 as the pressure was increased from 500 to 550 bar. The upper middle phase is believed to be more oil-rich relative to the lower middle phase. Thus, for both of the temperatures it is reasonable to assume that the system will move away from optimum, toward the WII(—) state, upon increasing the pressure further.

At 80 °C and at low pressures, 128 bar, close to the bubble point of the oil, the system is far from optimum, showing very large $S(w)$ and low $S(o)$, $S(w)/S(o) = 20$. Upon increasing the pressure the system appears to move toward the optimum, $S(w)$ decreases, and $S(o)$ increases. At 270 bar an upper middle phase appeared in the cell. A further increase in the pressure up to 400 bar resulted in an increase in $U(m)$ and a decrease in $L(m)$, leaving the total middle phase volume nearly constant. This is illustrated by Fig. 6. The linearity in the plot of solubilization factors against the volume % of the upper middle phase supports that the solubilization is governed by the relative phase volumes of the middle phases (Fig. 7). The linear relation is valid up to 90% of the upper middle phase. From Fig. 7 it is further seen that the solubilization parameters of the lower and the upper middle phases are $S(w) = 15$, $S(o) = 2.2$ and $S(w) = S(o) = 9.5$, respectively. Thus, the system is very close to optimum at 400 bar. By increasing the pressure above 400 bar, $S(w)$ is slowly increased and $S(o)$ is slowly decreased, and the system is believed to move towards the WII(—) state.

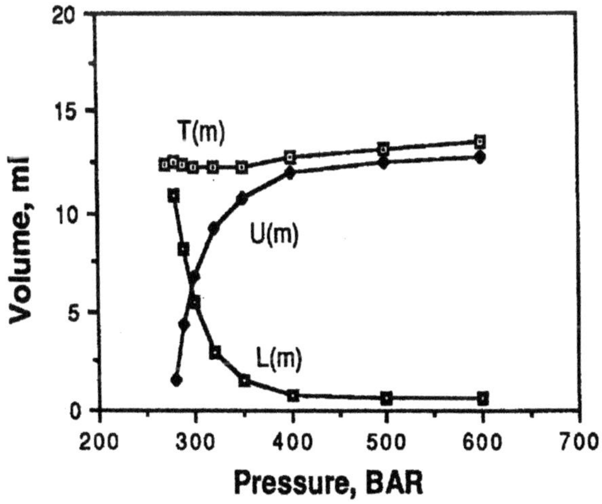

Fig. 6. The middle-phase volumes as the function of pressure at 80 °C. Surfactant: RL-3011, 2.0 wt%; Oil: 55.5 mole% C10 and 44.5 mole% C1; WOR: 1.0 at lab. conditions $T(m)$: total middle-phase volume; $U(m)$: upper middle-phase volume; $L(m)$: lower middle-phase volume

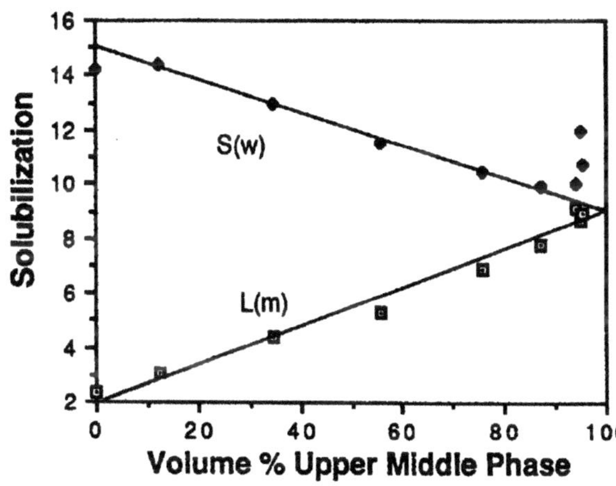

Fig. 7. The solubilization parameters as the function of the volume % of the upper middle phase at 80 °C and pressure range 170 to 600 bar

At 85 °C the system showed a similar phase behavior as that at 80 °C. Two middle phases were detected, but no optimum was observed. The solubilization of water into the middle phase(s) was always higher than the solubilization of oil.

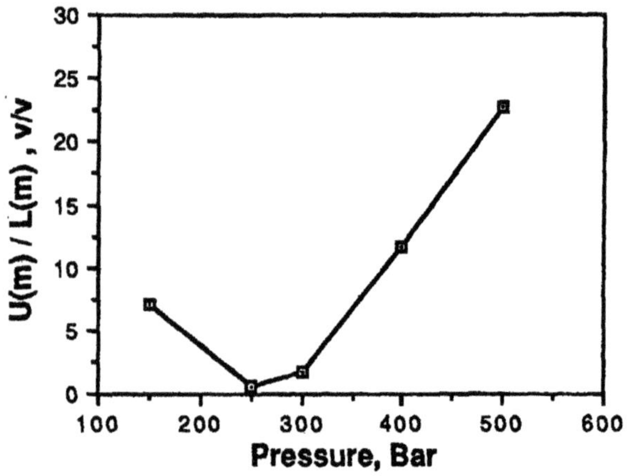

Fig. 8. The volume ratio between the upper and lower middle phase as the function of pressure at 90 °C

At 90 °C two middle phases are observed over the entire pressure range studied. The volume ratio between the upper and the lower middle phases changed in the manner shown in Fig. 8. By increasing the pressure the volume ratio will decrease to a certain value and then start to increase again. Thus, at this temperature the relative phase volumes of the middle phases are very sensitive to changes in pressure around 250 bar.

At 100 °C two middle phases were only observed at pressures below 200 bar. The volume ratios $U(m)/L(m)$ at 135 and 150 bar were calculated to 7.3 and 16.7, respectively. Thus, by comparing Fig. 5 at 90° and 100 °C, it is noticed that increase in the temperature will depress the two middle phase regions to lower pressure. The solubilization of water is always higher than the solubilization of oil, and the system appears to move further towards the WII(—) state as the pressure is increased.

The solubilization parameters of oil and brine in the middle phase(s) as a function of temperature at different pressures are shown in Fig. 9. Large changes in the solubilization parameters are observed in the temperature range 70° to 90 °C for the different pressures studied. The formation of two middle phases with different solubilization may, to some extent, explain these phase changes.

It is interesting to note that between 70° and 80 °C a drastic increases in $S(w)$ takes place at the different pressures, which can result in the formation of a second and more oil-rich upper middle phase, causing a significant drop in $S(w)$. This happens at 300 to 600 bar (Fig. 9). Increase in the temperature will increase the volume ratio between the upper and the lower middle phases causing a decrease in $S(w)$ and an increase in $S(o)$. Again, between 80° and 85 °C a drastic increase in $S(w)$ and a lowering of $S(o)$ appears in the pressure range 150 to 350 bar. Above 350 bar the variation in the solubilization parameters in this temperature range is low. It should be noted that at 300 and 500 bar and the temperatures 85° and 75 °C, respectively, only one middle phase was observed. In the case of 500 bar, this means that at 70 °C a lower middle phase is formed, and at 75 °C this middle phase is the entire middle phase. Above 75 °C an upper middle is formed, and it is increased in volume as the temperature increases.

At 150, 250, and 350 bar optimums in the phase behavior have been determined with temperatures equal to 55°, 60°, and 62 °C. The lack of an optimum in the three-phase region at 300 bar, within the temperature range studied, may indicate that the relationship between T(opt) and P(opt) at the present salinity is not a continuous function.

In general then, according to Figs. 5 and 9, a salinity of 3.80 wt% NaCl will give a phase behavior toward the WII(—) state relative to optimum. Recalling that we did observe an optimum in phase behavior for the n-decane system at 60 °C and 80 bar at the same salinity. Thus, adding methane to n-decane, resulting in an average molecular weight equal to hexane, the system will probably require a salinity even higher than 3.8 wt% NaCl in order to obtain optimum. This is in line with Nelson [10] and Pueto and Reed [9], who showed that live oil systems deviated from the optimum salt — molar volume correlation.

Even though unexpected phase behaviors were observed in the temperature range 70°-90 °C, it is a general trend that the multiphase system moves towards the WII(—) state as the pressure is increased. Regarding the temperature effects, it is difficult from the present results to draw any general conclusions concerning the effects of the phase behavior.

Multiphase behavior close to optimum

The salinity of the brine has been increased from 3.80 wt% NaCl to 4.30 wt%, related to the aqueous phase; otherwise the experimental conditions were the same as described previously. The relation between the solubilization parameters of oil and water in the middle phase(s) microemulsion(s) and the

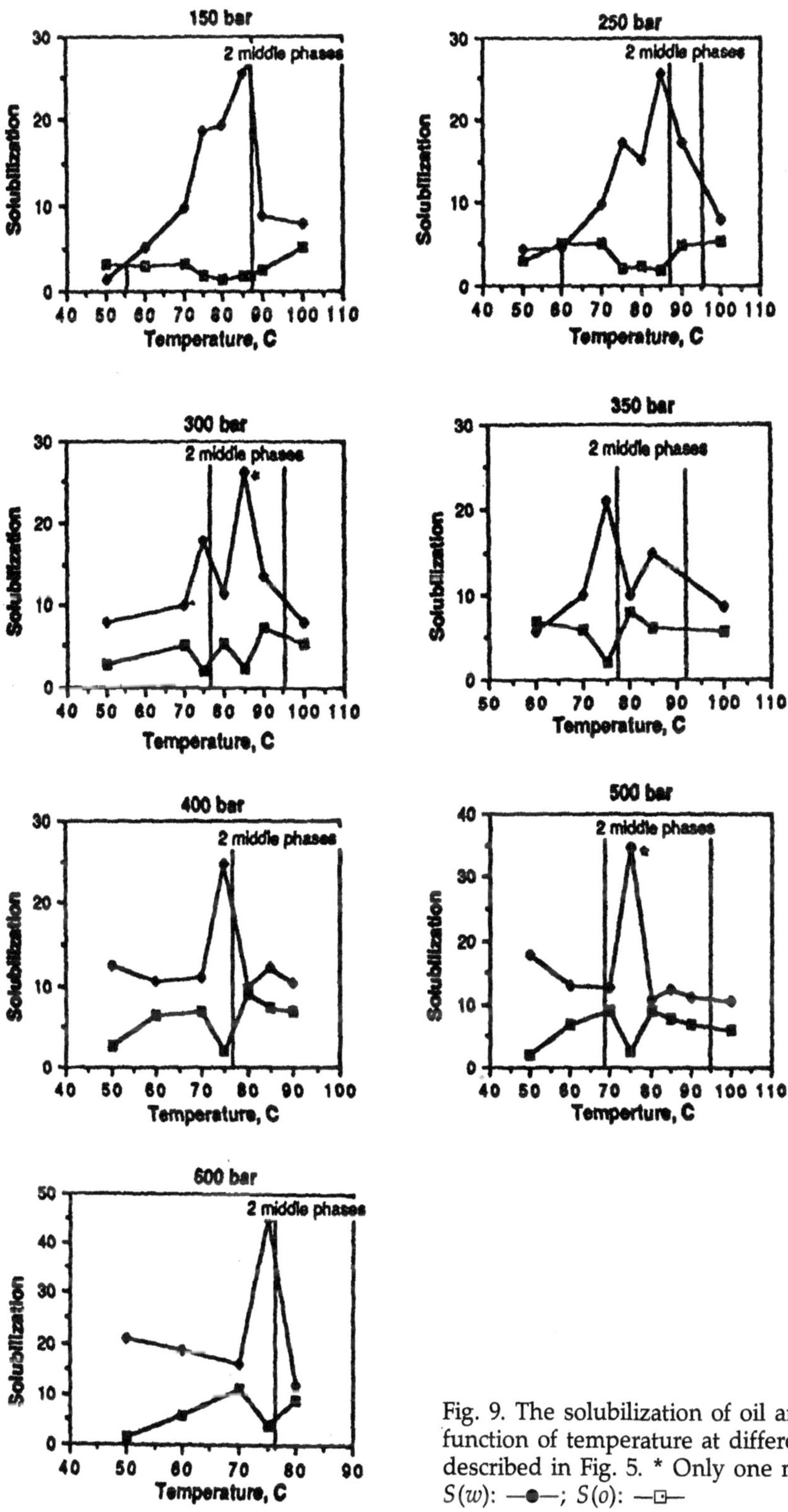

Fig. 9. The solubilization of oil and water in the middle phase(s) as the function of temperature at different pressures. Surfactant system as described in Fig. 5. * Only one middle phase observed.
$S(w)$: —●—; $S(o)$: —□—

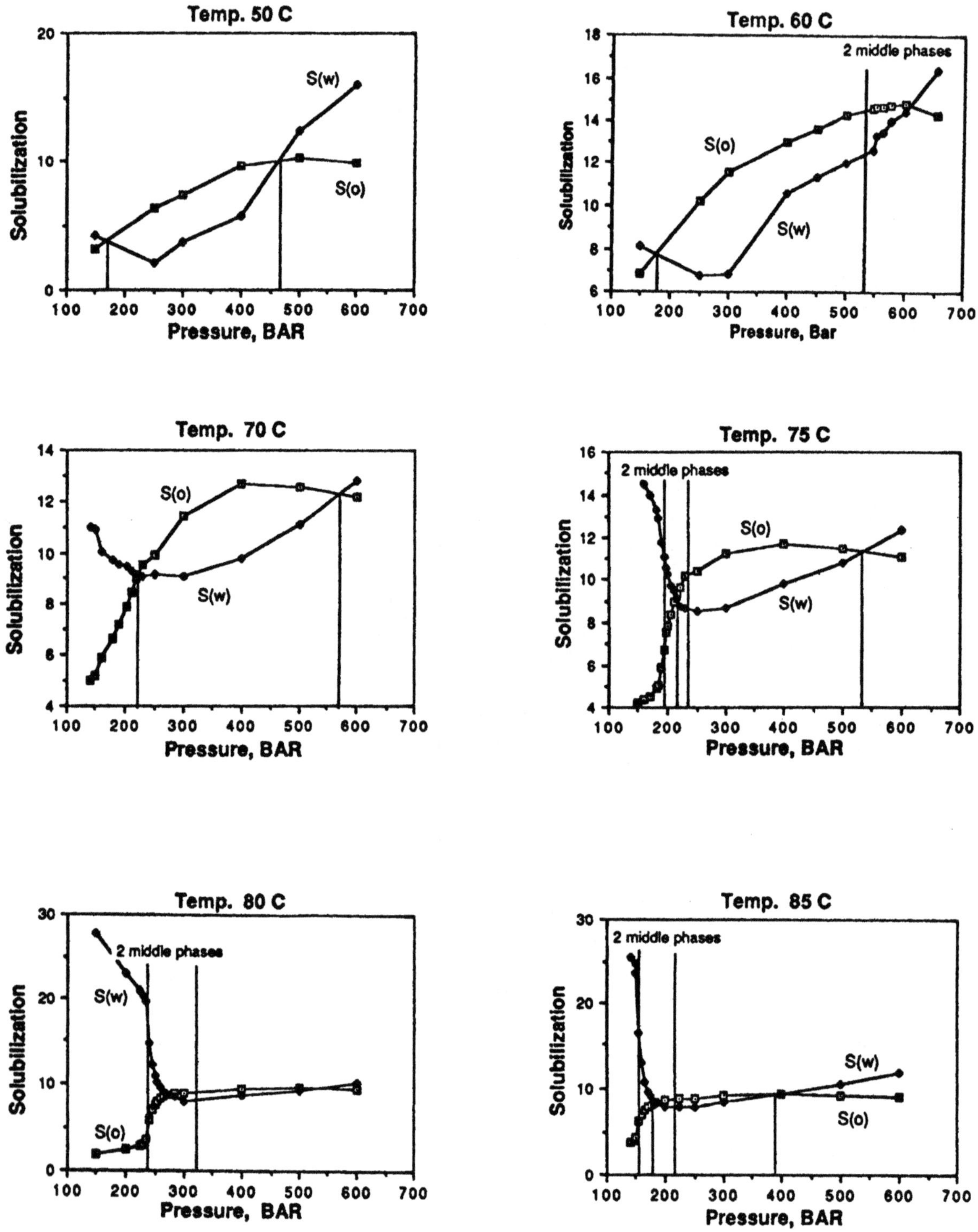
Temp. 50 C
S(w)
S(o)
Solubilization
Pressure, BAR
Temp. 60 C
2 middle phases
S(o)
S(w)
Solubilization
Pressure, Bar
Temp. 70 C
S(o)
S(w)
Solubilization
Pressure, BAR
Temp. 75 C
2 middle phases
S(o)
S(w)
Solubilization
Pressure, BAR
Temp. 80 C
2 middle phases
S(w)
S(o)
Solubilization
Pressure, BAR
Temp. 85 C
2 middle phases
S(w)
S(o)
Solubilization
Pressure, BAR

Fig. 10

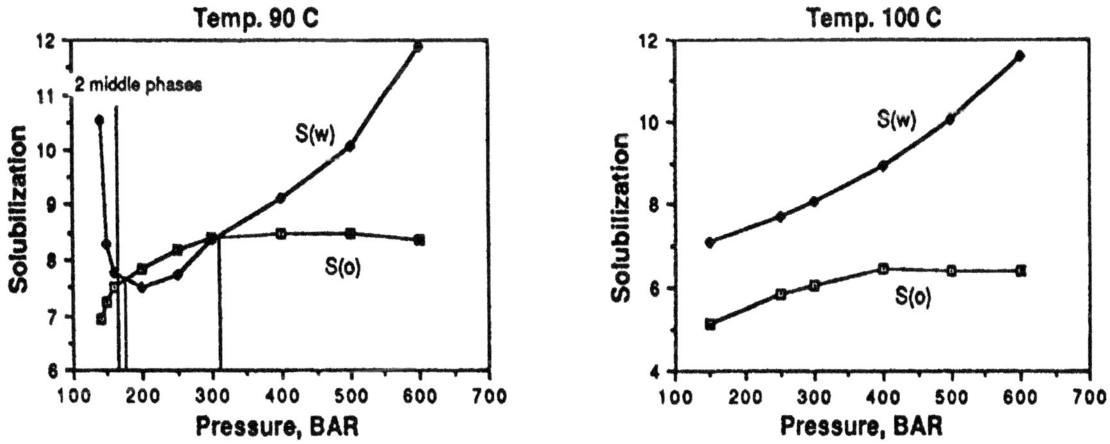

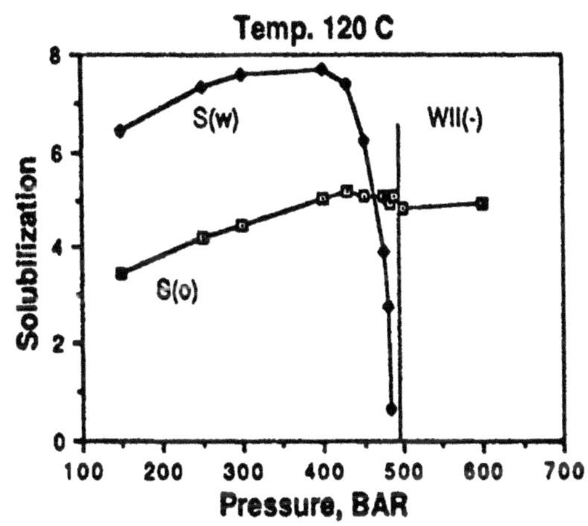

Fig. 10. The solubilization of oil and water in the middle phase(s) as the function of pressure at different temperatures. Surfactant: RL-3011, 2.0 wt%; Oil: 55.5 mole% n-C10 and 44.5 mole% C1. Brine: 4.3 wt% NaCl; WOR: 1.0 at lab. conditions

pressure at different temperatures is shown in Fig. 10. It is obvious by comparing with Fig. 5 that the relations between $S(o)$, $S(w)$, T, and P are different at a salinity closer to optimum.

At 50° and 70°C only one middle phase was observed. The systems showed optima in the pressure at 170 and 465 bar at 50°C, and at 220 and 570 bar at 70°C. It should be noted that at 50°C the excess water phase was clouded at all pressures up to 500 bar.

At 60°C and rather high pressure, about 540 bar, a lower middle phase was formed that increased in volume as the pressure increased. The result is an increase in $S(w)$ and a decrease in $S(o)$, indicating that the system will move towards a WII(—) state with a further increase in pressure. An optimum in the three-phase region is observed at about 180 bar.

At 75°, 80°, 85°, and 90°C two middle phases were observed at pressures below 320 bar. The pressure range at which two middle phases occurs is rather limited. The solubilization parameters in the two middle phase areas are average values of both the middle phases. It is interesting to note that at low pressures, just before the middle phases occur, the variation in $S(w)$ and $S(o)$ with pressure is very large. After the two middle phases region, $S(w)$ and $S(o)$ are rather similar, however, in all cases $S(o) > S(w)$. At higher pressures it appears that the systems move towards the WII(—) state, resulting in optimum pressures at 540, 530, 390, and 310 bar at 75°, 80°, 85°, and 90°C, respectively.

A closer look into the pressure regions where two middle phases occur are illustrated in Fig. 11. At 75°C the total middle-phase volume is nearly constant, while the water-rich lower middle phase decreases, and the oil-rich upper middle phase increases to nearly the same extent as the pressure increases. The linearity in the solubilization param-

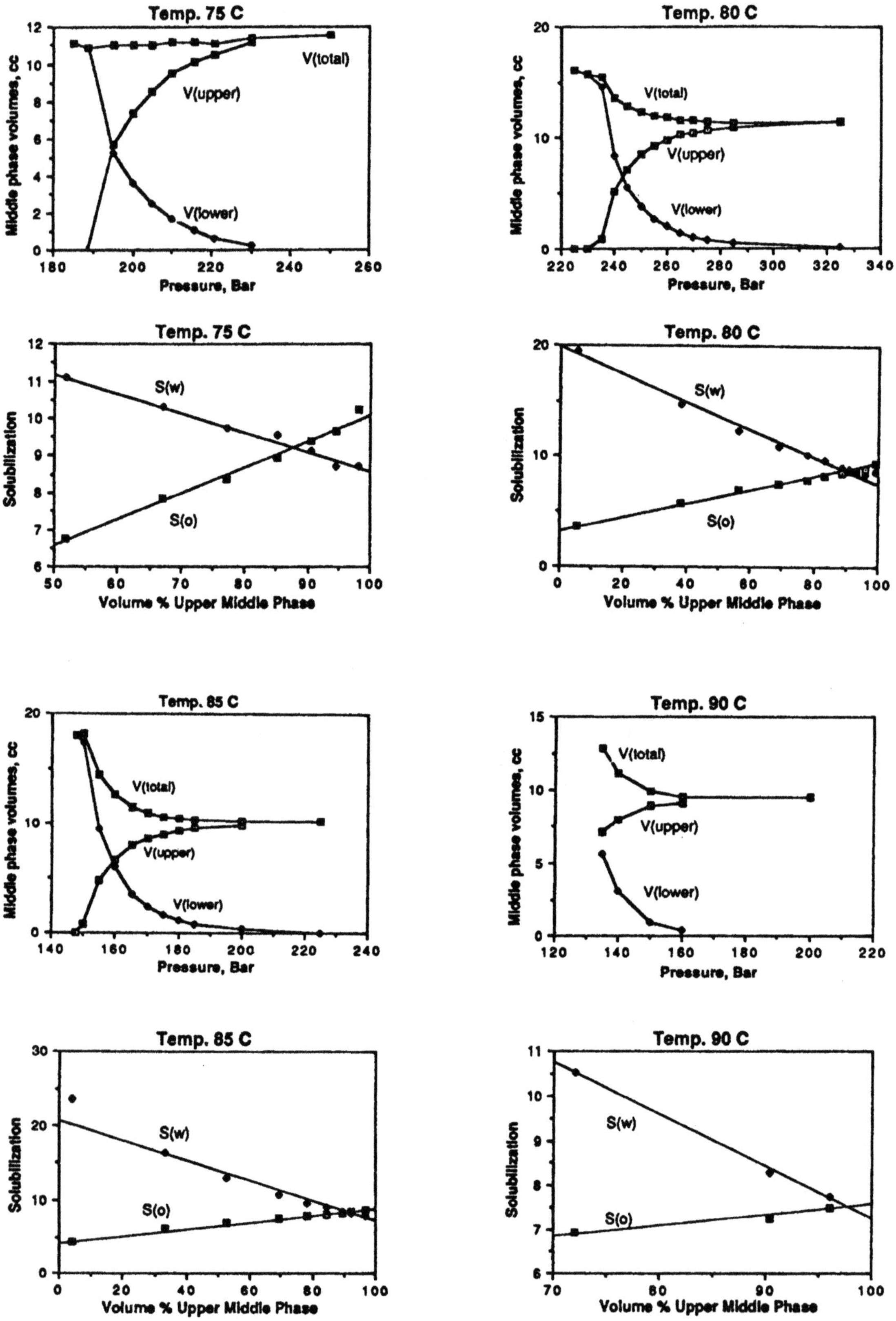

Fig. 11. Middle-phase volumes and solubilization parameters as the function of pressure at different temperatures. Surfactant system as described in Fig. 10

eters vs the relative volume of the two middle phases (Fig. 11) strongly suggests that the solubilization of oil and water are determined by the relative amount of the two middle phases. At the other temperatures, 80°, 85°, and 90°C, the total middle-phase volume decreases as the upper middle phase is formed. This is probably due to the strong solubilization of water in the lower middle phase in these cases. However, it appears that the linear relation between solubilization and relative middle-phase volumes is maintained. Thus, the solubilization in the two middle phase regions is determined by the relative middle-phase volumes. In other words, at a certain temperature, $S(o)$ and $S(w)$ in each of the two middle phases are believed to be constant during the pressure region where two middle phase occur. The solubilization parameters in the two middle phases at the different temperatures have been determined, and the values are shown in Fig. 12. it is seen that the lower middle phase is water-rich while the upper middle phase is slightly towards the WII(+) state relative to optimum.

The upper middle phase is believed to have a bicontinuous structure with a viscosity different from the assumed water-continuous lower middle phase. The flow rate of these two middle phases in a porous media will depend on the relative saturation and the viscosity of the phases. Thus, in certain pressure regions a "dispersion" of the one component surfactant, due to different partition between the two middle phases, may take place in a chemical flooding process.

The fact that the upper middle phase is close to optimum, Figure 12 further indicates that the optimum solubilization decreases as the temperature increases.

We have not performed any light-scattering measurements on the two middle phases, and we do not have any final proof that both of the two middle phases are real microemulsions. However, we feel that the data shown in Fig. 11 strongly support that the two middle phases are in equilibrium with each other, and that both of the middle phases are a middle phase in a three-phase microemulsion system, depending on P and T. This means that there exist certain temperatures and pressures at which the chemical potential of the oil in a water-rich microemulsion phase and in a microemulsion phase close to optimum are equal.

At 100°C (Fig. 10) and $S(w) > S(o)$ in the pressure range 150—600 bar, no optimum in the pressure

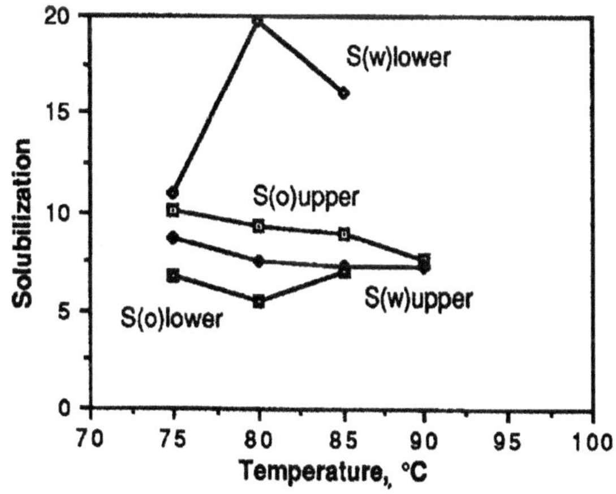

Fig. 12. Solubilization of oil and water in the upper middle phase in the four-phase pressure region as the function of temperature (system as described in Fig. 10)

was observed; the system moves towards the WII(—) state with increase in pressure.

At 120°C the solubilization of water into the middle phase decreases rapidly as the system moves towards the WII(—) state at about 500 bar. At this time the excess water phase will also contain surfactants, and the assumption that all the surfactant is in the middle phase is not valid. Above 500 bar only two phases occur, and the values of $S(o)$ represent the solubilization of oil in the lower microemulsion phase.

In Fig. 13 the solubilization parameters have been plotted against the temperature at different pressures. The two middle-phase regions and the optimum temperatures are shown. Large variations in $S(o)$ and $S(w)$ are observed below and in the two middle-phase regions. However, increasing the temperature above the two middle-phase regions the systems move towards the WII(—) state. In general then, at rather high temperatures and pressures the system will move towards the WII(—) state as these two intensive variables further increase.

The relationship between P(opt) and T(opt) in the three-phase region is plotted in Fig. 14; the stippled line indicates the general trend. Above 250 bar the correlation is similar to the behavior of the dead-oil system (Fig. 4), i.e., an inverse linear relationship. Also the optimum solubilization parameters appear to increase as the pressure is increased. Below 250 bar, T(opt) increases with increase in P(opt), and the optimum solubilization increases with increase

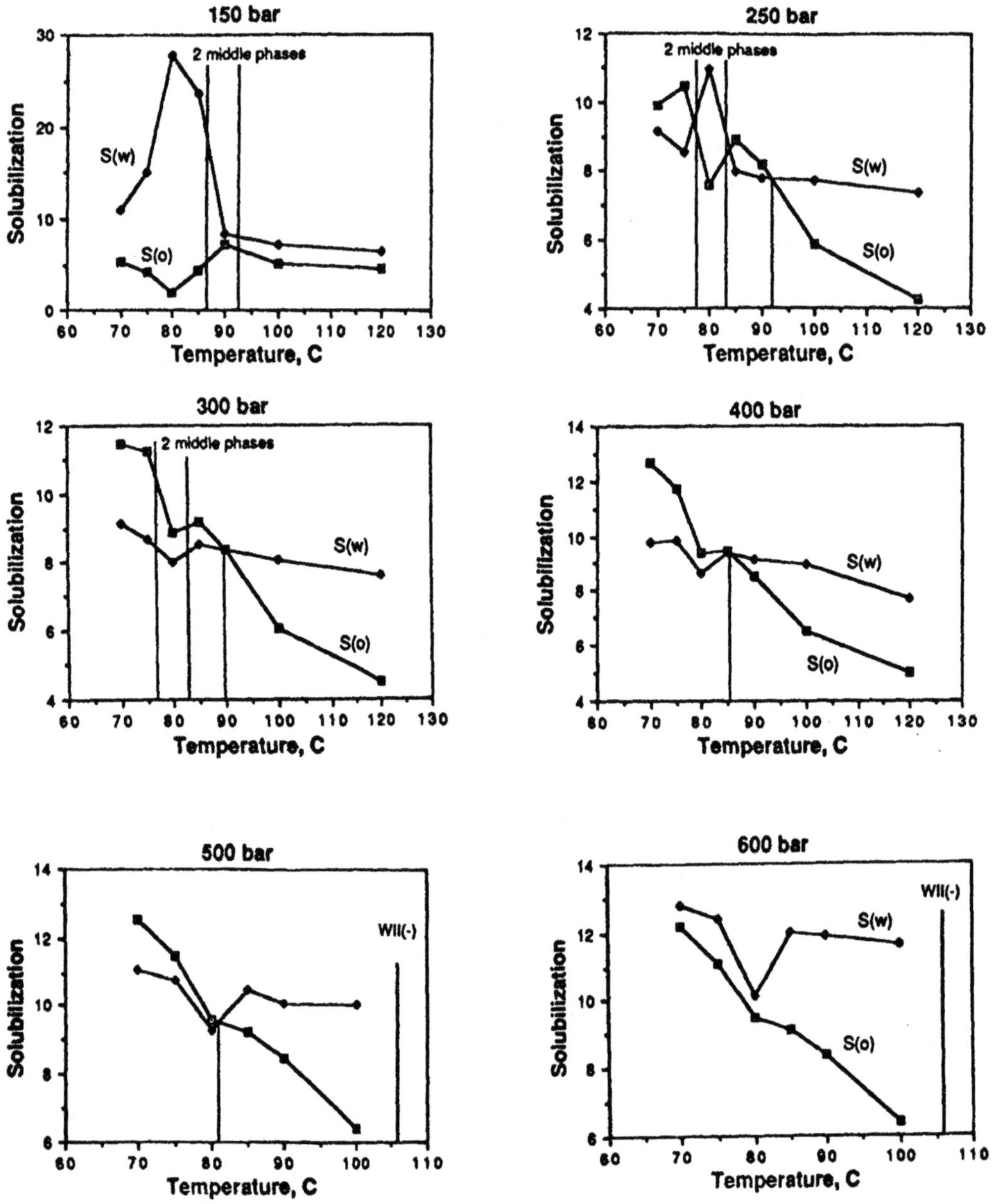

Fig. 13. The solubilization of oil and water into the middle phase(s) as the function of temperature at different pressures (system as described in Fig. 10)

in *T*(opt) and *P*(opt). In a reservoir-flooding situation the surfactant system should be at the optimum. The slope of the lines indicates that below 250 bar a change in *P*(opt) of 50 bar corresponds to a change in *T*(opt) of 26 °C, while above 250 bar the corresponding change in *T*(opt) is only 3 °C. In practical situations, at given flooding conditions, it is important to know the relationship between *T*(opt) and *P*(opt), because these two intensive variable will always change a little during the oil production.

The high salinity multiphase behavior

The salinity of the brine was 4.78 wt% NaCl related to the aqueous phase, and otherwise the ex-

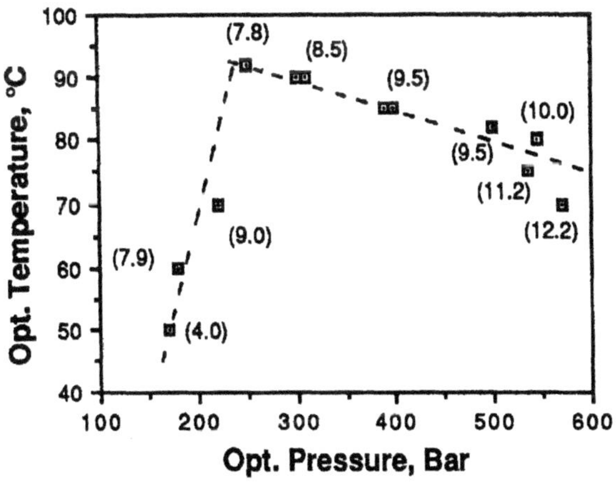

Fig. 14. The relationship between P(opt) and T(opt) for the microemulsion system described in Fig. 10. The optimum solubilization parameters are listed

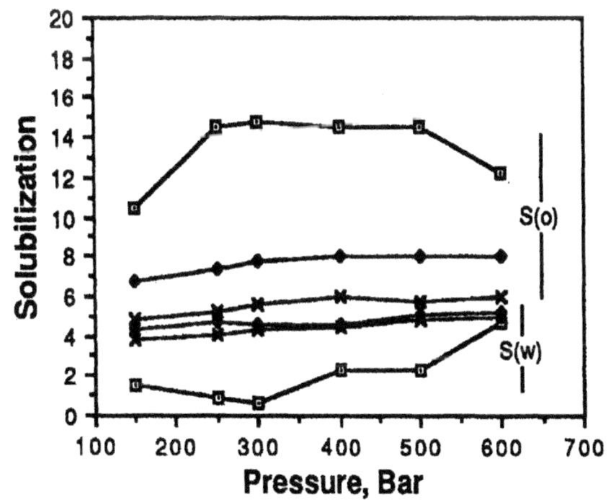

Fig. 15. The solubilization of oil and water in the middle phase as the function of pressure at different temperatures. Surfactant: RL-3011, 2.0 wt%; Oil: 55.5 mole% n-C10 and 44.5 mole% C1; Brine: 4.76 wt% NaCl; WOR: 1.0 at lab. conditions. 80°C: —□—, 100°C: —◆—, 120°C: — x —

perimental conditions were the same. The system was only studied at three different temperatures: 80°, 100°, and 120°C, and the solubilization parameters as the function of pressure are shown in Fig. 15. At each temperature $S(o) > S(w)$ over the entire pressure range. For the two highest temperatures a nearly constant ratio between $S(o)$ and $S(w)$ is observed at the different pressures. At 80°C the data indicate that the system moves towards the WII(—) state as the pressure is further increased. At 100° and 120°C both $S(o)$ and $S(w)$ increase slightly with the increase in pressure, and it is impossible from the present data to determine whether the system moves towards the WII(—) or WII(+) state.

Comparing Figs. 5, 10, and 15, it is obvious that the phase behavior of the high salinity system, toward the WII(+) state relative to optimum, is quite different from the lower salinity systems regarding variations in P and T. Only one middle phase was observed in the high salinity case, and the variation in $S(o)$ and $S(w)$ is much smaller.

The solubilization parameters at the different pressures have been plotted against the temperatures in Fig. 16. Between 80°—100°C the three-phase microemulsion system showed a drastic decrease in $S(o)$ and an increase in $S(w)$. A further increase in temperature to 120°C will result in a further decrease in $S(o)$, while $S(w)$ is nearly constant for all the pressures. Thus, it appears that the system will move towards an optimum and further toward the WII(—) state as the temperature is increased beyond 120°C.

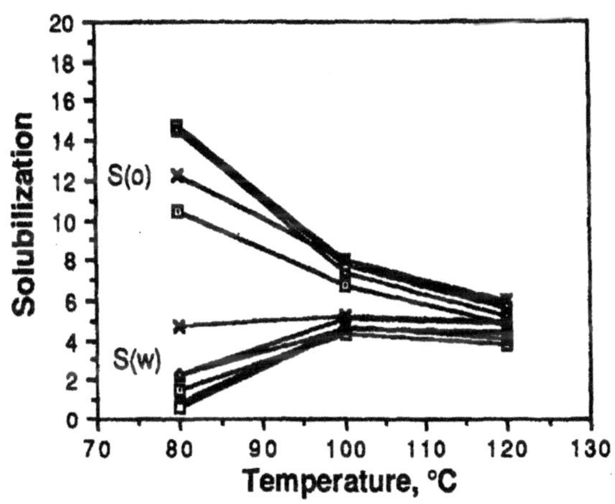

Fig. 16. The solubilization of oil and water in the middle phase as the function of temperature at different pressures (system as described in Fig. 15). 150 bar: —□—; 250 bar: —■—; 300 bar: —□—; 400 bar: —▲—; 500 bar: —◇—; 600 bar: — x —

Conclusions

The following conclusions can be made concerning the phase behavior of a multiphase microemulsion system consisting only of an anionic surfactant, model oil, and NaCl-brine:

The dead-oil system

1) The phase behavior with respect to changes in *P* and *T* of a "dead" oil multiphase microemulsion system containing no alcohol is similar to anionic surfactant systems containing alcohol.
2) The system moves towards the WII(—) state when pressure and temperature increases.
3) At a specific salinity *T*(opt) appears to show an inverse linear relationship with *P*(opt).
4) The *P* and *T* relation for a phase transition from a WIII to a WII(—) state shows that the higher the temperature is, the lower is the pressure at whcih the phase transition takes place.

The live-oil system

1) The phase behavior with respect to changes in *P* and *T* is strongly related to the salinity of the multiphase system.
2) At low salinities complicated phase behavior was observed, resulting in two middle phases at certain *P* and *T*.
3) At higher salinities, toward the WII(+) state relative to optimum, only one middle phase was observed.
4) In the four-phase regions the solubilization of oil and water into the middle phases is determined by the relative volume of two middle phases.
5) The solubilization of oil and water in the upper middle phase, which is close to optimum, decreases slightly with increase in temperature.
6) In general, the data suggest that the systems move toward the WII(—) state upon increasing the pressure beyond 400 bar, and the same appears to occur at temperatures above 100°C.
7) In the three-phase region, and salinities close to optimum, there appears to be an inverse linear relationship between *P*(opt) and *T*(opt).

Acknowledgement

This project was financed by The Norwegian Petroleum Directorate as part of the state supported SPOR-Program on enhanced oil recovery.

References

1. Abrams A (1975) Soc of Petr Eng J Oct 437—447
2. Huh C (1979) J Colloid Interface Sci 71:408—426
3. Skauge A, Fotland P (1986) Paper SPE/DOE 14932, 5th SPE/DOE Symposium on Enhanced Oil Recovery, Tulsa, April 20—23, II:109
4. Puerto MC, Reed RL (1985) Paper 14290, 60th Annual Technical Conference and Exhibition of the Soc of Petr Eng, Las Vegas, Sept 22—25
5. Puig JE, Scriven, Davis HT, Miller WG (1982) AIChE Symposium Series 78:p1
6. Noronha JC, Shah DO (1982) AIChE Symposium Series 79:p42
7. Offen HW (1980) The Rev of Phys Chem of Japan, The 50th Commemorative Volum 50:97—118
8. O'Connell JP, Walker RD (1982) Paper presented at The AIChE Annual Spring Meeting, Houston, April 5—9
9. Puerto MC, Reed RL (1982) Paper SPE/DOE 10678, 3th Symposium on Enhanced Oil Recovery, Tulsa, April 4—7, p 51
10. Nelson RC (1982) Paper SPE/DOE 10677, 3th Symposium on Enhanced Oil Recovery, Tulsa, April 4—7, p 35
11. Rossen WR, Kohn JP (1982) Paper SPE 11210, 57th Annual Fall Technical Conference and Exhibition of The Soc of Petr Eng of AIME, New Orleans, Sept 26—29
12. Fotland P, Skauge A (1986) J Dispersion Science and Technology 7(5):563—579
13. Long G, Chierici G (1961) Petr Eng July: B25—B31
14. Sage BH, Berry VM (1971) Hand Book Am Petrol Institute, Section IV p 66 and VI p 92

Authors' address:

Dr. Tor Austad
Rogaland University Center
Ullandhaug
4001 Stavanger, Norway

Surface Modification and Interactions

Progress in Colloid & Polymer Science

Progr Colloid Polym Sci 82:313—320 (1990)

Effects of particle concentration, ionic strength, pH and temperature on the microelectrophoretic mobility of cationic polystyrene latex. I

F. Galisteo, F. J. de las Nieves López, M. Cabrerizo, R. Hidalgo-Alvarez

Department of Applied Physics, Interface Physics and Colloidal System Group, Faculty of Sciences, University of Granada, Spain

Abstract: The microelectrophoretic behavior of dilute dispersions of positively charged polystyrene particles has been studied. The effects of particle concentration, ionic strength, pH, and temperature on microelectrophoretic mobility have been analyzed. Change in mobility on the addition of different co-ions and counterions was investigated. The pH at which the isoelectric condition is established was found to be 8.5. The charge-determining groups on the positively charged polystyrene particles were slightly basic with a pK_a of 8.2. Mobilities increase as does temperature. The dependence of mobility on temperature is mainly due to the variation of viscosity and dielectric constant of liquid medium with temperature.

Key words: Microelectrophoretic mobility; cationic polystyrene latex

Introduction

Monodisperse spherical polystyrene lactices have proven to be very useful model systems for studying various colloidal phenomena. Recently, some studies have been made on the electrokinetic behavior of negative polystyrene latices [1—6]. However, papers considering the electrokinetic behavior of positive polystyrene lattices are very scarce [7]. In both cases, the results raise a number of questions, for instance, concerning the occurrence of a maximum in the zeta-potential as a function of ionic strength.

Furthermore, the surface of a colloidal model has to be smooth and rigid, however, a number of recent studies [4, 6] indicate this to not be the case when negative polystyrene latex is used. In order to establish whether the reported results are typically for negatively charged latices only, or if they are representative of positively charged particles as well, the present electrokinetic study was undertaken.

Among the various electrokinetic techniques available for characterization of electrical double layers in colloidal dispersions, one usually reverts to measurements of the electrophoretic mobility, because this technique is well established and easily applicable to dilute dispersions. For that reason, microelectrophoresis is the most commonly used method for obtaining the zeta-potential of colloid particles. Furthermore, the theory of microelectrophoresis has been dealt with extensively, so that the mobility values may be converted into zeta-potential whatever the electrokinetic radius. O'Brien and White [8] have recently published a theoretical treatment of electrophoresis, which takes into account any combination of ions in solution and the possibility of very high zeta-potentials (up to 250 mV), far enough from the values to be expected in most experimental conditions. Nevertheless, the theory of O'Brien and White [8] assumes the absence of ionic conduction inside the plane of shear, and Midmore and Hunter [4] have very recently shown that this condition is not obeyed by the latex/electrolyte system below electrolyte concentrations of about 0.001 M. According to these authors, the most likely explanation for the mobility/electrolyte concentration maximum seems to be the shear plane moving away from the surface as the electrolyte concentration decreases. This phenomenon results in two effects: firstly, it lowers the zeta-potential in the usual way, and second it lowers the mobility by a much greater percentage by introducing ionic conduction in the diffusse layer, but inside the plane of shear.

In principle, the mobility of colloids in dilute dispersions should be independent on particle concentration, since the overlap between diffuse layers is extremely weak; however, cationic colloids are easily contaminated by anionic materials from the glass, e.g., silicates [9], and hence, mobility is apparently dependent on particle concentration [7]. Consequently, one must first find the most appropriate conditions to reduce contamination of the cationic latex with silicates dissolved from glass.

It is well-known that the mobility of latex particles depends on ionic strength; however, the effect of different co-ions and counterions on mobility of positively charged latex particles has not been studied exhaustively yet. Furthermore, the mobility of this kind of latex is pH-dependent [7]. In its turn, the pH in the preparation of a latex exerts a decisive influence on the shape of the curve expressing such pH-dependence, so that the determination of the electrokinetic properties of a positively charged latex requires a detailed study of the effect of pH on mobility.

Goodwin et al. [9] found that prolonged heating of the cationic latices at a temperature above 90 °C caused coagulation. This was considered to be a consequence of hydrolysis of the amidine group at elevated temperature. Hence, temperature plays an important role in the stability of cationic latices. For that reason, some experiments have been accomplished at different temperatures (20—40 °C).

Materials and methods

Positively charged monodisperse latices were prepared according to Goodwin et al. [9]. Some slight modifications were introduced to adapt the process for using the initiator azo — N,N'-dimethyleneisobutyramidine-hydrochloride (ADMBA) for the polymerization process. The initiator concentration was 2.7×10^{-3} M, the ionic strength during preparation was 3.27×10^{-2} M, and the reaction temperature was maintained constant at 50 °C.

The latex was dialized extensively and subsequently cleaned by ion-exchange over a mixed bed. The ion-exchange resins used were Dowex 50W-X4 in the H^+ form and Dowex 1-X4 in the OH^- form. These resins had been rigorously purified by the method of Van den Hul and Vanderhoff [10]. Surface charge density (σ_0 = 327 mC m^{-2}) was obtained by conductometric and potentiometric titration. Although positively charged latices in general are prone to hydrolysis, in this case σ_0 proved to be constant for a period of time longer than 2 years. This behavior demonstrates the usefulness of ADMBA as the initiator, confirming the suggestion made by Goodwin et al. [9]. The latex sample was stored in Pyrex glassware with a solid content of 6.9%, and maintained at 4 °C.

The average particle diameter, as obtained by means of electron microscopy, was (712 ± 12) nm.

All chemicals used were of A.R. quality. Water was purified by reverse osmosis, followed by percolation through charcoal and mixed bed, ion-exchange resins (Millipore).

Electrophoretic mobilities were obtained with a Zeta-Sizer IIc (Malvern Instruments, Malvern, England), by taking the average of four measurements at the stationary level in a cylindrical cell. Zeta-Sizer IIc is an advanced instrument which allows rapid and accurate measurements of electrophoretic mobility of colloids. Measurements are based on an analysis of the frequency spectrum of laser light scattered by the particle suspension using a photon counting detector and digital correlator.

Electrophoretic mobilities were converted into zeta-potentials by simply using the Smoluchowski equation

$$\mu_e = \frac{\varepsilon_0 \varepsilon_r}{4\pi\eta} \zeta , \qquad (1)$$

where $\sigma_0\sigma_r$ and η are, respectively, the dielectric permittivity and absolute viscosity of the liquid medium. This equation, however, does not consider the so-called "relaxation effect" due to the deformation of the ionic atmosphere during electrophoretic motion, which takes place mainly at high zeta-potential and electrokinetic radius around 10. Since the relaxation effect plays an important role in the electrokinetic behavior of cationic latex at low ionic strength, we have also used a computer program for the conversion of mobility measurements to zeta-potential designed by O'Brien and White [8]. This program is far more flexible than Wiersema graphs which have traditionally been used for the interpretation of mobility data. Zeta potentials obtained by O'Brien and White method should be taken only as a first approximation to zeta potentials of polystyrene/electrolyte interface, since this method assumes the absence of ionic conduction between the plane of shear and the surface. If this conduction is not taken into account, the calculated zeta potential will be too low. This is particularly a problem at low electrolyte concentration.

Results and discussion

1. Effect of latex particle concentration

The latex dispersions were prepared at different particle concentration between 3 and 50 mg latex per 100 cm^3 of solution. In principle a slight μ_e-dependence on particle concentration might be expected [11]. In fact, as can be seen in Fig. 1 the latex

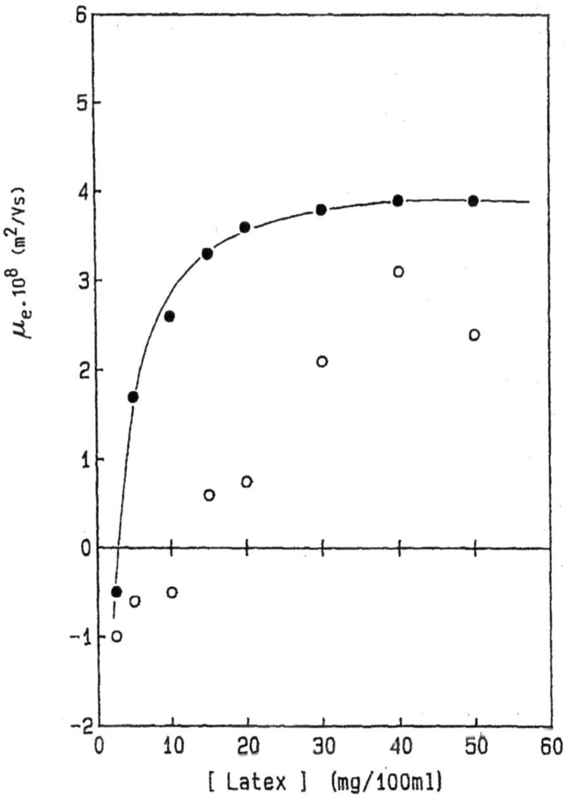

Fig. 1. Electrophoretic mobility of cationic polystyrene particles against latex concentration for two storage times: ● = 0.5 h; ○ = 24 h

concentration is important when the dispersions are very dilute (<30 mg latex per 100 cm^3 of solution), even making μ_e negative. When the latex concentration is higher than 30 mg/100 cm^3, however, the mobility values reach a plateau. Hence, according to the results shown in Fig. 1, latex concentration plays an important role in the electrokinetic behavior of cationic polystyrene latex. To check any possible influence of storage time of dispersion on the anomalous dependence of μ_e on latex particle concentration, some experiments were made with dilute cationic polystyrene dispersion stored in Pyrex glass for 24 h. In this case more negative mobilities were found and the results were erratic (see Fig. 1).

In a recent careful study of the effect of a trace impurity on isoelectric point (i.e.p.), Furlong et al. [12] have shown that even at pH 6 a shift of 0.2 pH units (from 5.95 to 5.75) occurs in the i.e.p. of TiO_2 over a period of 2 months. At pH 10, however, a shift of 0.8 pH units (from 9.6 to 8.8) occurred in the i.e.p. of Al_2O_3 in a mere 16 h when stored in a glass vessel. Because of the low i.e.p. of SiO_2 the effect is always to shift the i.e.p. of the substrate to lower the pH values. As Hunter [13] has pointed out, one impurity often ignored is the soluble silicate deriving from prolonged storage of electrolyte solutions in glass prior to use, and this problem is particularly severe when dealing with alkaline solutions. Thus, we consider that the anomalous dependence of μ_e on particle latex concentration is because cationic polystyrene is easily contaminated by anionic materials (silicates) from the glass. Hence, it was essential that the latex concentration always exceed 30 mg polystyrene per 100 cm^3 of solution, since otherwise the results would have been erratic. At lower latex concentrations the particle are probably contaminated by small amounts of dissolved silicates, as the effects are most pronounced at high pH values and prolonged storage of electrolyte solutions in glass vessel.

2. Ionic strength effect

We conducted experiments to see whether mobility changed on the addition of different monovalent co-ions and counterions. As can be seen in Fig. 2 the mobility depends non-linearly on ionic strength of three different monovalent co-ions (Li^+, Na^+, and K^+), and a maximum mobility occurs when ionic strength is around 5×10^{-3} M, whatever the co-ions of the electrolyte solution. It has already been remarked that the occurrence of the maxima in mobility values is well-established for negatively charged latices, but is now apparent that it is typical for polystyrene latices in general. The common origin of the effect is also obvious from the observation that the maximum for both types of polystyrene latex is situated in the concentration range 10^{-3} — 10^{-2} M. Its exact location is not sensitive to the character of the co-ion, as illustrated in Fig. 2.

In order to check any influence of monovalent counterions on the mobility of positively charged latex, some microelectrophoretic measurements were performed with NaCl, NaBr, and NaI (Fig. 3). The mobility-dependence on ionic strength is also non-linear, and the exact location of maximum mobility is somewhat sensitive to the ionic radius of counterion. The lower values for the maximum and for the mobilities in the high concentration range

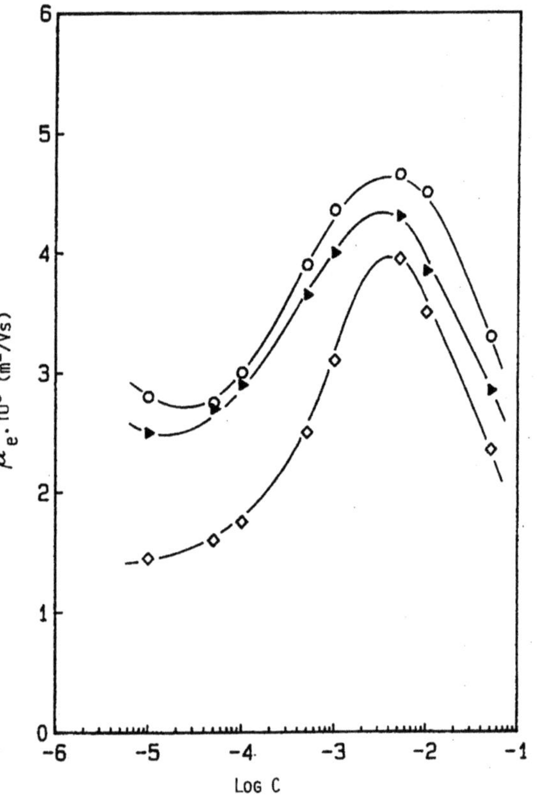

Fig. 2. Electrophoretic mobility of cationic polystyrene particles against electrolyte concentration for monovalent co-ions: ○ = KCl; ▲ = NaCl; ◇ = LiCl

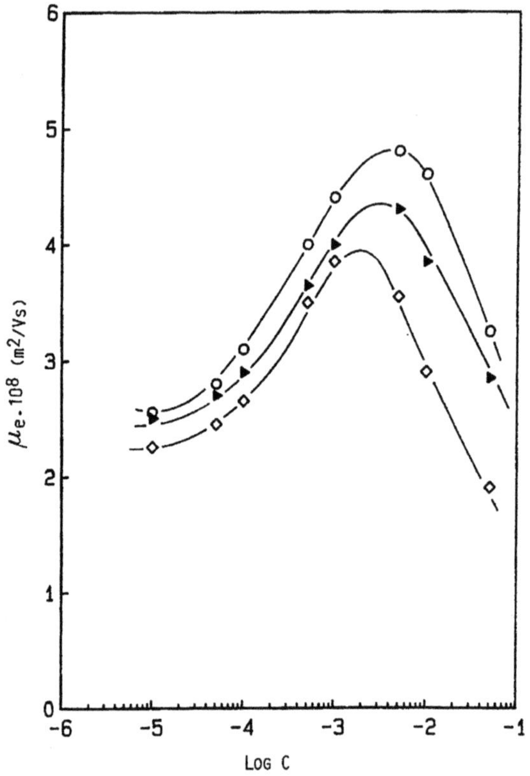

Fig. 3. Electrophoretic mobility of cationic polystyrene particles against electrolyte concentration for monovalent counterions: ○ = NaBr; ▲ = NaCl; ◇ = NaI

for I^- in comparison with Cl^- and Br^- are possibly due to some specific adsorption of the more polarizable I^- ions.

Also, the effect of various divalent co-ions (Ba^{2+}, Sr^{2+} and Ca^{2+}) on mobility has been studied. Electrophoretic mobilities for polystyrene latex particles with positive surface titratable groups are virtually identical for the series $CaCl_2$, $BaCl_2$, $SrCl_2$ and shows a maximum as the ionic strength of the supporting electrolyte decreases (see Fig. 4).

Midmore and Hunter [4] have recently shown that mobilities for negative polystyrene latex are very similar for the lyotropic series KF, KCl, KBr, KI; combining these results with the large differences in the physical properties of the anions in these series, they conclude that the maximum in mobility must be due to physical changes in particle surface properties rather than co-ion adsorption as suggested by Zukoski and Saville [3]. Recently, Chow and Takamura [6] have also suggested that surface roughness of latex particles is one of the most important factors in determining their zeta-potentials from electrophoresis. According to these authors, the surface roughness of latex particles gives rise to a large value for the location of the shear plane, resulting in a smaller value for the zeta potential at low ionic strength.

Counterion valencies can play an important role in the electrokinetic behavior of latex particles. Besides, a marked effect of the charge of the counterions on location of the maximum in the mobility/log electrolyte concentration might be expected if this maximum were due to ion adsorption on the latex surface. For that reason, we conducted some experiments with electrolytes of different valencies: $NaNO_3$, Na_2SO_4, and Na_3PO_4. As can be seen in Fig. 5, the mobility maximum appears again when ionic strength is around 5×10^{-3} M. Positive mobilities have been found when $NaNO_3$ and $NaSO_4$ were used as electrolytes, although with Na_2SO_4 mobility values were lower than those for $NaNO_3$. Both electrolytes are indifferent and they

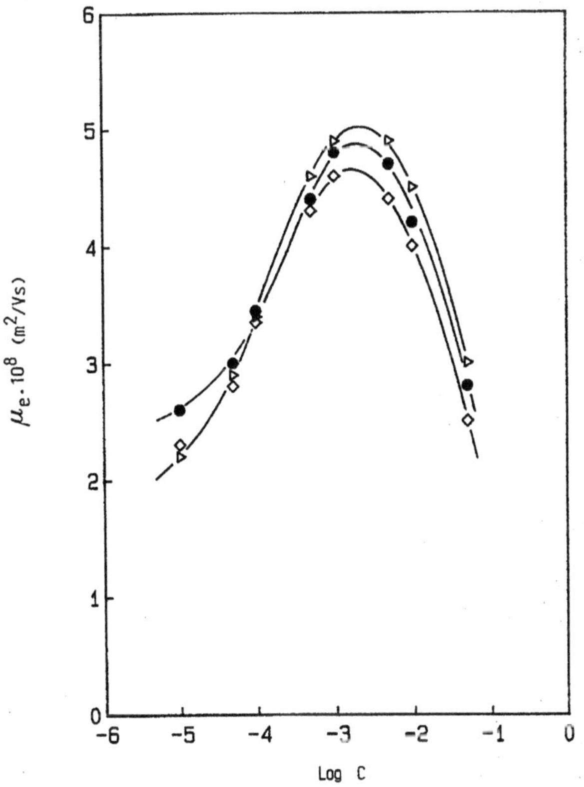

Fig. 4. Electrophoretic mobility of cationic polystyrene particles against electrolyte concentration for divalent coions: ● = $SrCl_2$; △ = $BaCl_2$; ◇ = $CaCl_2$

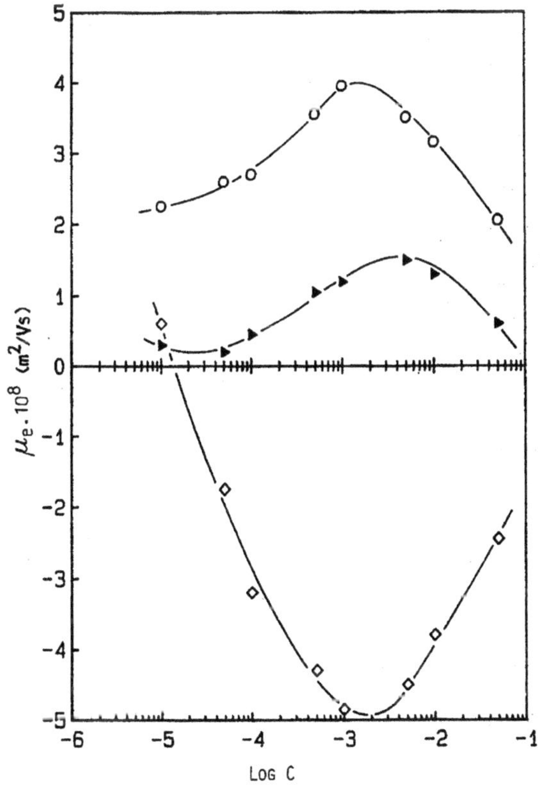

Fig. 5. Electrophoretic mobility of cationic polystyrene particles against electrolyte concentration for counterions of different valence: ○ = $NaNO_3$; ▲ = Na_2SO_4; ◇ = Na_3PO_4

provoke only a decrease in electrical double-layer thickness. The phosphate counterions, however, are specifically adsorbed on the surface and they confer to cationic polystyrene a negative charge which is now electrokinetically mobilized. It must be note that the electrolyte concentration which corresponds to maximum mobility is not sensitive to the counterion valencies, which suggests again that this anomalous behavior of cationic polystyrene is mainly related to particle surface properties.

3. *pH effect*

Figure 6 shows the effect of pH on the electrophoretic mobility. The experiments indicated that the latex particles had a high positive mobility at acid pH values, particularly at pH $\leqslant$ 4. As the pH became alkaline the mobility decreased sharply and was negative when pH > 8.5. The amphoteric behavior of this cationic polystyrene sample is very peculiar and quite different to that shown in [7]. Bearing in mind that the samples used in [7] was prepared at acid pH, whereas the sample studied here was prepared at neutral pH, we can infer that the pH plays an important role in the final characteristics of the cationic polystyrene latex prepared with the initiator ADMBA-2ClH. This initiator has a fast rate of decomposition, thus increasing the radical flux at the beginning of the reaction and thereby more effectively stabilizing the particles formed. The charge-determining groups on the latex surface were amidine groups and their pK_a is around 10.2 for latex prepared using ABA-2ClH as initiator [14]. This indicates that these surface basic groups function as weak bases, and the presence of H^+ or OH^- ions would clearly influence the dissociation of the amidine groups and it could be anticipated that the pK_a of the surface groups would depend on pH used during the preparation process of cationic polystyrene latex.

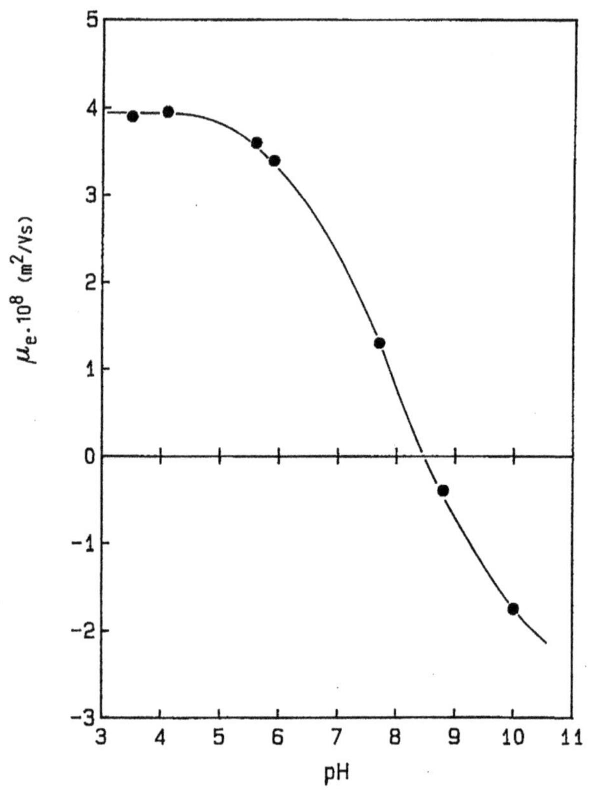

Fig. 6. Electrophoretic mobility of cationic polystyrene particles against pH of the solution (ionic strength: 10^{-3} M KBr)

The isoelectric point of this cationic polystyrene latex is around pH 8.5. Determination of the surface pK_a can be made by zeta-potential measurements [15]

$$pK_a = (pH)_{\zeta_{max/2}} + \frac{e\zeta_{max/2}}{2.303\ kT} . \qquad (2)$$

According to Eq. (2), the charge-determining groups on the cationic polystyrene latex surface were slightly basic with a pK_a of 8.2. This value roughly agreed with that of amidine groups obtained by potentiometric titration and using ABA-2ClH as initiator [14]. The latices prepared using ADMBA-2ClH are less prone to hydrolysis than those prepared with ABA-2ClH; in fact, it shows that the heterocyclic ring is more stable towards hydrolysis than the butyramide group. Likewise, it seems that the amidine groups have different basicity, depending on the location of such groups in the initiator structure.

4. *Temperature effect*

A prolonged heating of the cationic polystyrene latices may cause coagulation. Temperature is an important factor in the colloidal stability of this latex. The stability of a lyophobic colloidal sol, according to the DLVO theory, is determined by the balance between the repulsive and attractive forces of the approaching particles. It has long been recognized that the zeta-potential is a very good index of the magnitude of the repulsive interaction between colloidal particles. This parameter is commonly used to assess the stability of a colloidal sol. In order to know whether the coagulation of the cationic latex was due to a decrease in the magnitude of the repulsive forces between the latex particles, some microelectrophoresis experiments were made with latex dispersions at different temperatures.

As can be seen in Fig. 7, the mobilities of cationic latex particles increase as temperature does within

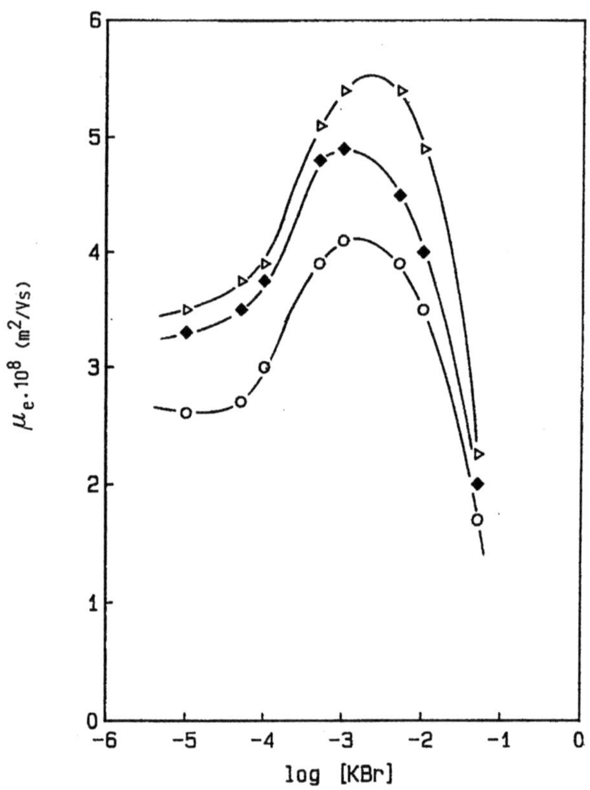

Fig. 7. Electrophoretic mobility of cationic polystyrene particles against KBr concentration for different temperatures: ○ = 20°C; ◆ = 30°C; △ = 40°C

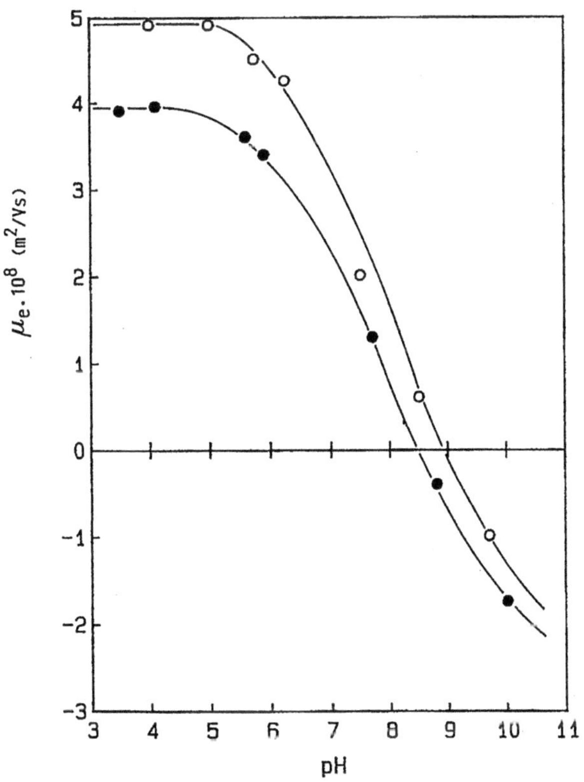

Fig. 8. Electrophoretic mobility of cationic polystyrene particles against pH of the solution at two different temperatures: ● = 20 °C; ○ = 30 °C

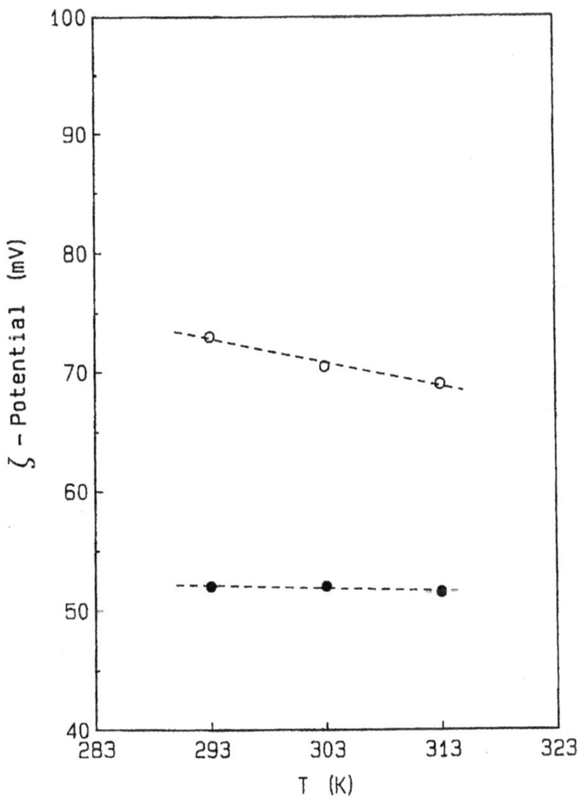

Fig. 9. Zeta-potential of cationic polystyrene particles against absolute temperature at two different ionic strengths as (O'Brien and White treatment): ○ = 10^{-3} M KBr; ● = 10^{-2} M KBr

the range of ionic strength studied (from 10^{-5} to 5×10^{-2} M). Figure 8 shows the dependence of μ_e with pH at two different temperatures. Also here, mobility increases as temperature does over the pH range studied (pH from 3 to 10).

The increase in mobility upon increasing temperature might also involve an increase of the magnitude of the repulsive interaction between latex particles; however, mobility depends on dielectric permittivity and the viscosity of the liquid media as well. Thus, mobilities have to be converted into zeta-potentials since the effect of temperature on repulsive interactions is related to the effect of thermal agitation upon the structure and thickness of the electrical double layer; that is to say, any effect of temperature on the double layer has to bring out a change in the zeta-potential. The zeta-potential values obtained by the theory of O'Brien and White [8] do not depend on temperature over the range 293—313 K (see Fig. 9); only when the ionic strength is relatively low (ca. 10^{-3} M KBr) does the zeta-potential tend to decrease slightly as temperature increases. This decreasing trend of zeta-potential is consistent with the compression of the double layer since its thickness decreases somewhat when temperature increase (κ^{-1} = 86.3 nm at 293 K, and κ^{-1} = 85.2 nm at 313 K). Nevertheless, in general, the dependence of mobility on temperature is mainly due to the effect of temperature on the viscosity and dielectric constant of the medium. Therefore, the charge determining groups on latex surface are not affected by hydrolysis when temperature is lower than 313 K, and probably this is true even at somewhat higher temperatures.

Figure 10 shows the zeta-potential values obtained at 293 K when mobilities are converted into this parameter according to the Smoluchowski equation, and O'Brien and White treatment. As can be seen, both methods provide identical values of zeta-potential at high ionic strength ($c > 10^{-2}$ M KBr),

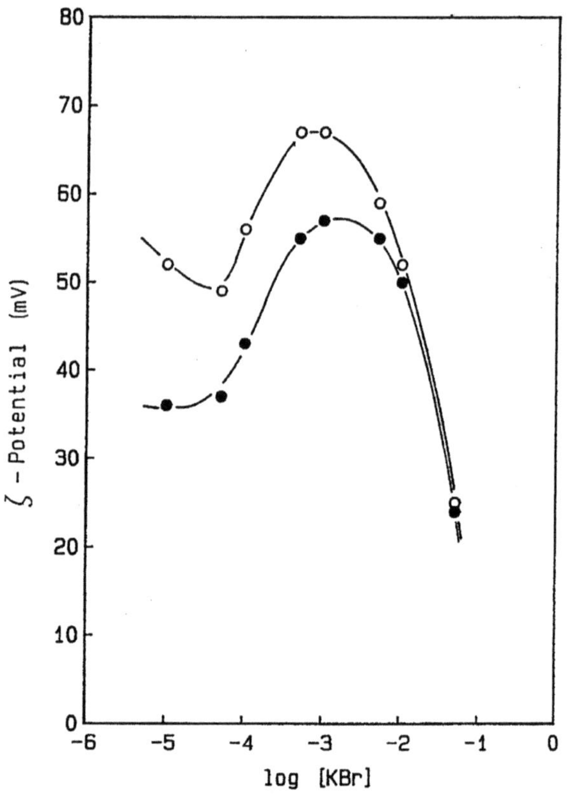

Fig. 10. Zeta-potential of cationic polystyrene particles against electrolyte concentration at 20°C: ● = Smoluchowsky equation; ○ = O'Brien and White treatment

but significant differences appear when ionic strength is lower than 10^{-2} M. It demonstrates that these differences are actually due to the relaxation effect, and that the Smoluchowski equation is not adequate for converting mobilities into zeta-potential values at low ionic strength.

Acknowledgements

This work was supported by Dirección General de Investigación Científica y Técnica (Spain), project no. PA86-0235, and by Junta de Andalucía. We thank our colleague Dr. Juan Salcedo Salcedo for the computer program. Also, we thank Mr. A. Van der Linde of the Department of Physical and Colloidal Chemistry, Agricultural University of Wageningen (The Netherlands), for kindly supplying the initiator ADMBA.

References

1. Goff JR, Luner P (1984) J Colloid Interface Sci 99:468—483
2. van der Put AG, Bijsterbosch BH (1983) J Colloid Interface Sci 92:499—507
3. Zukosky CF, Saville DA (1985) J Colloid Interface Sci 107:322—333
4. Midmore BR, Hunter RF (1988) J Colloid Interface Sci 122:521—529
5. Delgado A, Gonzalez Caballero F, Salcedo J (1986) Acta Polym 37:361—364
6. Chow RS, Takamura K (1988) J Colloid Interface Sci 125:226—236
7. Hidalgo Alvarez R, de las Nieves FJ, van der Linde AJ, Bijsterbosch BH (1986) Colloids Surfaces 21: 259—266
8. O'Brien RW, White LR (1978) J Chem Soc Faraday Trans II 74:1607—1626
9. Goodwin JW, Ottewill RH, Pelton R (1979) Colloid and Polymer Sci 257:61—69
10. van den Hul HJ, Vanderhoff JW (1968) J Colloid Interface Sci 92:336
11. Anderson JL (1981) J Colloid Interface Sci 82:248
12. Furlong DN, Freeman PA, Lau ACM (1981) J Colloid Interface Sci 80:20
13. Hunter RJ (1981) Zeta Potential in Colloid Science. Principles and Applications, London: Academic Press
14. Napper DH, Netschey A (1971) J Colloid Interface Sci 37:528
15. Jacobasch HJ (1984) Die Angew Makromol Chem 128:47

Authors' address:

R. Hidalgo-Alvarez
Dept. of Applied Physics
Interface Physics and Colloidal Systems Group
Faculty of Sciences
University of Granada
18071 Granada, Spain

Resistance to fouling accomplished by modification of ultrafiltration membranes

M. Nyström, M. Laatikainen, K. Turku and P. Järvinen

Department of Chemical Technology, Lappeenranta University of Technology, Finland

Abstract: In this study polysulfone (PSU) membranes or smooth PSU surfaces were modified by adsorption of macromolecules from solution. A cationic polyelectrolyte poly(ethylene imine) (PEI) and a neutral block copolymer (PEOB) were used as modifying agents. The effects of modification were studied with ultrafiltration experiments using ovalbumin (OA), bovine serum albumin (BSA) or lysozyme (LYS) as model substances and with contact-angle measurements. Moreover, titrations of the proteins, streaming potential measurements of some membranes, and adsorption studies with the quartz crystal microbalance were carried out. — The enhanced electrostatic repulsion between the protein molecules and the membrane and an increased hydrophilicity of the membrane surface decreased fouling and the importance of these two factors was pH-dependent. The results showed that modification, both with PEI and PEOB decreased fouling by about 25% for LYS at pH 6.1. For OA modification with PEI gave positive results at nearly all pH values studied. The best results were obtained near or below the isoelectric point of OA. Modification with PEOB improved flux for BSA at pH 6.1 but not at pH 3.1. Flux reduction and the amount of adsorption on the PSU surface correlated for BSA and LYS. Moreover, the data imply adsorption of LYS molecules also into the pores of the membrane.

Key words: Fouling; ultrafiltration; modification; proteins; poly(ethylene imine)

Introduction

In ultrafiltration (UF), which is a separation process in which macromolecules are separated from the solvent or small molecules with membranes where the pore sizes are between 1 and 100 nm, one of the greatest problems is the fouling of the membranes. This phenomenon can be caused by various factors. Sometimes small polyvalent ions tend to precipitate the macromolecules, resulting in accumulation of the precipitate on the membrane surface [1], even if the charges of both the macromolecule and the membrane have the same sign. Moreover, the macromolecules themselves interact with the surface of the membrane as the repulsive energies between the surface of the membrane and the macromolecular species are weaker than the sum of their mutual attractive energies. This results in adsorption on the surface and in the pores of the UF membrane. Both phenomena cause either reversible or irreversible fouling, which decreases the solvent flux through the membrane.

Adsorption of proteins on solid surfaces has been studied with several different methods including ellipsometry [2], ESCA [3], and microcalorimetry [4, 5]. For the measurement of adsorption on UF membranes radioactive labelling technique has been used [6]. One of the most studied proteins is bovine serum albumin (BSA). An ellipsoid protein molecule like BSA can adsorb on the membrane in different ways, making a more densely packed layer if adsorbed with its edge against the solid surface than when adsorbed in a side-on manner [3]. When proteins adsorb on a hydrophobic surface the system tries to minimize the free energy[2]. Consequently, stronger interactions between the protein molecules and the surface are formed with time and conformational changes may occur [2]. The entropic and enthalpic driving forces connected with protein adsorption have been studied by Norde [5]. Adsorption of a protein molecule is also accompanied by water desorption from the surface [4]. Fouling in ultrafiltration can be compared to adsorption on solids, but in UF the driving pressure and the flux through the heterogeneous pores complicates the situation.

In this study an effort was made to relate the electrochemical properties of the model proteins and the membrane surfaces with the ultrafiltration flux results. For this purpose the proteins were titrated potentiometrically and the amount of charged groups at different pH values was evaluated. The zeta potentials of the membranes studied were calculated from streaming potential data. Different types of membranes were obtained by surface modification of commercial polysulfone membranes. The properties of the modified surfaces were also characterized by contact-angle measurements. Adsorption of the model macromolecules on the modified or unmodified membranes was studied by means of UF experiments and, in some cases, by *in situ* adsorption measurements with a quartz crystal microbalance (QCM). Finally, the factors governing membrane fouling due to adsorption of the model substances are discussed.

Experimental

Materials

The pieces of polysulfone (PSU) membrane used were of type GR 61 PP delivered from De Danske Sukkerfabrikker (DDS). This type of UF membrane has a nominal cut-off of 20000, which means that it retains 90% of the molecules having this molar mass. These membranes have a relatively wide pore size distribution [7] and the pore diameters can vary between 2 and 6 nm [8]. Smooth PSU model surfaces were prepared from PSU of GR type obtained from DDS.

Poly(ethylene imine) (PEI), was supplied by Aldrich Chemie (Cat. No. 18,197-8, M = 50000—60000) as a 50% aqueous solution. Before use PEI was dialyzed against distilled water in order to remove low molecular weight fractions.

The block copolymer of ethylene oxide and vinyl acetate (PEOB) was synthesized according to Napper [9]. 7.5 g of polyethylene oxide (PEO) (M = 35000) and 0.4 g of recrystallized dibenzoyl peroxide were dissolved in 60 g of toluene and during a period of 30 min, 3.75 g of freshly distilled vinyl acetate was added to the refluxed solution. The resulting PEOB was precipitated by the addition of n-heptane and the precipitate was filtered and washed with n-heptane and dried in vacuum at 50 °C. The product was dissolved in water, and filtered through a 0.22-μm membrane in order to remove the polyvinyl acetate homopolymer, then evaporated and dried. Finally, it was dissolved in dichloromethane and precipitated by n-hexane, filtered, and dried in vacuum.

The proteins used as model macromolecules were all supplied by Sigma Chemical Co. The bovine serum albumin (BSA) (grade V, containing 96—99% albumin) was obtained from no:s A-4503 and A-6003, the ovalbumin (OA) was of grade V, no. A-5503, and the lysozyme (LYS) was of grade I, no. L-6876.

In the preparation of the solutions p.a. grade electrolytes and RO-filtered water (Millipore) with a conductivity less than 1 μS/cm were used.

Methods

Preparation of the surfaces

For the UF experiments the PSU membranes were modified by adsorption. The membranes were exposed to an aqueous solution of the modifying macromolecule (PEI, PEOB) for 30 min at a fixed pH and ionic strength. The concentrations were 1 g/m^3 for PEI and 5 g/m^3 for PEOB. If the concentrations were lower no effects in the UF experiments could be detected and if they were higher the net increases of the flux due to surface modification were more than counterbalanced by the extensive blocking of the pores by the modifying material. Typically, modification with the stated concentrations resulted in a flux decrease of about 20% [10].

For contact-angle measurements and for QCM adsorption studies smooth polysulfone surfaces were cast, either on a glass rod used for the contact-angle studies or on the quartz crystal to be used in QCM experiments. Cyclohexanone solutions containing 5 and 12 w-% of PSU, respectively, were used. The different concentrations were used, because the glass rods were cast by dipping, and the quartz crystals by spreading a drop of solution on the surface when spinning the crystal. The polysulfone surfaces were carefully dried, first in air and then in vacuum before use so that no traces of solvent were left. The absence of solvent was verified by contact-angle measurements. These surfaces were modified in a similar way as described above for the membranes.

Potentiometric titrations

The electrical properties of the proteins and the PEI were measured by means of potentiometric titrations. In these experiments a known amount of the macromolecule was dissolved in a KCl solution of a fixed ionic strength (0.015 M or 0.15 M). First, the solution was titrated to pH 3 with 0.1 M HCl and after that, to pH 11 with 0.1 M KOH. The proton demand or release of the macromolecule was calculated as the difference between the sample experiment and a blank experiment in which the macromolecule was not present.

Streaming potential measurements

Zeta potentials of the membranes were calculated from measured streaming potentials according to a method described in detail elsewhere [11]. In this method the potential difference (ΔE), the so-called streaming potential, generated by the flow of saline solution through the pores of the membrane under the pressure difference (ΔP) is measured. This method, however, suffers from several difficulties and, consequently, the calculated zeta potential must be taken as a relative value. This apparent zeta potential is usually lower than the true zeta potential, because it is calculated using the Helmholtz-Smoluchowski equation, which assumes that the ratio of the pore radius to the double layer thickness is large. This condition is not fulfilled for the saline solution in the pores of the UF membrane. Moreover, the measured streaming potentials of the modified membranes depend on the ratio of the adsorption layer thickness to the "skin" layer thickness of the UF membrane, which means that they only partly describe the surface where adsorption takes place [11].

Contact-angle measurements

The contact angles between pure water, air, and the model surfaces were measured by a method similar to the Wilhelmy plate method, except for that a rod with a radius of 10 mm was used instead of a plate. The contact angles were recorded with a video camera provided with magnifying optics. The advancing (Θ_A) and receding (Θ_R) contact angles were measured for the rod descending into the water and then ascending from the water, respectively. The movement of the rod was controlled to give a constant speed of 1 mm/min. Both contact angles were measured from a television screen by visual inspection during constant movement after a stabilization period of 3 min. No correction in the values of the contact angles were made because the radius of the rod used was large compared with the radius of the meniscus [12]

Ultrafiltration experiments

The UF experiments were carried out at 25°C in a flat plate module described in detail elsewhere [13]. The membrane area was 19.3 cm^2. The membranes were stabilized for 2.5 h by filtration of pure water at a pressure of 3 bar. After this period the flux was essentially constant and the experiments with the model macromolecular solutions were started. At first a saline solution (KCl) of a fixed ionic strength and pH was filtered through the membrane at a pressure of 1.5 bar and the permeate flux was measured gravimetrically. Then the modification, if any, was carried out by contacting the membrane to the modifying solution for 30 min. After that the permeate flux of the same saline solution as above was measured. Then the lowest protein concentration in the feed solution was adjusted and the flux was measured after a stabilization period of five min. Flux reduction was calculated from:

$$\text{Flux red.} = 1 - J/J_0 \,, \tag{1}$$

where J is the flux after protein addition and J_0 is the flux of the electrolyte solution after modification. This flux J_0 is usually smaller than the flux before modification (mostly about 20% smaller). After this the protein concentration was increased with subsequent additions of a stock solution, and the flux was registered for each concentration. The bulk concentration of protein c_b was corrected for concentration polarization to obtain the surface concentration c_m according to:

$$c_m = c_b \exp(J/k) \,, \tag{2}$$

where k is the mass transfer coefficient calculated from Eq. (3), according to Jonsson [14]:

$$k = 0.023(D/d_H)\mathrm{Re}^{0.8}\mathrm{Sc}^{0.33} \,. \tag{3}$$

The hydraulic diameter d_H of the present system was 1.9 mm and the Reynolds number Re was 5300. The Schmidt number Sc varied, depending on the diffusion coefficients D of the proteins. In Table 1 are collected the diffusion coefficients used and the calculated mass transfer coefficients.

Adsorption measurements

Adsorption studies were carried out with the quartz crystal microbalance (QCM) method as described in detail elsewhere [16]. The method is based on the high sensitivity of an oscillating piezoelectric crystal to mass changes taking place on the crystal surface. The quartz crystal with gold electrodes was coated with PSU. The experiments were made in a deaerated 0.015 M KCl solution; 0.02% of sodium azide was added in order to prevent bacterial contamination. At the beginning of each experiment the frequency difference Δf_0 between the coated crystal and a stable reference crystal was measured after a stabilization period with pure salt solution at pH 6.0. The protein stock solution was prepared by dissolving 0.01 g of the protein in 1 cm^3 of 0.015 M KCl-solution. From this solution subsequent additions were made to the stirred solution and after each addition (after a stabilization period of 20 min) the Δf value was determined. Then a new addition was made and the cor-

Table 1. Diffusion coefficients D and calculated mass transfer coefficients k for bovine serum albumin BSA, ovalbumin OA and lysozyme LYS, t = 25°C

	D [m²/s]		k [m/s]
BSA	6.8×10^{-11}	[15]	1.86×10^{-5}
OA	7.6×10^{-11}	[10]	1.92×10^{-5}
LYS	11.9×10^{-11}	[15]	2.70×10^{-5}

responding Δf was determined. This sequence was repeated until the adsorption isotherm was completed. The surface concentration Γ in mg/m² was calculated from the measured frequency shift $\Delta f - \Delta f_0$ in Hz by means of Eq. (4), where the proportionality factor depends only on the properties of the crystal:

$$\Gamma = -0.28(\Delta f - \Delta f_0) \ . \quad (4)$$

Results and discussion

Characterization experiments

The amount of charged groups of the proteins studied has been evaluated from titration data. In Fig. 1 are depicted results from potentiometric titrations in 0.015 M KCl. The titrations at other ionic strengths (not shown here) made it possible also to estimate the isoionic points (IIP) of the proteins. The proteins BSA and OA have their IIP's rather close to each other, BSA at pH 4.8 and OA at pH 4.6. Values given in literature for their isoelectric points (IP) are 5.0 and 4.7 for BSA and OA, respectively [17], which seem to be rather close to the IIP's measured here. BSA has about 60% more titratable groups as compared to OA [18]. According to Fig. 1 a somewhat higher ratio, i.e., 1.8 is obtained when the groups titrated between the pH values 3.5 and 10.5 are taken into account. LYS has its IP at pH 11 [17] so only its behavior below the IP could be studied here.

At their IP's the proteins are zwitterions with both positively and negatively charged groups, but with a net charge of zero. At the IP's the protein molecules have smaller volumes, because of intramolecular electrostatic attraction [19], and they contain less water and are, therefore, more easily precipitated or adsorbed. At pH values far away from their IP's the proteins carry only negative charges above, and only positive charges below their IP's. The region in between the net charge is either negative or positive and is composed of both positive and negative charges. The coexistence of both charges is probably of great importance for their adsorption behavior.

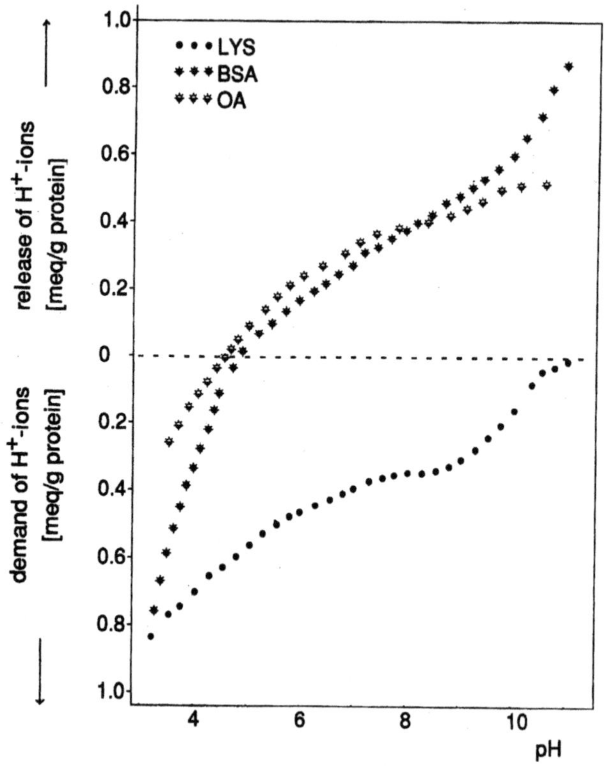

Fig. 1. Release or demand of H^+-ions in milliequivalents per g protein in the titration of the proteins lysozyme (LYS), bovine serum albumin (BSA), and ovalbumin (OA) in a solution of 0.015 M KCl at 25°C

The GR membrane made of PSU is non-ionic in nature. However, zeta potential data, which have been reported earlier [11], show that the membrane is increasingly negatively charged above pH 3.5, probably due to adsorption of either OH^- ions or Cl^- ions. Below this IP the charge of the membrane is either zero or slightly positive, but the measured data were too scattered to give a conclusive result. In this pH range the ionic strength becomes too high for making accurate measurements.

Poly(ethylene imine) (PEI) is a cationic polyelectrolyte with its point of zero charge (pzc) at pH 10.8 [20] and its association with H^+ ions is completed at pH 3. Its positive charge between pH 3 and 8 has been verified by streaming potential measurements of a PEI modified membrane [11]. Titration data (not shown here) also showed that at pH 3 its total amount of charged groups was 18.3 meq/(g PEI) and that its pzc is at pH 10.8. At high ionic strengths the PEI molecule is slightly more dissociated than in dilute electrolyte solutions [10]. According to viscosity measurements [10] the PEI molecule has a slight tendency to expand in acidic solutions, especially at low ionic strengths.

The PEOB is non-ionic, because it is a block copolymer of uncharged monomers. Zeta potentials of the PEOB modified membranes have not been measured here nor has it been titrated, so it has only been presumed to be neutral.

The results from the titration experiments and from streaming potential measurements are collected in Fig. 2. The IP of lysozyme is taken from literature [17].

The results from the contact-angle measurements are shown in Table 2. It can be seen that the pure PSU membrane is hydrophobic in nature. The large difference between Θ_A and Θ_R can be due to inhomogeneity and roughness of the surface [21]. When PEI or PEOB has adsorbed on the PSU surface the contact angles decrease a little, indicating an increase in hydrophilicity.

The hydrphilicity of the proteins has not been measured in this study, but values taken from literature show that their hydrophilicity increases as OA < BSA < LYS [17].

ESCA measurements of the membranes carried out at The Institute of Surface Chemistry (YKI) in Stockholm showed an increase in the ratio of oxygen to carbon for the modified membranes, compared to the unmodified PSU membrane. The ESCA results also verified that the modification agents had permanently adsorbed by showing an increased nitrogen to carbon ratio due to PEI and a shielding of the sulphur signal of the PSU membrane, due to PEOB.

Table 2. Advancing (Θ_A) and receding (Θ_R) contact angles of an unmodified polysulfone (PSU) surface and of surfaces modified with poly(ethylene imine) (PEI) and a block copolymer (PEOB). 90° was the upper limit for the measurements

Contact angles	PSU	PSU ± PEI	PSU ± PEOB
Θ_A	>90°	88 ± 1°	84 ± 1°
Θ_R	68 ± 1°	62 ± 2°	46 ± 1°

Ultrafiltration experiments

The effect of surface modification was studied by ultrafiltration of solutions containing the model proteins. In Fig. 3 results are given from experiments where lysozyme solutions were ultrafiltered with the unmodified GR 61 polysulfone

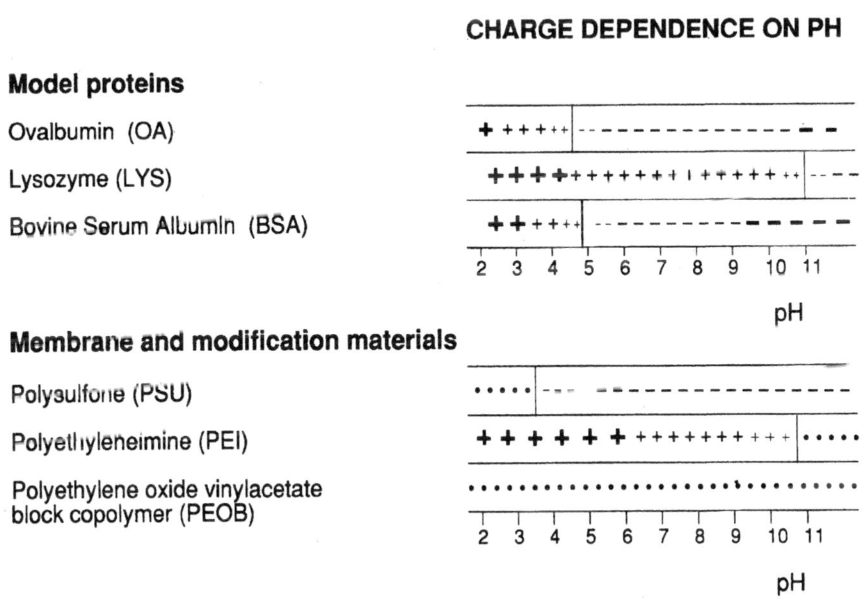

Fig. 2. Charge dependence on pH of BSA, OA, and LYS, and of the PSU membrane, and the materials with which it has been modified

membrane and with membranes modified with PEI and PEOB.

At pH 6.1 LYS is positively charged, according to the results discussed above. The unmodified polysulfone membrane is negatively charged, according to the zeta potential measurements. This results in electrostatic attraction between the protein and the membrane surface leading to a high flux reduction; flux reduction is high even at very low bulk concentrations of LYS.

If the membrane is modified with PEI, which possesses a positive charge at this pH, electrostatic repulsion is achieved, which results in a decreased flux reduction. However, the difference between the results corresponding to the conditions of repulsion (PEI-modified membrane) and attraction (unmodified membrane) is rather small. One can notice that this difference develops already at small bulk concentrations of the protein, where also probably electrostatic repulsion is most effective, due to the low coverage of the surface. According to the contact-angle results the PEI-modified membrane also is more hydrophilic than the PSU membrane, a matter which should further prevent protein adsorption. The high values for flux reduction at higher bulk concentrations can be due to penetration of the pores by the lysozyme molecules. The molar mass of LYS is 14300 g/mol, which is actually less than the cut-off value for the membrane.

The modification with the hydrophilic uncharged PEOB gives an even better result than the modification with PEI. The PEOB molecule consists of one

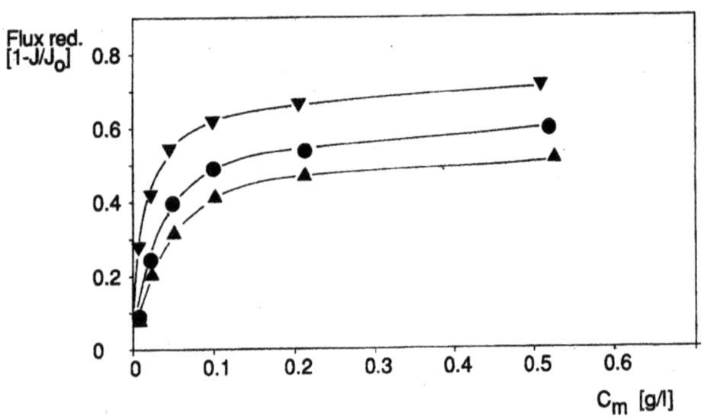

Fig. 3. Flux reduction as a function of lysozyme surface concentration (c_m) in ultrafiltration with a GR 61 polysulfone membrane without (▼) and with modification with PEI (●) and PEOB (▲). ΔP = 1.5 bar, pH = 6.1, I = 0.015 M KCl, and t = 25°C

block of PVAc, which has been shown [9] to be anchored to the hydrophobic surface in a latex system, and one block of PEO which was supposed to entangle into the solution outwards from the surface. This configuration is assumed to give an essentially hydrophilic surface at full coverage. This effect is clearly seen in the contact angles (Table 2), although the changes in numerical values are relatively small, indicating only partial coverage. The decrease in flux reduction due to the PEOB modification (Fig. 3) can thus be attributed to the increasing hydrophilicity. Gölander and Kiss [3[have studied the

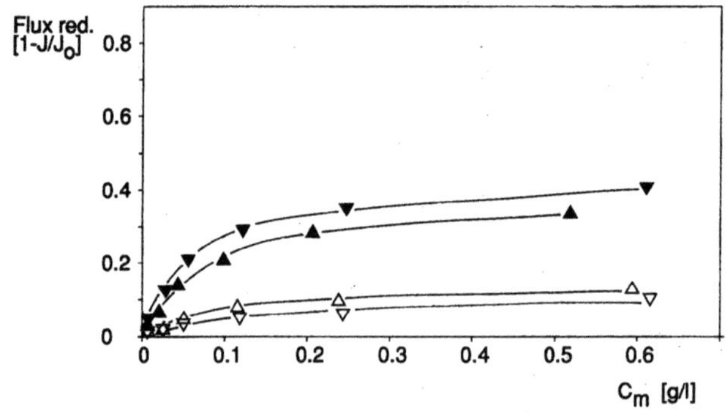

Fig. 4. Flux reduction as a function of bovine serum albumin surface concentration (c_m) in ultrafiltration with a GR 61 polysulfone membrane without (pH 6.1 ▼, pH 3.1 ▽) and with modification with PEOB (pH 6.1 ▲, pH 3.1 △). ΔP = 1.5 bar, I = 0.015 M KCl, and t = 25°C

adsorption of proteins on PEO-coated surfaces and found that, among other proteins, BSA is much less adsorbed on such a hydrophilic surface than on a hydrophobic one. Similar adsorption studies were made by Lundström [22] with human fibrinogen and human fibronectin on hydrophilic and hydrophobic silicon surfaces. Lundström explains the good results obtained with a hydrophilic surface with the fact that it attracts so much water that adsorption of proteins is prevented.

In order to see if the positive effect of modification with a hydrophilic substance was independent of pH some ultrafiltration experiments at different pH values with BSA and membranes modified with PEOB were carried out. In Fig. 4 are given the results from experiments at pH 6.1 and 3.1. At pH 6.1 both the BSA molecules and the membrane are slightly negatively charged. At pH 3.1, however, the BSA molecules are positively charged and the membrane either not charged at all or weakly positively

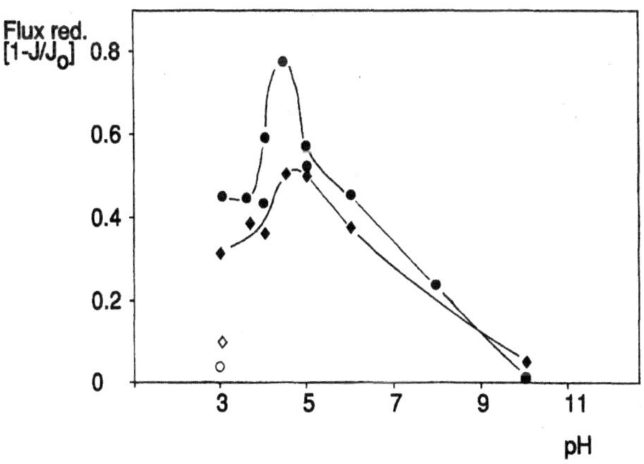

Fig. 5. Flux reduction as a function of pH in the ultrafiltration of ovalbumin with a GR 61 polysulfone membrane with or without modification with PEI at a surface concentration (c_m) of 0.5 g OA/dm^3. ΔP = 1.5 bar and t = 25°C. I = 0.15 M KCl; without PEI (●) and with PEI (◆), I = 0.015 M KCl; without PEI (○) and with PEI (◇)

charged. At pH 6.1 the modification improves the UF result even though electrostatic repulsion is probably shielded, which shows the importance of the hydrophilicity of the membrane. At pH 3.1 flux reduction, even without modification, is so low that the modification with PEOB does not improve it at all.

Ultrafiltration experiments with ovalbumin as the model substance were carried out over the pH range 3—10 at different ionic strengths. Both unmodified membranes and membranes modified with PEI were used. The results for an ionic strength of 0.15 M are depicted in Fig. 5. At pH values near the IIP of OA (4.6) or below it, the modification with PEI gives somewhat better results than without PEI. At pH 3 electrostatic repulsion, which decreases adsorption, is achieved and near the IIP of OA the hydrophilicity of PEI is favorable for the result. At pH 10 PEI does not improve the result as the flux reduction is low anyway, nor does it considerably increase flux reduction, as it is only slightly charged at this pH.

In Fig. 5 are also included results obtained at an ionic strength of 0.015 M at pH 3. They show that at lower electrolyte concentration flux reduction is considerably decreased both for the mofidied and the unmodified membrane, which is in accordance with results obtained by Fane et al. for unmodified membranes [23]. One explanation for this is that the charges of the molecules are partially shielded by the higher electrolyte concentration. However, the high flux reduction at the higher ionic strength can possibly be interpreted as a result from the electroviscous effect arising when cations (H^+ and K^+) penetrate the pores through a possibly thin, but highly positively charged adsorption layer or concentration polarization layer consisting of OA molecules [10, 24]. This hypothesis is corroborated by the observed variations in pH and conductivity of the feed and filtrate solutions, when various polyelectrolyte systems have been studied in our laboratory.

Correlation between flux reduction data and adsorption data

It was interesting to study whether the flux reductions from ultrafiltration experiments correlated with the data from adsorption on a smooth PSU surface. The model surface was meant to simulate the surface properties of the membrane. UF data were taken from experiments with BSA and LYS and an unmodified PSU membrane at pH 6 and the results are shown in Fig. 6, where flux reduction is plotted against surface concentration. It can be seen that a rather good correlation is achieved but flux reduction is continuously higher for LYS than for BSA. This can result from the fact that the LYS molecule (M = 14300) is smaller than the BSA molecule (M = 68000) and, therefore, gets more densely packed on the membrane. It is also small enough to penetrate the pores to some extent and also foul the inside of the pores, making them smaller. The immediate rise in flux reduction at small concentrations can also be taken as a sign of this phenomenon (Fig. 3).

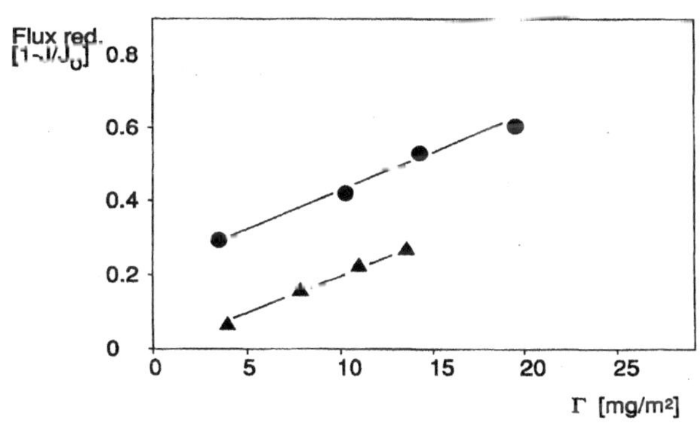

Fig. 6. Correlation of flux reduction in ultrafiltration with the surface concentration Γ of the proteins on a PSU coated quartz crystal. (●) LYS, (▲) BSA, ΔP = 1.5 bar (UF), pH = 6.1, I = 0.015 M KCl, and t = 25°C

Aimar et al. [25] have studied the relationship between flux reduction and BSA adsorption on IRIS 3038 membranes at different pH. They obtained a linear relationship between the logarithm of hydraulic resistance of the fouled membrane to the logarithm of the amount of adsorbed protein on the membrane.

Conclusions

From the results above can be concluded that there are many factors governing the fouling of the membranes, one of which undoubtedly is electrostatic repulsion. This repulsion was achieved between the membrane surface and the protein molecules in solution, when an appropriate pH and modification were used. Consequently, modification with PEI decreased flux reduction in the pH region 3 to 4.6 for OA about 30%.

In the cases where attraction between the membrane and the protein molecules exists, but where intermolecular repulsion is strong, very little fouling takes place. This situation was attained at pH 6 with the negatively charged OA and the positively charged PEI-modified membrane. An explanation for this behavior can be that, at first, a thin layer of the protein is attracted and adsorbed onto the membrane, but then this layer (being of the same charge as the molecules in solution) repels further adsorption.

When the membrane is rendered more hydrophilic by modification with an uncharged molecule, such as PEOB, the hydrophobic nature of the membrane is shielded and adsorption due to hydrophobic interaction becomes less favorable. This effect was established at pH 6.1 with a PEOB modified membrane and LYS. The combination gave an improvement of about 30% at small concentrations of LYS, and an improvement of 20% at higher concentrations.

The pH-dependence of flux reduction seems to result from variations in the intermolecular electrostatic repulsion between the protein molecules in solution. At the IP the net charges of the proteins are zero and they cannot repel each other and, therefore, they get more effectively adsorbed and foul the membrane. The strong fouling tendency near the IP was observed above for BSA and OA and has been frequently reported in literature [23, 6, 24]. Far away from the IP the electrostatic repulsion between like molecules is high and fouling is much less severe. The favorable effect of both high and low pH values can be seen from the UF results at pH 10 and 3 for BSA and OA. The less satisfactory results for LYS, even when electrostatic repulsion is established, probably result from the fact that the molecules are small enough to partly penetrate the pores and then get stuck there.

Acknowledgements

The authors are indebted to Nordisk Industrifond, The Finnish Academy of Sciences, and the Maj and Tor Nessling Foundation for financial support. The authors also thank Prof. Matti Lindström for valuable, inspiring discussions.

References

1. Hanemaaijer JH, Robbertsen T, van den Boomgaard Th, Gunnink JW (1989) J Membr Sci 40:199—217
2. Lundström I (1985) In: Glantz P-O, Leach SA, Ericson T (eds) Oral Interfacial Reactions of Bone, Soft Tissue and Saliva. IRL Press, Oxford, pp 9—23
3. Gölander C-G, Kiss E (1988) J Colloid Interface Sci 121:240—253
4. Norde W, Fraaye JGEM, Lyklema J (1987) In: Brash JL, Horbett TA (eds) Proteins at Interfaces. Physicochemical and Biochemical Studies. ACS Symposium Series 343, American Chemical Society, Washington, DC, pp 36—47
5. Norde W (1986) Adv Colloid Interface Sci 25:267—340
6. Matthiasson E (1984) Macromolecular Adsorption and Fouling in Ultrafiltration and their Relationship to Concentration Polarization. Thesis, Lund University, Lund, Sweden
7. Nilsson J-L (1989) A Study of Ultrafiltration Membrane Fouling, Thesis, Lund University, Lund, Sweden
8. Jonsson G (1986) Personal communication
9. Napper DH (1970) J Colloid Interface Sci 32:106—114
10. Nyström M (1989) J Membr Sci 44:183—196
11. Nyström M, Lindström M, Matthiasson E (1989) Colloids Surfaces 36:297—312
12. Padday JF (1969) In: Matijevic' E (ed) Surface and Colloid Science, vol 1. Wiley-Interscience, New York, pp 39—151
13. Lindström M, Nyström M, Laatikainen M (1988) Sep Sci Technol 23:703—717
14. Jonsson G (1984) Desalination 51:61—77
15. Tanford C (1961) Physical Chemistry of Macromolecules. John Wiley & Sons, Inc. New York: pp 357—358
16. Laatikainen M (1989) J Colloid Interface Sci 132:451—461
17. Burton WG, Nugent KD, Slattery TK, Summers BR, Snyder LR (1988) J Chromatogr 443:363—379
18. White A, Handler P, Smith EL (1964) Principles of Biochemistry, third edition. Mc Graw-Hill Book Company, New York, p 123

19. Fruton JS, Simmonds S (1958) General Biochemistry. J Wiley & Sons, Inc, New York, pp 85—113
20. Lindquist GM, Stratton RA (1976) J Colloid Interface Sci 55:45—59
21. Hiemenz PC (1977) Principles of Colloid and Surface Chemistry. Marcel Dekker, Inc, New York
22. Lundström I (1983) Physica Scripta T4:5—13
23. Fane AG, Fell CJD, Suki A (1983) J Membr Sci 16:195—210
24. Wahlgren M, Sivik B, Nyström M (1990) Acta Polytech Scand, Ch-series 194:1—18
25. Aimar P, Baklouti S, Sanchez V (1986) J Membr Sci 29:207—224

Authors' address:

Acting assoc. prof. Marianne Nyström
Laboratory of Polymer Chemistry
Department of Chemical Technology
Lappeenranta University of Technology
P.O. Box 20
53851 Lappeenranta, Finland

Functionalized mica surfaces obtained by a cold plasma process

P. M. Claesson[1]), D. L. Cho[2]), C. G. Gölander[2]), E. Kiss[3]), J. L. Parker[4])

[1]) The Surface Force Group, Department of Physical Chemistry, The Royal Institute of Technology, Stockholm and The Institute for Surface Chemistry, Stockholm, Sweden
[2]) The Institute for Surface Chemistry, Stockholm, Sweden
[3]) L. Eötvös University, Department of Colloid Science, Budapest, Hungary
[4]) Department of Applied Mathematics, Research School of Physical Sciences, Institute for Advanced Studies, Australia

Abstract: A surface modification procedure based on the cold plasma technique has been utilized in order to introduce new functional groups onto muscovite mica surfaces. In the first reaction step mica surfaces are exposed to a water vapor plasma. The interaction between mica surfaces and reactive species in the plasma results in formation of surface hydroxyl groups. These groups are reactive toward chlorosilanes in the gas phase, which provides a versatile way to obtain mica surfaces with a variety of different surface chemical groups. For instance, poly(ethylene oxide) chains have been covalently attached to mica surfaces. — In a different surface modification route, mica surfaces have been coated with thin plasma polymer layers. Mechanical and surface properties of plasma polymer films depend, of course, on the type of monomer, but also on the condition used during polymerization (e.g., flow rate, pressure and discharge power). The surface force technique has, together with spectroscopic analyses, provided information about how the adhesion force and mechanical strength of different plasma polymers of acrylic acid depend on the polymer structure.

Key words: Surface modification; mica; cold plasma; plasma polymer; surface force

Introduction

The muscovite mica basal plane has become a very important substrate surface in fundamental research. The smoothness and the known chemical composition, in particular the known number of exchangeable potassium ions, has made it possible to quantitatively investigate adsorption of ions, surfactants or proteins by means of photoelectron spectroscopy [1—2]. Muscovite mica is also the preferred substrate for direct measurement of intermolecular forces with the type of surface force apparatus (SFA) originally developed by Tabor and co-workers [3, 4] and later perfected at the Australian National University by Israelachvili and co-workers [5, 6]. Other surfaces that have been used in various devices for direct measurement of surface forces include silica [7—10], sapphire [11] and platinum [12].

Muscovite mica is built up of about 1-nm thick aluminosilicate layers. The atoms within each layer are held together by strong covalent silicon-oxygen and aluminium-oxygen bonds. Each layer is negatively charged and this charge is compensated exactly by potassium, and, to a lesser extent, sodium, ions located between the layers. The bond between each layer is due to relatively weak electrostatic and van der Waals forces. The difference in strength between interlayer and intralayer forces is the reason why mica is easily cleaved into large, molecularly smooth sheets. The ions located on the mica basal plane are readily exchanged for other types of cations which forms the basis for modifying the mica surface by adsorption from solution or by Langmuir-Blodgett deposition. The hydroxyl groups that do exist in the mica structure are buried 2 Å below the surface, and they are not accessible for reactions with chlorosilanes.

The range of systems that may be investigated with the SFA is greatly enhanced by modifying the mica basal plane without roughening the surface. This is traditionally done by adsorption from solution [13] or by deposition of insoluble surfactants using the Langmuir-Blodgett technique [14]. Adsorption from solution is the simplest method, but it is hard to have a precise control of the orientation of adsorbed molecules and the chemical composition of the modified surface. In contrast, the Langmuir-Blodgett technique facilitates an excellent control of surface composition and molecular orientation. However, it has been found that the stability of deposited layers, particularly bilayers or multilayers, in many cases limits their use in fundamental research [15, 16] as well as in technical applications [17].

Recently, it has been shown that the cold plasma technique can be used as an alternative way to chemically alter the surface composition of the mica basal plane [18–20] without roughening the surface significantly. In this paper we present some routes to obtain novel functionalized surfaces by the plasma technique and some results obtained from measurements of the interaction between such surfaces.

Experimental

Surface modification by cold plasma

A cold (room-temperature) plasma can be created by exposing a low pressure gas (0.01—1 torr) to a radiofrequency electromagnetic field. Interactions between the electromagnetic field and gas molecules create a range of reactive species including high energy electrons, free radicals, excited molecules and ions. The degree of surface modification can be controlled by varying the conditions used during plasma treatment.

The design of the glass plasmareactor (inner diameter 15 cm, length 65 cm) has been described previously [18]. The reactor is evacuated with a rotary vacuum pump and the radio frequency field (125—375 kHz) is supplied by a power generator (ENI HPG-2) via two external copper band electrodes. The reactor is also equipped with pressure transducers and flow control systems.

Surface modification of muscovite mica by exposure to a water vapor plasma was carried out at a pressure of 20 mtorr. Subsequent gas phase reactions with chlorosilanes were performed in a desiccator saturated with silane vapor.

Acrylic acid plasma polymerization was carried out at a pressure of 17 mtorr and a flow rate of 4 standard cubic centimeters per minute.

The surface force technique

Crack formation, long-range and adhesion forces were studied by means of a surface force apparatus [5]. Two molecularly smooth mica surfaces were first silvered on one side and then glued (silvered side down) onto cylindrical (radius of curvature about 2 cm) silica discs. The two surfaces were, after surface modification, mounted in a crossed cylinder geometry inside the stainless steel measuring chamber. When white light is directed perpendicular to the surfaces a standing wave pattern (fringes of equal chromatic order) is created. The wavelengths of the standing waves, which depend on the optical path length between the silver layers, facilitate measurement of the surface separation to within 0.2 nm [21]. The fringe pattern may also be used for measuring surface deformation [22], capillary condensation [23], crack formation [20], and refractive index of adsorbed layers [21].

The force is measured with a double variable spring supporting one of the surfaces. The deflection of the spring is determined interferometrically to within 1 nm, which allows the force acting between the surfaces to be determined from Hooke's law.

Chemicals

Muscovite mica was obtained from Mica Supplies Limited, Christchurch, England. Acrylic acid monomer (purity >99%) was received from Merck, and all the silanes were from Petrarch Systems. Water was purified with a MilliQ unit and deaerated for several hours before being used in the surface force apparatus. Poly(ethylene oxide) with a mean molecular weight of 1900 g/mol was purchased from Sigma. All chemicals were used without further purification.

Results and discussion

Mica treated with water vapor plasma

A water vapor plasma has been utilized in order to introduce surface hydroxyl groups on muscovite mica surfaces. The extent of surface modification with respect to the plasma power and the exposure time was optimized in order to maximize the number of hydroxyl groups and minimize the surface roughness. The chemical changes on the surface caused by water vapor plasma treatment were investigated by means of ESCA. Analyses of photoeletron spectra showed that potassium and aluminium atoms were depleted from the surface region [18], thereby creating a surface composed of mainly silicon and oxygen atoms. The sputtering of potassium and aluminium atoms increased with the plasma discharge power and with the exposure time [19].

The degree of surface modification caused by water vapor plasma treatment was also tested by first reacting the modified surface with a hydrophobic, monofunctional chlorosilane, and then measuring the advancing and receding water contact angle. Unmodified sheets of muscovite mica have no hydroxyl groups on the basal plane and do not, except possibly at the edges, react with chlorosilanes. However, after exposure to water vapor plasma chlorosilanes are easily covalently bonded to the basal plane. The optimal surface modification procedures was taken as the lowest power and the shortest treatment time that, after a gas phase reaction with chlorosilanes, resulted in a strongly hydrophobic surface with a small contact angle hysteresis. A satisfying result was obtained for mica surfaces treated at a power of 20 W for 2.5 min [18, 19]. ESCA analyses of these surfaces showed that about 1.7×10^{14} potassium ions per cm^2 and 3.7×10^{14} aluminium atoms per cm^2 had been removed. This corresponds to essentially all potassium ions, and is significantly more than all aluminium ions, in the top layer. That mica surfaces modified by water-vapor-plasma treatment (20 W, 2.5 min) remain highly smooth is supported by the fact that oscillating forces, due to the molecular packing of liquid molecules, have been observed between such surfaces immersed in octamethylcyclotetrasiloxane [18].

Forces between water vapor plasma treated mica surfaces

The adhesion force in air between water-vapor-plasma-treated mica surfaces was found to be 500 mN/m. This value is slightly lower than the adhesion between the same pair of sheets before plasma treatment, which was 600 mN/m. The long-range forces between water-vapor-plasma-treated mica in conductivity water and in 10^{-3} M KBr is illustrated in Fig. 1. The measured forces were consistent with predictions based on the DLVO theory down to a separation of 1—2 nm. At large separations a repulsive double-layer force with the expected Debye-length dominated the interaction. The double-layer force was weaker than the corresponding force between unmodified mica surfaces. At a separation of 3—4 nm a van der Waals force pulled the surfaces into an attractive minimum located about 1 nm from the position reached when the surfaces was pressed firmly together. The adhesion force was 5 mN/m in water

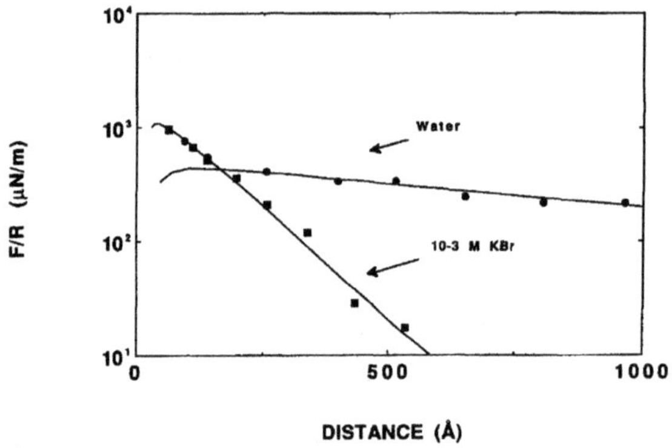

Fig. 1. Forces between water-vapor-plasma-treated mica surfaces immersed in pure water (●) and in 10^{-3} M KBr (■) as a function of surface separation. The force is normalized by the radius of the surfaces and plotted on a logarithmic scale. The solid lines are theoretical force curves calculated by assuming a double-layer interaction at constant potential and a van der Waals attraction characterized by a non-retarded Hamaker constant of 2.2 $\cdot 10^{-20}$ (J). The surface potential and the Debye-length were 110 mV and 1500 Å in pure water. The corresponding values in 10^{-3} M KBr was 60 mV and 100 Å

and 0.6 mN/m in 10^{-3} M KBr. This is significantly lower than the value (40 mN/m in pure water) observed for unmodified mica. The repulsive force observed at distances below 1—2 nm is due to surface roughness and dehydration of the surfaces. It was noted that water plasma treated mica surfaces were easily damaged when squeezed together. After damage the increased surface roughness resulted in a stronger repulsive force at short separations and the disappearance of the adhesive minimum.

Forces between silanated mica surfaces

After a gas-phase reaction with a fluorocarbon-based silane [(tridecafluoro-1,1,2,2-tetrahydro-octyl)-1-dimethylchlorosilane] for 3 min the advancing and receding contact angles were found to be 95° and 75°, respectively. ESCA analyses of these surfaces showed that the surface area per silane is about 45 $Å^2$. Hence, the silane layer is not tightly packed, which rationalizes the rather high contact angle hysteresis.

The forces measured between two types of fluorocarbon-coated mica surfaces are compared in Fig. 2. The forces acting between water vapor

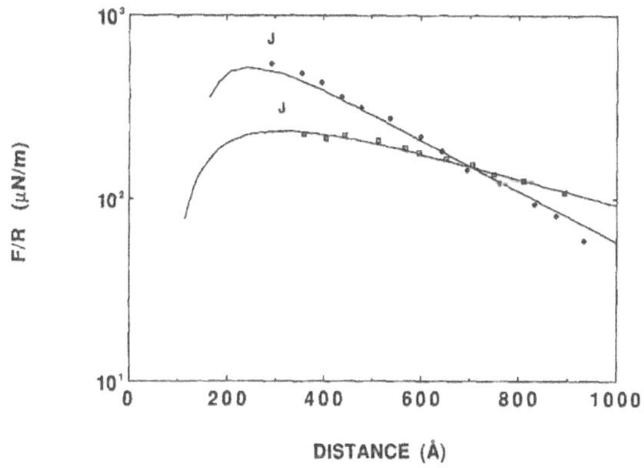

Fig. 2. The normalized force between water-vapor-treated mica reacted with (tridecafluoro-1,1,2,2-tetrahydrooctyl)-1-dimethylchlorosilane immersed in water (◙). The solid line is calculated by assuming a double-layer interaction at constant charge (surface potential 73 mV, Debye-length 580 Å), a nonretarded van der Waals attraction (Hamaker constant 2.2 · 10—20 (J)) and an exponential hydrophobic attraction (pre-exponential factor 0.8 mN/m, decay constant 120 Å). The forces between mica-carrying adsorbed layers of [N-(α-trimethylammoniumacetate)-O-O′-bis-(1H,1H,2H,2H-perfluorodecyl)-L-glutamate chloride] in a sonicated surfactant solution (2 · 10^{-4} M) is represented by (◆). The corresponding solid line is calculated assuming a surface potential of 100 mV, a Debye-length of 300 Å, the same Hamaker constant as above, and an exponential attraction with a preexponential factor of 8 mN/m, and a decay-length of 7 nm

plasma treated and silanated mica surfaces, prepared as described above, were measured in one set of experiments [18—19]. The other type of fluorocarbon surface was obtained by adsorbing a cationic, double-chained fluorocarbon surfactant [N-(α-trimethylammoniumacetate)-O-O′-bis(1H,1H,2H,2H-perfluorodecyl)-L-glutamate chloride] from a sonicated surfactant solution (2 · 10^{-4} M) [16]. In both cases a repulsive double-layer force dominates the long-range interaction, whereas a strong hydrophobic attraction is present at distances below 20—30 nm. In the case of water-vapor-treated and silanated mica the surface charge originates from the mica substrate. In contrast, the surface charge observed in the surfactant system originates from adsorption of a partial bilayer. Qualitatively similar long-range attractive forces have been observed between a range of different hydrophobic surfaces prepared by silanation [9, 18, 19], Langmuir-Blodgett deposition [24] or adsorption from solution [13]. The molecular mechanism for the *long-range* hydrophobic attraction between macroscopic surfaces is still under debate, even though some suggestions have been presented in the literature [25, 26].

Mica surfaces with grafted poly(ethylene oxide) chains

Water-vapor-plasma-treated mica surfaces may also be used as a starting point for grafting various polymer chains to the surface. For instance, mica surfaces with covalently attached poly(ethylene oxide) chains have been obtained. This is important since the interaction between poly(ethylene oxide) chains is of fundamental interest, because of a strong temperature dependence with a less repulsive interaction at higher temperatures [27, 28]. Further, it has been demonstrated that the adsorption of proteins is very low on poly(ethylene oxide)-coated surfaces [29], and its application as a biocompatible material has been suggested.

In the second reaction step surface hydroxyl groups on water-vapor-plasma-treated mica were reacted with 3-isocyano dimethyl propyl chlorosilane. This reaction was carried out in the gas phase for 30 min. Finally, terminal hydroxyl groups on poly(ethylene oxide) chains were reacted with the isocyanate group by covering the surface with a poly(ethylene oxide) melt (T = 70 °C, reaction time 2 h). Non-covalently attached poly(ethylene oxide) was removed by excessive rinsing in pure water. From ESCA analyses of the modified surfaces the area per polymer chain (molecular weight 1900 g/mol) was determined to be around 3.0 nm^2.

The investigation of the interaction between this type of poly(ethylene oxide) coated mica surfaces has just been initialized; so far, only the interaction at room-temperature has been studied (Fig. 3). At large separation a repulsive double-layer force dominates the interaction (the surface charge originates from the mica surface). At separations below about 30 nm a repulsion due to interactions between poly(ethylene oxide) segments becomes apparent. The repulsion measured on approach is larger than on separation. The reason for this hysteresis is not clear, but may indicate the presence of non-covalently attached polymer chains. This hypothesis will be tested by varying the preparation method. In comparison to the forces measured between mica surfaces immersed in a 10^{-3} M poly-

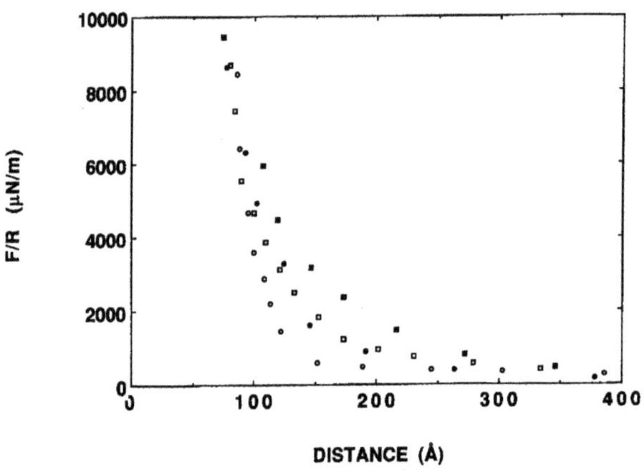

Fig. 3. The normalized force in water between mica surfaces carrying grafted poly(ethylene oxide) chains (unfilled symbols) and between mica-carrying adsorbed layers of poly(ethylene oxide)-lysine in a 10^{-3} M poly(ethylene oxide)-lysine solution (filled symbols). The forces measured on approach (squares) are more repulsive than those measured on separation (circles)

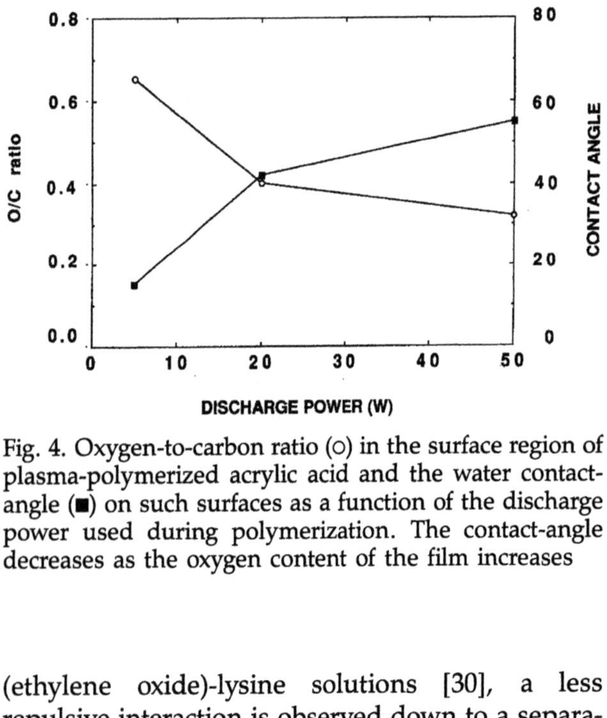

Fig. 4. Oxygen-to-carbon ratio (o) in the surface region of plasma-polymerized acrylic acid and the water contact-angle (■) on such surfaces as a function of the discharge power used during polymerization. The contact-angle decreases as the oxygen content of the film increases

(ethylene oxide)-lysine solutions [30], a less repulsive interaction is observed down to a separation of 90 Å (Fig. 4). This supports the suggestion made in a previous paper [30] that the most long-range part of the interaction between mica surfaces immersed in poly(ethylene oxide)-lysine (molecular weight 1900 g/mol) solutions is largely due to the presence of polymers in solution. The compressed thickness of the grafted poly(ethylene oxide) layer is about 7—8 nm. The same layer thickness was found for electrostatically attached poly(ethylene oxide)-lysine adsorbed from a 10^{-3} M solution. This is consistant with ESCA analyses that show that the amount of poly(ethylene oxide) is the same on the two types of surfaces.

Mica coated with plasma polymers

Plasma polymerization is a simple, dry, and versatile way to functionalize or coat surfaces with thin polymer layers. The plasma polymerization technique can utilize a range of monomers, including normally unreactive molecules like, for example, methane or ethane. The properties of the polymer formed by plasma polymerization naturally depend on the type of monomer, but also of the conditions in the plasma (e.g., pressure, monomer flow rate and discharge power). Hence, a plasma polymer of a given monomer is likely to differ structurally, mechanically, and in interfacial properties compared with a polymer formed by the same monomer using a conventional polymerization method. The combination of spectroscopic studies (ESCA, FTIR) and measurement with a surface-force apparatus has provided valuable information concerning the properties of plasma polymer coatings [20]. We have coated mica with a range of plasmapolymers including those formed from some alkanes, alkenes, tetrafluorethylene, siloxanes, and acrylic acid.

In the case of thin (20—50 nm) plasma polymers of acrylic acid the mechanical and interfacial properties vary considerably with the discharge power used during polymerization. This is correlated to a change in chemical structure of the polymer: with increasing discharge power the oxygen content of and the number of carboxylic acid groups in the film decreases, whereas the degree of crosslinking increases [20]. Figure 4 illustrates how the oxygen-to-carbon ratio (calculated from the carbon 1 s and oxygen 1 s photoelectron signal from ESCA analysis) and the water contact-angle vary with the discharge power. As expected, the wettability increases with increasing oxygen content.

The pull-off force normalized by the radius of the surface (the adhesion force) increases with decreasing discharge power. There are two factors contributing to this effect. First the number of hydrogen bonds formed between the surfaces decreases with

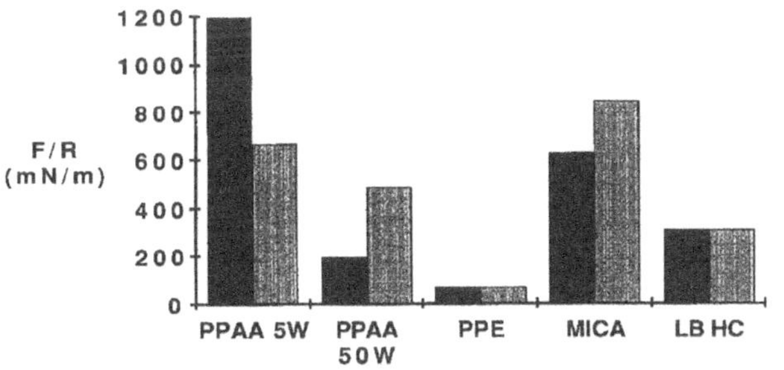

Fig. 5. The pull-off force normalized by the radius of the surfaces for some different types of surfaces. PPAA is plasma-polymerized acrylic acid, PPE is plasma polymerized ethylene, and LB HC is a hydrocarbon surface obtained by Langmuir-Blodgett deposition of a monomolecular film. Black bars indicate measurements in any air and grey bars in humid air

increasing power. Secondly the degree of crosslinking increases with increasing power, which counteracts chain entanglement and adaptive deformation in the contact region. Hence, the effective contact area decreases with increasing discharge power. The adhesion force in dry and humid air for some systems are compared in Fig. 5. Note the high adhesion force in dry air for mica coated with an acrylic acid layer prepared at a discharge power of 5 W.

When water vapor is present in the measuring chamber a capillary condensate form between hydrophilic surfaces in contact (Fig. 6). The width (W) and the thickness (T) of the condensate increases with decreasing water contact-angle. For instance, at close to 100% humidity the capillary condensate formed between plasmapolymers of acrylic acid deposited at 5 W (contact-angle 15°) had the dimensions T = 200 nm and W = 200 μm, whereas the corresponding values for polymers prepared at 50 W (contact-angle 55°) was T = 20 nm and W = 100 μm.

The adhesion force is related to the interfacial energy between the surface and vapor (γ_{sv}) and the interfacial energy between two surfaces in contact in vapor (γ_{ssv}) according to:

$$F_c/R = C(\gamma_{sv} - \gamma_{ssv}) ,$$

where γ_{ssv} is zero for perfect surfaces. C is a constant which has a value between 4π and 3π [22, 23]. As the humidity increases more water vapor adsorbs on hydrophilic surfaces and this reduces the surface tension according to the Gibbs adsorption

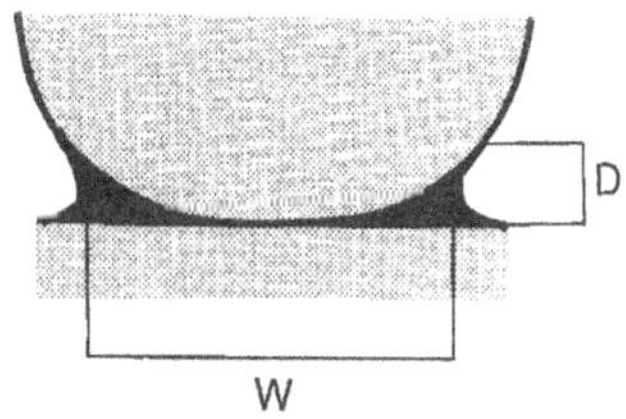

Fig. 6. In humid air a capillary condensate forms around the contact region between hydrophilic surfaces. In our system the width of the capillary (W) is typically 100 μm and the thickness is (T) 10—300 nm

equation. Hence, for smooth hydrophilic surfaces where changes in γ_{ssv} are negligible, one expects the adhesion force to decrease with increasing relative humidity. This was observed for plasma polymerized acrylic acid prepared at 5 W. The interfacial energy between a rough hydrophilic surface and water vapor also decreases with increasing humidity. Nevertheless, the adhesion force may increase in the presence of a capillary condensate. The reason is that the adhesion in dry air is low, due to a small effective contact area [31]. The adhesion force in the presence of a capillary condensate is, to a first approximation, equal to $4\pi\gamma_{lv}\cos\theta$.

In contrast, due to negligible water adsorption, the humidity is not expected to influence the adhesion between hydrophobic surfaces. This has been confirmed for smooth hydrocarbon surfaces prepared by Langmuir-Blodgett deposition and for molecularly rough surfaces prepared by plasma polymerization of ethylene (Fig. 6).

That adsorption of water vapor reduces the interfacial tension means that the cohesion within a material is lowered. This was observed experimentally in the case of plasma polymerized acrylic acid prepared at 50 W. In dry air the coated mica surfaces could be brought into adhesive contact several times without disruption of the coating. However, in humid air, when a capillary condensate formed between the surfaces in contact, cracks appeared in the surface coating as soon as the adhesive contact was broken. The cracks were most numerous in the surface area covered by the capillary condensate.

Conclusion

Hydroxyl groups have been introduced on the basal plane of muscovite mica by exposure to a low-temperature, water-vapor plasma. The resulting surface, which readily can be further modified by a gas-phase reaction with chlorosilanes, is sufficiently smooth for accurate surface-force measurements. Such measurements have been carried out between water-plasma-treated mica, between silanated mica, and between mica surfaces carrying covalently attached poly(ethylene oxide) chains.

Adhesion and mechanical properties of plasma polymers deposited on mica surfaces have been investigated with the surface-force technique. The adhesion force between plasma polymerized acrylic acid decreases with decreasing number of free acid groups. At the same time the wettability decreases and the degree of crosslinking increases. Adsorption of water vapor significantly lowers the mechanical stability of crosslinked films of plasma-polymerized acrylic acid.

References

1. Claesson PM, Herder P, Stenius P, Eriksson JC, Pashley RM (1986) J Colloid Interface Sci 109:31
2. Herder PC, Claesson PM, Herder CE (1987) J Colloid Interface Sci 119:155
3. Tabor D, Winterton RHS (1969) Proc Roy Soc A312:435
4. Israelachvili JN, Tabor D (1972) Proc Roy Soc A331:19
5. Israelachvili JN, Adams GE (1978) J Chem Soc Faraday Trans I 74:975
6. Parker JL, Christenson HK, Ninham BW, Rev Sci Instrum, (1989) 60:3135
7. Derjaguin BV (1960) Scientific American 203:47
8. Peschel G, Belouschek P, Muller MM, Muller MR, König R (1982) Colloid Polymer Sci 260:444
9. Rabinovich YI, Derjaguin BV (1988) Colloids Surf 30:243
10. Horn RG, Smith DT, Haller W (1989) Chem Phys Lett 162:404
11. Horn RG, Clarke DR, Clarkson MT (1988) J Mater Res 3:413
12. Smith CP, Maeda M, Atanasoska HS, White DJ, McClure (1988) J Phys Chem 92:199
13. Israelachvili JN, Pashley RM (1984) J Colloid Interface Sci 98:500
14. Marra J (1985) J Colloid Interface Sci 107:446
15. Claesson PM, Berg JB, Thin Solid Films, (1989) 176:157
16. Christenson HK, Claesson PM, Herder PC, Berg JM (1989) J Phys Chem 93:1472
17. Roberts G, Barton J, Flanagan M, Grey G, Pitt C, Shanks I, Schildt M (1984) "Langmuir-Blodgett Films — Current Status and Prospects for Further Development". The report of the Langmuir-Boldgett working party to the Science and Engineering Research Council, UK
18. Parker JL, Cho DL, Claesson PM, J Phys Chem, in press
19. Parker JL, Claesson PM, Cho DL, Ahlberg A, Tidblad J, Blomberg E, J Colloid Interface Sci, in press
20. Cho DL, Claesson PM, Gölander CG, Johansson K, submitted J Applied Polymer Sci, in press
21. Israelachvili JN (1973) J Colloid Interface Sci 44:259
22. Horn RG, Israelachvili JN, Pibac F (1987) J Colloid Interface Sci 115:480
23. Christenson HK (1988) J Colloid Interface Sci 121:170
24. Claesson PM, Christenson HK (1988) J Phys Chem 92:1650
25. Eriksson JC, Ljunggren S, Claesson PM (1989) J Chem Soc Faraday Trans 2 85:163
26. van Oss CJ, Good RJ (1988) J Dispersion Sci Tech 9:355
27. Claesson PM, Kjellander R, Stenius P, Christenson HK (1986) J Chem Soc Faraday Trans 1 82:2735
28. Carvell M, Hall DG, Lyle IG, Tiddy GJT (1986) Faraday Disc Chem Soc 81:223
29. Gölander CG, Kiss EJ (1988) Colloid Interface Sci 121:240
30. Claesson PM, Gölander CG (1987) J Colloid Interface Sci 117:366
31. Israelachvili JN (1985) "Intermolecular and Surface Forces with Applications to Colloidal and Biological Systems", Academic Press, London

Authors' address:

Dr. P. Claesson
The Surface Force Group
The Institute for Surface Chemistry
Box 5607
114 86 Stockholm, Sweden

Progress in Colloid & Polymer Science Progr Colloid Polym Sci 82:337—344 (1990)

Applications of the hard-soft acid-base (HSAB) principle to solid adhesion and surface triboInteractions

L. H. Lee*)

Webster Research Center, Xerox Corporation, Webster, New York, USA

Abstract: We attempt to extend the Hard-Soft Acid-Base (HSAB) principle for the reactions in solutions to interactions in solids. First we point out the important link between the absolute hardness of acid-and-base and the average energy gap. Then we discuss the electronic band structures of various solids, e.g., metals, semimetals, semiconductors, and insulators. On the basis of energy gaps, we elaborate various consequences of the acid-base interactions in solids. The applications of the HSAB principle and the frontier orbital concept to the solid adhesion and surface interactions between metals and polymers will be verified by experimental results reported in the literature. The new findings reported in this paper should be beneficial to those who are carrying out research in or processing thin-film microelectronic devices or thick-film multilayer structures.

Key words: Acid-base interaction; ahesion; HSAB; insulator; metal; polymer; semiconductor; semimeta solid; triboInteraction

1.0 Introduction

The donor-acceptor interaction [1, 2] and the acid-base interaction [3] have been reviewed. In many instances the two terms, though different, have been used interchangeably to describe the interactions involving the exchange of electrons between a donor and an acceptor. For polymer adhesion, Fowkes [4, 5] and Bolger et al. [6] have pointed out the important role of the acid-base interaction in the formation of an adhesive bond.

For the acid-base interaction in solutions, Pearson [7] proposed the hard-soft acid-base (HSAB) principle to describe some basic rules about the kinetics and equilibrium of the reaction. In this paper, we attempt to apply the HSAB principle to solid interactions with the aid of the frontier orbital method. We shall first describe the HSAB principle as it has evolved in recent years [8—11] and shall then describe the band structures of solids. After we demonstrate the compatibility between the HSAB principle and the band structures in the solid state, we then illustrate with several examples of adhesion and tribointeractions between metals and polymers. We are hoping that, in so doing, we can further understand the dynamics and energetics of the tribointeractions at the molecular level.

*) Honorary professor of the Chinese Academy of Sciences.

2.0 HSAB Principle

2.1 HSAB Principle for inorganic reactions

In an inorganic reaction the generalized Lewis acid-base interaction can be represented:

$$\underset{\text{(Acid)}}{A} + \underset{\text{(Base)}}{:B} \rightleftharpoons \underset{(AB\ \text{Complex})}{A\!:\!B} \ . \quad (1)$$

Pearson [7] has proposed for the above interaction to classify acids and bases according to the absolute hardness, which will be defined later in Fig. 1:

— A hard acid contains an acceptor atom of high positive charge and small size. It does not have easily excitable outer electrons and is not polarizable.

- A soft acid contains an acceptor atom of low positive charge and large size. It has several easily excitable outer electrons and is polarizable.
- A hard base contains a donor atom of low polarizability that is hard to reduce. It is associated with empty orbitals of high energy and hence inaccessible.
- A soft base contains a donor atom of high polarizability and low electronegativity. It is easily oxidized and associated with unoccupied low-lying orbitals.

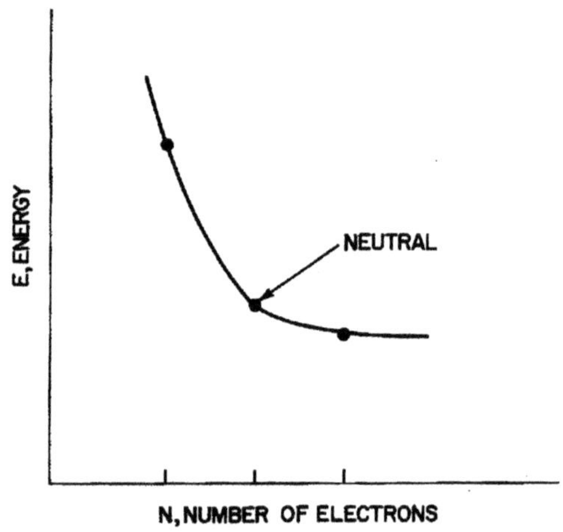

Fig. 1. Plot of electronic energy vs number of electrons for a fixed collection of nuclei (after [8], reproduced with permission)

There are two rules governing the acid-base interactions in solutions:

1) Regarding equilibrium: Hard acids prefer to coordinate with hard bases, and soft acids to soft bases.
2) Regarding kinetics: Hard acids react readily with hard bases, and soft acids with soft bases.

The above principle has been well tested in inorganic reactions [11].

2.2 HSAB Principle for organic reactions and the frontier orbital approach

For an organic reaction the HSAB principle has also been applied for the study of kinetics and equilibrium, and the frontier orbital method has been used to illustrate the electrophilic and nucleophilic interactions [12] (Fig. 2):

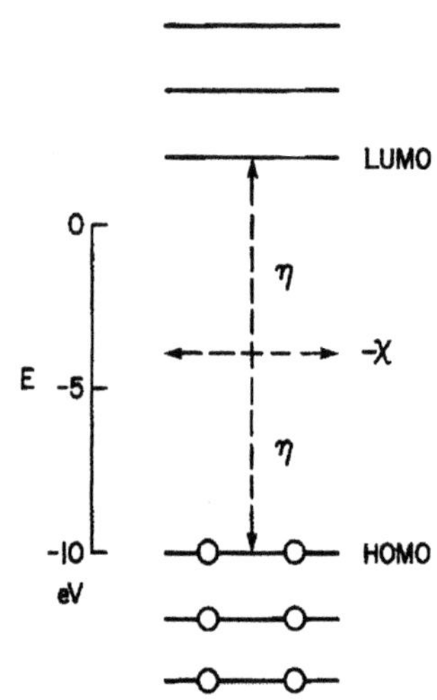

Fig. 2. Orbital energy diagram for a molecule. HOMO, highest occupied molecular orbital (MO); LUMO, lowest unoccupied MO, (after [8], reproduced with permission)

- A hard electrophile (or acid) has a high-energy lowest unoccupied molecular orbital (LUMO) and usually has a positive charge.
- A soft electrophile has a low-energy LUMO, but does not necessarily have a positive charge.
- A hard nucleophilic (or base) has a low-energy (highest occupied molecular orbital) HOMO and usually has a negative charge.
- A soft nucleophile has a high-energy HOMO, but does not necessarily have a negative charge.

Furthermore, according to Koopmans' theorem [13] the frontier orbital energies are given by

$$-E_{\mathrm{HOMO}} = I \; ; \quad -E_{\mathrm{LUMO}} = A \; , \tag{2}$$

where I is the ionization potential, and A is the electron affinity.

2.3 Perturbation equation

Hudson and Klopman [14] have derived an equation to describe the effect of orbital perturbation of two molecules on chemical reactivity. Their equation for the interaction energy ΔE can be simplified by using the HOMO of a nucleophile and the LUMO of an electrophile:

$$\Delta E = \underbrace{-\frac{Q_N Q_E}{\varepsilon R^2}}_{\text{The Coulombic term}} + \underbrace{\frac{2(C_N C_E \beta)^2}{E_{HOMO} - E_{LUMO}}}_{\text{The frontier orbital term}}, \quad (3)$$

where Q_N, Q_E are the total charges for the nucleophile and electrophile, respectively; C_N and C_E are the coefficients of the atomic orbitals N and E, respectively, in the molecular orbital; β is the resonance integral, ε is the permittivity, and R is the distance between N and E. E_{HOMO} is E_V, the energy of the valence band edge, and E_{LUMO} is E_C, the energy of the conduction band edge.

On the basis of the above equation, the HSAB rules can be restated as follows:

- A hard-hard interaction (or reaction) is fast because of a large Coulombic attraction, as described by the first term of Eq. (2).
- A soft-soft interaction (or reaction) is fast because of a large orbital interaction between the HOMO of the nucleophile and the LUMO of the electrophile, as described by the second term of Eq. (2).

2.4 Electronegativity and absolute hardness

When the HSAB principle was first introduced, the meaning of hardness was not easily understood. Certainly, it is not the hardness that measures the resistance against deformation. Then, what is it? It was not until the last several years that the real definition of absolute hardness [8, 9] received a quantitative and theoretical backing [10].

Firstly, we must define absolute electronegativity X. According to Mulliken [15], for atomic species, the electronegativity χ^M is the negative of chemical potential μ:

$$\chi^M = 1/2(I + A) = -\mu . \quad (4)$$

However, this relation is only applicable to a single atom, but not to a molecule. If the total electronic energy of an atom is plotted against the total number of electrons, a curve similar to that in Fig. 1 is obtained. The first ionization potential for the species will be much larger than the electron affinity and much smaller than the second ionization potential. Assuming a differential curve, the electronegativity may be defined as

$$\chi^M = -(\partial E/\partial N)_Z , \quad (5)$$

where Z is the total number of nuclear charges in the atom.

The second derivative of the same is defined to be absolute (or chemical) hardness, η, and the quantity:

$$\eta = 1/2(\partial^2 E/\partial N^2)_Z . \quad (6)$$

The corresponding operation definition is the corresponding finite difference formula

$$\eta = 1/2(I - A) . \quad (7)$$

Hence, the absolute hardness η is the resistance of the chemical potential to change in the number of electrons. The opposite of η is the absolute softness, σ.

According to the frontier orbital method [16], the relationship of η with the energies of LUMO and HOMO is shown in Fig. 2. Thus, from Fig. 2,

$$\eta = -1/2(E_{HOMO} - E_{LUMO}) . \quad (8)$$

Here χ^M is the horizontal broken line in the middle of the energy gap. Hence the energy gap E_g is twice the absolute hardness η; this is an important link between chemistry and physics in the solid state.

3.0 Extension of the HSAB principle to solid interactions

In the literature, we have not found any formal application or extension of the HSAB principle to solid interactions*). In this paper, we shall demonstrate that the extension of the HSAB principle to solid interactions is feasible in view of the electronic band structures of solids. We shall discuss the physical meaning of the absolute hardness in terms of the average energy gap.

*) The work by Yang W, Parr RC on metals and catalyst (Proc. Natl. Acad. Sci. USA (1985) 82, 67234) could be the earliest work related to this aspect.

3.1 Electronic band structures of solids

Solids can be classified as metals, semimetals, intrinsic semiconductors, and insulators. The band structures of solids are illustrated in Fig. 3. Monovalent metals [17], e.g., Na, have a partially filled valence band, the lower half of which is occupied. The Fermi level is in the valence band, but is at the top of the occupied orbitals. Furthermore, there is still an energy gap between the valence band and the conduction band (unoccupied MO). In some metals, such as the bivalent metals, the valence band is full but it overlaps a higher unoccupied conduction band. In this case, the Fermi level is in the conduction band and the overlapped valence band. Thus, the electrons close to the Fermi level are still free to move as the extra bands supply the unoccupied states. In the latter case, there appears to be no minimum energy gap E_g^0, which is generally reported in the literature. However, it is not obvious that there will always be an average energy gap E_g^{Av}, especially in metal atoms, which will be discussed later.

In addition to metals, there are semimetals [18], such as graphite, whose valence band and conduction band can overlap. In general, their minimum energy gaps are very narrow. The third class of solids is the intrinsic semiconductor; its minimum energy gap E_g^0 is generally below 3 eV. Thus, thermal excitation alone can create an electron-hole pair to enhance conduction. The Fermi level of the intrinsic semiconductor [17] lies between the valence band (HOMO) and the conduction band (LUMO). Hence,

$$E_F = 1/2(E_{\text{HOMO}} + E_{\text{LUMO}}) = 1/2(E_c + E_v) = -\chi^M , \quad (9)$$

where E_F is the energy of the Fermi level. The fourth class of solids is the insulator, which includes most polymers. The minimum energy gap of an insulator is generally above 3 eV. Therefore, thermal excitation alone cannot enhance the conduction of electricity.

3.2 Average energy gap and absolute hardness

The energy gaps generally reported in the literature are minimum energy gaps E_g^0. On the other hand, the average energy gap E_g^{Av} [19, 20] is defined as the difference in the energy between the conduction and valence bands:

$$E_g^{Av} = -(E_{\text{HOMO}} - E_{\text{LUMO}}) = -(E_v - E_c) = 2\eta . \quad (10)$$

According to Phillips [19], E_g^{AV} should contain both covalent (homopolar) component E_h and ionic (heteropolar) component C:

$$(E_g^{Av})^2 = (E_h)^2 + (C)^2 . \quad (11)$$

For many $A^N B^{8-N}$ compounds, E_g^{Av} can be determined from dielectric measurements.

For comparison, both E_g^0 and E_g^{Av} for several elements are listed in Table 1. In general, E_g^{Av} is

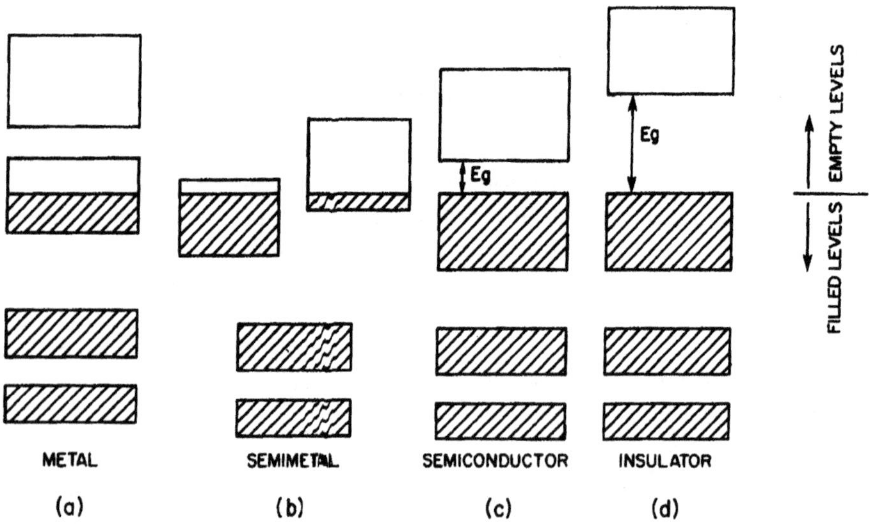

Fig. 3. Band structures of solids: a) metal, b) semimetal, c) semiconductor, and d) insulator [35], adapted with permission

Table 1. Minimum and average energy gaps and absolute hardness values for atoms (eV)

	E_g^0 1)	E_g^{Av} 1)	η 2) (cal'ed)	η 3)
C	5.4	13.6	6.8	5.0
Si	1.1	4.8	2.4	3.4
Ge	0.7	4.3	2.2	3.4
Sn	0	3.1	1.6	3.1

1) E_g^0 and E_g^{Av} from [20]; 2) η^{Av} is calculated from E_g^{Av}; 3) η is the value obtained in [9] (where hardness values quoted are for isolated metal atoms and not for bulk metals).

always larger than E_g^0. From Table 1 the η values calculated from E_g^{Av} are rather close to those calculated from the $A - I$ data; however, the agreement is still not good enough.

We also attempt to calculate E_g^{Av} from η for several transition metal atoms (Table 2). The η values that fall between 3 to 4 indicate that most of these metals are rather "soft". Since there are not many E_g^{Av} values in the literature for comparison, we assume that they reasonably good. We shall discuss the applications of the HSAB principle based on these η values.

Table 2. Average energy gap and hardness values for transition metal atoms (eV)

Metal	I 1)	A 2)	χ^M	η 3)	E_g^{Av} 4) (cal'ed)
Ti	6.82	0.08	3.45	3.4	6.8
V	6.7	0.5	3.6	3.1	6.2
Cr	6.77	0.66	3.72	3.1	6.2
Mn	7.44	0	3.72	3.7	7.4
Fe	7.87	0.25	4.06	3.8	7.6
Co	7.8	0.7	4.3	3.6	7.2
Ni	7.64	1.15	4.4	3.3	6.6
Cu	7.73	1.23	4.48	3.3	6.6
Mo	7.10	0.75	3.9	3.2	6.4
Ru	7.40	1.5	4.5	3.0	6.0
Pd	8.34	0.56	4.45	3.9	7.8
Ag	7.58	1.30	4.44	3.1	6.2
Pt	9.0	2.1	5.6	3.5	7.0
Au	9.23	2.31	5.77	3.5	7.0

1) [33]; 2) [34]; 3) [9]; 4) E_g^{Av} is calculated from η obtained from $A - I$ values.

3.3 Absolute hardness of solid

Now let us examine again the band structure of solids in Fig. 3. In Fig. 3a the structure of metals with zero or low E_g signifies that all metals are "soft". However, they do not appear to be as "soft" as the η values calculated from the $A - I$ data (Table 2). Indeed, metals have been classified as amphoteric materials. Most of the metals are "soft" acids, and some of them "soft" bases. When two metals are brought together into close contact, one of them assumes to be an acid while the other is a base. In a recent paper by Cain et al. [21], the interaction at the Cu/Cr interface has been treated as an acid-base interaction. In this case a soft base (Cu) and a soft acid (Cr) react preferentially. In fact, the Cr atom (η = 3.1 eV) is as soft as the Cu atom (η = 3.3 eV) (see Table 2).

Figures 3b and c show that both semimetals and semiconductors should be rather "soft" because of relatively low E_g^0 and E_g^{Av} as well (most of η's < 3.0 eV). Semiconductors, generally, react readily with metals even at ambient temperatures [22]. Thus, the interaction takes place as a "soft" acid (metal) with a "soft" base (semiconductor). Both chemical reaction and interdiffusion [23] jointly create a diffused "interphase" instead of an abrupt "interface".

In Fig. 3d it is noted that the band structure of an insulator contains a large gap. Thus, most insulators, regardless of organic or inorganic, are considered to be rather "hard" on the basis of the energy gap alone. Generally, they are harder than metals, semimetals, and semiconductors.

It is important to remember that the hardness of one material with respect to the other depends on its η. The one with a lower η value is "softer" than the one with a higher η, though they both can be "hard"; even in the same material one functional group can be harder than the other. In this case, a new term, local hardness $\tilde{\eta}$ [11] has been introduced to signify the relative hardness of each functional group.

4.0 Applications of the HSAB principle to adhesion and tribointeractions between metals and polymers

The interactions between metals and polymers are good examples of the applications of the HSAB principle to solids. When Cr is deposited on pyromellitic dianhydride-oxydianiline polyimide (PMDA-ODA PI) there appears to be some chemical bonding between the Cr atom and PI. In this case, presumably, Cr is a "soft" acid and PI is a "hard" base. How can the bonding take place? According to Ho et al. [24, 25] the reaction, indeed, does not proceed in the manner of the Cr atom attacking one of the carbonyl groups (nucleophiles). What really happens is that the Cr atom delocalizes and forms a charge transfer (or acid-base) complex with the PMDA unit of the PI, as shown in Fig. 4. In frontier orbital terminology, the stabilization of the complex is achieved through the transfer of electronic charge from the metal d-states of Cr (HOMO) to the LUMO of the π system of the PMDA unit of PI. In this manner, PI acts as a "soft" acid by accepting the electron and Cr (η = 3.1 eV) becomes a "soft" base by donating the electron. Other transition metals, e.g., the Ti atom (η = 3.4 eV) [26], are also very reactive with PI. The general reaction presumably follows the same path of the acid-base interaction through a charge-transfer complex.

One of the exceptions is the Cu atom [27], which does not interact with PI after the deposition and gives rise to a much weaker complex. Recently, Cain and Matienzo [28] have used tight-binding calculations of the extended Hückel type to find the relative acid strengths of Cu compounds in the following decreasing order: CuF_2 > CuO > CuF ≈ Cu_2O > Cu. Indeed, the Cu ions react faster than Cu with functional polymers. The Cu^{2+} ion has a Fermi level lying below the top of the Cu 3d band, and some of the Cu 3d orbitals are unoccupied and able to accept electron pairs by acting as an acid. In the HSAB terminology Cu becomes a "hard" acid in the form of Cu^{2+} ion, because the hardness for Cu is 3.3 eV, and that for Cu^{2+} is 8.3 eV.

Generally, many transition metals react well with polymers, ceramics [29], etc., partly because of the ease of oxidation and partly because of the availability of d-orbitals. Buckley [30], with pin-and-disc experiments has found that metals react with PTFE and PI under high vacuum. With PTFE, the adhesive forces are three times the applied load. Now, we can also attribute these reactions to the HSAB interactions. In the literature there should be numerous examples of the surface interactions that can be classified as the acid-base interaction; however, we are unable to discuss them all in this paper. One of the reasons is that the chemical hardness values for polymers have yet to be determined indirectly. (We plan to publish the calculated values in the future.)

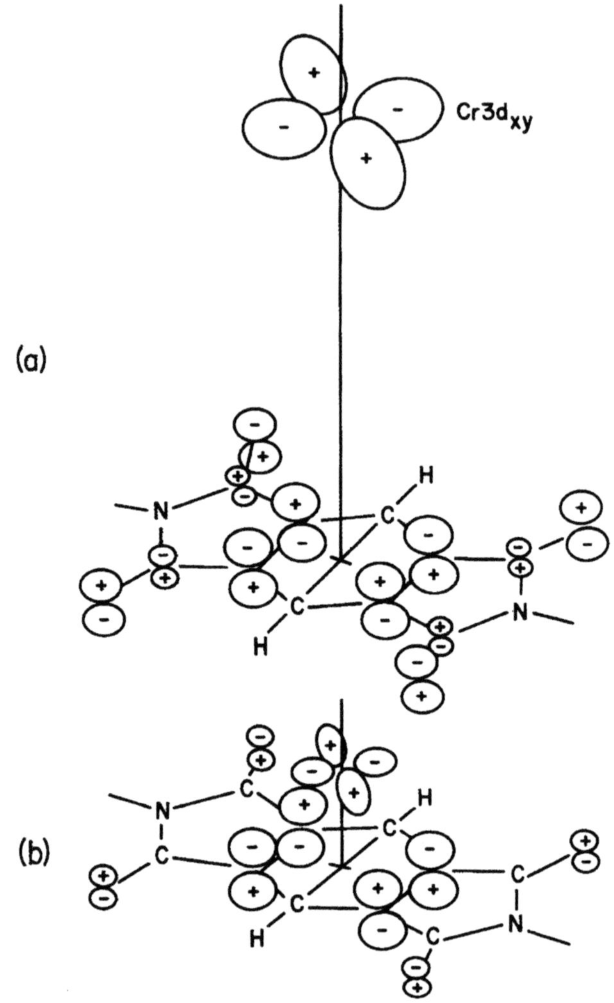

Fig. 4. Scheme of Cr-polyimide complex formation a) LUMO of PMDA fragment and Cr 3 d_{xy} orbital (occupied) when Cr is at a distance; and b) Cr over the six-member central ring of the PMDA (after [24] reproduced with permission)

It is important to point out that though chemical hardness values are rather similar for all metals, not all metals react in the same way or at the same rate with polymers. Since all metals are "soft" and most polymers are "hard", the metal-polymer reactions or adhesion, unlike the exception Cr-PI reaction mentioned earlier, are generally difficult to achieve. To circumvent the situation, most metals have to be transformed into harder counterparts, e.g., related oxides of higher chemical hardness values. This is exemplified in the cited case about Cu^{2+} ions [28] which are harder than the metal atoms Cu. Consequently, the reactivities of the metal-polymer interaction are determined by the ease of the formation of oxides, and that ease is, in turn, controlled by the heat of formation H_f of oxides [31]. In theory, the more negative the H_f, the more reactive the interaction. For the same reason, Au and Ag with high electronegativities are always non-reactive or inert toward polymers. We shall explore, in detail, the metal-polymer adhesion in a forthcoming paper on the chemistry and physics of solid adhesion [32].

5.0 Conclusions

The extension of the HSAB principle and the frontier orbital concept to solid interactions has been demonstrated. For the HSAB principle, the absolute hardness has been discussed along with the average energy gaps in solids. The mechanisms of interactions between different types of solids have been elaborated in light of the HSAB principle. Both the HSAB principle and the frontier orbital concept have been found to be beneficial in explaining the mechanisms for several metal-polymer adhesion and surface interactions. Though we have not supplied any direct experimental evidence, we believe that our findings should be helpful to those in carrying out research in or processing thin-film microelectronic devices and thick-film multilayer structures.

Acknowledgements

The author thanks Professor R. G. Pearson of the University of California at Santa Barbara for his comments and for providing a reference. The author also appreciates the comments on the original manuscript made by Professor R. J. Good of the University of New York at Buffalo. Several excellent comments made by the referee for this paper are gratefully acknowledged.

Nomenclature

A electron affinity
C ionic (heteropolar) component of the average energy gap
C_E coefficient of the atomic orbital of the electrophile
C_N coefficient of the atomic orbital of nucelophile
ΔE interaction energy
E electronic energy
E_c energy of the conduction band edge
F_F energy of the Fermi level
E_g^0 minimum energy gap
E_g^{Av} average energy gap
E_h covalent (homopolar) component of the average energy gap
E_{HOMO} energy of the highest occupied molecular orbital
E_{LUMO} energy of the lowest unoccupied molecular orbital
E_V energy of the valence band edge
I ionization potential
N number of electrons for a fixed collection of nuclei
Q total electron charge
R distance between the electrophile and nucleophile
Z total number of nuclear charge in the atom
β resonance integral
ε permittivity
η absolute hardness
$\underline{\eta}$ local hardness
μ chemical potential
σ absolute softness
χ^M electronegativity (Mulliken)

References

1. Gutman V (1978) The Donor-Acceptor Approach to Molecular Interactions. Plenum Press, New York
2. Deryagin BV, Krotova NA, Smilga VP (1978) Adhesion of Solids. Engl Ed Translated by R. K. Johnson, Plenum, New York
3. Jensen WB (1978) Chem Rev 78:(1)1
4. Fowkes FM (1972) J Adhes 4:155, also in Recent Advances in Adhesion, Lee LH (ed) Gordon and Breach, New York and London (1973), p 39
5. Fowkes FM (1987) J Adhes Sci Technol 1:(1)7 (1987)
6. Bolger JC, Michaels AS (1969) in Interface Conversions, Weiss P, Cheevers D (eds) Chapt 1, Elsevier, New York
7. Pearson RG (1963) J Am Chem Soc 85:3533
8. Pearson RG (1987) J Chem Educ 64:(7)563
9. Pearson RG (1988) Inorg Chem 27:734, also J Am Chem Soc 110:7684 (1988)
10. Parr RG, Pearson RG (1983) J Am Chem Soc 105:7512
11. Pearson RG (1973) Hard and Soft Acids and Bases, Dowden, Hutchinson and Ross, Stroudsburg, PA
12. Isaacs NS (1987) Physical Organic Chemistry, Longman Scientific and Technical, with Wiley, New York
13. Koopmans T (1934) Physica 1:104
14. Hudson RF, Klopman G (1967) Tetrahedron Lett 12:1103; also Theo Chim Acta 8:165 (1967)
15. Mulliken RS (1953) J Chem Phys 3:573, also (1953) 3:586
16. Fukui K (1982) Science 218:(4574)747
17. Gutmann F, Lyons LE (1967) Organic Semiconductor, Wiley, New York, p 17
18. Carter DL, Bate RT (eds) (1971) The Physics of Semimetals and Narrow-Gap Semiconductors, Pergamon Press, Oxford
19. Phillips JC (1970) Rev Mod Phys 42(3):317
20. Burstein E, Pinzuk A, Wallis RF (1971) The Physics of

Semimetals and Narrow Gap Semiconductors, Carter DL, Bate RT (eds) Pergamon Press, Oxford, p 251
21. Cain SR, Matienzo LJ, Emmi F (1989) J Phys Chem Solids 50:(1)87
22. Braicovich L (1988) The Chemical Physics of Solid Surfaces and Heterogeneons Catalysts, King DA, Woodruff DP (eds) Chapt 6, Elsevier, Amsterdam
23. Brillson LJ (1982) Surf Sci Rep 2:123
24. Ho PS, Silvermann BD, Haight RA, White RC, Sanda PN, Rossi AR (1988) IBM J Res Develop 32:(5)658
25. Ho PS, Haight RA, White RC, Silverman BD, Faupel F, to be published in 1990 in Fundamentals of Adhesion, Lee LH (ed) Plenum, New York
26. Ohuchi FS, Freilich SC (1986) J Vac Sci Technol A4:(3)1039
27. Haight RA, White RC, Silverman BD, Ho PS (1988) J Vac Sci Technol A6:(4)2188
28. Cain SR, Matienzo LJ (1988) J Adhes Sci Technol 2:(5)395
29. Buckley DH (1981) Surface Effects in Adhesion, Friction, Wear and Lubrication, Elsevier, Amsterdam
30. Buckley DH, Brainard WA (1975) in Advances in Polymer Friction and Wear, Lee LH (ed) Vol A, Plenum, New York, p 315
31. Mattox DM (1973) Thin Solid Film 18:173
32. Lee LH (1990) Fundamentals of Adhesion, Lee LH (ed) Chapter 1, to be published by Plenum, New York (1990)
33. Moore CE (1970) Natl Stand Ref Data Ser NSRDS-NBS 34
34. Hotop H, Lineberger WC (1985) J Phys Chem Ref Data 14:731
35. Epstein AJ, Miller JS (1979) Sci Amer 241(4):52
36. Lee LH (1990) to be published in Journal of Adhesion Science and Technology

Author's address:

Dr. Lieng-Huang Lee
Webster Research Center
Xerox Corporation
Webster, New York 14580, USA

Appendix

To complement existing acid-base theories, we further demonstrate the theoretical relevance of the hard-soft acid-base (HSAB) principle for solutions to surface interactions and solid adhesion. On the basis of density-functional theory, two basic parameters, i.e., chemical potential μ and absolute hardness η, will be shown to govern an acid-based interaction. From them, the number of transferred electrons ΔN between A and B of the system can be estimated

$$\Delta N \approx -(\Delta\mu/\Sigma\eta) .$$

By extending the HSAB principle for solutions to solids, we have found two important properties of solids, i.e., work function φ and the average energy gap E_g^{Av}, which are equivalent to the two chemical parameters. Thus, we propose to estimate the number of transferred electrons for solid interactions by the derived equation [36]:

$$\Delta N \approx \Delta\varphi/\Sigma E_g^{Av} .$$

Progr Colloid Polym Sci 82:345—348 (1990)

Immobilization of anti-hCG on gold and aluminum surfaces

S. Airikkala and W. M. Albers

Technical Research Centre of Finland, Medical Engineering Laboratory, Tampere, Finland

Abstract: Thin films of gold and aluminum were silanized with γ-glycidoxypropyltrimethoxysilane in vacuum. The modified metal substrates were coated with monoclonal antibody against human Chorionic Gonadotropin (hCG). The binding of hCG and polyclonal anti-mouse IgG was measured with surface plasmon resonance (SPR). The antigen hCG gave only a small shift in resonance angle, while the anti-mouse IgG gave a stronger shift. Quantitation of the amount of hCG binding to the surface was done by a radioimmunoassay (RIA) of similar surfaces as used in the SPR experiments.

Key words: Protein immobilization; silanization; surface plasmon resonance; gold; aluminum

1. Introduction

Direct optical monitoring of antigen binding to an antibody-coated surface has recently received considerable attention, due its potential as a reagentless immunoassay technique. One problem is to attain stable attachment of highly specific molecules on the surface of the measuring device. Proteins can be immobilized in a number of different ways: covalent binding to a reactive insoluble support, physical adsorption at a solid surface, entrapment in polymeric gels or within microcapsules or crosslinking by means of bifunctional reagents [1]. Covalent coupling for the immobilization of proteins is based on the formation of a covalent bond between the protein molecules and support material.

Chemical modification using organofunctional silanes is an important technique. The most usual method to silanize surfaces is to let the reaction beween the surface and organosilane occur in a liquid phase [2]. However, the presence of water will cause irreproducible silane layers as a result. These problems can be circumvented by chemical vapor deposition of the silane [3]. Most of the work involving silanization has been done by depositing silanes from solution onto glass surfaces and less information is available on the use of coupling agents on metal surfaces. Daniels et al. [4] have immobilized biotin on the surface of a metal coating of either gold or silver using solution of aminopropyltriethoxysilane in toluene. A thin silica film was used as an additional layer between metal and silane. Glucose oxidase has been attached to platinum-platinum oxide screens via alkylamine silane-glutaraldehyde coupling [5].

The aim of this study was to immobilize monoclonal antibody against human Chorionic Gonadotropin (hCG) on to gold and aluminum surfaces for further use in surface plasmon resonance (SPR) studies. Gold and aluminum coated glass and silicon slides were epoxylated with γ-glycidoxypropyltrimethoxysilane. The resulting modified metal substrates were coated with a monoclonal antibody against hCG. HCG is a reproductive hormone, which is produced in the placenta and by certain tumors. Its physiological effects are probably related to maintenance of the corpus luteum and to the development of fetal gonadal function. With SPR the binding of hCG was measured, as well as the binding of polyclonal anti-mouse IgG.

A surface plasmon is a collective motion of electrons on the surface of the metal layer, excited by the impact of light of appropriate wavelength at a particular angle. For monochromatic light, the SPR is observed as a sharp minimum in light reflectance when the angle of incidence is varied. The angle at which the resonance occurs is extremely sensitive to the refractive index of the medium adjacent to the metal surface. The binding of proteinaceous antigens to antibodies immobilized on the metal surface perturbs the local refractive index and produces a change in the resonance conditions [4, 6]. This can be observed as a change in the intensity of the reflected light if the angle of incidence is kept at near-resonance conditions. A radioimmunoassay (RIA) of similar surfaces, as in the SPR measurements, was used as a means of standardization. Estimates of the affinity and capacity constant of the immobilized antibody for hCG were obtained by fitting the RIA results to a single-site, equilibrium-binding model. In this way the maximum amount of hCG coupling to the antibody could be determined and related to the SPR experiments.

Experimental

Gold and aluminum films, with thickness of 48.5 nm and 16.5 nm, resp., were sputtered on 25 × 50 mm glass slides (type BK7). Depositions were also made on silicon wafers, which were cut in pieces of 6 × 9 mm for use in RIA measurements. The γ-glycidoxypropyltrimethoxysilane was obtained from Aldrich (Steinheim, FRG). Bovine serum albumin (BSA, RIA-grade) and polyclonal anti-mouse IgG antibody were from Sigma Chemical Co. (St. Louis, USA). The monoclonal antibody against hCG was from Medix Biochemica (Kauniainen, Finland; Prod. Nr. 5008). The antibody was β-chain specific and had an affinity for hCG or 9×10^{10} l/mol. Radiolabeled hCG (^{125}I-hCG) was taken from the hCG-RIA test of Medix Biochemica. The hCG standard, used in all measurements, was obtained from Boehringer (Mannheim, FRG). The standard had an immunological potency of 9 IU/µg and a molecular weight of 46000. Phosphate-buffered saline (PBS) contained 150 mM NaCl and 20 mM KH_2PO_4 (pH 7.4). SPR assay buffer was made dissolving BSA in PBS at a concentration of 0.01% followed by filtration. For SPR measurements the hCG and the anti-mouse IgG were reconstituted with SPR assay buffers: hCG at a concentration of 90000 IU/l, the anti-mouse IgG at a concentration of 0.1 mg/ml. The RIA assay buffer consisted of 0.33% BSA, 3 mM EDTA, 13 mM $NaH_2PO_4 \cdot 2\,H_2O$ (pH = 7.5). For RIA-assay the hCG standard was reconstituted and further diluted in RIA-assay buffer.

Gold- and aluminum-coated glass and silicon slides were epoxylated with γ-glycidoxypropyltrimethoxysilane containing a glycidoxy organic group and a trimethoxysilyl inorganic group. Gold slides were first cleaned with chromosulfuric acid and rinsed with isopropanol and water. Aluminum slides were rinsed with isopropanol and water. The slides were dried in nitrogen stream and placed in a vacuum oven together with a Petri dish containing 1 ml of freshly distilled epoxysilane. The oven was evacuated and the silane deposited at 140 °C in 2 h. The oven was thereafter cooled and the slides kept in vacuum. Anti-hCG was coated on the slides by adding 4.4 µl (4.4 µg) of the antibody solution per cm^2 of epoxylated surface and placing the slides in a moist chamber at room temperature for 2 h. To block nonspecific adsorption the slides were preserved at 4 °C in 0.5% BSA in PBS. Radioimmunoassay of slides coated with antibody was carried out in microwells: hCG standard and radiolabeled hCG (in 300 µl of assay diluent) were pipetted to each microwell, after which a coated slide was added. The slides were mildly shaken at room temperature for 4 h. After binding of the hCG the slides were washed three times with PBS containing 0.1% Tween 20 and counted for radioactivity on a gamma counter.

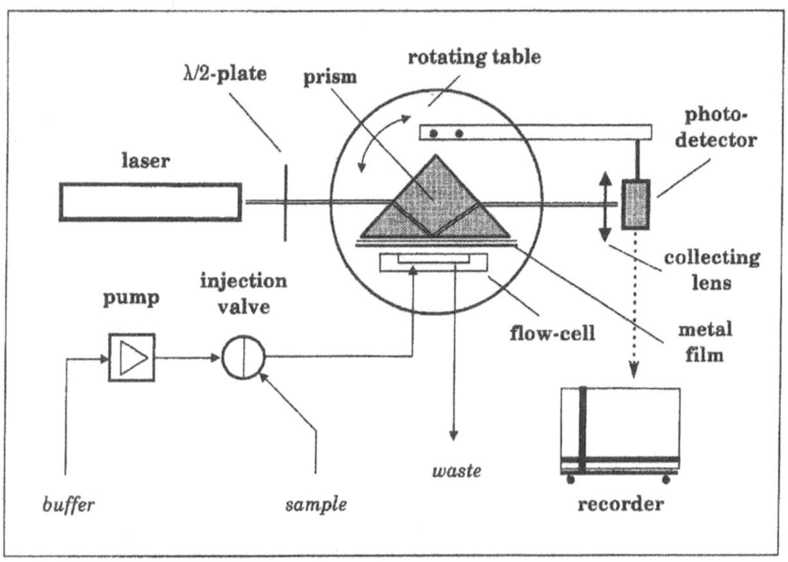

Fig. 1. Set-up for surface plasmon resonance (SPR) studies

In the surface plasmon resonance experiments the Kretschmann configuration was used [7]. The surface plasmon was generated by means of a prism placed on a rotating table (Fig. 1). The light source was a He-Ne laser (Siemens, 20 mW) with a polarized output. The light beam of the laser was conducted through a $\lambda/2$ plate, for accurate rotation of the plane of polarization, and an attenuator before entering the prism. Glass slides with deposited metal films were attached to the prism with index matching immersion oil. The reflected light was collected by lenses and directed upon a photodetector (UDT 61 Optometer). The signal from the photodetector was recorded on a x,y-recorder. The light intensity was measured as a function of incidence angle. Kinetic response curves and intensity changes at a constant angle in time were also recorded. The immunological reactions were studied by means of a flow-cell with a volume of approx. 60 µl. After rinsing with ethanol and drying in ambient air, the flow-cell was placed over the sensing area and PBS buffer was pumped through the cell with a flow rate of 0.5 ml/min. Protein solutions of 110 µl volume were introduced by means of an injection valve. The following solutions were successively injected: 0.01% BSA, 90000 IU/l hCG in 0.01% BSA and 0.1 mg/ml anti-mouse IgG in 0.01% BSA.

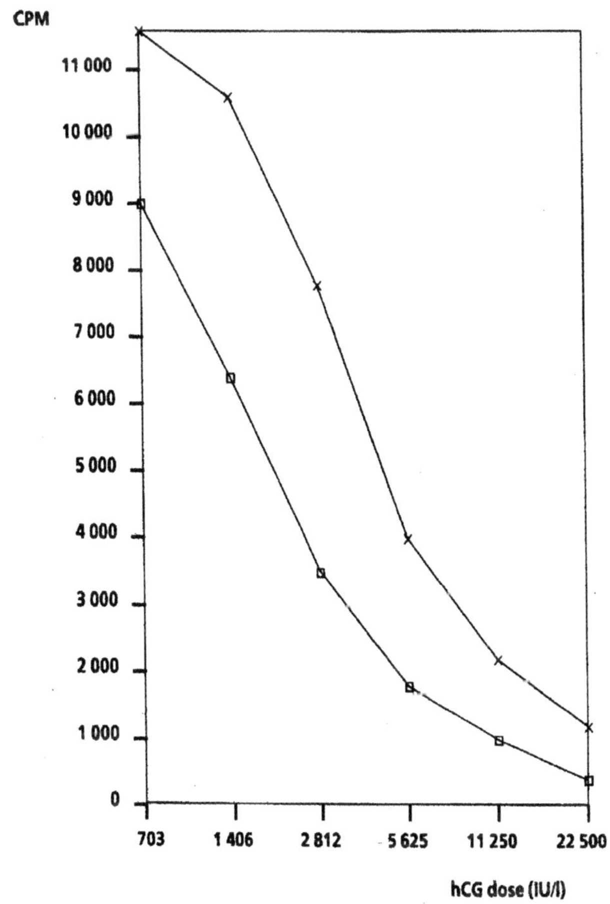

Fig. 2. RIA curves of gold (-X-) and aluminum (-□-)

3. Results and discussion

The obtained standard curves from RIA are depicted in Fig. 2. The antibody-coated aluminum slides appeared to be less effective in hCG binding than antibody coated gold slides. The RIA standard curves were fitted to a single-site binding model by non-linear regression using the "LIGAND" program [8], which gave the results of Table 1. The monoclonal anti-hCG gave the highest affinity constant on aluminum, while the binding capacity was highest on gold. From the values of binding capacity (156 fmole on epoxylated gold and 60 fmole on epoxylated aluminum) the saturation surface concentrations of hCG were calculated: 13.3 ng/cm^2 on gold and 5.1 ng/cm^2 on aluminum.

Table 1. Binding constants of the immobilized anti-hCG as calculated with "LIGAND"

Metal	Affinity constant		Capacity constant	
	K_a (l/mole) $\times 10^{-10}$	c.v.%	Q (femtomole)	c.v.%
gold	0.49	18	156	4
aluminum	1.58	59	60	7

Figure 3 shows the SPR curves of gold and aluminum. Epoxylated gold and aluminum slides, precoated with antibody, showed response to hCG in SPR, and the response was stronger on gold. The poor SPR response of antibody-coated aluminum towards hCG can be explained from the observed lower hCG binding capacity. Considering the surface area sampled by the laser (4 mm^2), the amount of hCG, that was actually detected with SPR was approx. 0.5 ng/cm^2 for gold slides and 0.2 ng/cm^2 for aluminum slides. A polyclonal antibody against mouse IgG gave a much stronger response than that of hCG on both surfaces. We find this an obvious effect of both the higher molecular weight of the anti-IgG in comparison with hCG and the polyclonal nature of the anti-IgG. Non-specific adsorption of BSA was hardly noticed at a concentration of 0.01%.

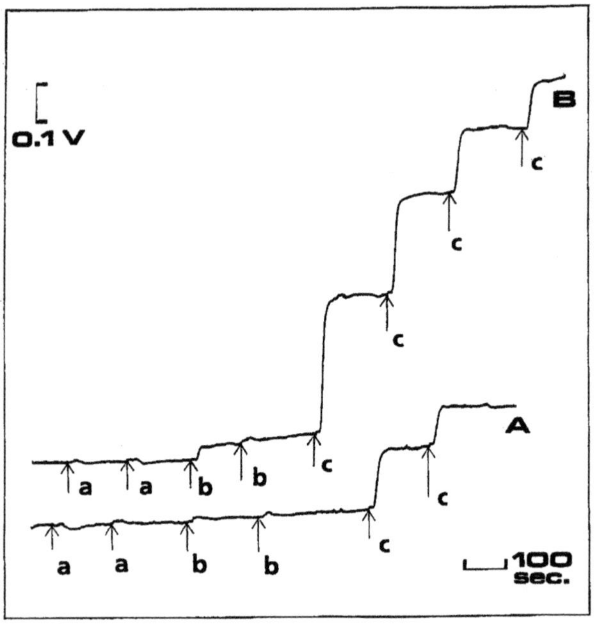

Fig. 3. SPR response curve of anti-hCG immobilized on epoxylated aluminum (A) and gold (B). Injections: a) 0.01% BSA, b) 90000 IU/l hCG in 0.01% BSA, and c) 0.1 mg/ml anti-mouse IgG in 0.01% BSA

Orientation problems with antibodies to the surface of the support can be reduced by the use of spacers like protein A, protein G or other Fc receptors. Some of these compounds have a strong affinity for only the Fc portion of immunoglobulins. If the Fc receptor is immobilized on the surface first, then the antibody will be bound in a fashion that will leave the antigen binding parts of the antibody free in solution. Sensitivity in SPR experiments can in this way be improved. Other means to achieve higher sensitivity, from the instrumental side, are the use of a differential measurement mode and another type of flow-cell that enables faster reaction kinetics. We propose those types of improvements in further experiments.

4. Conclusion

SPR is a suitable, direct detection method for polyclonal antibodies, as found here in the case of IgG/anti-IgG. For the determination of clinically interesting concentrations of hCG, however, the system in the present setup was not sensitive enough; the sensitivity can be improved as discussed above.

Acknowledgements

Dr. Janusz Sadowski is gratefully acknowledged for technical assistance in the SPR measurements.

References

1. Kennedy JF, White CA (1980) Int Polym Sci Tech 7 (11):52—56
2. Weetall HH (1976) In: Mosbach K (ed) Methods Enzymol XLIV. Immobilized Enzymes. Academic Press, Orlando, pp 134—148
3. Jönsson U, Olofsson G, Malmqvist M, Rönnberg I (1985) Thin Solid Films 124:117—123
4. Daniels PB, Deacon JK, Eddowes MJ, Pedley DG (1988) Sensors and Actuators 15:11—18
5. Wingard LB, Ellis D, Yao SJ, Schiller JG, Liu CC, Wolfson SK, Drash AL (1979) J Solid-Phase Biochem 4(4):253—262
6. Cullen DC, Brown RGW, Lowe CR (1987/88) Biosensors 3(4):211—225
7. Kretschmann E (1971) Z Physik 241:313—324
8. Munson PJ, Rodbard D (1980) Anal Biochem 107:220—239

Authors' address:

W. M. Albers
Technical Research Centre of Finland
Medical Engineering Laboratory
P.O. Box 316
33101 Tampere, Finland

Theoretical analysis of the forward reaction of antibody binding to surface-immobilized antigen

M. Werthén[1,2], M. Stenberg[3] and H. Nygren[1]

[1] Dept. of Histology, University of Göteborg
[2] Dept. of Biochemistry and Biophysics, University of Göteborg
[3] Dept. of Solid-State Electronics, Chalmers University of Technology, Göteborg, Sweden

Abstract: The binding of monoclonal antibodies to surface-adsorbed antigen was measured by ellipsometry. Mouse IgG-antibodies directed against dinitrophenyl were used. The different binding affinities to free hapten in solution had previously been determined. — The experimental findings indicate that the mechanism of heterogeneous antibody-antigen reactions is more complex than the corresponding reaction mechanism in solution. The initial binding of antibodies easily becomes diffusion-rate-limited. The progressive, forward-reaction rate is continuously retarded. The reverse reaction is slow enough to make the total reaction practically irreversible. Experimental data were considered, especially in order to analyze the decreasing forward reaction rate. — It was found that the forward rate constant has a dependence on time, such that $k(t) \propto k_1 t^{-h}$, where $0 < h < 1$. This scaling property of the binding reaction can be used in calculation of intrinsic reaction rate coefficients. The analysis reveals a concentration-independent, intrinsic, forward-rate constant. The magnitudes of the calculated initial reaction rates are comparable to reaction rates measured in homogeneous antibody-antigen reactions.

Key words: Monoclonal antibody; antigen; solid phase; scaling properties; fractal kinetics

Introduction

Experimental studies of antibody binding to surface-immobilized antigen have shown that the reaction mechanism is more complex than that of homogeneous antibody-antigen reactions. The initial reaction is often diffusion-rate-limited [1], the progressive forward-reaction rate is 1000-fold lower than the corresponding reaction rate when both reactants are in solution, and it is continuously decreasing with time [2]. The reverse reaction rate is low and the total reaction can be considered as effectively irreversible [3]. The heterogeneous antibody-antigen reaction cannot be interpreted using classical kinetics, assuming the reaction to establish an equilibrium between the forward- and reverse-reaction rates.

This study was undertaken in order to analyze the forward-reaction rate of antibody binding to surface-immobilized antigen using ellipsometry measurements of two monoclonal IgG-antibodies bound to surface-adsorbed antigen. Attempts are made to interpret the experimental results with the formalism of fractal reaction kinetics.

Experimental

Antigen and antibodies

Monoclonal IgG-antibodies (Mab1 and Mab2) directed against dinitrophenyl (DNP) were a generous gift from Prof. M. Steward, London. The characteristics of the antibodies used have previously been described [4]. The equilibrium constants of the antibodies used, measured in solution, are $0.9 \cdot 10^6\ M^{-1}$ and $1.1 \cdot 10^7\ M^{-1}$, respectively, [4]. DNP coupled to bovine serum albumin (BSA/DNP) was used as antigen [3], with an epitope density of 33 epitopes per protein molecule.

Experimental procedure

Methylsilanized silicon wafers were used as substrate [5]. The antigen was adsorbed to the surface by immersion for 2 h of the silicon wafers into phosphate-buffered saline (PBS), pH 7.2, containing antigen (0.1 mg protein/ml). The wafers were then rinsed in PBS, followed by a short rinse in water, blown dry, and placed in a humified chamber. Drops of PBS with different antibody content were placed on the wafers for various periods of time. The reactions were stopped by rapid rinsing with PBS, followed by a short rinse in water, and drying in an air current. The dissociation of bound antibodies (with known amount of bound antibodies) was measured after rinsing the wafers in PBS for timeperiods ranging from 15 min to 20 h. Control incubations of antibodies were made on wafers coated with BSA without DNP. All of the values presented are adjusted with respect to any background seen in controls. The wafers were examined in a comparison ellipsometer (SagaX, Göteborg, Sweden) and the amount of bound antigen and bound antibody was determined as described previously [6]. The error of measurement was $0.15 \cdot 10^{-9}$ mmol/cm^2.

Theory

The homogeneous antibody-hapten reaction has a well-known reaction mechanism, that being a rapid reaction with forward-reaction rate constants k_1, typically in the range 10^7—10^8 M^{-1} s^{-1}. It is an equilibrium reaction with reverse-rate constants k_{-1}, typically in the range 10—100 s^{-1}. Antibody binding to proteins is a somewhat slower reaction with a forward reaction rate constant in the range 10^5—10^6 M^{-1} s^{-1} [7]. The reaction is described by

$$Ab + Ag \underset{k_{-1}}{\overset{k_1}{\rightleftarrows}} AbAg \; , \quad k_1/k_{-1} = K_a \; . \tag{1}$$

K_a is the equilibrium constant of the reaction, also defined as the affinity of the antibody for the free hapten.

Assuming the heterogeneous reaction to be an equilibrium reaction, the reaction rate can be described by

$$\frac{dS_{ab}}{dt} = k_1 C_{ab}(S_{\max} - S_{ab}) - k_{-1} S_{ab} \; , \tag{2}$$

[2, 5], where S_{ab} is the surface concentration of bound antibodies, C_{ab} is the antibody concentration in solution, and $S_{\max}$ is the maximal amount of bound antibody at the surface. As the dissociation of bound antibodies from the surface is very low, the rate equation can be simplified to

$$\frac{dS_{ab}}{dt} = k_1 C_{ab}(S_{\max} - S_{ab}) \; . \tag{3}$$

The forward reaction rate constant can be calculated using experimental S_{ab} and $S_{\max}$ values. $S_{\max}$ can also be estimated theoretically by assuming a tightly packed layer of antibodies bound to the surface.

Antibody-antigen reactions at solid/liquid interfaces are often diffusion-rate-limited in the initial phase of the reaction, due to depletion of antibodies close to the surface. The diffusion-limited surface concentration of bound antibodies can be calculated using the relation for diffusion limitation to plane surfaces. The diffusion-limited surface concentration is described by

$$S_{ab} = (2/\sqrt{\pi}) C_{ab} \sqrt{Dt} \; , \tag{4}$$

where D is the diffusion constant of IgG-antibodies ($4.0 \cdot 10^{-7}$ cm^2/s) [1]. This relation can be used to make sure that the experiments are performed in time and concentration intervals where the reaction is not limited by the diffusion rate of antibodies to the surface.

In classical kinetics the rate constants have not time dependence. However, in experimental studies of the reaction kinetics at solid/liquid interfaces it has been shown that the forward-reaction rate constants are dependent on time. In recent studies of different types of heterogeneous reactions, both experimental and computer simulations, the reaction data has been interpreted by a theory often referred to as fractal reaction kinetics [8], to describe the time-scaling phenomena of these reactions.

The forward-reaction rate constant is defined as a function of time described by

$$k(t) \propto k_1 t^{-h} \; , \quad 0 < h < 1 \; ,$$

where $k(t)$ should rather be referred to as reaction-rate coefficient, and k_1 is the intrinsic forward-reaction rate constant.

In homogeneous reactions, where the reaction-rate constant is time-independent, $h = 0$. In addition to time dependence, reactions with geometric constraints often have anomalous reaction orders [8].

Results and discussion

In this study two monoclonal antibodies were chosen (which are not diffusion-rate-limited) using concentrations and incubation times presented here. The binding of Mab 1 and Mab 2 was measured with close intervals for times up to 1000 s, and then with longer intervals for times up to 3 days.

The surface concentration continues to increase for incubation times longer than 24 h. The amount of bound antibodies shows a linear relation to $\log(t)$ in this time interval [5]. This slow reaction rate is not due to dissociation since experimental studies have shown that the amount of bound antibodies does not decrease after 20 h of rinsing in PBS [5]. Neither is the slow reaction rate caused by saturation. The maximal surface concentration of bound antibodies, S_{max}, was determined experimentally by prolonged incubation for up to 7 days. The experimentally found value $8 \cdot 10^{-9}$ mmol/cm^2 is in accordance with a theoretical value for a tightly packed layer of antibodies bound to the surface antigen with both Fab-fragments.

The forward-reaction rate constant can be calculated (Eq. (3)) using the experimentally measured values of surface concentration of bound antibodies.

For both Mab 1 and Mab 2 there is a rapid initial decrease of the reaction-rate constant which then continues to decrease more slowly. The forward-reaction-rate constants for binding times up to 1000 s are shown in relation to time in Fig. 1. During the time interval presented the reaction-rate constant decreases logarithmically in relation to time, and the time dependence can be described by $k_1 \propto t^{-h}$; h is the slope (in Fig. 1) with values in the range 0.52—0.70. The time dependence of the forward-reaction-rate constant has been suggested to stem from the ordered distribution of reacted molecules at the surface. According to fractal reaction kinetics, h is related to the geometrical distribution of reactants, the fractal dimension D at the surface, such that $D = 1 + h$ [9].

The kinetics and isotherms of antibody binding to surface-immobilized antigen are similar to kinetics and isotherms of protein-adsorption to a hydrophobic surface. Recent studies of protein adsorption to a hydrophobic surface [10] have shown that the protein molecules are not adsorbed randomly at the surface, but are heterogeneously ordered in clusters. This gives us reason to believe that the antibody binding is not a random process, but rather that binding occurs in an ordered heterogeneous distribution, i.e., as growing clusters of antibodies at the antigen surface. This assumption is supported by electromicroscopic studies of the binding of gold-antibody particles to surface-immobilized antigen. The antibody-gold particles bind in clusters, with cluster sizes dependent on antibody affinity [11].

The intrinsic forward reaction rate constant for the binding reaction can be evaluated in Fig. 1. The

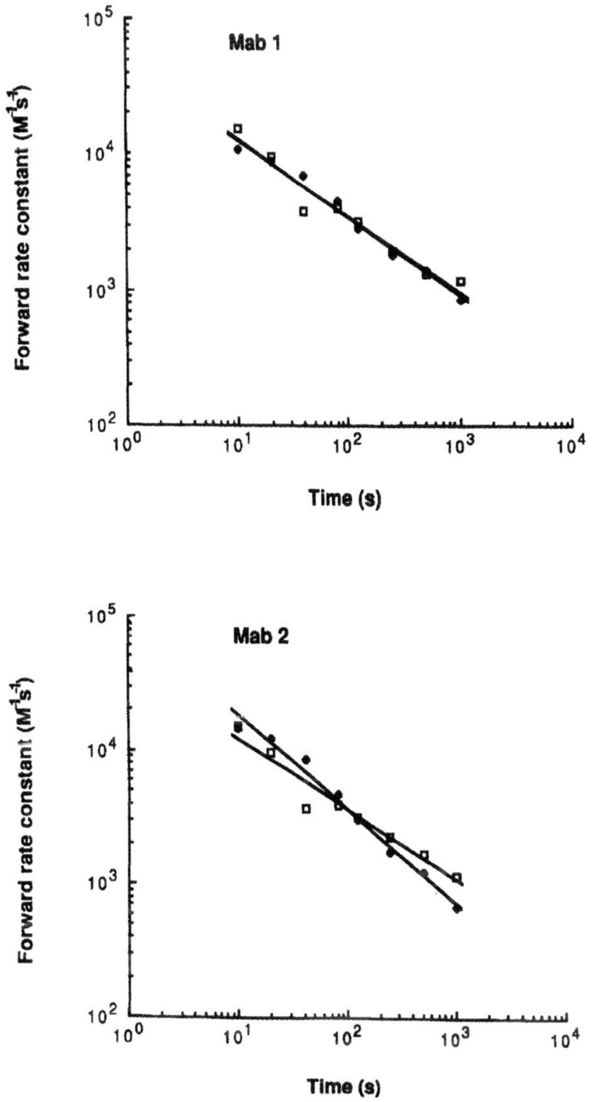

Fig. 1a, b. The forward reaction rate constants (M^{-1} s^{-1}) vs time (s) for times ranging from 10 to 1000 s. Values for two antibody concentrations are shown: □ 30 μg/ml and ◆ 100 μg/ml. 1a) Mab 1; 1b) Mab 2

reaction rate constant $0.5 \cdot 10^5$ M^{-1} s^{-1} is of the same order of magnitude as the forward-reaction-rate constants in homogeneous reactioans of antibodies and protein antigen. This fact implies that the fractal kinetics of the solid phase reaction is due to secondary effects and not to a change in the intrinsic reaction rate.

Although the exact reaction mechanism of antibody-antigen reactions at interfaces is still unknown, the formalism for fractal reaction kinetics seem to be a fruitful way of describing experimental data on antibody binding to surface-immobilized antigen.

References

1. Nygren H, Stenberg M (1985) J Colloid Interface Sci 107:560
2. Nygren H, Stenberg M (1989a) Immunology 66:3
3. Nygren H, Werthén M, Stenberg M (1987) J Immunol Methods 101:63
4. Stanley C, Lew AM, Steward M (1983) J Immunol Methods 64:119 (The Mabs 1 and 2 are here labelled 47 and 51 respectively)
5. Nygren H, Kaartinen M, Stenberg M (1986) J Immunol Methods 92:219
6. Stenberg M, Nygren H (1983) J Phys (Paris), Colloq 10:12
7. Mason DS, Williams AF (1986) In: Weir DM (ed) Handbook of Experimental Immunology, Blackwell, Oxford, p 38:1
8. Kopelman R (1988) Science 241:1620
9. Liebovitch LS, Fischbarg J, Koniarek JP, Todorova I, Wang M (1987) Biochimica et Biophysica Acta 896:173
10. Nygren H, Stenberg M (1989b) Progr in Colloid and Polymer Sci, 81
11. Werthén M, Nygren H (1988) J Immunol Methods 115:71

Authors' address:

H. Nygren
Dept of Histology
University of Göteborg
POB 33031
40033 Göteborg, Sweden

 Progr Colloid Polym Sci 82:353—359 (1990)

Immobilization of membrane enzymes into magnetizable, phospholipid bilayer-coated, inorganic colloids

M. De Cuyper and M. Joniau

Interdisciplinary Research Centre, Katholieke Universiteit Leuven — Campus Kortrijk, Kortrijk, Belgium

Abstract: Nanometer-sized magnetic particles, which can be efficiently captured in a magnetic filter device during high-gradient magnetophoresis, were encased by a phospholipid bilayer made of the zwitterionic dimyristoylphosphatidylcholine. Evidence for the bilayer configuration of the lipid coat was mainly derived from equilibrium adsorption experiments. Lipid-depleted, beef-heart cytochrome *c*-oxidase, used as a model enzyme, was incorporated into these so-called magnetoliposomes by sonication. As a result, the enzyme demonstrates a 15-fold enhancement of its activity, which is comparable with the degree of reactivation measured in the presence of small unilamellar vesicles. The suitability of this type of catalytic biocolloids for biotechnological purposes is further demonstrated in a miniature pilot, fixed-bed bioreactor in which the enzyme-phospholipid-magnetite complex is first trapped by magnetic forces on the magnetic filter and subsequently overflowed with a solution of reduced cytochrome *c*. In this continuous mode of operation, the conversion efficiency remains almost constant on a time-scale of hours. Consequently, magnetoliposomes carrying membrane enzymes show up as promising entities to be used in magnetically controlled bioreactors.

Key words: Bioreactor; enzymes (membrane-bound); immobilization; magnetic fluids; magnetoliposomes; magnetophoresis (high-gradient); phospholipid vesicles

Introduction

Magnetic particles have important applications in numerous industrial processes such as in printing practices using magnetic ink, in the development of magnetic joints, in the clean-up of oil spills, etc. [1]. For biotechnological purposes, the particles are very frequently covered with reactive polymers so that proteins or enzymes can be anchored onto them in a covalent way [2]. As far as the immobilization of enzymes is concerned, only water-soluble types have been used and exploited as such in bioreactors. Conversely, enzymes which in vivo are located in membranes, i.e., membrane enzymes, have been largely neglected. One of the main reasons for this lack of interest is that the availability of purified, membrane-bound enzymes in quantities sufficient for commercial purposes is severely restricted by the laborious purification schemes, which, moreover, result in very low recoveries. Also, as compared with soluble enzymes, membrane ones need to be manipulated in a special manner [3]. Indeed, the latter are highly hydrophobic by nature and, consequently, they need to be surrounded at all times by a lipidic matrix in order to prevent spontaneous aggregation, which, in turn, very often causes inactivation. In biological membranes this environment consists of a phospholipid bilayered structure.

In our efforts to study intermembraneous interactions we recently constructed such a phospholipid bilayered coat around small magnetite particles [4]. In a high-gradient magnetic field these so-called magnetoliposomes can be easily withdrawn from solution by trapping them on a magnetic filter device. These two unique features of magnetoliposomes (presence of a bilayer structure and ease of recovery) encouraged us to check whether magnetoliposomes could be used as valuable carriers for membrane-bound enzymes. As the phospholipid component we took the zwitterionic dimyristoylphosphatidylcholine (DMPC) which has been studied in depth as far as its bilayer-forming capacity is concerned [5]. As a model membrane enzyme we chose beef heart cytochrome *c*-oxidase. In vivo, this protein penetrates into the inner mitochondrial membrane where it participates in the final stage of cell respiration in which electrons originating from oxidized foodstuffs are transferred from cytochrome *c* to molecular oxygen [6].

In the present paper we report i) on the way in which DMPC-magnetoliposomes are generated, ii) on the influence of these stabilized colloids on the catalytic activity of the incorporated enzyme, and iii) on the application of this type of catalytic biocolloid in a miniature, fixed-bed bioreactor as a model for further applications.

Materials and methods

Materials

Chemicals: DMPC (>99% pure; molecular weight: 677) was obtained from two suppliers: Avanti Polar Lipids (Birmingham, Alabama, USA) and Sigma Chemie (Deisenhofen, FRG). No difference in quality was observed by analytical thin-layer chromatography or was experienced during the experiments. All other chemicals used in the present study were procured from sources mentioned in our previous reports [4, 7].

Proteins: Cytochrome *c* (Type III) was obtained from Sigma. For the kinetic experiments the stock solution [42 μM in 5 mM 2-(*tris*(hydroxymethyl)methylamino)-1-ethanesulfonic acid, (TES)] was reduced for approximately 95% by adding solid $Na_2S_2O_4$ (Merck, Darmstadt, FRG). Lipid-depleted cytochrome *c*-oxidase was purified from beef-heart mitochondria according to the method of Yu et al. [8] and stored at −70°C at a concentration of 11.4 mg/ml in a phosphate buffer containing 0.25 M sucrose and 1.5% cholate. The heme *a* concentration (11 nmol/mg protein) and the amount of residual phospholipid molecules (0.03 mg/mg protein) were determined as described previously [7].

Methods

Preparation recipe of DMPC vesicles and DMPC magnetoliposomes: Small, unilamellar DMPC vesicles were prepared at temperatures exceeding their gel-to-liquid phase transition (23°C; [5]) at a concentration of 5–10 mg/ml in 5 mM TES, pH 7.0, by ultrasonication.

The procedure for the generation of DMPC-magnetoliposomes is analogous to that previously developed for the production of magnetizable phosphatidylglycerol structures [4]. The protocol is represented schematically in Fig. 1 and further outlined in the accompanying legend. As judged by transmission electron microscopy (Zeiss EM10C), the modal size of the iron cores equals 14 nm.

High-gradient magnetophoresis: Magnetophoresis was done with a (tap) water-cooled Bruker electromagnet (Type BE 15), equipped with 15-cm diameter plane pole pieces. With a pole gap of 3.1 nm and an electric current of 30 A (corresponding electric field: 80 V), the magnetic

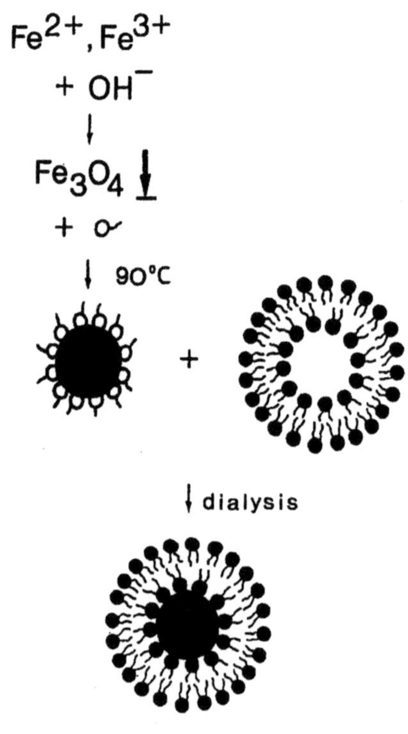

Fig. 1. Stylized illustration of the preparation recipe of magnetoliposomes. Colloidal Fe_3O_4 was generated by coprecipitation of 3 g $FeCl_2 \cdot 4$ aq and 6 g $FeCl_3 \cdot 6$ aq (dissolved in 25 ml water) with an excess (12.5 ml) of concentrated ammonia. The resultant precipitate was washed two times with 50 ml ammonia/water (5/95) and heated to 90°C for 4 min, meanwhile adding 1 g of lauric acid (o—) as a dispersing agent. The resulting coated particles were diluted with water and stored at a concentration of 28.6 mg Fe_3O_4/ml. Incubation and dialysis of this water-based magnetic fluid with performed DMPC sonicated vesicles (the requisite amount is described in the Results) produces lauric acid-free Fe_3O_4 particles covered with a phospholipid bilayer [4]. Non-adsorbed phospholipids were separated from the magnetoliposomes by high-gradient magnetophoresis. Magnetite recovery occurs with an efficiency of 98–99%

field intensity equals approx 1.5 Tesla. Local, high-gradient magnetic fields were generated by putting a plug of magnetizable stainless steel fibers (Bekaert, Belgium) in the conduit tubings (Silastic Mecial Grade Tubing, Dow Corning; inner and outer diameters: 0.078 and 0.125 inch, respectively). Unless otherwise stated, the samples were pumped through this magnetic filter device at a rate of 12 ml/h.

Enzymatic activity measurements: Cytochrome *c*-oxidase activity measurements were done on a Beckman DU-70 spectrophotometer at 550 nm with reduced cytochrome *c* (see above) as a substrate. After approx. 5 min the end point of the reaction was induced by adding a few grains of the oxidant $K_3Fe(CN)_6$ (UCB, Belgium). Using the GraphPAD software program (San Diego, USA), first-order rate constants, k_1, were calculated by the following formula:

$$k_1 = \frac{\ln(A_{t_2} - A_{t_\infty}) - \ln(A_{t_1} - A_{t_\infty})}{t_1 - t_2},$$

where A_{t_1}, A_{t_2}, and A_{t_∞} are the extinctions at time t_1, t_2, and after addition of ferricyanide, respectively.

Other dosage methods: Phosphate and iron analyses were done as mentioned in our previous report [4].

Results

DMPC-magnetoliposomes

The association of DMPC with Fe_3O_4 was studied in detail by constructing the adsorption isotherm. Increasing amounts of sonicated DMPC vesicles (from 0 to 24 mg dissolved at a concentration of 10 mg/ml) were added to 0.12 ml of the magnetic fluid stock solution (Fe_3O_4 concentration: 28.6 mg/ml) and adjusted to a final volume of 2.52 ml. After dialysis during 2 days, the DMPC-Fe_3O_4 complexes were captured in the magnetic filter device as described in the legend of Fig. 1, and further washed with buffer (1 ml) to remove non-adsorbed phospholipids remaining in the interstitial liquid.

The adsorption profile (Fig. 2) shows two distinct regions: at low total DMPC-concentrations virtually all the lipid molecules are adsorbed, whereas at higher concentrations the isotherm has a rounded shape and reaches a plateau value of approx 0.9 mmol DMPC/g Fe_3O_4. The latter part of the isotherm can be accurately fitted by the Langmuir adsorption equation (see [4]). In its linearized form (Fig. 3) the *y*-intercept is inversely proportional to the association constant, which in this case equals 1340 M^{-1} (r^2 = 0.99_0).

As shown in Fig. 2, saturation of the Fe_3O_4 colloid starts at a free DMPC-concentration of 4.3 mmol/l. For a final Fe_3O_4 concentration in the incubation mixture of 1.36 g/l, this point is obtained at a total DMPC-concentration of 5.48 mmol/l, which corresponds to a weight ratio of DMPC/Fe_3O_4 = 2.73. These conditions were applied to routinely prepare stock solutions of DMPC magnetoliposomes.

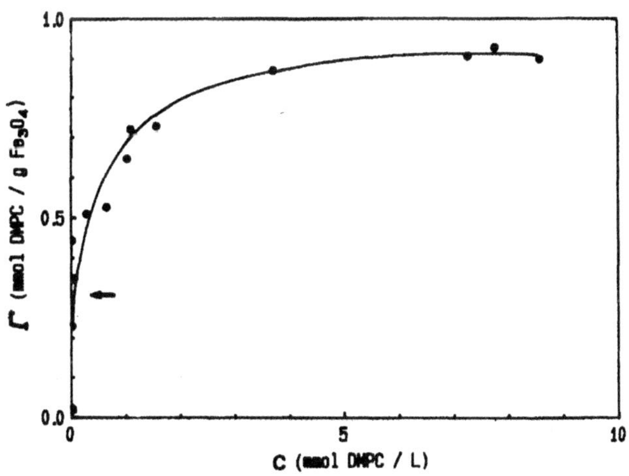

Fig. 2. Isotherm for the adsorption of DMPC from sonicated vesicles onto Fe_3O_4 in TES buffer (pH 7.0, 5 mM) at 37 °C. The arrow indicates the change from the high-affinity to the Langmuir-type adsorption curve

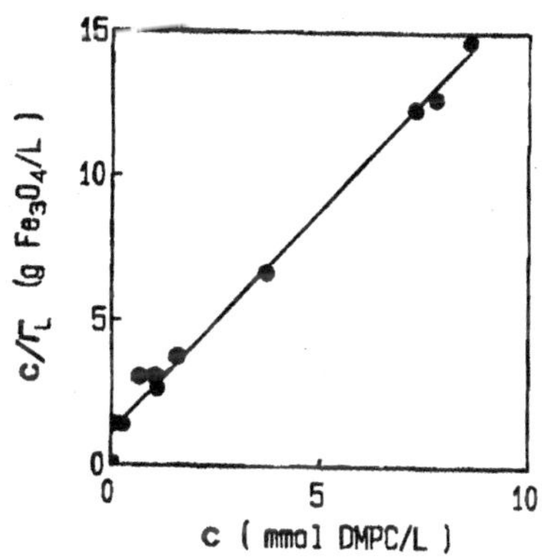

Fig. 3. Linearized Langmuir adsorption isotherm for the binding of DMPC onto magnetite (see Fig. 2)

Catalytic activity of cytochrome c-oxidase

The activity of cytochrome *c*-oxidase was followed at pH 7.0 in a 5 mM TES buffer under three different conditions: either as the lipid-depleted enzyme, or after incorporation in DMPC small unilamellar vesicles, respectively, in DMPC-magnetoliposomes (see above). In both of the latter cases incorporation was effected by cosonication of the lipid structures (3 ml at a concentration of 0.472 mmol DMPC/l) in the presence of 8.4 µl of the cytochrome *c*-oxidase stock solution (11.4 mg/ml). The total sonication

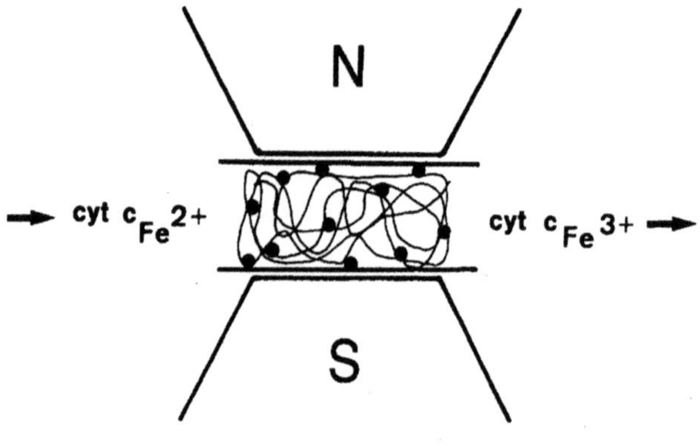

Fig. 5. Diagram of the flow reactor in which ferrocytochrome *c* is continuously oxidized. The catalytic biocolloids (●) are fixed by magnetic forces onto the magnetic filter

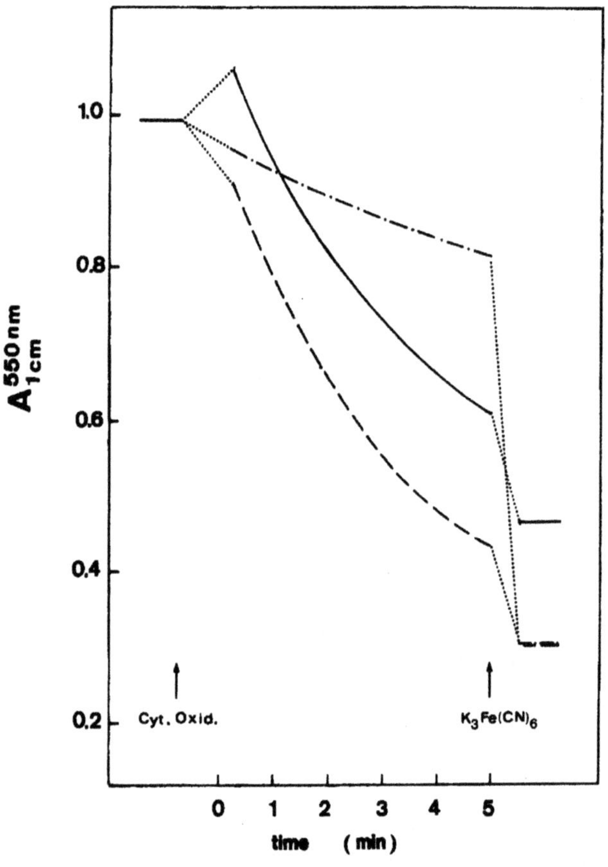

Fig. 4. Time-dependent decrease in the optical density at 550 nm of ferrocytochrome *c* (42 µM) after addition of three different cytochrome *c*-oxidase preparations: either in a lipid-deprived form (—•—•—), or after its incorporation in small sonicated DMPC vesicles (– – –), or in DMPC-magnetoliposomes (———). Fully oxidized cytochrome *c* is obtained by addition of $K_3Fe(CN)_6$. Note that in the case of the enzyme-magnetoliposome preparation the absorption values are somewhat higher as a result of the dark color of the iron oxide colloids

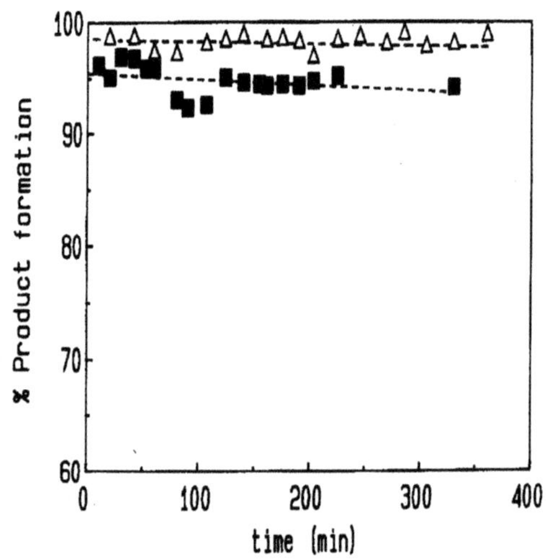

Fig. 6. Degree of ferrocytochrome *c* conversion achieved as a function of time. The flow rate of substrate equals 6 ml/h (△) and 12 ml/h (■)

time equaled 2 min, divided into four periods of 30 s, each separated by a 30-s interval to allow for thermal equilibration. To equalize for possible direct effects on the enzyme structure brought about by the sonication action, the lipid-poor enzyme underwent the same treatment.

The time-course of the reaction of the three different cytochrome *c*-oxidase preparations (60 µl) with reduced cytochrome *c* (3 ml of a 42 µM solution) is shown in Fig. 4. With the lipid-depleted enzyme, a k_1 of only 0.02_1 min^{-1} was found. Considerably higher activities were obtained with the enzymatic DMPC vesicles (k_1 = 0.33_2 min^{-1}) and DMPC-magnetoliposomes (k_1 = 0.29_5 min^{-1}).

Miniature, fixed-bed bioreactor

The possible application of cytochrome *c*-oxidase-DMPC-Fe_3O_4 complexes in a fixed-bed bioreactor (schematically represented in Fig. 5) was investigated using a miniaturized set-up. Three ml of the enzyme-magnetoliposome preparation (see above) were first captured by magnetic forces in the magnetic wire matrix and subsequently overflowed in a continuous way with the cytochrome *c* substrate solution. At regular time intervals the absorbance of the eluate was measured (Fig. 6).

At a fixed flow rate of 12 ml/h, approximately 95% of the substrate was converted, whereas at a slower rate the percentage of product formation increased to about 98%. Furthermore, the conversion efficiency scarcely dropped within the time scale of the experiment (6 h).

Discussion

Magnetoliposome formation

The use of magnetizable particles has been gradually introduced in many biotechnological applications. Mostly, the size of the colloid particles used is in the µm-range [9]. Due to their relatively large dimensions, they can be easily withdrawn from solution by conventional magnetic forces.

However, compared with these structures, smaller particles (diameter ⩽50 nm) have significant advantages. They are indeed characterized by a much more favorable surface-to-volume ratio and they do not settle spontaneously, because the gravity force is largely overshadowed by Brownian motions. Furthermore, quantitative capturing from an aqueous solution can also easily be achieved in high-gradient magnetic fields. By virtue of their small size, monodomain Fe_3O_4 particles extracted from magnetotactic bacteria are excellent candidates to attain these goals [10], but, unfortunately, they cannot be recovered in sufficient quantities for commercial purposes. Reimers and Khalafalla [11] were pioneers in synthesizing in situ nanomer-sized magnetite particles which demonstrate superparamagnetic properties. In the initial phase of development steric and electrostatic stabilization of these particles was achieved by oleic acid, thereby creating an organic solvent-based magnetic fluid. With shorter chain length surfactants, however, water compatible magnetic fluids can be created. In this work, we used lauric acid as a dispergent. We did not investigate in depth the exact arrangement of the laurate molecules on the iron oxide surface, but due to the high affinity of iron for carboxylate compounds (see below), most probably the fatty acid is adsorbed by its COO^- group onto the Fe_3O_4 surface, thereby forming a monolayer. In the case of dodecylsulfate adsorption onto iron oxide particles, however, Meguro et al. [12] claim that some additional SDS molecules are needed (with their polar headgroups directed towards the aqueous medium) in order to obtain a stabilized dispersion. Whether, in our case, a similar additional association of some laurate molecules is necessary to further improve the electrostatic stabilization cannot be completely excluded. However, from molecular shape considerations, the formation of a complete bilayer or admicelle is highly improbable [13].

At any rate, addition of sonicated DMPC vesicles to the laurate-coated Fe_3O_4 particles causes a displacement of the fatty acid molecules [4]. Undoubtedly, this exchange process is promoted by the stronger van der Waals interaction forces, occurring between the longer chain length diacylphospholipid molecules [14]. Alternatively, as far as the binding strength due to polar interactions is concerned, some guidance may be obtained from the Paneth-Fajanse rule which states that ionic components will strongly adsorb on the condition that they form compounds not easily soluble or weakly dissociated compounds with the oppositely charged ions of the solid lattice [15]. In this respect, a qualitative idea about the difference in phosphate-iron and carboxylate-iron affinity can be obtained by comparing the values of the solubility products for simple iron salts which equal 3.5×10^{-11} and 1.3×10^{-22} for $FeCO_3$ and $FePO_4$, respectively [16].

Regarding the deposition of the outer leaflet, which starts at a DMPC/Fe_3O_4 (mmol/g) ≈ 0.3, the isotherm shows Langmuirian characteristics, pointing to a significant role of hydrophobic adsorption forces [17]. In analogy with observations which we made using different types of phosphatidylglycerols (unpublished), we propose the following mechanistic model for the build-up of the outer layer: first, a phospholipid molecule escapes from the donor vesicles; subsequently, it spontaneously moves through the aqueous phase; and finally, it adsorbs onto the monolayer-covered magnetizable structures. In previous experiments we found that for DMPC the half-time value for such a spontaneous lipid transfer process between different types of vesicles equals approx. 100 min [18]).

Enzymatic magnetoliposomes

Some preliminary observations dealing with the unique possibilities that magnetoliposomes offer in the field of enzyme immobilization are described here. In the literature, only a few recent reports deal with the use of non-magnetizable, phospholipid-coated solid supports for fixation of peptides or enzymes. For instance, with the objective to develop HPLC methods for studying interactions between cell membranes and some peptides, Pidgeon and Venkataram [19] produced silica particles which were enveloped with a single layer of DMPC molecules. Alternatively, in the elegant work of Rothe and Aurich [20] phospholipid bilayers were first constructed around large spherical polymer particles and subsequently tagged with different types of enzymes. However, a key difference with our system is that in both investigations the phospholipids need to be first chemically modified either in their apolar [19] or polar [20] part in order to afford a strong covalent anchoring onto the solid support.

In the present study, the incorporation of the membrane-bound enzyme, cytochrome *c*-oxidase, into performed DMPC-magnetoliposomes is facilitated by a short sonication step. The lipid-poor enzyme with which we start, contains only 7—8 residual phospholipid molecules, whereas approximately 50 molecules are needed to construct a complete annulus around the intramembraneous region of the enzyme [21]. Apparently, the DMPC molecules present in the bilayer coat of the magnetoliposomes can restore the lipid environment of the enzyme, thereby inducing a reactivation which is as adequate as that found with the more "classical" small unilamellar DMPC vesicles. On a molecular scale, however, neither the exact configuration of the phospholipid molecules nor their precise role in the reactivation process are fully understood.

Besides their reactivation capacity, the DMPC molecules are also responsible for a firm fixation of the enzyme in the solid-bilayer complex. The results of Fig. 6 indeed show that — within the time-frame of the experiments — there is practically no drop in the conversion efficiency. Furthermore, by carefully choosing the flow rate of substrate, the percentage of product formation can be adjusted as desired. These features, together with the fact that our active biocolloids can be efficiently trapped on the magnetic filter device by high-gradient magnetic forces, make these entities valuable structures to be used in magnetically controlled bioreactors. Further work to evaluate the contribution of different parameters on the immobilization and performance conditions is now being carried out.

Acknowledgements

We gratefully appreciate the excellent technical contributions of Mrs. T. Derycke and Ms. C. Declercq. This work was supported by the "Belgische Fonds voor Geneeskundig Wetenschappelijk Onderzoek" (Contract # 3.0063.86) and "Nationaal Fonds voor Wetenschappelijk Onderzoek (Krediet aan Navorsers 89—90).

References

1. Rosensweig RE (1979) Advances in Electronics and Electron Physics 48:103—199
2. Adalsteinsson O, Lamotte A, Baddour RF, Colton CK, Pollak A, Whitesides GM (1979) J Mol Catal 6:199—255
3. Madden T (1986) Chem Phys Lipids 40:207—222
4. De Cuyper M, Joniau M (1988) Eur Biophys J 15:311—319
5. Janiak MJ, Small DM, Shipley GG (1979) J Biol Chem 254:6068—6078
6. Denis M (1986) Biochimie 68:459—470
7. De Cuyper M, Joniau M (1980) Eur J Biochem 104:397—405
8. Yu C-A, Yu L, King TE (1975) J Biol Chem 250:1383—1392
9. Halling PJ, Dunnill P (1980) Enzyme Microb Technol 2:2—10
10. Matsunaga T, Kamiya S (1987) Appl Microbiol Biotechnol 26:328—332
11. Reimers GW, Khalafalla SE (1974) US Patent Specification # 1439031
12. Meguro K, Tomioka S, Kawashima N, Esumi K (1983) Progr Colloid Polymer Sci 68:97—100
13. Israelachvili JN, Marčelja S, Horn RG (1980) Quart Rev Biophys 13:121—200
14. Tanford C (ed) (1980) The Hydrophobic Effect: Formation of Micelles and Biological Membranes, John Wiley and Sons, New York Chichester Brisbane Toronto
15. Skoog DA, West DM (eds) (1963) Fundamentals of Analytical Chemistry, Holt, Rinehart and Winston, New York Chicago San Francisco Toronto London, p 167

16. Meites L (ed) (1963 — Reissue 1982) Handbook of Analytical Chemistry, Mc Graw-Hill Book Co, New York San Francisco Toronto London, Sydney, p 1—16
17. Attwood D, Florence AT (1983) Surfactant Systems: Their Chemistry, Pharmacy and Biology. Chapman and Hall, London New York, p 22
18. De Cuyper M, Joniau M, Engberts JBFN, Sudhölter EJR (1984) Coll Surfaces 10:313—319
19. Pidgeon C, Venkataram UV (1989) Anal Biochem 176:36—47
20. Rothe U, Aurich H (1989) Biotechnol Appl Biochem 11:18—30
21. Jost PC, Griffith HO, Capaldi RA, Vanderkooi G (1973) Proc Natl Acad Sci USA 70:480—484

Authors' address:

Prof. Dr. Marcel De Cuyper
Katholieke Universiteit Leuven — Campus Kortrijk
Interdisciplinary Research Centre
Universitaire Campus
8500 Kortrijk, Belgium

Author Index

Subject Index

GPSR Compliance

The European Union's (EU) General Product Safety Regulation (GPSR) is a set of rules that requires consumer products to be safe and our obligations to ensure this.

If you have any concerns about our products, you can contact us on ProductSafety@springernature.com

In case Publisher is established outside the EU, the EU authorized representative is:

Springer Nature Customer Service Center GmbH
Europaplatz 3
69115 Heidelberg, Germany

Batch number: 10578403

Printed by Printforce, the Netherlands